Fundamental Theories of Physics

Volume 196

The international monograph series "Fundamental Theories of Physics" aims to stretch the boundaries of mainstream physics by clarifying and developing the theoretical and conceptual framework of physics and by applying it to a wide range of interdisciplinary scientific fields. Original contributions in well-established fields such as Quantum Physics, Relativity Theory, Cosmology, Quantum Field Theory, Statistical Mechanics and Nonlinear Dynamics are welcome. The series also provides a forum for non-conventional approaches to these fields. Publications should present new and promising ideas, with prospects for their further development, and carefully show how they connect to conventional views of the topic. Although the aim of this series is to go beyond established mainstream physics, a high profile and open-minded Editorial Board will evaluate all contributions carefully to ensure a high scientific standard.

More information about this series at http://www.springer.com/series/6001

Dirk Puetzfeld · Claus Lämmerzahl
Editors

Relativistic Geodesy

Foundations and Applications

Springer

Editors
Dirk Puetzfeld
ZARM
University of Bremen
Bremen, Germany

Claus Lämmerzahl
ZARM
University of Bremen
Bremen, Germany

ISSN 0168-1222 ISSN 2365-6425 (electronic)
Fundamental Theories of Physics
ISBN 978-3-030-11499-2 ISBN 978-3-030-11500-5 (eBook)
https://doi.org/10.1007/978-3-030-11500-5

Library of Congress Control Number: 2018967737

This Springer imprint is published by the registered company Springer Nature Switzerland AG
The registered company address is: Gewerbestrasse 11, 6330 Cham, Switzerland

Preface

Recent years have seen the advent of high precision measuring methods, in particular, modern clocks reached an unprecedented level of accuracy and stability. This was accompanied by important developments in the fields of atom and laser interferometry. Laser interferometers improved by several orders of magnitude and interferometry in space nowadays is a mature technology ready for practical applications. All this is of direct importance for many fields of physics, and consequently for geodesy.

The high precision of these new experimental capabilities made clear that geodesy can no longer rely solely on Newtonian concepts, which are still used within the field. Geodetical models and the interpretation of data within these models therefore inevitably require concepts which go beyond the Newtonian picture of space and time. The theoretical underpinning of geodesy should therefore be based on the special and the general theory of relativity, the latter still represents the most successful gravity theory to the present date. This new "relativistic geodesy" is the topic of the present volume.

In 2016, we organized[1] an international conference in Bad Honnef (Germany) on the *Relativistic Geodesy: Foundations and Applications*. The conference brought together the leading experts in their respective fields and was very well received by the speakers as well as by the audience. We would like to thank the WE-Heraeus Foundation for the generous support of this conference. Our thanks also go to the Physikzentrum Bad Honnef where the conference took place.

The positive reception and the feedback after the conference made clear that there is a strong demand for an up-to-date volume, covering the methods employed in current research in the context of the relativistic geodesy. This book intends to give such a status report. It hopefully is of value for the experts working in this field and may also serve as a guideline for students. At the same time, we should warn potential readers that it is not intended to serve as a replacement for a textbook on either of the subjects of gravitational physics or geodesy. But we hope that it bridges some of the gaps between the relativity and the geodesy communities, in

[1]http://puetzfeld.org/relgeo2016.html.

particular, when it comes to implementation and application of relativistic concepts and methods in the field of geodesy.

The present volume is based on the lectures given at the conference and gives an overview over the following topics:

- Time and frequency metrology
- Chronometric geodesy
- (Clock) gradiometry
- Satellite experiments
- Navigation systems
- Tests of gravity by means of geodetic measurements

In covering these topics, definitions and methods from relativistic gravity are introduced. Emphasis is put on the coverage of the geodetically relevant concepts in the context of Einstein's theory (e.g., role of observers, use of clocks, and definition of reference systems). Furthermore, fundamental questions in the context of the measuring process, as well as approximation methods which make certain calculations feasible, are discussed in detail.

We as editors are deeply indebted to the contributors to this volume, who made great efforts to present their respective areas of research in an accessible way to a broader audience. We hope that the material presented in here will prove to be useful as a reference for experienced researchers, as well as serve as an inspiration for younger researchers who want to enter the exciting emerging field of relativistic geodesy.

Bremen, Germany

Dirk Puetzfeld[2]
Claus Lämmerzahl

[2]D. P. acknowledges the support by the Deutsche Forschungsgemeinschaft (DFG) through the grant PU 461/1-1.

Contents

Contributions

- A. Bauch *Time and frequency metrology in the context of relativistic geodesy*
- P. Delva, H. Denker, G. Lion *Chronometric geodesy: methods and applications*
- Y. N. Obukhov, D. Puetzfeld *Measuring the gravitational field in General Relativity: From deviation equations and the gravitational compass to relativistic clock gradiometry*
- P. A. Hogan *A Snapshot of J. L. Synge*
- B. Mashhoon *General Relativistic Gravity Gradiometry*
- S. Kopeikin *Reference-ellipsoid and normal gravity field in post-Newtonian geodesy*
- E. W. Grafarend *Anholonomity in Pre and Relativistic Geodesy*
- B. Coll *Epistemic relativity: An experimental approach to physics*
- A. Hees, A. Bourgoin, P. Delva, C. Le Poncin-Lafitte, P. Wolf *Use of geodesy and geophysics measurements to probe the gravitational interaction*
- B. Hartmann *Operationalization of basic relativistic measurements*
- H. Quevedo *Can spacetime curvature be used in future navigation systems?*
- J.-W. van Holten *World-line perturbation theory*
- D. Philipp, D. Puetzfeld, C. Lämmerzahl *On the applicability of the geodesic deviation equation in General Relativity*
- R. König, I. Ciufolini *Measurement of frame dragging with geodetic satellites based on gravity field models from CHAMP, GRACE and beyond*
- I. Ciufolini *Tests of General Relativity with the LARES satellites*

Contributors

Andreas Bauch Physikalisch-Technische Bundesanstalt, Braunschweig, Germany

Adrien Bourgoin Dipartimento di Ingegneria Industriale, University of Bologna, Bologna, Italy

Ignazio Ciufolini Dip. Ingegneria dell'Innovazione, Università del Salento, Lecce, Italy;
Centro Fermi, Roma, Italy;
Centro Fermi - Museo Storico della Fisica e Centro Studi e Ricerche "Enrico Fermi", Rome, Italy

Bartolomé Coll Relativistic Positioning Systems, Department of Astronomy & Astrophysics, University of Valencia, Burjassot, Valencia, Spain

Pacome Delva SYRTE Observatoire de Paris, Université PSL, CNRS, Sorbonne Université, LNE, Paris, France

Heiner Denker Institut für Erdmessung, Leibniz Universität Hannover (LUH), Hannover, Germany

Erik W. Grafarend Department of Geodesy and Geoinformatics, Faculty of Aerospace Engineering and Geodesy, Faculty of Mathematics and Physics, Stuttgart, Germany

Vahe Gurzadyan Center for Cosmology and Astrophysics, Alikhanian National Laboratory, Yerevan, Armenia

Bruno Hartmann Humboldt University, Berlin, Germany

Aurélien Hees SYRTE, Observatoire de Paris, Université PSL, CNRS, Sorbonne Université, LNE, Paris, France

Peter A. Hogan School of Physics, University College Dublin, Belfield, Dublin 4, Ireland

Rolf König Helmholtz-Zentrum Potsdam Deutsches GeoForschungsZentrum GFZ, Wessling, Germany

Sergei Kopeikin Department of Physics and Astronomy, University of Missouri, Columbia, MO, USA

Claus Lämmerzahl Center of Applied Space Technology and Microgravity (ZARM), University of Bremen, Bremen, Germany

Christophe Le Poncin-Lafitte SYRTE, Observatoire de Paris, Université PSL, CNRS, Sorbonne Université, LNE, Paris, France

Guillaume Lion LASTIG LAREG IGN, ENSG, Univ Paris Diderot, Sorbonne Paris Cité, Paris, France

Bahram Mashhoon Department of Physics and Astronomy, University of Missouri, Columbia, MO, USA;
School of Astronomy, Institute for Research in Fundamental Sciences (IPM), Tehran, Iran

Richard Matzner Theory Group, University of Texas at Austin, Austin, TX, USA

Yuri N. Obukhov Theoretical Physics Laboratory, Nuclear Safety Institute, Russian Academy of Sciences, Moscow, Russia

Antonio Paolozzi Scuola di Ingegneria Aerospaziale, Sapienza Università di Roma, Rome, Italy

Claudio Paris Centro Fermi - Museo Storico della Fisica e Centro Studi e Ricerche "Enrico Fermi", Rome, Italy

Erricos C. Pavlis Joint Center for Earth Systems Technology (JCET), University of Maryland, Baltimore County, MD, USA

Dennis Philipp Center of Applied Space Technology and Microgravity (ZARM), University of Bremen, Bremen, Germany

Dirk Puetzfeld Center of Applied Space Technology and Microgravity (ZARM), University of Bremen, Bremen, Germany

Hernando Quevedo Instituto de Ciencias Nucleares, Universidad Nacional Autónoma de México, Mexico, DF, Mexico;
Dipartimento di Fisica and ICRANet, Università di Roma "La Sapienza", Rome, Italy;
Department of Theoretical and Nuclear Physics, Kazakh National University, Almaty, Kazakhstan

John Ries Center for Space Research, University of Texas at Austin, Austin, TX, USA

Giampiero Sindoni Scuola di Ingegneria Aerospaziale, Sapienza Università di Roma, Rome, Italy

Jan-Willem van Holten Nikhef, Amsterdam, The Netherlands

Peter Wolf SYRTE, Observatoire de Paris, Université PSL, CNRS, Sorbonne Université, LNE, Paris, France

Time and Frequency Metrology in the Context of Relativistic Geodesy

Andreas Bauch

Abstract A status report is given on current practice and trends in time and frequency metrology. Emphasis is laid on such fields of activity that are of interest in the context of relativistic geodesy. In consequence, several topics of a priori general relevance will not be dealt with. Clocks and the means of comparing their reading are equally important in practically all applications and thus dealt with in this contribution. The performance of commercial atomic clocks did not change significantly during the last 20 years. Progress is noted in the direction of miniaturization, leading to the wide-spread use of chip-scale atomic clocks. On the other hand, research institutes invested considerably into the perfection of their instrumentation. Cold-atom caesium fountain clocks realize the SI-second with a relative uncertainty of close to 1×10^{-16}, and with a relative frequency instability of the same magnitude after averaging over a few days only. Optical frequency standards are getting closer to being useful in practice: outstanding accuracy combined with improved technological readiness can be noted. So one necessary ingredient for relativistic geodesy has become available. Satellite-based time and frequency comparison is here still somewhat behind: Time transfer with ns-accuracy and frequency transfer with 1×10^{-15} per day relative instability have become routine. Better performance requires new signal structures and processing schemes, some appear on the horizon.

1 Introduction

From the author's perspective, relativistic geodesy requires the following actions: Two "super-clocks" have to be operated simultaneously. Their frequency accuracy and stability, and the frequency difference or frequency ratio prevailing when both are operated side-by-side need to be determined at the outset. Then one of them is kept at a known "height", the other one at an unknown "height". Now they need to be

A. Bauch (✉)
Physikalisch-Technische Bundesanstalt, Bundesallee100,
38116 Braunschweig, Germany
e-mail: andreas.bauch@ptb.de
URL: http://www.ptb.de/time

D. Puetzfeld and C. Lämmerzahl (eds.), *Relativistic Geodesy*,
Fundamental Theories of Physics 196, https://doi.org/10.1007/978-3-030-11500-5_1

compared with sufficient accuracy in order to determine the height-difference to the desired accuracy, based on the frequency difference or ratio observed while they are operated at their new locations. Validity of General Relativity is taken for granted. For an exercise of this kind to be meaningful, the comparisons at the two sites should be made with an uncertainty of 10^{-17} or below, which indeed requires "super" clocks and comparison means and remains a real challenge as we will see. In the literature one finds a few early citations of "relativistic geodesy" or "chronometric leveling", one referred to quite often is [1]. At the time of writing a very detailed elaboration on the matter is underway, based on campaigns running during 2011–2015 [2].

Coordinated and disciplined operation of atomic clocks and of time transfer systems is on the other hand integral part of time and frequency metrology, comprising in general:

(a) Development of quartz oscillators and atomic clocks and their operation;
(b) Characterization of the properties of oscillators and clocks;
(c) Realization of time scales, e.g. national legal time;
(d) Time and frequency comparison of clocks and time scales, locally and remotely;
(e) Dissemination of time-of-day, time interval, and standard frequency to the public.

In the context of relativistic geodesy, and for economy of writing, it is admissible to skip some subjects. But it is essential to understand the status and the progress to be expected in several aspects of (a), (c) and (d). The subjects will be laid down in four sections before some conclusive statements are given. From the onset it should be clear that a certain simplification is unavoidable. Optical frequency standards and fiber-based frequency transfer for the comparison of optical clocks will be dealt with in separate articles in this volume and covered in a few words only in this one.

2 Characterization of Time and Frequency Signals

This brief section is intended to introduce the vocabulary used further on and to introduce the two quantities used in the characterization of frequency standards and clocks. The devices will then be described subsequently. Let us start with statistical signal properties. The frequency of oscillators and clocks is subject to systematic and random variations with respect to their intended nominal output value. Many measures for quantitative characterization are extensively covered in the literature [3, 4],[1] but all are based on the following formal description of the observed signal.

---

[1]Note: Although this is a review article, my intention was not to provide an exhaustive list of references, but rather to limit myself to text books and previous review articles of other authors with few exceptions.

A frequency standard outputs a (nearly) sinusoidal signal voltage described by

$$V(t) = [V_0 + e(t)] \times \sin\{2\pi \nu_0 t + \phi(t)\}, \tag{1}$$

where $\nu_0, \phi(t), V_0$, and $e(t)$ are the nominal frequency, the instantaneous phase fluctuations, the nominal signal amplitude, and its temporal variations, respectively. Further practical quantities are the instantaneous phase-time variations, $x(t) = \phi(t)/(2\pi \nu_0)$, and the instantaneous normalized frequency departure $y(t) = (d\phi/dt)/(2\pi \nu_0)$. Both can be analyzed in the time domain and in the frequency domain. For the remainder of this article the restriction to time-domain quantities is justified, which are based on mean frequency values $\bar{y}(\tau)$ measured during an averaging time τ. The most popular measure is the Allan variance

$$\sigma_y^2(\tau) = \left\langle (\bar{y}_{k+1}(\tau) - \bar{y}_k(\tau))^2 \right\rangle /2. \tag{2}$$

Here the $\bar{y}_k$-values are understood as a contiguous (no dead time) series of data, and the brackets $\langle\rangle$ signify an infinite time average, including normalization. In practice, a finite sum of terms is only available. Ideally the number of samples at the longest averaging time τ should be ten or larger. A double-logarithmic plot of $\sigma_y(\tau)$ versus τ helps to discriminate among some causes of instability in the clock signal because they lead to a different slope of the plot. If shot noise of the detected atoms is the dominating noise source, the frequency noise is white and $\sigma_y(\tau)$ decreases like $\tau^{-1/2}$. In this case, $\sigma_y(\tau)$ agrees with the classical standard deviation of the sample. Long-term deviations from this $\tau^{-1/2}$-behaviour are quite common and indicate that parameters defining ν_0 are not stable. In such a case the classical standard deviation would diverge with increasing τ and increasing observation time. In Fig. 1 I show schematic examples of the frequency instability expected or observed for a variety of atomic frequency standards. More detailed plots of that kind are shown subsequently as Figs. 2 and 3.

With the exemption of the active hydrogen maser, the following expression relates the observed $\sigma_y(\tau)$ to operational parameters of a frequency standard typically for a wide range of averaging times τ:

$$\sigma_y(\tau) = \frac{\eta}{Q \times (S/N)} \times \frac{1}{\sqrt{\tau/s}}. \tag{3}$$

Here η is a numerical factor of the order of unity, depending on the shape of the resonance line and of the method of frequency modulation to determine the line center. Q is the line quality factor (transition frequency / line width of the observed transition), and S/N is the signal-to-noise ratio for a 1 Hz detection bandwidth. The frequency standards discussed subsequently differ in the combination of the quantities Q, S and N. To understand the leap from microwave to optical frequency standards seen in Fig. 1, a look at Q is helpful.[2] In a caesium fountain clock the

---

[2]In order to understand the word "leap": Only very few fountain clocks achieve an instability as shown, see discussion in Sect. 3.2.

10 GHz transition is detected with a 1 Hz line width, whereas, to give orders of magnitude, a 500 THz optical transition is detected with a line width of 5 Hz, so the Q-values differ by 10^4.

Now we turn to the characterization of systematic effects. The term *accuracy* is often used in a broad sense to express the agreement between the clock's average output frequency and its nominal value conforming to the SI second definition. But according to the rules of metrology [5], *accuracy* should not be combined with a quantitative statement. Nevertheless, product manufacturers often state the accuracy of their devices as a range of output frequencies to be expected, usually without giving details about the causes of potential frequency deviations. A detailed uncertainty estimate, on the other hand, is provided for so-called primary clocks and optical frequency standards. It reflects the quantitative knowledge of all (known) effects which may cause the output frequency to deviate from the transition frequency of unperturbed atoms (or ions) at rest. The components of the *uncertainty* due to individual effects and finally the combined uncertainty are stated. The combined uncertainty reported for a primary clock expresses the potential deviation of its second-ticks from the SI-second. For all other devices it reflects the state of knowledge of systematic effects, whereas the absolute value of the transition frequency involved can be given only with the (sometimes larger) uncertainty of the primary clock that served as reference for its measurement.

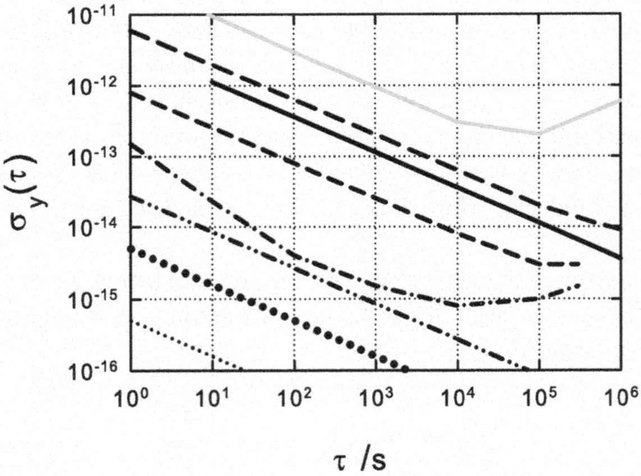

Fig. 1 Relative frequency instability $\sigma_y(\tau)$ of different atomic frequency standards, from top to bottom: (typical) rubidium standard (grey); commercial caesium standard type 5071A, high-performance option (long dash); PTB primary clock CS2 (solid); passive hydrogen masers (short dash); active hydrogen maser (dash-dot); PTB CSF2, state 2016 (dash-dot-dot); single ytterbium ion optical frequency standard (bold dots); strontium optical lattice clock (dots)

3 Atomic Clocks

Atomic properties such as energy differences between atomic eigenstates and thus atomic transition frequencies are believed to be natural constants and thus not to depend on space and time (apart from relativistic effects). They are governed by fundamental constants which describe the interactions among particles and fields. This basic principle governs all kinds of atomic clocks. The most detailed treatment of the underlying physics is given in the books of Vanier and Audoin [6] and Vanier and Tomescu [7], in less depth in Audoin and Guinot [8] and in [9], where the reader can also find detailed explanations of their function.

3.1 Commercial Clocks

3.1.1 Rubidium Gas Cell and Miniaturized Frequency Standards

For completeness they have to be mentioned here, as they are produced in large quantities and are indispensible in the fields of telecommunication, power grid management, navigation, just wherever the performance of a quartz oscillator is insufficient. The atomic reference transition is the 6.84 GHz hyperfine splitting frequency in ^{87}Rb. Several manufacturers share the large market. The devices differ in performance, size, power consumption, and we give in the following some performance figures as a guideline: As we will see, they have little importance in the context of relativistic geodesy, maybe just in the background to keep the infrastructure functioning.

Rubidium gas cell frequency standards come in packages between half a liter and less than 100 cm^3 and have a power consumption between 5 and 20 Watt. The relative deviation of the output frequency (typically 10 MHz) from its nominal value is of the order 10^{-9} and difficult to predict. During a month the value may change by 1 to 30×10^{-11} due to aging. The relative frequency instability is of order 10^{-11} at $\tau = 1$ s and white noise characteristics prevail up to 1000 s or even 10000 s of averaging, depending very much on the stability of the environment. Unless very special care in the packaging is taken, the devices are sensitive to external magnetic fields and temperature changes. A very important application is their use as so-called GNSS-disciplined oscillators: The offset as well as the long-term aging and sensitivity to external perturbations is suppressed by steering the output frequency to a reference signal received from a GNSS, today still most common from the US Global Positioning System GPS (see Sect. 5.1). Because of their low weight and power consumption rubidium clocks appeared particularly suited for use on board of satellites. Space qualified versions are today operated in the navigation satellites of all global navigation satellite systems (GNSS), serving as the source for the synthesis of the GNSS signals, i.e. carrier frequency and modulation.

So-called chip-scale atomic clocks are on the market since a number of years and represent an attractive alternative to the rubidium clocks. The atomic reference transition is the 9.19 GHz hyperfine splitting frequency in ^{133}Cs that has traditionally been used in the caesium beam clocks discussed in the following section. In a very compact package and at a power consumption of order 100 mW they outperform quartz oscillators and almost reach the rubidium performance [10]. Their main application is in battery-powered and hand-held devices. Their development was sponsored by the US military for future use in hand-held GPS receivers. But in fact, they are now deployed in thousands as part of undersea reflection seismology sensor installations deployed by oil exploration companies.

3.1.2 Commercial Caesium Clocks

Caesium atomic beam clocks have been produced commercially since the late 1950s, starting with the so-called Atomichron of the National Company [11]. When designing commercial clocks, a compromise between weight, volume, power consumption, and performance and cost is unavoidable. Several manufacturers have participated in this business over the years [12], but today essentially all production of instruments for civil use is in hands of Microsemi (www.microsemi.com). 25 years since its first appearance on the market, the model 5071A, initially developed by Hewlett-Packard, then branded as Agilent, later produced by Symmetricom, a firm taken over by Microsemi recently, is the work horse in the timing community. Standard and high-performance versions of this clock are on the market. Part of the improved specifications of the latter versus the former are due to a larger atomic flux employed which entails a larger S/N ratio. The price to be paid (literally) is a faster depletion of the caesium reservoir, thus a reduced period of warranty. Recently Oscilloquartz (ADVA Optical Networking, www.oscilloquartz.com) announced the forthcoming release of a commercial beam clock, using the technique of optical pumping for state selection and detection. An instability lower by a factor of three than for conventional commercial caesium clocks was reported at conferences, but no experience on accuracy and long-term performance has been published yet.

I give examples of the observed performance of clocks of type 5071A operated at PTB in laboratory environment during 2015. In Fig. 2 (left) records of the clock rate with reference to UTC(PTB) are shown. The clocks designated C1, C8, and C9 are high-performance versions, C6 is a standard performance version. In this context, UTC(PTB) can be regarded as an ideal reference, its scale-unit being very close to the SI-second (see Sect. 4). The maximum rate we note is that of the clock C9 of about 4400 ns/360 days, corresponding to a relative frequency difference of 142×10^{-15}. This is a typical value for this type of clock, for which the manufacturer specifies the magnitude of the offset from the nominal frequency (*accuracy*) as below 500×10^{-15}. We note in case of clock C8 that its rate (slope of the plot) changed during the year. The relative frequency instability values of the four clocks are shown in the right part of Fig. 2. The clocks' frequency instability is governed by white frequency noise for averaging times up to a few days of averaging. The so-called flicker floor is substantial for the device C8 and also noticeable for others. The specs shown are from a current sales brochure whereas the clocks C8 and C9 are more than

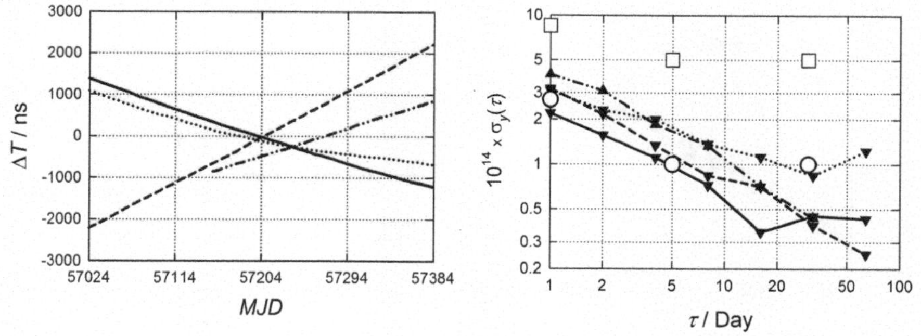

Fig. 2 Left: rate of four 5071 commercial caesium atomic clocks with reference to UTC(PTB) during 2015 (Modified Julian Day number MJD 57384 corresponds to 2015-12-28): C1 (solid line), C8 (dotted line), C9 (dashed line), C6 (dash-dot-dot line). Right: relative frequency instability of the clocks derived from the data shown left, specifications from the 2015 brochure of Microsemi for the standard performance clock (open square) and the high-performance clock (open circle)

20 years old and each already needed beam tube replacement twice. So the slight violation of the current specs is not surprising. The standard performance clock C6 is substantially more stable than the current specs predict. In summary one can say that the performance of these devices is remarkable and very useful in general time-keeping activities, but nevertheless their instability is prohibitive to use them in serious quantitative tests of relativity and also in the context of relativistic geodesy.

PTB continues to operate its legacy CS1 and CS2 caesium atomic beam clocks as the last ones world-wide of a previously larger ensemble of that kind. They were developed with the intention to surpass the limitations of commercial clocks and are each unique specimen. Their uncertainty for the realization of the SI second has been well developed and published. It amounts to 8×10^{-15} for CS1 and 12×10^{-15} for CS2 [13]. Their relative frequency instability is not so different from those of the commercial devices at short averaging times, but in the long term no flicker level above $(1\text{-}2)\times 10^{-15}$ can be noted. CS1 and CS2 constitute a back-up reference for the realization of PTB's time scale UTC(PTB), see Sect. 4.

3.1.3 Hydrogen Masers

The ground state hyperfine splitting of the hydrogen atom corresponds to a transition line at a frequency of 1.4 GHz. Research into the use of this atomic transition in a frequency standards started at Harvard University in the 1950s. In the active maser, as it is called, stimulated emission inside a high-Q cavity which encloses the hydrogen atoms kept in a storage bulb is used to detect the atomic transition [14]. In the passive maser the transition is probed by injecting radiation into the cavity and observing the effect on the atoms. Limited by the difficulty to control a variety of perturbing effects, the maser output frequency reflects the unperturbed hyperfine splitting frequency of hydrogen atoms only with an uncertainty of order 10^{-11}. But, as already shown

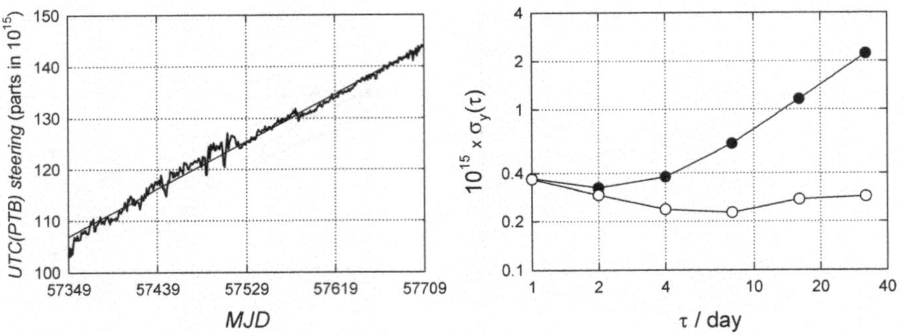

Fig. 3 Left: Frequency steering applied to the PTB active hydrogen maser used for generation of UTC(PTB) during one year, ending at MJD 57709 (2016-11-17); right: Relative frequency instability of the data from the second half of the period shown in the plot left; original data (solid symbols), frequency drift removed (open symbols)

in Fig. 1, both types of standards share a remarkably low frequency instability at averaging times up to one day, the lowest for any commercially available frequency standard at present.

Based on a long tradition, masers are today produced in Russia and in the US. A Swiss firm combines Russian physics packages with Swiss electronics. Small scale production of masers is reported from China and Japan, but the products are not used outside the respective country. Only one Russian manufacturer currently offers a passive maser commercially. However a space qualified variant serves as local frequency source in satellites of the European satellite navigation system Galileo. In the future also satellites of the Russian counterpart GLONASS shall be equipped with passive masers. Traditionally, active masers have served as frequency references in very-long baseline interferometry observatories since their existence. Nowadays several National Metrology Institutes (NMI) realize their reference time scales based on an active hydrogen maser, steered in frequency with respect to a superior "primary" reference. The same strategy is followed for the physical realization of the system time of Galileo [15]. These applications are not hampered by the frequency drift typically associated with masers. Without precautions, the drift caused by the aging of the mechanical cavity structure typically is of order 10^{-15}/day. Cavity auto-tuning reduces the frequency drift to the order 10^{-16}/day which is then caused by other effects [14]. In anticipation of the next paragraph and Sect. 4, Fig. 3 illustrates the performance of the active maser that has been used as physical source for the generation of PTB's reference time scale UTC(PTB) during one year until mid November 2016. The UTC(PTB) steering is derived from daily comparisons with one or both caesium fountain clocks of PTB. The regression line represents the linear part of the maser frequency drift of 1.03×10^{-16}/day. During the second half of the period shown, the steering was based on the average of the two fountains, with particularly higher data availability of fountain CSF2, which clearly improved the day-to-day stability. For this period the combined frequency instability is shown in the right plot, based on the original data and with the linear drift removed, respectively.

Active masers constitute also an important infrastructure in laboratories operating optical frequency standards. Direct comparisons of remote masers, however, cannot answer questions in the context of relativistic geodesy because of the lack of accuracy.

3.2 Cold-Atom Fountain Clocks

Laser cooling to μK temperature is the key to the success of the fountain concept [16]. In a fountain the laser cooled cloud of atoms is launched upwards with a velocity v_s and the microwave excitation is performed during the ballistic flight, as illustrated in Fig. 4. The atoms come to rest under the action of gravity at a height of $H = v_s^2/(2g)$. With a height of the fountain setup of about 1 m and $v_s = 4.4\ m/s$ the total time of flight, back to the starting point, is about 0.9 s. On their way the atoms interact twice with the field sustained in the microwave cavity, on their way up and then on their way down, separated in time by the so-called interaction time. This is typically 0.5 s, leading to a width of the observed resonance of 1 Hz. During clock operation, the transition probability is determined changing the probing frequency f_p from cycle to cycle alternately on either side of the central resonance feature. The difference of successive measurements is numerically integrated and represents the difference

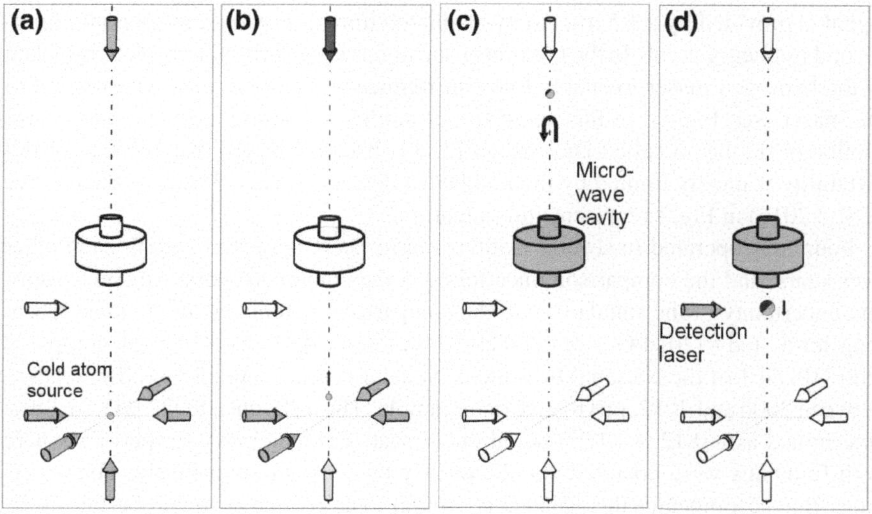

Fig. 4 Operation of a fountain frequency standard, illustrated in a time sequence from left to right. Arrows represent laser beams (white if they are blocked); **a** Loading of a cloud of cold atoms; **b** Launch of the cloud by de-tuning of the frequency of the vertical lasers; **c** Cloud expansion during the ballistic flight; **d** Second passage of the atoms through the microwave cavity and probing of the state population by laser irradiation and fluorescence detection

between the frequency source that drives the synthesis electronics producing the probing signal and the observed caesium transition frequency.

The relative difference between the observed and the unperturbed transition frequency due to several systematic effects amounts to about 10^{-13} only, much less than the 2×10^{-10} in beam clocks, and evaluation of several caesium fountains proved that they realize the SI second with an uncertainty in the low 10^{-16} range. See [17] for an overview on fountains and [18] for a detailed uncertainty evaluation for PTB's second fountain clock CSF2. In details, its uncertainty depends on the operational conditions that slightly change from period to period. During September 2016, CSF1 and CSF2 were operated with a stated uncertainty of 3.5×10^{-16} and 2×10^{-16}, respectively. During the last 24 months, including October 2016, data from 10 caesium fountain frequency standards were published in the context of collaboration with the Bureau International des Poids et Mesures (BIPM), see next section, with stated uncertainties ranging between 0.17×10^{-15} and 2×10^{-15}. Data are shown in Fig. 5 in the next section.

The frequency instability of caesium fountain clocks depends largely on the operational parameters, mostly on the atom number in the cloud, but also on the source of microwave radiation that irradiates the atoms. A few fountains are operated intentionally with very low atom numbers, thereby minimizing the frequency shift due to cold-atom collisions [7, 17]. On the other hand, low frequency instability is desirable, in particular when the frequency of reference transitions of optical clocks shall be measured in SI Hz as realized with the fountains. Here PTB has pioneered the routine use of an optically stabilized microwave oscillator [19] instead of a quartz-oscillator based microwave synthesis. The short term stability of the microwave signal is provided by a 1.5 μm cavity-stabilized fiber laser via a commercial femtosecond frequency comb. In the long-term, the microwave oscillator involved is locked to the hydrogen maser to enable fountain frequency measurements with respect to the maser (see Fig. 3). In this setup the instability contribution of the microwave oscillator via the so-called Dick-effect [7, 17] becomes negligible and the overall instability is mostly limited by the number of detected atoms. The respective curve (CSF2 2016) in Fig. 1 represents this situation.

Fountains operated in several institutes have been compared among each other over years, and the comparison uncertainty is the combined statistical and systematic uncertainty of the standards and the comparison techniques. In the most recent long term study [20] FO-2 of SYRTE/Observatoire de Paris (OP) was compared with NIST-F1 of the National Institute of Standards and Technology (NIST), USA, between 2006 and 2012, just to give one example. The difference FO2 - NIST-F1 was determined as -0.12×10^{-15}, based on 24 scattered data sets over periods where both fountains were operated simultaneously or quasi-simultaneously. The uncertainty that accompanies this value is not trivial to determine as the properties of the fountains and of the time transfer changed over the years. From [20], Table 1, I estimate 0.51×10^{-15} for the systematic part and 0.22×10^{-15} for the statistical part, where I assume the statistical contribution to decrease with the square-root of the number of comparisons. Each fountain frequency was corrected for the gravitational red shift, so as if both were operated at zero height. The corrections amounted to

6.54×10^{-15} for FO-2 and 179.95×10^{-15} for NIST-F1 (Boulder, Colorado, height some 1600 m above sea level). So the comparisons represent an - even quite crude - type of relativistic geodesy experiment, although it was never named like that. Determination of the red-shift correction at the Boulder site is quite challenging and was described in [21].

In fact, the French FO-2 is a double-fountain in which also rubidium atoms (^{87}Rb) are launched and their ground-state hyperfine transition at 6.8 GHz is observed. The use of Rb atoms instead of Cs is motivated by the fact that at a given number density of atoms in the cloud, the frequency shift due to cold atom collisions is considerably smaller. This would allow a better frequency stability, and this was the impetus for the United States Naval Observatory (USNO) to build a group of Rb-fountains that has shown indeed excellent performance [22].

3.3 *Optical Frequency Standards*

Frequency standards in the infra-red and in the visible range of the electromagnetic spectrum have been developed and used since decades. The most prominent use is as wavelength standards in practical length metrology and for the realization of the meter. Since the new definition of the meter became effective in 1983, this SI unit should be realized according to a mise-en-pratique. One method is the use of radiations whose wavelength in vacuum or whose frequency is stated in a list that is periodically updated and that can be retrieved from the BIPM web-site and found in the Appendix 2 of the BIPMs SI-brochure [23]. Detailed account of frequency standards used for this purpose can, e.g., be found in [24] .

It has been predicted for quite some time that the performance - in terms of accuracy and frequency instability - of a laser as a frequency standard, when stabilized to a suitable and narrow optical transition between a metastable state and the ground state, might surpass that of a frequency standard in the radio-frequency region. Reasons for that are at hand and were mentioned before (Sect. 1). The increase in Q-factor leads to a much reduced frequency instability at short averaging times which paves the way to explore systematically frequency shifting effects in a conveniently short measurement time. In addition, the magnitude of systematic energy level shifts is of the same order as in microwave atomic frequency standards, so that the relative uncertainty is dramatically reduced.

Three ingredients for an optical frequency standard to become feasible have been perfected during the last years: the required stable interrogation oscillator (clock laser), the optical frequency comb for counting the cycles of optical frequencies (glorified with the Nobel prize given to T. Hänsch in 2005 [25]), and confinement of laser-cooled atomic species to a range whose dimensions are smaller than the wavelength in radio-frequency or laser traps. The basic technology was described in the textbooks [7, 24], and a detailed survey on optical frequency standards with more than 200 references included was recently provided by Ludlow et al. [26].

Two distinct kinds of optical frequency standards have been developed, differing in the kind of trap used. Charged atoms (ions) can easily be trapped with electric fields without significantly disturbing their atomic energy levels and thus the resonance (clock) frequency. A "single ion at rest in free space" [27] represents the ultimate isolation of a spectroscopic object and thus the smallest systematic frequency shifts. But the frequency control of the clock laser is inevitably based on the (weak) signal obtained from one ion only. Neutral atoms need to be trapped using forces exerted by attacking the charge distribution inside the atom, which inevitably has an influence on the electronic structure. Seminal work of H. Katori showed how to set up a trap made from standing laser fields which shifts the two energy levels defining the clock transition equally [28]. In such an optical lattice ensembles of several thousands of atoms can be cooled and stored.

The storage and laser cooling methods are applicable to a great variety of atoms and ions in different charge states. When selecting an atom for an optical clock, the properties of the reference transition therefore play an important role as is described in detail in [26]. The most significant results from trapped ion frequency standards up to now have involved the ions $^{27}Al^{+}$, $^{40}Ca^{+}$, $^{88}Sr^{+}$, $^{115}In^{+}$, $^{171}Yb^{+}$ (Q), $^{171}Yb^{+}$ (O) (see explanations below), and $^{199}Hg^{+}$. Neutral-atom based frequency standards employed ^{87}Sr, ^{88}Sr, ^{171}Yb, and ^{199}Hg. Measurements of the optical transition frequencies of these species with respect to caesium fountain clocks in the relative uncertainty range of 10^{16} have been reported (except for ^{115}In). In PTB, two different experiments (with the ions Yb^{+} and with Sr atoms) succeeded in reaching this accuracy range. The $^{171}Yb^{+}$ possesses two suitable transitions, the $^{2}S_{1/2}$ - $^{2}D_{3/2}$ quadrupole transition at 436 nm (Q) and the $^{2}S_{1/2}$ - $^{2}F_{7/2}$ octupole transition (O) at 467 nm. At the time of writing, the octupole transition frequency can be realized with an uncertainty of 3×10^{-18} [29], and the ratio between the two transition frequencies is known with a relative uncertainty of about 10^{-16}. The uncertainty for the realization of the clock transition frequency in ^{87}Sr was estimated as 2×10^{-17}, and measurements of the frequency ratio $^{171}Yb^{+}(O)/^{87}Sr$ have been performed several times during the last years. This kind of ratio measurements are instrumental in fundamental physics studies, as discussed at the end. A transportable variant of the Sr optical frequency standard has been developed at PTB [30] and was recently used in a kind of demonstration exercise for chronometric leveling. Publication of results is pending.

This subsection was written as the last one of the manuscript, but nevertheless it will inevitably fail to be up-to-date at the time of its publication. The rate of progress is very high and the number of groups actively involved is quite large. Whether 1×10^{-18} relative uncertainty will have been reached already? I dare not make a prediction.

4 International Time Scales and Their Local Realizations

International Atomic Time (Temps Atomique International, TAI) and Coordinated Universal Time (UTC) are maintained and disseminated by the BIPM Time Department. About 75 NMIs and astronomical and scientific laboratories that operate atomic clocks of different kind and time transfer equipment are the players in the activity. Each of them, designated "k", realizes an approximation to UTC, denoted UTC(k), which is used as the reference for local clock comparisons. Data of the kind $[UTC(k) - clock_i(k)]$ are provided to the BIPM where i designates an individual clock (total number involved in 2016 about 400). Time transfer using calibrated equipment provides the differences between the UTC(k) time scales with ns-uncertainty (see next section), and respective data are provided to the BIPM as well.

Almost all atomic clocks involved are commercial caesium clocks and hydrogen masers, as described above. In addition, about 12 laboratories develop and maintain nowadays caesium fountain clocks. By combining the clock and time transfer data using the algorithm ALGOS, an averaged time scale, called free atomic time scale, is calculated. The algorithm was designed to provide a reliable scale with optimized frequency instability for one month averaging time [31]. Individual clocks contribute with statistical weights that are based on their performance during the last 12 months. In a second step, the relative departure of the scale of the free atomic time scale from the SI second is determined from data of the primary frequency standards. The departure is ideally brought to zero by a very gentle frequency adjustment, and the resulting scale is TAI. In Fig. 5 the comparison of the 10 fountain clocks for which data were made public during the last two years, including October 2016, with TAI is shown. Ideally, the data points would scatter around zero. The dispersion of the points mostly reflects the instability introduced by the comparison between the respective fountain, its local intermediate reference (hydrogen maser, UTC(k)), and TAI. The individual uncertainty contributions are reported in the monthly Circular T and explained in an explanatory supplement for which a link is provided on the web [32].

Some atomic transitions were recognized as so-called secondary representations of the second by a Working Group of the Consultative Committee (CC) for Length (CCL) and the CC for Time and Frequency [33]. The idea behind is to draw upon stable and accurate frequency standards for the monitoring of TAI and at the same time to prepare a solid data base for a decision about a future re-definition of the second. The 6.8 GHz ground-state hyperfine transition frequency in ^{87}Rb is one of them. Data from the French FO-2 have been reported in Circular T since 2012. In 2016 the first data from an optical frequency standard having the transition $5s^2\ ^1S_0 - 5s5p\ ^3P_0$ in ^{87}Sr at 429 THz as reference were submitted by SYRTE / OP to the BIPM. It can be expected that other research teams will follow this example during the coming years.

UTC, the final product, has the same scale unit as TAI, but differs from TAI by an integer number of seconds, introduced as "leap seconds" on request from the International Earth Rotation and References Systems Service (IERS) [8, 31].

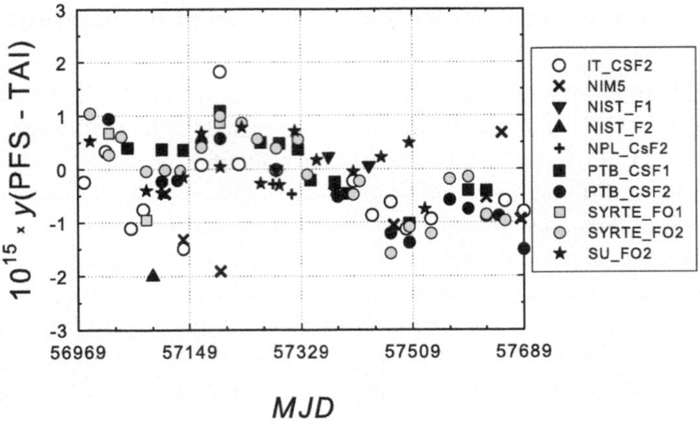

Fig. 5 Comparison of primary frequency standards with TAI during two years, ending MJD 57689 (2016-10-28), data taken from Circular T [32]. The standards are operated at Istituto Nazionale di Ricerca Metrologica, Torino, Italy (IT), LNE-SYRTE / Paris Observatory, NIST, National Institute of Metrology (NIM), Beijing, PR China, National Physical Laboratory (NPL), Teddington, UK, PTB, and VNIIFTRI, Mendeleevo, Russia (SU)

Dissemination of UTC by the BIPM happens in the form of time series $[UTC - UTC(k)]$ for selected dates in the past month being published in the Circular T [32]. The Circular T provides the traceability to international standards (time unit, frequency, and time scale) for each participating institute. The NMIs in turn are responsible for the dissemination of their UTC(k) in their respective countries and thus provide traceability to calibration services, academic institutes, and to the common public.

For some applications the publication of Circular T only once per month with the last reported value from typically more than ten days in the past was felt as an inconvenience. The BIPM Time Department implemented a rapid realization of UTC, called UTCr, which has been published every week since July 2013 [34]. UTCr gives daily values of $[UTCr - UTC(k)]$ for a subset of laboratories which committed to submit data daily (and thus fully automatic) to the BIPM. The difference $[UTC - UTCr]$ is at the few-nanoseconds level, but inevitably not zero: the clock ensemble is different (smaller) and also the time link data used are not the same as for UTC generation. So typically once per month a small time step aligns UTCr with UTC, which is considered as a nuisance for other applications for which a monotonous $[UTCr - UTC(k)]$ would be desirable.

There is no strict rule or recommendation governing the offset between UTC and UTC(k). For many years an offset of 100 ns was felt appropriate, more recently several institutes strive to stay within 10 ns. End of October 2016 the (absolute) difference $[UTC - UTC(k)]$ was below 20 ns for 39 institutes. To stay within 10 ns is definitely facilitated if the time scale is built from the average of a large ensemble of atomic clocks - as it used to be the situation at USNO for many years and continues to be today - or from a hydrogen maser steered towards a long-term stable, accurate

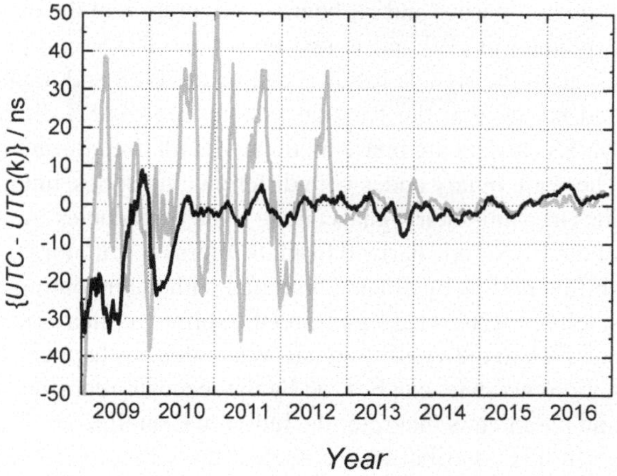

Fig. 6 Comparison of UTC with UTC(PTB) (black) and UTC(OP) (grey) during eight years

reference, such as a fountain clock. PTB was the first institute to go this way [35], starting in 2010. Figure 6 illustrates the success of this strategy that was adopted two years later also by the French laboratory LNE-SYRTE / OP.

5 Satellite-Based Time and Frequency Transfer

The comparison of distant clocks has always been an important part of time metrology. A comparison on a local and regional scale can be achieved with electrical signals transported in cables. Unsurpassed accuracy could be demonstrated during recent years by using optical fibers to transport either stabilized laser radiation or modulated laser signals [36, 37] even over 1000 km distances (see the contribution by G. Grosche in these proceedings.). On a global scale, however, the use of radio signals from or via satellites remains the first choice [38, 39]. Subsequently, two satellite-based methods are presented, the reception of signals of Global Navigation Satellite Systems (GNSS), and Two-Way Satellite Time and Frequency Transfer (TWSTFT).

5.1 GNSS-Based Time and Frequency Transfer

The primary purpose of all Global Navigation Satellite Systems (GNSS) is to serve as a positioning and navigation system. But each system relies on accurate timing, more precise, the satellite ranges used to calculate position are derived from propagation time measurements of the signals transmitted from each satellite in view. The signals broadcast by GNSS satellites are derived from onboard atomic clocks (caesium beam, rubidium gas cell frequency standards, passive hydrogen masers). Details of signal properties and on-board configuration of the existing GNSS are inter alia well explained in [39] and in text books on GNSS, e.g. [40]. A brief description of the usage of the signals follows, adapted to GPS and the European counterpart Galileo, and noting that the Russian system GLONASS differs in many details [39]. The GNSS carriers are phase modulated with pseudo-random noise codes (PRN-codes). These are binary codes with different chip codes, unique for each satellite. All satellites transmit their signals on the same frequencies. A receiver generates a local copy of the PRN-code derived from its internal oscillator. This local copy is electronically shifted in time by δt and correlated with the incoming antenna signal. If the received satellite PRN-coded signal and the shifted replica match, the receivers tracking loops can lock to the satellite signal. When this has happened data, usually called the navigation message, can be read by the receiver, reporting the almanac, orbit parameters and parameters that refer the individual satellite clock to the underlying GNSS time.

In this contribution we neglect the type of receiver that has de facto the widest general use, but that is inappropriate for accurate clock comparisons. It combines the recorded value δt with the navigation message to discipline the frequency of its inbuilt quartz oscillator to GNSS time and delivers standard frequency output and a one pulse per second electrical signal (1 PPS) representing GNSS time. Called GNSS disciplined oscillator (GNSSDO)[3] it is the common instrumentation in calibration laboratories, industry, wherever such signals are needed. Another variety of such instruments outputs the time-of-day information, converted from the navigation message, either in a clock display, in standard electrical time codes like IRIG, or for distribution in the Internet or in local area networks using the Network Time Protocol (NTP).

In what might be called "scientific" timing receivers, the measured time offset δt for each satellite in view with respect to the local reference signal connected to the receiver is stored. The information contained in the navigation message is used to provide output data in the form of local reference (local time scale) minus GNSS time. Modern receivers are capable to measure also the phase difference between the received carrier signal and the local reference once code lock had been established. Such GNSS carrier phase measurements are two orders of magnitude more precise than the code data. Code and carrier phase measurement results are usually output in the so-called receiver independent exchange format (RINEX) [41]. The current version that is adapted to the multitude of GNSS and transmitted signals is 3.03, but older versions (GPS + GLONASS only) are still common.

Precise point positioning (PPP) is a code and carrier phase-based analysis technique that has become very popular and most often combines GPS observations at the two transmit frequencies f_1 = 1575.42 MHz (L1) and f_2 = 1227.60 MHz (L2). PPP builds on the precise satellite orbits and clock products generated by the International GNSS Service IGS (see www.igs.org). The position of the antenna of an isolated GPS receiver is provided by PPP with high accuracy on a global scale. At the same time, the difference between local reference clock and IGS time, a time

[3]For improved hold-over capability, some models include a Rb frequency standard (see Sect. 3.1) that is steered with a few hours time constant.

scale generated by IGS, is calculated and reported in the output data. A software package in frequent use has been developed by National Resources Canada and the software has been generously made available to several timing laboratories for local installations, and an online service is also available [42].

Code-based time transfer in the popular common-view (CV) method has been used already for decades and has still its merits. It is built upon simultaneous reception of the transmitted signal from the same satellite by two receivers on Earth. Thereby the impact of common errors in the GPS signals caused by errors in the satellite position, instabilities of the satellite clocks, and the effects of the intentional degradation (known as "selective availability") that had been applied to the GPS signal until May 2000 are strongly reduced. Receivers of the first generation used for time comparison were single-channel, single-frequency (L1) receivers.[4] The propagation of GNSS signals are affected by atmospheric effects. The ionosphere provokes delays that can be modeled on a global scale only to a limited extent. Substantial errors occur, particularly during periods of high solar activity and when the receiver is at low latitude. As the ionosphere shows dispersion, group velocity and phase velocity are affected with opposite sign and depend on the carrier frequency. This property is used in advanced receivers that receive and process signals on both frequencies f_1 and f_2 to determine the ionospheric delay in situ. Data generated in this way are labeled as L3P-data.

With increasing availability and accuracy of IGS products, the common-view method has almost been replaced by GPS all-in-view (AV), which is in practice simpler to implement. After exchange of the (standardized [43]) data files among the laboratories, the individual observation data are corrected for the above mentioned effects based on IGS products before averages over convenient intervals are formed. Subtraction of corresponding data allows the comparison of the local time scales or frequency standards. Comparisons within Europe practically give the same results in CV and AV, even without the use of external products. AV is, however, particularly useful in intercontinental comparisons and thus widely used today by BIPM in its undertaking to realize TAI. Directives on a common format and standard formulae and parameters for code-based data evaluation were developed jointly by the BIPM and the CCTF [43].

Figure 7 illustrates the advantage of dual-frequency reception in a comparison between PTB Braunschweig and IMBH Sarajevo, Bosnia-Herzegovina. Data collection happened during 2014 in support of the operations establishment and receiver calibration of the time laboratory at the Bosnian NMI. L3P data are more noisy on short averaging times, but are free from daily variations seen in the single frequency data. Daily patterns can be explained with insufficient modeling of the ionosphere.

[4]As an aside, I remember well that my start as PhD student at PTB in 1983 coincided with the installation of the first receiver of this kind in the PTB time-unit laboratory, a single-channel GPS receiver for time transfer provided by the then National Bureau of Standards (now NIST). This receiver is now at display in the Deutsche Uhrenmuseum (German clock museum), Furtwangen, Black Forest.

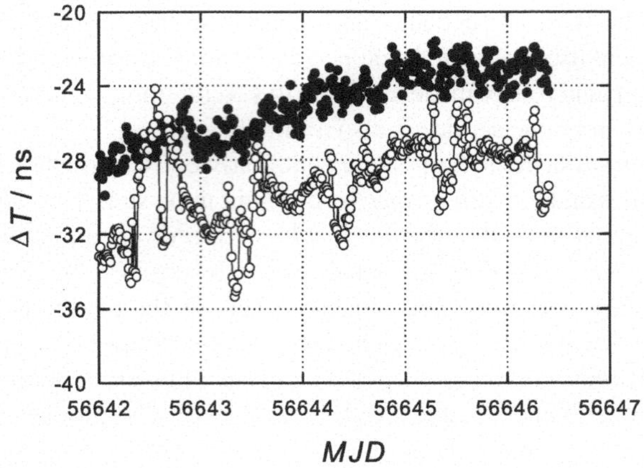

Fig. 7 GPS CV time comparisons between the NMI of Bosnia-Herzegovina, IMBH, Sarajevo, and PTB; open symbols: L1C single frequency data, full symbols: dual frequency L3P data obtained from the same receivers. The offset is due to incomplete delay calibration of the IMBH receiver at the time of data taking in 2014

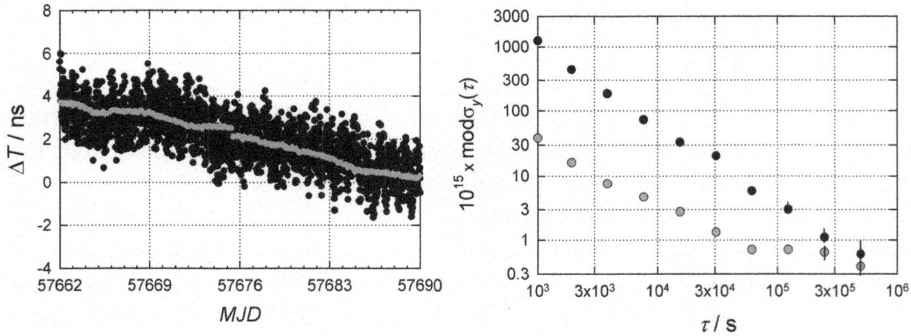

Fig. 8 GPS comparison $\Delta T = UTC(OP) - UTC(PTB)$, evaluated by BIPM, GPS P3 (black) and PPP (grey) data (left), frequency comparison instability (right)

The second example,[5] shown as Fig. 8 contrasts L3P (code-based) and PPP evaluation of data collected with receivers at Observatoire de Paris and PTB during October 2016. The kink in the PPP data around the middle of the period points to an interruption of continuous recording of observations at one (or both) receivers. The PPP solution then continues with a new estimate of the integer phase offset which is affected by the more noisy code-based measurements. The instability (kink removed) achieved with both methods converges at averaging times exceeding one day which reflects the instability of the two time scales that are compared (see also Fig. 6). The so-called Modified Allan variance (modσ_y) is a variant of the Allan variance

[5] Source of data: BIPM ftp server at ftp://ftp2.bipm.org/pub/tai/timelinks/lkc/, data in monthly files, Read Me file provided.

explained in Sect. 2 [4]. In general, it has been found that frequency comparisons between distant clocks using PPP links show an instability of about 1×10^{-15} at 1 day averaging time and a few 10^{-16} at 5–10 days averaging.

Joint work of BIPM and the French Centre National d'Etudes Spatiales (CNES) has led to improved capabilities of PPP for longer averaging times than one day. Their approach is called Integer PPP (IPP) and avoids the treatment of a priori unknown number of carrier cycles between the space craft antenna and the receiver on Earth (the "phase ambiguities") as floating point numbers as it is done in the "standard" PPP [44]. They demonstrated comparisons with an instability of 2×10^{-15} at 5 h averaging and 10^{-16} at 4 days. The data analysis is currently still quite laborious, but work is ongoing to make IPPP a useful tool, potentially also in the context of relativistic geodesy. As a demonstration of IPPP performance, a comparison between two fountains during 60 days could reveal whether the 10^{-17} region could be reached. But this has not happened yet.

5.2 *Two-Way Satellite Time and Frequency Transfer*

Two-way satellite time and frequency transfer (TWSTFT) is based on the exchange of signals between two active terminals A and B. Signals propagate between A and B in both directions, simultaneously, and at each terminal the time-of-arrival of the signal from the other side is recorded. All propagation delays cancel to first order when the two measurement results are combined to provide the time difference between the clocks connected to the two terminals. To connect terminals on Earth, TWSTFT is made using fixed satellite services in the Ku-band and the X-band, and geo-stationary telecommunication satellites serve as relay [38, 39, 45]. There are still several effects causing non-reciprocities which are discussed in detail in the literature [46]. Most of the propagation related effects can, however, account for small non-reciprocities only, typically not exceeding 0.1 ns, depending on the geometry of the locations of the stations and the satellite, and the transmission frequencies.

A brief description of the established services follows. Pseudorandom noise (PRN) binary phase-shift keying (BPSK) modulated carriers are transmitted. The phase modulation is synchronized with the local clock's 1 PPS output. Each station uses a dedicated PRN code for its BPSK sequence in the transmitted signal. The receiving equipment is capable to generate the BPSK sequence of each remote station and to reconstitute a 1 PPS tick from the received signal. This is measured by a time-interval counter (TIC) with respect to the local clock. Following a pre-arranged schedule, both stations of a pair lock on the code of the corresponding remote station for a specified period, measure the signal's time of arrival, and store the results. After exchanging the data records the difference between the two clocks is computed. Within Europe, both stations are within the same antenna footprint of the satellite, and signals are routed through the same transponder electronics. In this favourable case, the link delays can be calibrated and time transfer with uncertainty of 1 ns or even slightly below is feasible. Satellites rarely have an antenna footprint that is wide enough

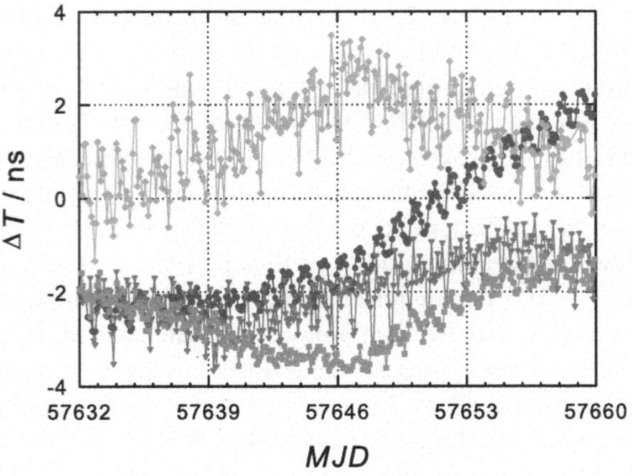

Fig. 9 TWSTFT comparisons UTC(PTB)-UTC(k) during September 2016, looking at the plots from top downwards: "k" = INRIM (Torino, Italy), NIST, SYRTE/OP and USNO

to cover both Europa and US or Europe and Asia, respectively, so that the above condition is not fulfilled. Here another method has to be used, and for many links between the Europe, the US and Asia GPS data have been used for delay calibration. TWSTFT has as well proven to provide a relative uncertainty for frequency transfer of about 1×10^{-15} at averaging times of one day [47]. TWSTFT is therefore used in the international network of time keeping institutions supporting the realization of TAI [32].

On the other hand, admittedly, TWSTFT as used today has some weakness, as can be identified by looking at some time transfer results shown in Fig. 9. Each of the data points represents the result of a two-minute data collection of time scale comparison between PTB and a remote institute, two in Europe and two in the US. Nominally there are 12 such measurements per day per link. In the plots we note different levels of noise, and apparently systematic variations with daily period, the strength of which is not constant with time. A lot of studies went into the cause of such "diurnals", and the above mentioned non-reciprocities were suspected as causes. But recently evidence was found that they are likely caused by the receiving electronics which cannot always cope with the changing receive frequency. It is modulated in a daily rhythm due to the classical Doppler effect proportional to the (small, order m/s) line-of-sight velocity of the geostationary satellite with respect to the receiving antenna. This effect and a good part of measurement noise could be suppressed or reduced if a larger phase-shift keying rate would be used. This would spread the signal power in a wider band, but would require a larger portion of the transponder bandwidth reserved - and paid for - for the application. But the cost is currently prohibitive to establish a continuous all-year service of such kind.

In addition to the routine comparisons with institutes in Europe and the USA, PTB supported recently two experiments aiming at improvements of the performance of TWSTFT links. One experiment involved European institutes in a collaboration funded by the European Commission, and consisted of the transmission/reception of signals with a 20-fold wider spectral distribution in the Ku-band region for a few weeks, thereby circumventing some limitations mentioned before. The results showed a good part of the expected reduction in short-term measurement noise, but no significant improvement in the long term. The other one was managed and evaluated by National Institute of Communications Technology (NICT), Japan. Here it could be demonstrated that transmission of signals in a quite small spectral band, but using the carrier phase as the measurement quantity provides frequency comparison with less than 10^{-15} instability when averaging longer than 20 000 s at quite favorable operational cost, which is considered as important as the performance itself. Research on that subject is ongoing.

6 Look Ahead and Conclusions

The "Atomic Clock Ensemble in Space" (ACES) unfortunately remains "on the horizon" only. The launch of the scientific mission that relies on the availability of the International Space Station (ISS) experienced once more delay into 2018. The ACES project will involve a space segment and a complex ground segment [48, 49]. The space segment will comprise PHARAO (Projet d'Horloge Atomique par Refroidissement d'Atomes en Orbite), a primary frequency standard based on laser-cooled caesium atoms, and an active hydrogen maser. An on-board time scale will be generated that should reflect the short-term instability of the maser and the long-term characteristics and accuracy of PHARAO. Connection to the ground is going to be provided by the so-called Microwave Link (MWL) and the European Laser Timing (ELT) space terminals. All that will be installed on the Columbus External Payload Facility. The MWL follows the principle of TWSTFT, now between space and ground, but the transmitted signals are going to occupy a hundred times wider bandwidth than those used routinely in these days. The MWL shall be used to compare the space clocks with high-performance ground clocks at seven sites worldwide where ground terminals are going to be installed. The ACES campaign of 18 months duration shall be used for a couple of fundamental physics studies and international clock comparisons. The projected frequency transfer capabilities are competitive with the uncertainty of a few optical frequency standards and could be used to verify the relativistic red shift of the frequency standards at the sites at the 10^{-17} level.

In general, satellite-based time and frequency transfer serves plenty of applications, and in particular with the advent of new freely available signals and newer modulation schemes on GNSS some improvement can be expected. But it seems unlikely that the gap in performance (stability) between optical frequency standards and the GNSS comparison techniques can be significantly reduced. The GNSS all-in-view technique is unique: it allows comparisons among laboratories wherever they are located on Earth, and installation of receiver and antenna is quite simple. TWSTFT can bridge approximately 10 000 km because both sites must simultaneously be in the field of view of the same satellite. The synchronization of the ground stations of the deep space tracking networks maintained by NASA and ESA has practically to rely on GNSS comparisons as they are separated by about 120 degree in longitude on the globe. For sure, TWSTFT offers high potentials in terms of achievable measurement noise and accuracy, but it remains open to see who is going to pay the bill.

"More accurate clocks - What are they needed for?" has been partially answered in [50], and many publications demonstrate the interdependence of metrology and fundamental science. The research into atomic frequency standards and time and frequency dissemination can help improving our understanding of the laws of physics in general. Without any pretension of completeness let me first mention pulsar timing as a fascinating branch of radio-astronomy [51]. Although I doubt that pulsar time scales [52] are going to have properties adequate for replacing atomic time scales, we remember that pulsar timing was the basis for the indirect proof of the existence of gravitational radiation emitted by the binary system of two neutron stars [53, 54]. The emission of gravitational waves was predicted and the accompanying orbital energy loss became observable as an orbital period change. A very active field of study in these days is search for variations of fundamental constants. Several parameters that are usually designated as "constants", such as charge and mass of the electron as well as the fine structure constant α are predicted to vary on cosmological time scales, and laboratory searches of the - if at all - tiny temporal variations today involve atomic frequency standards [55]. As explained in [55], atomic transition frequencies depend in a different way on these constants. Measurement of the ratio of atomic transition frequencies of different atomic species (or of transition frequencies in the unit Hertz as realized with caesium fountain clocks) repeatedly over time allows under certain circumstances to determine limits on the temporal variations in our days. The uncertainty of such ratio measurements got lower and lower with time during recent years, mostly due to the dramatic improvement in the performance of optical frequency standards, as laid out in Sect. 3.3. Just to give one example, at the time of writing (April 2017), the tightest limit of variation of α is in the low 10^{-18} per year, and not statistically significant [56]. Complemented by searches involving astrophysical data, such laboratory searches may in the future point to new physics beyond the standard model.

Disclaimer

The mentioning of individual products and their manufacturers is not to be understood as endorsement by PTB. Data obtained at PTB reflect the properties of the selected equipment and its installation conditions and may deviate from observations made at other sites.

Acknowledgements This review paper reports mostly on achievements of colleagues from all over the world. The fruitful collaboration belonged to the pleasures of the author's business life. Special thanks go to Ekkehard Peik, Dirk Piester and Stefan Weyers of PTB for critical reading of the manuscript.

References

1. A. Bjerhammar, On relativistic geodesy. Bull. Geod. **59**, 207 (1985)
2. H. Denker et al., Geodetic methods to determine the relativistic redshift at the level of 10^{-18} in the context of international timescales - a review and practical results. J. Geod. **92**, 487 (2018) to be published
3. *Standard definitions of Physical Quantities for Fundamental Frequency and Time Metrology, IEEE-Std 1189–1988* (IEEE, 1988)
4. W.J. Riley, *Handbook of Frequency Stability Analysis, NIST Special Publication*, vol. 1065 (NIST, 2008)
5. Joint Committee for Guides in Metrology, *Evaluation of Measurement Data - Guide to the Expression of Uncertainty in Measurement*. JCGM 100:2008 (2008)
6. J. Vanier, C. Audoin, *The Quantum Physics of Atomic Frequency Standards* (Adam Hilger, Bristol, 1989)
7. J. Vanier, C. Tomescu, *The Quantum Physics of Atomic Frequency Standards - Recent Developments* (CRC Press, Taylor & Francis Group, Boca Raton, 2016)
8. C. Audoin, B. Guinot, *The Measurement of Time* (Cambridge University Press, Cambridge, 2001)
9. E.F. Arias, A. Bauch, *Metrology of time and frequency, in Handbook of Metrology* ed. by M. Gläser, M. Kochsiek (WILEY-VCH Verlag, Weinheim, 2010), p. 315
10. R. Lutwak et al., The MAC - a miniature atomic clock. *Proceedings 2005 International Frequency Control Symposium and Exhibition* (2005), p. 752
11. P. Forman, Atomichron®: the atomic clock from concept to commercial product. Proc. IEEE **73**, 1181 (1985)
12. L.S. Cutler, Fifty years of commercial caesium clocks. Metrologia **42**, S90 (2005)
13. A. Bauch, The PTB primary clocks CS1 and CS2. Metrologia **42**, S43 (2005)
14. N.F. Ramsey, Experiments with separated oscillatory fields and hydrogen masers (Nobel Lecture). Rev. Mod. Phys. **62**, 541 (1990)
15. R. Piriz et al., The time validation facility (TVF): an all-new key element of the Galileo operational phase. *Proceedings of the Joint IEEE International Frequency Control Symposium and the European Frequency and Time Forum, Denver* (2015), p. 320
16. H.J. Metcalf, P. van der Straten, *Laser-Cooling and Trapping* (Springer, New York, 1999)
17. R. Wynands, S. Weyers, Atomic fountain clocks. Metrologia **42**, S64 (2005)
18. V. Gerginov et al., Uncertainty evaluation of the caesium fountain clock PTB-CSF2. Metrologia **47**, 65 (2010)
19. B. Lipphard, V. Gerginov, S. Weyers, Optical stabilization of a microwave oscillator for fountain clock interrogation. IEEE Trans. Ultrason. Ferroelectr. Freq. Control. **99**, 99 (2016)
20. G. Petit, G. Panfilo, Comparison of frequency standards used for TAI. IEEE Trans. Intrumentation Meas. **62**, 1550 (2013)
21. N.K. Pavlis, M.A. Weiss, The relativistic redshift with 3×10^{-17} uncertainty at NIST, Boulder, Colorado, USA. Metrologia **40**, 66 (2003)
22. S. Peil et al., Evaluation of long term performance of continuously running atomic fountains. Metrologia **51**, 263 (2014)
23. BIPM, Le Système international d'unités (The International System of Units), 8th edition. BIPM, Pavillon de Breteuil, F-92312 Sèvres Cedex, France, 2006, 2014 update)
24. F. Riehle, *Frequency Standards: Basics and Applications* (Wiley VCH, Weinheim, 2006)
25. T.W. Hänsch, Passion for precision (Nobel lecture). Rev. Mod. Phys. **78**, 1297 (2006)
26. A.D. Ludlow et al., Optical atomic clocks. Rev. Mod. Phys. **87**, 637 (2015)
27. H. Dehmelt, Coherent spectroscopy on single atomic system at rest in free space. J. Phys. (Paris) **42**, C8–299 (1981)
28. H. Katori, Spectroscopy of Strontium atoms in the Lamb-Dicke confinement, in *Proceedings of the 6th Symposium on Frequency Standards and Metrology* (2001). ((St. Andrews, Scotland), the proceedings were published by World Scientific, Singapore, in 2002, p. 323)

29. N. Huntemann et al., Single-ion atomic clock with 3×10^{-18} systematic uncertainty. Phys. Rev. Lett. **116**, 063001 (2016)
30. S.B. Koller et al., Transportable optical lattice clock with 7×10^{-17} uncertainty. Phys. Rev. Lett. **118**, 073601 (2017)
31. B. Guinot, E.F. Arias, Atomic time-keeping from 1955 to the present. Metrologia **42**, S20 (2005)
32. BIPM Time Department. Circular T. BIPM, http://www.bipm.org/en/bipm-services/timescales/time-ftp/Circular-T.html
33. P. Gill, F. Riehle, On secondary representations of the second, in *Proceeding of the 20th European Frequency and Time Forum* (2006), p. 282
34. G. Petit et al., UTCr: a rapid realization of UTC. Metrologia **51**, 33 (2014)
35. A. Bauch et al., Generation of UTC(PTB) as a fountain-clock based time scale. Metrologia **49**, 180 (2012)
36. C. Lisdat et al., A clock network for geodesy and fundamental science. Nat. Commun. **7**, 12443 (2016)
37. Ł. Śliwczyński et al., Fiber-optic time transfer for UTC-traceable synchronization for telecom networks. IEEE Commun. Stand. Mag. **1**, 66 (2017)
38. J. Levine, A review of time and frequency transfer methods. Metrologia **45**, S162 (2008)
39. ITU Study Group 7, *ITU Handbook: Satellite Time and Frequency Transfer and Dissemination* (International Telcommunication Union, Geneva, 2010)
40. E.D. Kaplan, C.J. Hegarty (eds.), *Understanding GPS, Principles and Applications*, 2nd edn. (Artech, Boston, London, 2006)
41. RINEX Working Group and RTCM-SC104, RINEX The Receiver Independent Exchange Format Version 3.02. International GNSS Service (2013), https://igscb.jpl.nasa.gov/igscb/data/format/rinex302.pdf
42. J. Kouba, P. Heroux, Precise point positioning using IGS orbit and clock products. GPS Solut. **5**(2), 12 (2002)
43. P. Defraigne, G. Petit, CGGTTS-Version 2E: an extended standard for GNSS time transfer. Metrologia **52**, G1 (2015)
44. G. Petit et al., 10^{-16} frequency transfer by GPS PPP with integer ambiguity resolution. Metrologia **52**, 301 (2015)
45. D. Kirchner, Two-way satellite time transfer. *Review of Radio Science 1996–1999* (1999), p. 27
46. A. Bauch, D. Piester, M. Fujieda, W. Lewandowski, Directive for operational use and data handling in two-way satellite time and frequency transfer (TWSTFT). BIPM, Rapport 2011/01, 2011 (2011)
47. A. Bauch et al., Comparison between frequency standards in Europe and the USA at the 10^{-15} uncertainty level. Metrologia **43**, 109 (2006)
48. L. Cacciapuoti, Chr. Salomon, Space clocks and fundamental tests: the ACES experiment. Eur. Phys. J. Spec. Top. **172**, 57 (2009)
49. L. Cacciapuoti, Chr. Salomon, Atomic clock ensemble in space. J. Phys. Conf. Ser. **327**, 012049 (2011)
50. E. Peik, A. Bauch, More accurate clocks - What are they needed for? PTB-Mitteilungen **119**(2), 16 (2009)
51. I.H. Stairs, Testing general relativity with pulsar timing. Living Rev. Relativ. (2003), http://relativity.livingreviews.org/open?pubNo=lrr-2003-5&=node5.html
52. G. Hobbs, Development of a pulsar-based time-scale. Mon. Not. R. Astron. Soc. **427**, 2780 (2012)
53. R.A. Hulse, The discovery of the binary pulsar (Nobel Lecture). Rev. Mod. Phys. **66**, 699 (1994)
54. J.H. Taylor, Binary pulsars and relativistic gravity (Nobel Lecture). Rev. Mod. Phys. **66**, 711 (1994)
55. N. Huntemann et al., Improved limit on a temporal variation of m_p/m_e from comparisons of Yb^+ and Cs atomic clocks. Phys. Rev. Lett. **113**, 210802 (2014)
56. C. Lisdat, Private communication (2017)

Chronometric Geodesy: Methods and Applications

Pacome Delva, Heiner Denker and Guillaume Lion

1 Introduction

The theory of general relativity (GR) was born more than one hundred years ago, and since the beginning has striking prediction success. Einstein proposed three effects for its experimental verification, all verified shortly after their prediction: the perihelion precession of Mercury's orbit, the deflection of light by the Sun, and the gravitational redshift of spectral lines of stars. Other predictions from GR had to wait decades before being confirmed experimentally. It is only in 1959 that the gravitational redshift is confirmed in a laboratory experiment by Pound, Rebka and Snider [1–4]. Two gamma-ray emitting iron nuclei at different heights were compared, verifying GR prediction with a relative accuracy of 10% (and later $<1\%$). In parallel, the era of atomic time began in 1955 with the caesium frequency standard built by Essen and Parry at the National Physical Laboratory (NPL) [5, 6]. Since then, the accuracy and stability of atomic clocks were constantly ameliorated, with around one order of magnitude gained every ten years (see Fig. 1).

In this context, the unit of time of the International System of Units (SI), the second, was officially defined with respect to a specified hyperfine transition of the caesium atom in the year 1967–1968.[1] Moreover, as local atomic timescales were

[1] Resolution 1 of the 13th General Conference on Weights and Measures (CGPM) [7].

P. Delva (✉)
SYRTE Observatoire de Paris, Université PSL, CNRS, Sorbonne Université,
LNE 61 avenue de l'Observatoire, 75014 Paris, France
e-mail: pacome.delva@obspm.fr

H. Denker
Institut für Erdmessung, Leibniz Universität Hannover (LUH),
Schneiderberg 50, 30167 Hannover, Germany

G. Lion
LASTIG LAREG IGN, ENSG, Univ Paris Diderot, Sorbonne Paris Cité,
35 rue Hélène Brion, 75013 Paris, France

D. Puetzfeld and C. Lämmerzahl (eds.), *Relativistic Geodesy*,
Fundamental Theories of Physics 196, https://doi.org/10.1007/978-3-030-11500-5_2

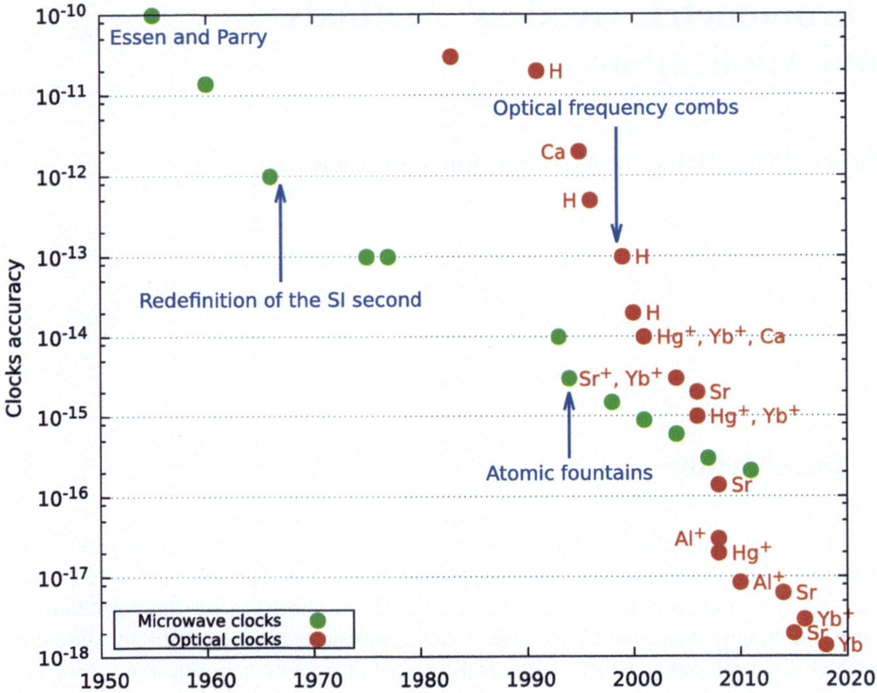

Fig. 1 Accuracy records for microwave and optical clocks. From the first Cs clock by Essen and Parry in the 1950's, an order of magnitude was gained every ten years. The advent of optical frequency combs boosted the performances of optical clocks, and they recently surpassed microwave clocks

developed thanks to commercial caesium clocks [8] as well as laboratory caesium standards, there was a need to compare these different timescales in order to build a mean international atomic time, which could be adopted by everyone. This was done at the beginning by the United States Naval Observatory (USNO) and the Bureau International de l'Heure (BIH) with radio time signals, which allowed time comparisons with uncertainties of the order of 1 ms [9]. A jump in accuracy occured with the first demonstration from the Hewlett-Packard company of the possibility of using commercial flights to transport its caesium clocks, allowing time transfer with an uncertainty of around 1 μs [9]. This eventually led to the famous test of Hafele and Keating [10, 11], who flew four caesium beam clocks around the world on commercial jet flights during October 1971. They predicted and measured the desynchronization of the proper times of these commercial atomic clocks with respect to the USNO atomic scale, and thus verified the gravitational redshift effect with an accuracy of around 12%.[2]

---

[2]This is based on the numbers given in table 1 of [10] and table 1 of [11]: the relative accuracy of the gravitational part of the relativistic shift effect is taken as $(\sqrt{18^2 + 10^2 + 7^2}\ \text{ns})/(179\ \text{ns})$.

Following the Hafele and Keating experiment, Briatore and Leschiutta did the first experimental measurements of the gravitational redshift with a direct comparison of ground caesium beam atomic frequency standards [12]. The two clocks were separated in heights by $\Delta h = 3250$ m, predicting a desynchronization of $\Delta t/t \approx g\Delta h/c^2 \approx 30.6$ ns d^{-1}, where g and c are the local gravity and the velocity of light in vacuum. The measurement gave (36.5 $\pm$ 5.8) ns d^{-1}, giving a relative accuracy of around 20%. Now, we can say that this experiment is amongst the first demonstrations of chronometric levelling (see Sect. 2.4): the clock comparison measured a difference of altitude between the two clocks of (3880 $\pm$ 620) m, to be compared with the otherwise measured value of 3250 m.

Now that the atomic clock accuracy reaches the low 10^{-18} in fractional frequency (see Fig. 1), and can be compared to this level over continental distances with optical fibres (see Sect. 3.3), the accuracy of chronometric levelling reaches the cm level and begins to be competitive with classical geodetic techniques such as geometric levelling and GNSS/geoid levelling. Moreover, the building of global timescales requires now to take into account these effects to the best possible accuracy. It is the topic of this chapter to explain how atomic clock comparisons and the building of timescales can benefit from the latest developments in physical geodesy for the modelization and realization of the geoid, as well as how classical geodesy could benefit from this new type of observable, which are clock comparisons that are directly linked to gravity potential differences.

In Sect. 2 we introduce fundamental concepts of GR concerning the measurement of time, relativistic reference systems and we review the recent literature of chronometric geodesy. In Sect. 3 we introduce the theory of frequency standard comparisons, beginning with the Einstein equivalence principle, followed by the description of the frequency techniques, and finally, we describe clock syntonization and the realization of timescales. Section 4 describes the geodetic methods for determining the gravity potential, namely the geometric levelling approach and the GNSS/geoid approach, as well as considerations about the uncertainties of these methods. In Sect. 5 we describe the European project International Timescales with Optical Clocks (ITOC) where unified relativistic redshift corrections were determined for several atomic clocks in European national metrology institutes. Finally, in Sect. 6 we present numerical simulations exploring what could be the contribution of clock comparisons for the determination of the geoid.

2 The Relativistic Framework

2.1 Observers and the Spacetime Manifold

The theoretical background of chronometric geodesy is general relativity (GR). In GR theory, space and time are bound together in a continuous entity named *spacetime*. Spacetime geometry specifies how matter and energy behave, while matter and energy distribution tells how spacetime geometry is *curved*. This is a non-linear process and the link between geometry on one side, and matter/energy on the other side is given by the Einstein equations. Gravitation is no longer a force as in Newtonian theory, but the manifestation of the variation of the background geometry. Variations of spacetime can be induced by a choice of coordinates, causing inertial effects, which act in a way similar but not equivalent to gravitation. The presence of energy/matter gives rise to curvature of the background geometry, and therefore gravity. However, gravitational effects can never be completely disentangled from inertial effects.

A spacetime is formally described by a four-dimensional differentiable manifold $\mathcal{M}$ endowed with a pseudo-Riemannian metric $\boldsymbol{g}$. A point of the manifold is called an *event*. Let us define an open subset $\mathcal{U} \in \mathcal{M}$ and an event P in this open subset. A *chart* or *coordinate system* $\{x^\alpha\}_{\alpha=0...3}$ can be defined in $\mathcal{U}$; it maps point $P \in \mathcal{U}$ to a point $\{x^\alpha_P\} \in \mathbb{R}^4$. The four real numbers x^α_P are the *coordinates* of event P. Generally a coordinate system cannot be defined on the whole manifold. An atlas is a collection of charts with some properties, which cover the whole manifold.

As the manifold is smooth, the difference vector $\mathrm{d}\vec{\boldsymbol{x}}$ between two infinitesimally close events may be defined. Then each event P is associated to a vector space $\mathcal{T}_P(\mathcal{M})$, called the *tangent space*, which contains the set of all possible four-vectors $\mathrm{d}\vec{\boldsymbol{x}}$. A basis $\{\vec{\boldsymbol{e}}_\alpha\}$ of the tangent space is usually called a *frame*. The introduced coordinates induce a coordinate basis $\vec{\boldsymbol{e}}_\alpha = \{\vec{\boldsymbol{\partial}}_\alpha\}_P \equiv \{\partial/\partial x^\alpha\}_P$. However, a frame does not need to be associated with any coordinate system.

The metric tensor $\boldsymbol{g}$ is a symmetric bilinear scalar function of two vectors. Given two vectors $\vec{\boldsymbol{v}}$ and $\vec{\boldsymbol{w}}$, the metric tensor returns a scalar called the dot product: $\boldsymbol{g}(\vec{\boldsymbol{v}}, \vec{\boldsymbol{w}}) = \vec{\boldsymbol{v}} \cdot \vec{\boldsymbol{w}} = \vec{\boldsymbol{w}} \cdot \vec{\boldsymbol{v}} = \boldsymbol{g}(\vec{\boldsymbol{w}}, \vec{\boldsymbol{v}})$. The metric can be characterized by its action on a basis of the tangent space. For example, $g_{\alpha\beta} \equiv \vec{\boldsymbol{\partial}}_\alpha \cdot \vec{\boldsymbol{\partial}}_\beta$ are the components of the metric tensor in the natural frame associated with the coordinate system $\{x^\alpha\}$. The infinitesimal interval

$$\mathrm{d}s^2 \equiv \mathrm{d}\vec{\boldsymbol{x}} \cdot \mathrm{d}\vec{\boldsymbol{x}} = g_{\alpha\beta}\mathrm{d}x^\alpha \mathrm{d}x^\beta \tag{1}$$

between two neighboring events is invariant under coordinate transformation. Trajectories of observers are defined by worldlines $\mathcal{C}$, which are parameterized 1-D curves in the manifold $\mathcal{M}$.

One of the most striking consequences of GR is the fact that coordinates do not have a direct physical interpretation as in the Newtonian theory. Indeed, one has to distinguish proper quantities from coordinate quantities. A proper quantity is the result of a physical measurement in a real or *Gedanken* experiment. It is mathematically described by a scalar, a quantity which is invariant under general coordinate transformations. However, a scalar quantity is not necessarily a physical measurement. The latter needs to be defined with respect to a reference frame adapted to the observer. The proper time τ of a clock is implicitly defined with the relation

$$\mathrm{d}s^2 = -c^2\mathrm{d}\tau^2 \,, \tag{2}$$

where c is a constant velocity characterizing the spacetime, which can be identified with the velocity of light in vacuum [13].

Proper time is invariant under general coordinate transformations. The proper times of two different clocks can be compared by means of a time transfer technique, while their proper frequencies can be compared with frequency transfer techniques (see Sect. 3). Time and frequency comparisons are linked, but both approaches lead to different formalisms and experimental techniques. Usually time transfer is more challenging, as it necessitates the knowledge of many instrumental delays with accuracy, while they can often be neglected in the frequency transfer.

When the spatial separation of two clocks is much smaller than the typical length of the background curvature of spacetime, then curvature effects can safely be neglected. This is a consequence of the Einstein equivalence principle (see e.g. [14]). This type of measurement will be termed *local comparison* of clocks. On the contrary, if the distance between both clocks is of the order or bigger than the typical length of the background curvature, the result of the comparison will have curvature perturbations, which depend on both the locations of the clocks and the particular time or frequency transfer technique. This type of measurement will be termed *non-local* or *distant comparison* of clocks.

The relation between proper time and coordinate time can be deduced from Eqs. (1)–(2):

$$\mathrm{d}\tau = \frac{1}{c}\sqrt{-\mathrm{d}\vec{x}\cdot\mathrm{d}\vec{x}}\ . \tag{3}$$

Let us integrate this relation along the worldline $\mathcal{C} : x^\alpha = f^\alpha(\lambda)$ parameterized by λ, between two events A and B belonging to $\mathcal{C}$ (see Fig. 2). The associated tangent vector is $\vec{\boldsymbol{v}} = \dot{f}^\alpha \vec{\boldsymbol{\partial}}_\alpha$, where $\dot{()} \equiv \mathrm{d}/\mathrm{d}\lambda$, and $\mathrm{d}\vec{x} = \vec{\boldsymbol{v}}\mathrm{d}\lambda$, such that:

$$\tau(A, B) \equiv \int_A^B \mathrm{d}\tau = \frac{1}{c}\int_{\lambda_A}^{\lambda_B} \mathrm{d}\lambda\sqrt{-g_{\alpha\beta}\dot{f}^\alpha\dot{f}^\beta}\ . \tag{4}$$

The parameter λ is usually chosen as the coordinate $x^0 = ct$ in the context of relativistic time and frequency transfer. It is clear from this formula that the proper time elapsed between two events A and B depends on the worldline $\mathcal{C}$, i.e. on the trajectory of the observer between these two events. One consequence of GR is that the parameter λ cannot be adapted such that both proper and coordinate time be equal everywhere. This is possible only in special relativity where there is no curvature in a well chosen reference system.

2.2 *Simultaneity and Synchronization*

Let us define an observer $\mathcal{O}$ with trajectory $\mathcal{C}$ parameterized by its proper time τ, and an event M which does not belong to $\mathcal{C}$ (see Fig. 2). How can we define an event on $\mathcal{C}$ which is simultaneous with event M. This is possible in the Newtonian space

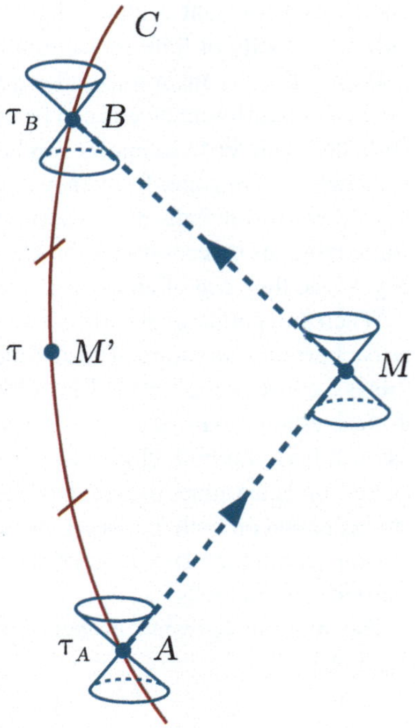

Fig. 2 Illustration of the Einstein synchronization convention

which is Euclidean, and therefore time is absolute, i.e. independent from the observer. However, in GR proper time is defined only along the worldline of the observer and is not a global property of spacetime. Only the light cone is a fundamental element of a given spacetime. The light cone is the collection of vectors $\vec{v} \in \mathcal{T}_P(\mathcal{M})$ which satisfies $\vec{v} \cdot \vec{v} = 0$. It is independent of the observer and divides the tangent space in three parts, *past* and *future* containing *time vectors* which satisfy $\vec{v} \cdot \vec{v} < 0$, and a third part containing *space vectors* which satisfy $\vec{v} \cdot \vec{v} > 0$. It is supposed here that the metric tensor has a signature $(-, +, +, +)$, i.e. at least one basis of the tangent space exists for which $\vec{v} \cdot \vec{w} = -v^0w^0 + v^1w^1 + v^2w^2 + v^3w^3$, where $\vec{v}, \vec{w} \in \mathcal{T}_P(\mathcal{M})$.

With the notion of the light cone, spacetime can be time-oriented, but it does not say which set of events can be considered simultaneous. Indeed in GR simultaneity can only be conventional and not an intrinsic property of spacetime. Einstein has suggested an operational definition of simultaneity. Suppose that an observer $\mathcal{O}$ is equipped with a clock and a system to send and receive electromagnetic signals. A signal is sent at event $A \in \mathcal{C}$, received and reflected with no delay at event M and finally received at event $B \in \mathcal{C}$ (see Fig. 2). The proper times along $\mathcal{C}$ corresponding to events A and B, respectively τ_A and τ_B, are measured with the clock. By convention, the event M' which is simultaneous to M along the observer trajectory corresponds to the proper time:

$$\tau = \frac{1}{2}(\tau_A + \tau_B) = \tau_A + \frac{1}{2}(\tau_B - \tau_A) . \tag{5}$$

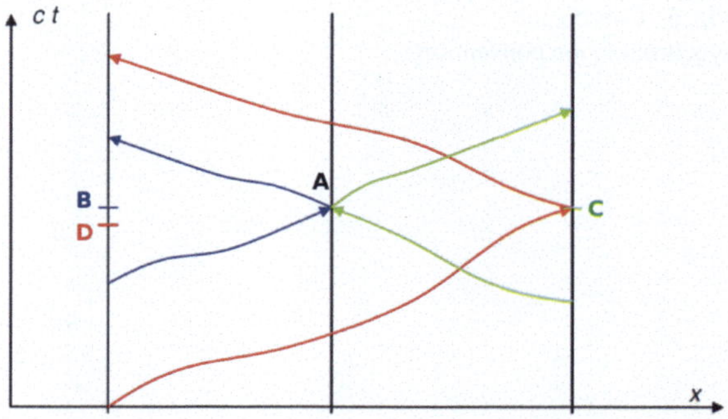

Fig. 3 Einstein synchronization convention is not transitive: events B and C are defined as simultaneous with A thanks to the convention, while D is defined as simultaneous to C. However, events B and D do not coincide

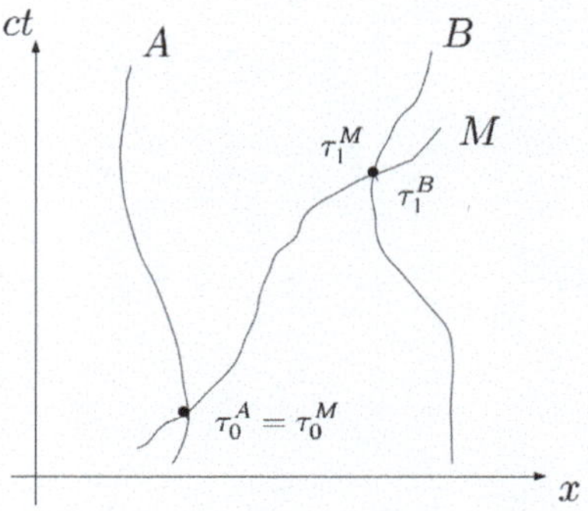

Fig. 4 Slow clock transport synchronization convention

This convention is usually called *Einstein synchronization*. It is a geometrical convention based on the concept of light cones, and an operational convention based on the exchange of electromagnetic signals. However this convention is not transitive. This is illustrated in Fig. 3. Events B and C are simultaneous to A (with Einstein synchronization); D is simultaneous to C, but D and B events do not generally coincide. Therefore this convention is not practical to define global timescales such as the TAI (Temps Atomique International). This problem was discussed in [15–17] in the context of satellite clock synchronization. However, in these articles the problem was thought as "synchronization errors", but it was in fact well understood in the context of GR, as noted in [18–20].

Another convention is the slow clock transport synchronization. Let us define three clocks A, B and M with corresponding worldlines $\mathcal{C}_A$, $\mathcal{C}_B$ and $\mathcal{C}_M$ (see Fig. 4). Clocks A and M are compared locally at event A_0 such that clock M proper time is set to $\tau^M = \tau_0^M = \tau_0^A$ at this event. Then clock M goes toward clock B and crosses worldline $\mathcal{C}_B$ at event B_1, where $\tau^M = \tau_1^M$ and $\tau^B = \tau_1^B$. The event B_0 on $\mathcal{C}_B$ is defined simultaneous to event A_0 on $\mathcal{C}_A$ by the slow clock transport synchronization convention with:

$$\tau_0^B = \lim_{v \to 0}[\tau_1^B - (\tau_1^M - \tau_0^M)] \,, \tag{6}$$

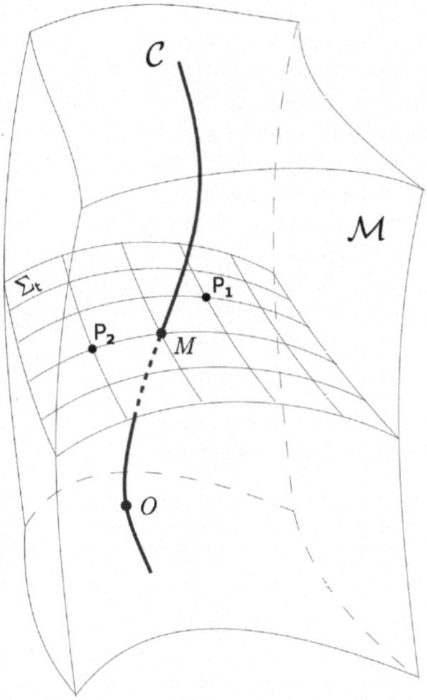

Fig. 5 Coordinate synchronization convention

where v is the coordinate velocity of clock M. The limit condition of null velocity is not feasible in a real experiment. Therefore, operationally, this convention depends on the particular trajectory of the mobile clock M, and reaches a different synchronization than the Einstein synchronization convention, as shown in [16]. However, in special relativity, i.e. with a null background curvature, it can be shown that both synchronization conventions are equivalent. If the spacetime geometry and the mobile clock trajectory are sufficiently known, then in the weak-field and low velocity approximation it is possible to use the coordinate time of clock M at events A_0 and B_1 instead of its proper time, so that the convention will not depend on the particular trajectory of the mobile clock. However it will depend on the relativistic coordinate system chosen to calculate the coordinate time. The inaccuracy of the time transfer operated with this convention can be assessed with the closing relation: $\tau_0^A = \lim_{v\to 0}[\tau_1^A - (\tau_1^M - \tau_0^M)]$.

Finally, we define the coordinate synchronization convention: two events P_1 and P_2 of coordinates $\{x_1^\alpha\}$ and $\{x_2^\alpha\}$ are considered to be simultaneous if the values of their time coordinates are equal: $x_1^0 = x_2^0$ (see Fig. 5). This definition follows the definition of simultaneity adopted in special relativity in [21, 22]. It is convenient to introduce three-dimensional hypersurfaces with constant time coordinate t: $\Sigma_t \equiv \{P \in \mathcal{M}, x_P^0 = ct\}$. By choosing a particular relativistic reference system we introduce a conventional foliation of spacetime, giving the hypersurfaces of simultaneity. The synchronization of clocks with this convention obviously depends on the chosen reference system. It is the most commonly used convention for the building of timescales such as TAI and Global Navigation Satellite System (GNSS) timescales.

Indeed, this convention is very similar to what is well known in Newtonian physics where the foliation of spacetime is absolute. For this convention to become operational it is necessary to define conventional relativistic reference systems. In special relativity, for clocks which are at rest with respect to an inertial reference system, the Einstein synchronization is a convenient procedure to achieve coordinate synchronization of clocks.

It was proposed in [23] to build a global "coordinate time grid" on and around the Earth, therefore realizing the idea of coordinate synchronization convention for clocks, without any problem of transitivity. The authors proposed to take as a reference a clock on the geoid,[3] i.e. to choose a conventional reference system $\mathcal{R}$ such that the proper time of a clock at rest on the geoid coincides with its coordinate time in $\mathcal{R}$. We will see later that this choice is convenient because it implies a simple link between the relativistic correction of a clock, in order to realize the coordinate time synchronization, and its altitude. The authors in [23] detailed several operational methods of time transfer using the coordinate synchronization convention: portable clocks, one-way and two-way synchronization with electromagnetic signals. The same authors in [24] estimated the main limitation on the determination of coordinate time: the knowledge of the geoid.

It is interesting to note that the question of synchronization of clocks in non-inertial reference systems raised a controversy in the 80's, driven by the development of Navstar Global Positioning System (GPS) and the need for a global timescale on Earth. This is reviewed in [25], where the author concludes: "In principle, the curved Schwarzschild space cannot be imbedded in a four-dimensional flat space without the addition of more dimensions. Thus the theoretical basis for the GPS navigational scheme would appear to be flawed, and a new algorithm would have to be constructed". Indeed, coordinate time synchronization can only be theoretically realized in approximation schemes, e.g. post-Newtonian approximation as reviewed in [26] for GPS. A different relativistic approach to this problem has been initiated in [27], where the idea is to give to a constellation of satellites the possibility to constitute by itself a primary and autonomous positioning system, without any need for synchronization of the clocks. Such a *relativistic positioning system* is defined with the introduction of *emission coordinates*, which have been re-introduced by several authors in the context of navigation systems [28–36].

A resolution concerning the global "coordinate time grid" was proposed by N. Ashby at the International Astronomical Union (IAU) Symposium No. 114, reported in [37]:

1. To adopt the coordinate time system (as approved by the Consultative Committee for the Definition of the Second (CCDS) and the International Radio Consultative Committee (CCIR)) as a global time scale for the Earth;
2. To continue further investigations for the determination and adjustment of the International Atomic Time (TAI) and the Terrestrial Dynamic Time (TDT).

---

[3] In the Newtonian sense, the geoid is the equipotential of the Earth's gravity (Newtonian) potential, which best coincides with the (mean) surface of the oceans.

The resolution was not adopted, but the chairman of the Scientific Organizing Committee, J. Kovalesky, considered that specialists in Celestial Mechanics and Astrometry needed more time to study the problem in competent commissions of IAU. Following this Symposium, several authors have contributed to the definition of global coordinate times [38–42].

As the definition of coordinate timescales necessitates the definition of a four dimensional relativistic reference system, the IAU working group had several complementary tasks in hand (Resolution C2 of the IAU General Assembly in 1985[4]):

1. the definition of the Conventional Terrestrial and Conventional Celestial Reference Systems,
2. ways of specifying practical realizations of these systems,
3. methods of determining the relationships between these realizations, and
4. a revision of the definitions of dynamical and atomic time to ensure their consistency with appropriate relativistic theories

Moreover, the President of the International Association of Geodesy (IAG) was invited to "appoint a representative to the working group for appropriate coordination on matters relevant to Geodesy". This work eventually led to the set of IAU Resolutions in 1991 and 2000 that define the present reference systems.

2.3 *Relativistic Reference Systems*

Several approaches have been considered for the definition of relativistic reference systems. Generalised Fermi coordinates were considered in [43–45]. However, the use of Fermi coordinates is not adapted to self-gravitating bodies for which mass-energy contributes to the determination of the initial metric $\boldsymbol{g}$ when solving the Einstein equations. For this reason, harmonic coordinates are preferred and recommended for the definition and realization of relativistic celestial reference systems [46–50], where the frame origin can be centered at the center-of-mass of a massive body. One drawback of harmonic frames is that the harmonic gauge condition does not admit rigidly rotating frames [51, chapter 8]. Other recent approaches are based on a perturbed Schwarzschild metric [52], or on the Kerr metric [53] in the different context of a slowly rotating astronomical object.

Following the pioneering works, a set of Resolutions was adopted at the IAU General Assembly in Manchester in the year 2000 [54]:

- B1.3: definition of the Barycentric and Geocentric Celestial Reference Systems (BCRS and GCRS)
- B1.4: form of the Earth post-Newtonian potential expansion

---

[4]All IAU Resolutions can be found at http://www.iau.org/administration/resolutions/general_assemblies/.

- B1.5: time transformations and realization of coordinate times in the Solar System (uncertainty $< 5 \times 10^{-18}$ in rate and 0.2 ps in phase amplitude for locations farther than a few Solar radii from the Sun)
- B1.9: definition of Terrestrial Time (TT)

We summarize here very briefly these resolutions. A relativistic reference system is implicitly defined by giving the components of the metric tensor in this reference system, in addition to a conventional spatial origin and orientation for the spatial part of the frame, and a conventional time origin for the time coordinate (the time orientation is trivial). The metric tensor is a solution of the Einstein equations in the low velocity and weak gravitational field approximation, for an ensemble of N bodies.

The Solar System Barycentric Celestial Reference System (BCRS), recommended by the IAU Resolutions, can be used to model light propagation from distant celestial objects and the motion of bodies within the Solar System. It is defined with:

$$g_{00} = -1 + \frac{2w(\vec{x})}{c^2} - \frac{2w(\vec{x})^2}{c^4} , \tag{7}$$

$$g_{0i} = -\frac{4}{c^3} w^i(\vec{x}) , \tag{8}$$

$$g_{ij} = \delta_{ij}\left(1 + \frac{2w(\vec{x})}{c^2}\right) , \tag{9}$$

where $\vec{x} \equiv \{ct, x^i\}$, with $i = 1\ldots3$, w and w^i are scalar and vector potentials. Its origin is at the barycenter of the Solar System masses, while the orientation of the spatial axes is fixed up to a constant time-independent rotation matrix about the origin (a natural choice is the International Celestial Reference System (ICRS) orientation which is fixed with respect to distant quasars). The coordinate time t is called Barycentric Coordinate Time (TCB). The origin of TCB is defined with respect to TAI: its value on 1977 January 1, 00:00:00 TAI (JD = 2,443,144.5 TAI) must be 1977 January 1, 00:00:32.184.

The unit of measurement of TCB should be chosen so that it is consistent with the SI second. An interesting discussion about timescale units can be found in [55]. As coordinate times such as TCB are not proper times, they cannot be directly measured by clocks. They are calculated using the corresponding metric components, e.g. Eqs. (7)–(9) for TCB, in combination with Eq. (4), which has to be inverted. Indeed, the basic observables to build timescales are the readings of proper times on clocks, which are local experiments. If the clocks are realizing the SI second, then the timescales calculated from these measurements are also in SI units, and the unit of such time coordinate could be named "SI-induced second".

The second relativistic reference system, recommended by the IAU Resolutions, is the Geocentric Celestial Reference System (GCRS). It can be used to model phenomenon in the vicinity of the Earth, such as its gravity field, artificial satellites orbiting the Earth or Earth rotation. It is defined with:

$$G_{00} = -1 + \frac{2V(\vec{X})}{c^2} - \frac{2V(\vec{X})^2}{c^4}\,, \quad (10)$$

$$G_{0i} = -\frac{4}{c^3}V^i(\vec{X})\,, \quad (11)$$

$$G_{ij} = \delta_{ij}\left(1 + \frac{2V(\vec{X})}{c^2}\right)\,, \quad (12)$$

where $(\vec{X}) \equiv \{cT, X^i\}$, and V and V^i are scalar and vector potentials. Note that we use notation V instead of usual notation W because W is commonly used in geodesy for the gravity potential. The frame origin is at the centre of mass of the Earth, and the orientation of the spatial axes is fixed with respect to the spatial part of the BCRS. The coordinate time T is called Geocentric Coordinate Time (TCG). It has the same origin and unit as TCB.

TCG is the proper time of a clock at infinity, and is not convenient because its rate differs from the one of clocks on the ground. Therefore IAU Resolutions introduced Terrestrial Time (TT), which differs from TCG by a constant rate $L_G = 6.969290134 \times 10^{-10}$:

$$\frac{\mathrm{d(TT)}}{\mathrm{d(TCG)}} = 1 - L_G. \quad (13)$$

The origins of TT and TCG are defined so that they coincide with TCB in origin: TT (resp. TCG) = TAI + 32.184 s on 1977 January 1 st, 0 h TAI. TT is a theoretical timescale and can have different realizations, e.g. TT(BIPM), or TT(TAI) = TAI + 32.184 s. (see e.g. [56]).

2.4 *Chronometric Geodesy*

Chronometric geodesy is the use of clocks to determine the spacetime metric. Indeed, the gravitational redshift effect discovered by Einstein must be taken into account when comparing the frequencies of distant clocks. Instead of using our knowledge of the Earth's gravitational field to predict frequency shifts between distant clocks, one can revert the problem and ask if the measurement of frequency shifts between distant clocks can improve our knowledge of the gravitational field. To do simple orders of magnitude estimates it is good to have in mind some correspondences:

$$1\,\mathrm{m} \leftrightarrow \frac{\Delta\nu}{\nu} \sim 10^{-16} \leftrightarrow \Delta W \sim 10\,\mathrm{m}^2\,\mathrm{s}^{-2}\,, \quad (14)$$

where 1 m is the height difference between two clocks, $\Delta\nu$ is the frequency difference in a frequency transfer between the same two clocks, and ΔW is the gravity potential difference (see Sect. 4.1) between the locations of these clocks.

From this correspondence, we can already recognize two direct applications of clocks in geodesy: if we are capable of comparing clocks to 10^{-16} accuracy, we can determine height differences between clocks with one meter accuracy (levelling), or determine geopotential differences with 10 m^2 s^{-2} accuracy.

To the knowledge of the authors, the latter technique was first mentioned in the geodetic literature by Bjerhammar [57] within a short section on a "new physical geodesy". Vermeer [58] introduced the term "chronometric levelling", while Bjerhammar [59] discussed the clock-based levelling approach under the title "relativistic geodesy", and also included a definition of a relativistic geoid. The term "chronometric" seems well suited for qualifying the method of using clocks to determine directly gravity potential differences, as "chronometry" is the science of the measurement of time. However the term "levelling" seems to be too restrictive with respect to all the applications one could think of using the results of clock comparisons. Therefore we will use the term "chronometric geodesy" to name the scientific discipline that deals with the measurement and representation of the Earth, including its gravity field, with the help of atomic clocks. It is sometimes also named "clock-based geodesy", or "relativistic geodesy". However this last designation is improper as relativistic geodesy aims at describing all possible techniques (including e.g. gravimetry, gradiometry, VLBI, Earth rotation, ...) in a relativistic framework. The natural arena of chronometric geodesy is the four-dimensional spacetime. At the lowest order, there is proportionality between relative frequency shift measurements – corrected from the first order Doppler effect – and (Newtonian) gravity potential differences. To calculate this relation one does not need the theory of general relativity, but only to postulate Local Position Invariance. Therefore, if the measurement accuracy does not reach the magnitude of the higher order terms, it is perfectly possible to use clock comparison measurements – corrected for the first order Doppler effect – as a direct measurement of (differences of) the gravity potential that is considered in classical geodesy. Comparisons between two clocks on the ground generally use a third reference clock in space, or an optical fibre on the ground (see Sect. 3.3).

In his article, Martin Vermeer explores the "possibilities for technical realisation of a system for measuring potential differences over intercontinental distances" using clock comparisons [58]. The two main ingredients are, of course, accurate clocks and a mean to compare them. He considers hydrogen maser clocks. For the links he considers a 2-way satellite link over a geostationary satellite, or GPS receivers in interferometric mode. He has also considered a way to compare proper frequencies of the different hydrogen maser clocks. Today this can be overcome by comparing primary frequency standards (PFS, see Sect. 3.2), which have a well defined proper frequency based on the transition of Caesium 133 used for the definition of the second. Secondary frequency standards (SFS), i.e. standards based on a transition other than the defining one, may also be used if the uncertainty in systematic effects has been fully evaluated, and the frequency measured against PFS.

With the advent of optical clocks, it often happens that the evaluation of systematics can be done more accurately than for PFS. This was one of the purposes of the European project[5] of "International timescales with optical clocks" [60], where optical clocks based on different atoms are compared to each other locally, and to

[5]projects.npl.co.uk/itoc.

PFS. Within this project, a proof-of-principle experiment of chronometric geodesy was done by comparing two optical clocks separated by a height difference of around 1000 m, using an optical fibre link [61].

Few authors have seriously considered chronometric geodesy in the past. Following Vermeer's idea, the possibility of using GPS observations to solve the problem of determining geoid heights has been explored in [62]. The authors considered two techniques based on frequency comparisons and direct clock readings. However, they leave aside the practical feasibility of such techniques. The value and future applicability of chronometric geodesy has been discussed in [63], including direct geoid mapping on continents and joint gravity-geopotential surveying to invert for subsurface density anomalies. They find that a geoid perturbation caused by a 1.5 km radius sphere with 20 percent density anomaly buried at 2 km depth in the Earth's crust is already detectable by atomic clocks with an achievable accuracy of 10^{-18}. The potentiality of the new generation of atomic clocks has been shown in [64], based on optical transitions, to measure heights with a resolution of around 30 cm. The possibility of determining the geopotential at high spatial resolution thanks to chronometric geodesy is thoroughly explored and evaluated in [65]. This work will be detailed in Sect. 6.

2.5 The Chronometric Geoid

Arne Bjerhammar in 1985 gives a precise definition of the "relativistic geoid" [59, 66]:

> The relativistic geoid is the surface where precise clocks run with the same speed and the surface is nearest to mean sea level.

This is an operational definition, which has been translated in the context of post-Newtonian theory [47, 67]. In these two articles a different operational definition of the relativistic geoid has been introduced based on gravimetric measurements: a surface orthogonal everywhere to the direction of the plumb-line and closest to mean sea level. The authors call the two surfaces obtained with clocks and gravimetric measurements the "u-geoid" and the "a-geoid", respectively. They have shown that these two surfaces coincide in the case of a stationary metric. In order to distinguish the operational definition of the geoid from its theoretical description, it is less ambiguous to give a name based on the particular technique to measure it. The term "relativistic geoid" is too vague as Soffel et al. [67] have defined two different ones. The names chosen by Soffel et al. are not particularly explicit, so instead of "u-geoid" and "a-geoid" one can call them "chronometric geoid" and "gravimetric geoid" respectively. There can be no confusion with the geoid derived from satellite measurements, as this is a quasi-geoid that does not coincide with the geoid on the continents [68]. Other considerations on the chronometric geoid can be found in [51, 69, 70].

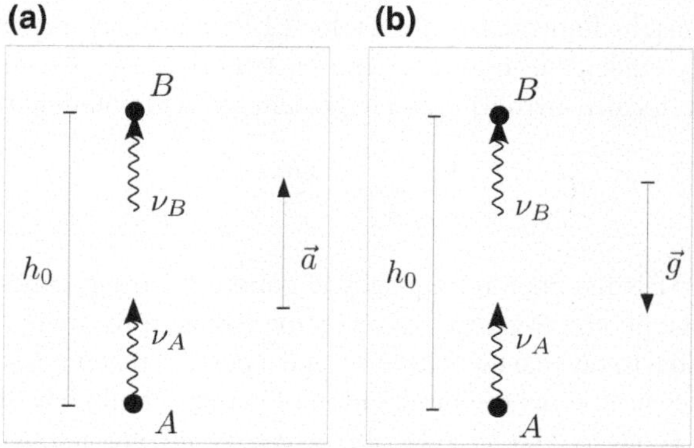

Fig. 6 A photon of frequency ν_A is emitted at point A toward point B, where the measured frequency is ν_B. a) A and B are two points at rest in an accelerated frame, with acceleration $\vec{a}$ in the same direction as the emitted photon. b) A and B are at rest in a non accelerated (locally inertial) frame in presence of a gravitational field such that $\vec{g} = -\vec{a}$

We notice that the problem of defining a reference isochronometric surface is closely related to the problem of realizing Terrestrial Time (TT). This is developed in more details in Sect. 3.5.

Recently, extensive work has been done aiming at the development of an exact relativistic theory for the Earth's geoid undulation [71], as well as developing a theory for the reference level surface in the context of post-newtonian gravity [72, 73]. This goes beyond the problem of the realization of a reference isochronometric surface and tackles the tough work of extending all concepts of classical physical geodesy (see e.g. [68]) in the framework of general relativity.

3 Comparisons of Frequency Standards

3.1 *The Einstein Equivalence Principle*

Let's consider a photon emitted at a point A in an accelerated reference system toward a point B, which lies in the direction of the acceleration (see Fig. 6). We assume that both points are separated by a distance h_0, as measured in the accelerated frame. The photon time of flight is $\delta t = h_0/c$, and the frame velocity during this time increases by $\delta v = a\delta t = ah_0/c$, where a is the magnitude of the frame acceleration $\vec{a}$. The frequency at point B (reception) is then shifted because of the Doppler effect, compared to the frequency at point A (emission), by an amount:

$$\frac{\nu_B}{\nu_A} = 1 - \frac{\delta v}{c} = 1 - \frac{ah_0}{c^2} . \tag{15}$$

Now, the Einstein Equivalence Principle (EEP) postulates that a gravitational field $\vec{g}$ is locally equivalent to an acceleration field $\vec{a} = -\vec{g}$. We deduce that in a non-accelerated (locally inertial) frame in presence of a gravitational field $\vec{g}$:

$$\frac{\nu_B}{\nu_A} = 1 - \frac{g h_0}{c^2} , \tag{16}$$

where $g = |\vec{g}|$, ν_A is the photon frequency at emission (strong gravitational potential) and ν_B is the photon frequency at reception (weak gravitational potential). As $\nu_B < \nu_A$, it is usual to say that the frequency at the point of reception is "red-shifted". One can consider it in terms of conservation of energy. Intuitively, the photon that goes from A to B has to "work" to be able to escape the gravitational field, then it looses energy and its frequency decreases by virtue of $E = h\nu$, with h the Planck constant.

If two ideal clocks are placed in A and B and the clock at A (strong gravitational potential) is used to generate the signal ν_A, then the signal received at B (weak gravitational potential) has a lower frequency than a signal locally generated by the clock at B.

3.2 Relativistic Frequency Transfer

Let's consider two atomic frequency standards (AFS) A and B which deliver the proper frequencies f_A and f_B. These two frequencies can be different if the two AFS are based on different atom transitions. Following the Bureau International des Poids et Mesures (BIPM) we name primary frequency standards (PFS) the AFS based on the atom of Caesium 133, more commonly named Caesium Fountains. The best PFS have a very low relative accuracy in the range 10^{-15} – 10^{-16} (see e.g. [74]). Then, we name secondary frequency standards (SFS) the AFS which are based on a different atom than the Caesium atoms. The Consultative Committee for Length (CCL)-Consultative Committee for Time and Frequency (CCTF) frequency standards working group is in charge of producing and maintaining a single list of recommended values of standard frequencies for the practical realization of secondary representations of the second.[6] SFS can have a relative accuracy down to the range 10^{-17} – 10^{-18} [75–77]. See also [78, 79], where a method is presented for analysing over-determined sets of clock frequency comparison data involving standards based on a number of different reference transitions.

The goal of a frequency comparison between two AFS A and B is to determine the ratio of their frequencies f_A/f_B. The most used technique for frequency comparison nowadays is the transmission of an electromagnetic signal between A and B, reaching the following formula:

$$\frac{f_A}{f_B} = \frac{f_A}{\nu_A}\frac{\nu_A}{\nu_B}\frac{\nu_B}{f_B} , \tag{17}$$

[6]See http://www.bipm.org/en/publications/mises-en-pratique/standard-frequencies.html.

where ν_A is the proper frequency of the photon at the time of emission t_A, and ν_B is the proper frequency of the same photon at the time of reception t_B. The ratio ν_A/f_A is known or measured, ν_B/f_B is measured, while ν_A/ν_B has to be modelled and calculated.

Let $S(x^\alpha)$ be the phase of the electromagnetic signal emitted by clock A. It can be shown that light rays are contained in hypersurfaces of constant phase. The frequency measured by A/B is:

$$\nu_{A/B} = \frac{1}{2\pi} \frac{\mathrm{d}S}{\mathrm{d}\tau_{A/B}} , \tag{18}$$

where $\tau_{A/B}$ is the proper time along the worldline of clock A/B. We introduce the wave vector $k_\alpha^{A/B} = (\partial_\alpha S)_{A/B}$ to obtain:

$$\nu_{A/B} = \frac{1}{2\pi} k_\alpha^{A/B} u^\alpha_{A/B} , \tag{19}$$

where $u^\alpha_{A/B} = \mathrm{d}x^\alpha_{A/B}/\mathrm{d}\tau$ is the four-velocity of clock A/B. Finally, we obtain a fundamental relation for the frequency transfer:

$$\frac{\nu_A}{\nu_B} = \frac{k_\alpha^A u_A^\alpha}{k_\alpha^B u_B^\alpha} . \tag{20}$$

This formula does not depend on a particular theory, and thus can be used to perform tests of general relativity. It is needed in the context of chronometric geodesy in order to calculate the gravity potential difference between two clocks for which the ratio f_A/f_B is well known.

Introducing $v^i = \mathrm{d}x^i/\mathrm{d}t$ and $\hat{k}_i = k_i/k_0$, it is usually written as:

$$\frac{\nu_A}{\nu_B} = \frac{u_A^0}{u_B^0} \frac{k_0^A}{k_0^B} \frac{1 + \frac{\hat{k}_i^A v_A^i}{c}}{1 + \frac{\hat{k}_i^B v_B^i}{c}} . \tag{21}$$

From Eq. (18) we deduce that:

$$\frac{\nu_A}{\nu_B} = \frac{\mathrm{d}\tau_B}{\mathrm{d}\tau_A} = \left(\frac{\mathrm{d}\tau}{\mathrm{d}t}\right)_A^{-1} \left(\frac{\mathrm{d}\tau}{\mathrm{d}t}\right)_B \frac{\mathrm{d}t_B}{\mathrm{d}t_A} . \tag{22}$$

The derivative $(\mathrm{d}t_B/\mathrm{d}t_A)$ is affected by processes in the frequency transfer itself and depends on the particular technique used for the frequency comparison. It is considered in more details in Sect. 3.3.

The derivatives $(\mathrm{d}\tau/\mathrm{d}t)$ in (22) do not depend on the frequency transfer technique but just on the state (velocity and location) of the emitting and receiving AFS. In Sects. 4 and 5 we focus on the best practical determination of these terms. Indeed, calculation of these terms is limited in accuracy by our knowledge of the Earth's gravitational field. We note that quantity (ν_A/ν_B) in (22) is a scalar, therefore invariant under general coordinate transformation (see Sect. 2.1). However the splitting of this quantity as written on the right-hand side of (22) is not invariant and depends on the particular relativistic reference system used for the splitting. The choice of a particular relativistic reference system gives a conventional meaning to simultaneity: two events are simultaneous if they have the same time coordinate t (see Sect. 2.2) for free and guided propagation.

For applications on the Earth, such as (ground) clock syntonization (Sect. 3.4) and the realization of a worldwide coordinate time (Sect. 3.5), a natural choice of a relativistic reference system is the spatial part of the geocentric celestial reference system (GCRS) together with the terrestrial time (TT) as a coordinate time (see Sect. 2.3). Following [80, 81], the coordinate to proper time transformation can be written down to a relative accuracy of 10^{-18} as:

$$\frac{\mathrm{d}\tau}{\mathrm{d}TT} = 1 + L_G - \frac{1}{c^2}\left[W^{\mathrm{static}} + W^{\mathrm{temp}}\right] . \tag{23}$$

where L_G is a constant defined in Sect. 2.3, and $W = V + Z$ is the gravity (gravitational plus centrifugal) potential, commonly used in geodesy (see e.g. [68, 82]). The gravity potential is split into a static part W^{static} and a part varying with time W^{temp}. Neglected terms in Eq. (23) are terms in c^{-4} or smaller as well as one term of order c^{-2} resulting from the coupling of higher order multipole moments of the Earth to the external tidal gravitational field. All the neglected terms in the transformation (23) amount in the vicinity of the Earth to a few parts in 10^{-19} or less [80].

The static part of the gravity potential W^{static} can be derived from geometric levelling or GNSS positions combined with a gravimetric geoid model. For instance, the best unified evaluation of the static gravity potential for several AFS in Europe was one of the main purposes of the ITOC project (see Sect. 5). The time varying part W^{temp} is dominated by solid Earth and ocean tides and further discussed in Sect. 3.6.

3.3 Frequency Transfer Techniques

We discuss in this section the foundations of two frequency transfer techniques widely used, based on the propagation of an electromagnetic signal either in free space or in an optical fiber. Free and guided propagation lead to different theoretical modelling approaches of the frequency transfer. We limit the presented results to one-way transfer and give appropriate references for two-way transfer techniques.

Free space time and frequency transfer can be realized using radiofrequency signals (of order 1–10 GHz) with well established techniques [83], and in the optical domain with lasers [84]. GNSS (Global Navigation Satellite Systems) [85–89] and TWSTFT (Two-Way Satellite Time and Frequency Transfer) [83, 90–93] have been widely used for years to perform clock comparisons and establish international timescales such as TAI [56]. The ACES MWL (MicroWave Link) [94] is being developed in the frame of the ACES (Atomic Clocks Ensemble in Space) experiment [95, 96]. New techniques using two-way laser links have been developed and operated, such as T2L2 (Time Transfer by Laser Light) [97–101], and others are in development, such as ELT [102, 103], which is part of the ACES experiment.

Existing free space frequency transfer techniques are in the range 10^{-15}–10^{-16} for the fractional frequency accuracy and stability, with the goal of being in the 10^{-17} range for the ACES experiment. However, they are not sufficient for the comparisons of optical clocks, which have fractional frequency accuracy and stability in the 10^{-17}–10^{-19} range [76, 77, 104–107]. Therefore, phase-coherent optical links have been developped using principally an optical fibre as a medium for the propagation [74, 75, 108, 109], attaining spectacular stability and accuracy in the range 10^{-19} and below. However, phase coherent free space optical links are also being developed [110–113]. It is not clear yet if these techniques will be able to be as good as optical fibre techniques, mainly because of the effect of atmospheric turbulence [114–116].

3.3.1 Free Space Propagation Comparisons

In the case of propagation in free space, if we suppose that the spacetime is stationary, i.e. $\partial_0 g_{\alpha\beta} = 0$, then it can be shown that k_0 is constant along the light ray, meaning that $k_0^A = k_0^B$. Then, from Eqs. (21) and (22) we deduce that

$$\frac{\mathrm{d}t_B}{\mathrm{d}t_A} = \frac{1 + \frac{\hat{k}_i^A v_A^i}{c}}{1 + \frac{\hat{k}_i^B v_B^i}{c}} \,. \tag{24}$$

The quantity $\mathrm{d}t_B/\mathrm{d}t_A$ in Eq. (22) can be computed with several methods. Two different approaches are presented in some detail in Appendix A of [117]: a direct integration of the null geodesic equations, and a simpler way, which is the differentiation of the time transfer function. This second method is quite powerful: a general method has been developed to calculate the time transfer function as a Post-Minkowskian (PM) series up to any order in G, the gravitational constant [118, 119]. See for example [120] for the calculation of the one-way frequency shift up to the 2-PM approximation. This method does not require the integration of the null geodesic equations. The frequency shift is expressed as integral of functions defined from the metric and its derivatives and performed along a Minkowskian straight line.

Let A be the emitting station, with GCRS position $\vec{x}_A(t)$, and B the receiving station, with position $\vec{x}_B(t)$. We use $t =$ TCG and hence the calculated coordinate time intervals are in TCG. The corresponding time intervals in TT are obtained by multiplying with $(1 - L_G)$. We denote by t_A the coordinate time at the instant of emission of an electromagnetic signal, and by t_B the coordinate time at the instant of reception. We define $r_A = |\vec{x}_A(t_A)|$, $r_B = |\vec{x}_B(t_B)|$ and $\mathrm{R}_{AB} = |\vec{x}_B(t_B) - \vec{x}_A(t_A)|$, as well as the coordinate velocities $\vec{v}_A = \mathrm{d}\vec{x}_A/\mathrm{d}t(t_A)$ and $\vec{v}_B = \mathrm{d}\vec{x}_B/\mathrm{d}t(t_B)$. Then the frequency ratio can be expressed as [117]:

$$\frac{\nu_A}{\nu_B} = \frac{1 - \frac{1}{c^2}\left[\frac{v_B^2}{2} + U_E(\vec{x}_B)\right] q_A}{1 - \frac{1}{c^2}\left[\frac{v_A^2}{2} + U_E(\vec{x}_A)\right] q_B}\,, \tag{25}$$

where U_E is the Newtonian potential of the Earth, and, if the desired accuracy is greater than 5×10^{-17},

$$q_A = 1 - \frac{\vec{N}_{AB}\cdot\vec{v}_A}{c} - \frac{4GM_E}{c^3}\frac{\mathrm{R}_{AB}\vec{N}_A\cdot\vec{v}_A + (r_A+r_B)\vec{N}_{AB}\cdot\vec{v}_A}{(r_A+r_B)^2 - \mathrm{R}_{AB}^2}\,, \tag{26}$$

$$q_B = 1 - \frac{\vec{N}_{AB}\cdot\vec{v}_B}{c} - \frac{4GM_E}{c^3}\frac{\mathrm{R}_{AB}\vec{N}_A\cdot\vec{v}_B - (r_A+r_B)\vec{N}_{AB}\cdot\vec{v}_B}{(r_A+r_B)^2 - \mathrm{R}_{AB}^2}\,, \tag{27}$$

where $\vec{N}_{AB} = (\vec{x}_B(t_B) - \vec{x}_A(t_A))/R_{AB}$, G is the gravitational constant, and M_E is the mass of the Earth.

Note that formulas (26) and (27) have been obtained by assuming that the field of the Earth is spherically symmetric. If an accuracy lower than 5×10^{-17} is required, it is necessary to take into account the J_2 terms in the Newtonian potential.

The terms of order c^{-1} correspond to the relative Doppler effect between the clocks. Terms of order c^{-2} in Eq. (25) are the classic second-order Doppler effect and gravitational redshift.[7] Terms of order c^{-3} amount to less than 3.6×10^{-14} for a satellite in Low-Earth Orbit and 2.2×10^{-15} for the ground. Terms of order c^{-4}, omitted in Eq. (25), can reach a few parts in 10^{-19} in the vicinity of the Earth [80].

3.3.2 Fibre Propagation Comparisons

If the signal propagates in an optical fibre, the term $(\mathrm{d}t_B/\mathrm{d}t_A)$ has been calculated up to order c^{-3} in [122] for one-way and two-way time and frequency transfers. The result for one-way frequency transfer is:

$$\frac{\mathrm{d}t_B}{\mathrm{d}t_A} = 1 + \frac{1}{c}\int_0^L \left(\frac{\partial n}{\partial t} + n\alpha\frac{\partial T}{\partial t}\right)\mathrm{d}l + \frac{1}{c^2}\int_0^L \frac{\partial \vec{v}\cdot\vec{s}_l}{\partial t}\mathrm{d}l\,, \tag{28}$$

where L is the total rest length of the fibre at time of emission t_A, n is the effective refractive index of the fibre, α is the linear thermal expansion coefficient of the fibre, T is the temperature of the fibre as a function of time and location, and $\vec{v}$ and $\vec{s}_l$ are the velocity and tangent vector fields of the fibre, respectively.

Up to the second order it does not depend on the gravitational field, as for the free propagation in vacuum. The first order term is due to the variation of the fibre length (e.g. due to thermal expansion) and of its refractive index. For a 1000 km fibre with refractive index $n = 1.5$ this term is equal to 2×10^{-13}, but this term cancels in a two-way frequency transfer. The second order term is the derivative of the Sagnac

[7] One can notice that the separation between a gravitational red-shift and a Doppler effect is specific to the chosen coordinate system. One can read the book by Synge [121] for a different interpretation in terms of relative velocity and Doppler effect only.

effect, which is of order 10^{-19} or less for a 1000 km fibre. The sign of this term depends on the direction of propagation of the signal in the fibre, such that it adds up when doing two-way frequency transfer. Finally the neglected third order term is of the order of 10^{-22} for a 1000 km fibre.

3.4 Clock Syntonization

Clock syntonization necessitates to calculate the derivatives $(\mathrm{d}\tau/\mathrm{d}t)$ from (22). Using (23) and neglecting all terms smaller than 10^{-18} we deduce:

$$\left(\frac{\mathrm{d}\tau}{\mathrm{d}TT}\right)_A^{-1}\left(\frac{\mathrm{d}\tau}{\mathrm{d}TT}\right)_B = 1 + \frac{W_A - W_B}{c^2}, \tag{29}$$

where $W = W^{\text{static}} + W^{\text{temp}}$. Therefore syntonization necessitates the knowledge of the difference of the gravity potential between locations A and B. Two widely used geodesy techniques can be used to determine the static part of this difference: geometrical levelling and GNSS positions combined with a gravimetric geoid model. Geometrical levelling has the advantage to be very accurate on short distances (typically 0.2–1.0 mm for a 1 km double run levelling) and should be preferred when comparing clocks within the same institute (local comparison). However, geometrical levelling accumulates errors with increasing distance (up to several dm over 1000 km distance) and hence the GNSS/geoid approach should be preferred for comparisons between different institutes.

Direct geometrical levelling between two points A and B does not necessitate a point of reference and leads to a high accuracy. However, when determining the height of the clocks with respect to the national height system, the reference point of the zero altitude can be very far away from the clocks and therefore the link to the zero altitude may lead to a bias in the determination of the height of the clock. Moreover, the reference point of the zero altitude can be different from one country to the other, because it can be based on different realizations of mean sea level. This leads to the problem of unifying national height systems [123]. The GNSS/geoid approach allows the derivation of the height system bias term for a particular country. It is therefore possible to correct for the bias in the geometrical levelling technique for international clock comparisons. However long distance errors cannot be avoided in geometrical levelling for distant comparisons, for which the GNSS/geoid approach is more adapted.

The GNSS/geoid method is based on the assumption that the gravitational potential is regular (zero) at infinity. This has the advantage that when using one gravimetric model, the zero origin of the gravitational potential is coherent between all locations covered by the model. High quality regional models exist for Europe and a new one was developed during the course of the ITOC project (EGG2015, see Sect. 5.3). Indeed, this technique is highly dependent on the quality and coverage of the ground gravimetric observables, and particular care should be taken in the use of the gravimetry dataset. This method allows the derivation of absolute potential values with about 2–3 cm accuracy in terms of heights (best case scenario, i.e., accurate GNSS positions, sufficient terrestrial data around sites of interest, and state-of-the-art global satellite geopotential utilized). Detailed considerations about the uncertainties of the two approaches, geometric levelling and GNSS/geoid, can be found in [81].

3.5 The Realization of Terrestrial Time (TT)

The realization of TT necessitates the knowledge of the absolute gravity potential. The TAI is the most commonly used realization of TT [56, 124]. First, comparisons of about 400 atomic clocks around the world in around 70 laboratories are used to calculate the free atomic scale (EAL), a fly-wheel timescale. In a second step, around 15 AFS are used to steer the unit of EAL such that its scale corresponds to the definition of the second. Direct comparisons between AFS are not necessary in this process. Instead, each laboratory compares its AFS to a (master) clock participating in EAL.

In 1980, the definition of TAI was given by the Consultative Committee for the Definition of the Second as:

> TAI is a coordinate time scale defined in a geocentric reference frame with the SI second as realized on the rotating geoid as the scale unit.

This reference to the geoid was very ambiguous. Indeed, the value of the gravity potential on the geoid, W_{geoid}, depends on the global ocean level which changes with time.[8] In addition, there are several methods to realize the geoid as "closest to the mean sea level" so that there is yet no adopted standard to define a reference geoid and W_{geoid} value (see e.g. a discussion in [125]). Several authors have considered the time variation of W_{geoid} , see e.g. [126, 127], but there is some uncertainty in what is accounted for in such a linear model. A recent estimate by Dayoub et al. over 1993–2009 gives $\mathrm{d}W_{\text{geoid}}/\mathrm{d}t = -2.7 \times 10^{-2}\,\mathrm{m}^2\,\mathrm{s}^{-2}\,\mathrm{yr}^{-1}$, mostly driven by the sea level change of +2.9 mm/yr. However, the rate of change of the global ocean level could vary during the next decades, and predictions are highly model dependent [128]. Nevertheless, to state an order of magnitude, considering a systematic variation in the sea level of order 2 mm/yr, different definitions of a reference surface for the gravity potential could yield differences in the redshift correction of the AFS of order 2×10^{-18} in a decade, which is of the same order than the best current SFS accuracies [76, 77].

However, this ambiguity disappeared with the new definition of TT adopted with IAU resolution B1.9 (2000) [54] (see Sect. 2.3). If TAI is a realization of TT then one has to apply a relative frequency correction, or redshift correction, to the AFS

[8] Here we use notation W_{geoid} instead of the commonly used W_0, in order to emphasize that there is no generally accepted conventional and unified value of the geoid gravity potential value.

frequency such that

$$\frac{\mathrm{d}\tau}{\mathrm{d}TT} = 1 , \tag{30}$$

This equation is exact in GR. Given a model of the spacetime metric and of the AFS worldlines, it implicitly defines an isochronometric hypersurface, i.e. an hypersurface where all clocks run at the same rate as TT. This hypersurface can be foliated using TT coordinate time as a collection of 3D hypersurfaces Σ_{TT} with constant TT (see Sect. 2.1). Following Eq. (23), the total correction (to be added) to the AFS relative frequency in order to realize TT is:

$$\epsilon = -\frac{W_0^{(\mathrm{IAU})} - W}{c^2} , \tag{31}$$

where $W_0^{(\mathrm{IAU})} = c^2 L_G = 62636856.000519\,\mathrm{m^2 s^{-2}}$, and W is the gravity potential at the clock location for the considered epoch. We have seen that it is usual in geodesy to separate the problem of modelization of the gravity field in a static part and a part varying with time (see Eq. (23) and Sect. 4). This splitting is conventional and it should be done with care as several conventions exist (see Sect. 5.2). Then the AFS correction ϵ can be split in a static part and a part varying with time:

$$\epsilon = \epsilon_{\mathrm{static}} + \epsilon_{\mathrm{temp}} = -\frac{W_0^{(\mathrm{IAU})} - W^{\mathrm{static}}}{c^2} + \frac{W^{\mathrm{temp}}}{c^2} . \tag{32}$$

Keeping only the static part of the gravity field, the problem becomes stationary and the isochronometric hypersurface Σ_{TT} is uniquely determined for clocks fixed on the Earth's surface. In the weak gravitational field and low velocity limit, it coincides at the lowest order with the Newtonian equipotential of the gravity field with exact value $W_0^{(\mathrm{IAU})} = c^2 L_G$. Higher order relativistic corrections (terms of order c^{-4} in ϵ) are of the order 2×10^{-19} or below in the AFS relative frequency [70].

We emphasize that the realization of TT does not necessitate any longer the realization of a geoid. The reference equipotential is just an equipotential with a well defined value, $W_0^{(\mathrm{IAU})}$, which is constant in time and exactly known: its value is a convention. However if this reference equipotential is defined theoretically with no ambiguity, it needs to be realized in the same way as the geoid, leading to inaccuracies in its realization, mainly due to the imperfect knowledge of the Earth's mass distribution. In this context, it is interesting to note that clocks in orbit around the Earth are less sensitive to the Earth's gravitational field, and thus to errors in its modelization.

As an illustration, let's take a clock in a satellite following an approximately circular orbit of radius a around the Earth. Approximating the Earth gravitational potential along the satellite trajectory with $V = GM/a$, where GM is the Earth gravitational parameter, then the velocity of the clock is $v = \sqrt{GM/a}$. In order for the clock to realize TT, one needs $V + v^2/2 \approx c^2 L_G$, i.e. $a \approx 9543\,km$, and a good knowledge of the trajectory of the satellite. It is shown in [80] that at this altitude the effect of solid Earth tides, ocean tide, polar motion, and changes in the atmospheric pressure are below 10^{-18} in fractional frequency. Moreover, tidal effects can also be calculated with uncertainties below 10^{-18} in fractional frequency.

The definition of the scalar potential in the context of relativistic reference frames, from which the redshift correction formula (31) is deduced, is coherent in the Newtonian limit with the assumption done in classical geodesy that the Newtonian potential is regular at infinity. Therefore the GNSS/geoid method is very well adapted to the determination of the redshift corrections in the context of relativistic reference frames. As discussed, when using national height systems one has to calculate corrections such that the assumption of regularity is fulfilled over the area covered by the clock comparisons. This will be illustrated in detail in Sect. 5.

Finally, according to Eqs. (29) and (31), syntonizing two AFS necessitates to determine the relative gravity potential between the locations of both clocks, while the realization of TT necessitates the determination of the absolute gravity potential at the location of the contributing AFS. If the redshift correction (31) is known for two clocks, it is easy to obtain Eq. (29) in order to syntonize them. Therefore, both the problem of syntonization and the realization of TT can be tackled by determining the absolute gravity potential at the locations of the contributing AFS.

3.6 *Temporal Variations of the Gravity Field*

For the temporal variations of the gravity field W^{temp}, one can refer to [129], where all corrections bigger than 10^{-18} in relative frequency are modelled and evaluated. The dominant effect is the gravity potential variation induced by solid Earth tides, which can be (at most) 10^{-16} for clock syntonization on international scales, and 10^{-17} for the realization of TT. The second major contributor is the induced signal of ocean tides. However, both solid Earth and ocean tide signals can be modelled down to an accuracy of a few parts in 10^{19}.

Several other time-variable effects can affect the clock comparisons at the 10^{-18} level, such as solid Earth pole tides, non-tidal mass redistributions in the atmosphere, the oceans and the continental water storage, as well as secular signals due to sea level changes and glacial isostatic adjustment. Non-tidal mass redistribution effects on the gravity field are strongly dependent on location and/or weather conditions. As clock comparisons now approach the 10^{-18} stability, it will be necessary to develop guidelines in order to include these effects for the syntonization of clocks and their contribution to the realization of TT. Recent analysis of optical clock comparisons have included temporal variations [130, 131].

4 Geodetic Methods for Determining the Gravity Potential

This section describes geodetic methods for determining the gravity potential, needed for the computation of the relativistic redshift corrections for optical clock observations. The focus is on the determination of the static (spatially variable) part of the potential field, while temporal variations in the station coordinates and the potential quantities (with the largest components resulting from solid Earth and ocean tide effects, see [129]) are assumed to be taken into account through appropriate reductions or by using sufficiently long averaging times. This is common geodetic practice and leads to a quasi-static state (e.g. by referring all quantities to a given epoch), such that the Earth can be considered as a rigid and non-deformable body, uniformly rotating about a body-fixed axis. Hence, all gravity field quantities including the level surfaces are considered in the following as static quantities, which do not change in time. On this basis, the static and temporal components of the gravity potential can be added to obtain the actual potential value at time t, as needed, e.g., for the evaluation of clock comparison experiments.

In this context, a note on the handling of the permanent (time-independent) parts of the tidal corrections is appropriate; for details, see, e.g., [132, 133], or [134]. The International Association of Geodesy (IAG) has recommended that the so-called "zero-tide system" should be used (resolutions no. 9 and 16 from the year 1983; cf. [135]), where the direct (permanent) tide effects are removed, but the indirect deformation effects associated with the permanent tidal deformation are retained. Unfortunately, geodesy and other disciplines do not strictly follow the IAG resolutions for the handling of the permanent tidal effects, and therefore, depending on the application, appropriate corrections may be necessary to refer all quantities to a common tidal system (see below and the aforementioned references).

In the following, some fundamentals of physical geodesy are given, and then two geodetic methods are described for determining the gravity potential, considering both the geometric levelling approach and the GNSS/geoid approach (GNSS – Global Navigation Satellite Systems), together with corresponding uncertainty considerations.

4.1 Fundamentals of Physical Geodesy

Classical physical geodesy is largely based on the Newtonian theory with Newton's law of gravitation, giving the gravitational force between two point masses, to which a gravitational acceleration (also termed gravitation) can be ascribed by setting the mass at the attracted point P to unity. Then, by the law of superposition, the gravitational acceleration of an extended body like the Earth can be computed as the vector sum of the accelerations generated by the individual point masses (or mass elements), yielding

$$\boldsymbol{b} = \boldsymbol{b}(\boldsymbol{r}) = -G \iiint_{\text{Earth}} \frac{\boldsymbol{r} - \boldsymbol{r}'}{|\boldsymbol{r} - \boldsymbol{r}'|^3} dm \ , \ dm = \rho dv \ , \ \rho = \rho(\boldsymbol{r}') \ , \tag{33}$$

where $\boldsymbol{r}$ and $\boldsymbol{r}'$ are the position vectors of the attracted point P and the source point Q, respectively, dm is the differential mass element, ρ is the volume density, dv is the volume element, and G is the gravitational constant. The SI unit of acceleration is ms^{-2}, but the non-SI unit Gal is still frequently used in geodesy and geophysics (1 Gal = 0.01 m s^{-2}, 1 mGal = 1 $\times 10^{-5}$m s^{-2}). While an artificial satellite is only affected by gravitation, a body rotating with the Earth also experiences a centrifugal force and a corresponding centrifugal acceleration $\boldsymbol{z}$, which is directed outwards and perpendicular to the rotation axis:

$$\boldsymbol{z} = \boldsymbol{z}(\boldsymbol{p}) = \omega^2 \boldsymbol{p} \ . \tag{34}$$

In the above equation, ω is the angular velocity, and $\boldsymbol{p}$ is the distance vector from the rotation axis. Finally, the gravity acceleration (or gravity) vector $\boldsymbol{g}$ is the resultant of the gravitation $\boldsymbol{b}$ and the centrifugal acceleration $\boldsymbol{z}$:

$$\boldsymbol{g} = \boldsymbol{b} + \boldsymbol{z} \ . \tag{35}$$

As the gravitational and centrifugal acceleration vectors $\boldsymbol{b}$ and $\boldsymbol{z}$ both form conservative vector fields or potential fields, these can be represented as the gradient of corresponding potential functions by

$$\boldsymbol{g} = \nabla W = \boldsymbol{b} + \boldsymbol{z} = \nabla V_E + \nabla Z_E = \nabla(V_E + Z_E) \ , \tag{36}$$

where W is the gravity potential, consisting of the gravitational potential V_E and the centrifugal potential Z_E. Based on Eqs. (33)–(36), the gravity potential W can be expressed as

$$W = W(\boldsymbol{r}) = V_E + Z_E = G \iiint_{\text{Earth}} \frac{\rho dv}{l} + \frac{\omega^2}{2} p^2 \ , \tag{37}$$

where l and p are the lengths of the vectors $\boldsymbol{r} - \boldsymbol{r}'$ and $\boldsymbol{p}$, respectively. All potentials are defined with a positive sign, which is common geodetic practice. The gravitational potential V_E is assumed to be regular (i.e. zero) at infinity and has the important property that it fulfills the Laplace equation outside the masses; hence it can be represented by harmonic functions in free space, with the spherical harmonic expansion playing a very important role. Further details on potential theory and properties of the potential functions can be found, e.g., in [134, 136, 137].

The determination of the gravity potential W as a function of position is one of the primary goals of physical geodesy; if $W(\boldsymbol{r})$ were known, then all other parameters of interest could be derived from it, including the gravity vector $\boldsymbol{g}$ according to Eq. (36) as well as the form of the equipotential surfaces (by solving the equation

$W(\boldsymbol{r}) = \text{const.}$). Furthermore, the gravity potential is also the ideal quantity for describing the direction of water flow, i.e. water flows from points with lower gravity potential to points with higher values. However, although the above equation is fundamental in geodesy, it cannot be used directly to compute the gravity potential W due to insufficient knowledge about the density structure of the entire Earth; this is evident from the fact that densities are at best known with two to three significant digits, while geodesy generally strives for a relative uncertainty of at least 10^{-9} for all relevant quantities (including the potential W). Therefore, the determination of the exterior potential field must be solved indirectly based on measurements performed at or above the Earth's surface, which leads to the area of geodetic boundary value problems (GBVPs; see below).

The gravity potential is closely related to the question of heights as well as level or equipotential surfaces and the geoid, where the geoid is classically defined as a selected level surface with constant gravity potential W_0, conceptually chosen to approximate (in a mathematical sense) the mean ocean surface or mean sea level (MSL). However, MSL does not coincide with a level surface due to the forcing of the oceans by winds, atmospheric pressure, and buoyancy in combination with gravity and the Earth's rotation. The deviation of MSL from a best fitting equipotential surface (geoid) is denoted as the (mean) dynamic ocean topography (DOT); it reaches maximum values of about ± 2 m and is of vital importance for oceanographers for deriving ocean circulation models [138].

On the other hand, a substantially different approach was chosen by the IAG during its General Assembly in Prague, 2015, within "IAG Resolution (No. 1) for the definition and realization of an International Height Reference System (IHRS)" [139], where a numerical value $W_0^{(\mathrm{IHRS})} = 62,636,853.4\ \mathrm{m^2s^{-2}}$ (based on observations and data related to the mean tidal system) is defined for the realization of the IHRS vertical reference level surface, with a corresponding note, stating that $W_0^{(\mathrm{IHRS})}$ is related to "the equipotential surface that coincides (in the least-squares sense) with the worldwide mean ocean surface, the most accepted definition of the geoid" [140]. Although the classical geodetic geoid definition and the IAG 2015 resolution both refer to the worldwide mean ocean surface, so far no adopted standards exist for the definition of MSL, the handling of time-dependent terms (e.g., due to global sea level rise), and the derivation of W_0, where the latter value can be determined in principle from satellite altimetry and a global geopotential model (see [141, 142]). Furthermore, the IHRS value for the reference potential is inconsistent with the corresponding value $W_0^{(\mathrm{IAU})}$ used for the definition of TT (see Sect. 3.5); Petit et al. [124] denote these two definitions as "classical geoid" and "chronometric geoid", respectively.

In this context, it is somewhat unfortunate that the same notation (W_0) is used to represent different estimates for a quantity that is connected with the (time-variable) mean ocean surface, but this issue can be resolved only through future international cooperation, even though it seems unlikely that the different communities are willing to change their definitions. In the meantime, this problem has to be handled by a simple constant shift transformation between the different level surfaces, associated with a thorough documentation of the procedures and conventions involved. It is clear that the definition of the zero level surface (W_0 issue) is largely a matter of convention, where a good option is probably to select a conventional value of W_0 (referring to a certain epoch) with a corresponding zero level surface, and to describe then the potential of the time-variable mean ocean surface for any given point in time as the deviation from this reference value.

4.2 The Geometric Levelling Approach

The classical and most direct way to obtain gravity potential differences is based on geometric levelling and gravity observations, denoted here as the geometric levelling approach. Based on Eq. (36), the gravity potential differential can be expressed as

$$dW = \frac{\partial W}{\partial x}dx + \frac{\partial W}{\partial y}dy + \frac{\partial W}{\partial z}dz = \nabla W \cdot \mathbf{d}s = \mathbf{g} \cdot \mathbf{d}s = -g\, dn\,, \tag{38}$$

where $\mathbf{d}s$ is the vectorial line element, g is the magnitude of the gravity vector, and dn is the distance along the outer normal of the level surface (zenith or vertical), which by integration leads to the geopotential number C in the form

$$C^{(i)} = W_0^{(i)} - W_P = -\int_{P_{0(i)}}^{P} dW = \int_{P_{0(i)}}^{P} g\, dn\,, \tag{39}$$

where P is a point at the Earth's surface, (i) refers to the selected zero level or height reference surface (height datum) with the gravity potential $W_0^{(i)}$, and $P_{0(i)}$ is an arbitrary point on that level surface. Thus, in addition to the raw levelling results (dn), gravity observations (g) are needed along the path between $P_{0(i)}$ and P, for details, see, e.g., [137]. The geopotential number C is defined such that it is positive for points P above the zero level surface, similar to heights. It should be noted that the integral in Eq. (39) and hence C is path independent, as the gravity field is conservative. Furthermore, the geopotential numbers can be directly linked to the redshift correction according to Eq. (32) if one takes $W_0^{(\mathrm{IAU})}$ as zero reference zero level reference potential.

However, regarding height networks, the zero level surface and the corresponding potential is typically selected in an implicit way by connecting the levelling to a fundamental national tide gauge, but the exact numerical value of the reference potential is usually unknown. As mean sea level deviates from a level surface within the Earth's gravity field due to the dynamic ocean topography (see Fig. 7), this leads to inconsistencies of more than 0.5 m between different national height systems across Europe, the extreme being Belgium, which differs by more than 2 m from all other European countries due to the selection of low tide water as the reference (instead of mean sea level).

Geometric levelling (also called spirit levelling) itself is a quasi-differential technique, which provides height differences δn (backsight minus foresight reading) with

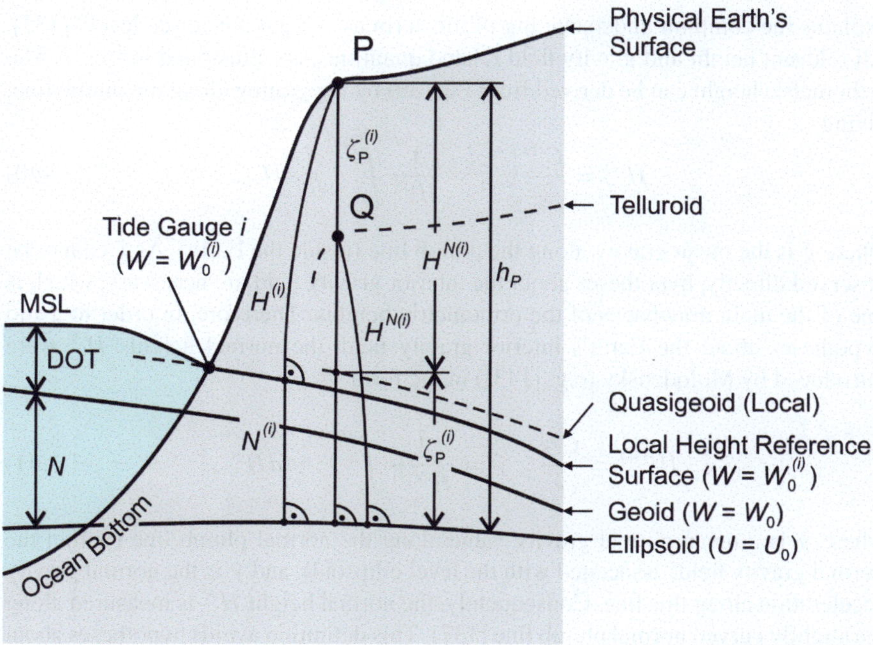

Fig. 7 Illustration of several quantities involved in gravity field modelling

respect to a local horizontal line of sight. The uncertainty of geometric levelling is rather low over shorter distances, where it can reach the sub-millimetre level, but it is susceptible to systematic errors up to the decimetre level over 1000 km distance (see Sect. 4.4). In addition, the non-parallelism of the level surfaces cannot be neglected over larger distances, as it results in a path dependence of the raw levelling results ($\oint dn \neq 0$), but this problem can be overcome by using potential differences, which are path independent because the gravity field is conservative ($\oint dW = 0$). For this reason, geopotential numbers are almost exclusively used as the foundation for national and continental height reference systems (vertical datum) worldwide, but one can also work with heights and corresponding gravity corrections to the raw levelling results (cf. [137]).

Although the geopotential numbers are ideal quantities for describing the direction of water flow, they have the unit m^2s^{-2} and are thus somewhat inconvenient in disciplines such as civil engineering. A conversion to metric heights is therefore desirable, which can be achieved by dividing the C values by an appropriate gravity value. Widely used are the orthometric heights (e.g. in the USA, Canada, Austria, and Switzerland) and normal heights (e.g. in Germany, France and many other European countries). Heights also play an important role in gravity field modelling due to the strong height dependence of various gravity field quantities.

The orthometric height H is defined as the distance between the surface point P and the zero level surface (geoid), measured along the curved plumb line, which explains the common understanding of this term as "height above sea level" [137]. All relevant height and gravity field related quantities, are illustrated in Fig. 7. The orthometric height can be derived from Eq. (39) by integrating along the plumb line, giving

$$H^{(i)} = \frac{C^{(i)}}{\bar{g}} \; , \; \bar{g} = \frac{1}{H^{(i)}} \int_0^{H^{(i)}} g \, dH \; , \tag{40}$$

where $\bar{g}$ is the mean gravity along the plumb line (inside the Earth). As $\bar{g}$ cannot be observed directly, hypotheses about the interior gravity field are necessary, which is one of the main drawbacks of the orthometric heights. Therefore, in order to avoid hypotheses about the Earth's interior gravity field, the normal heights H^N were introduced by Molodensky (e.g. [143]) in the form

$$H^{N(i)} = \frac{C^{(i)}}{\bar{\gamma}} \; , \; \bar{\gamma} = \frac{1}{H^{N(i)}} \int_0^{H^{N(i)}} \gamma \, dH^N \; , \tag{41}$$

where $\bar{\gamma}$ is a mean normal gravity value along the normal plumb line (within the normal gravity field, associated with the level ellipsoid), and γ is the normal gravity acceleration along this line. Consequently, the normal height H^N is measured along the slightly curved normal plumb line [137]. This definition avoids hypotheses about the Earth's interior gravity field, which is the main reason for adopting it in many countries. Indeed, the value $\bar{\gamma}$ can be calculated analytically, as the normal gravity potential of the level ellipsoid U is known analytically (see next section), but $\bar{\gamma}$ is slightly depending on the chosen reference ellipsoid. However, the normal height does not have a simple physical interpretation, in contrast to the orthometric height ("height above sea level"). Nevertheless, the normal height can be interpreted as the height above the quasigeoid, which is not a level surface and also has no physical interpretation (see [137]).

While the orthometric and normal heights are related to the Earth's gravity field (so-called physical heights), the ellipsoidal heights h, as derived from GNSS observations, are purely geometric quantities, describing the distance (along the ellipsoid normal) of a point P from a conventional reference ellipsoid. As the geoid and quasigeoid serve as the zero height reference surface (vertical datum) for the orthometric and normal heights, respectively, the following relation holds

$$h = H^{(i)} + N^{(i)} = H^{N(i)} + \zeta^{(i)} \; , \tag{42}$$

where $N^{(i)}$ is the geoid undulation, and $\zeta^{(i)}$ is the quasigeoid height or height anomaly; for further details on the geoid and quasigeoid (height anomalies) see, e.g., [137]. Equation (42) neglects the fact that strictly the relevant quantities are measured along slightly different lines in space, but the maximum effect is only at the sub-millimetre level (for further details cf. [134]).

Lastly, the geometric levelling approach gives only gravity potential differences, but the associated constant zero potential $W_0^{(i)}$ can be determined by at least one (better several) GNSS and levelling points in combination with the (gravimetrically derived) disturbing potential, as described in the next section. Rearranging the above equations gives the desired gravity potential values in the form

$$W_P = W_0^{(i)} - C^{(i)} = W_0^{(i)} - \bar{g}H^{(i)} = W_0^{(i)} - \bar{\gamma}H^{N(i)} , \tag{43}$$

and hence the geopotential numbers and the heights $H^{(i)}$ and $H^{N(i)}$ are fully equivalent.

4.3 The GNSS/Geoid Approach

The gravity potential W cannot be derived directly from Eq. (37) due to insufficient knowledge about the density structure of the entire Earth, and therefore it must be determined indirectly based on measurements performed at or above the Earth's surface, which leads to the area of geodetic boundary value problems. In this context, gravity measurements form one of the most important data sets. However, since gravity (represented as $g = |\boldsymbol{g}|$ = length of the gravity vector $\boldsymbol{g}$) and other relevant observations depend in general in a nonlinear way on the potential W, the observation equations must be linearized. This is done by introducing an a priori known reference potential and corresponding reference positions. Regarding the reference potential, traditionally the normal gravity field related to the level ellipsoid is employed, where the ellipsoid surface is a level surface of its own gravity field. The level ellipsoid is chosen as a conventional system, because it is easy to compute (from just four fundamental parameters; e.g. two geometrical parameters for the ellipsoid plus the total mass M and the angular velocity ω), useful for other disciplines, and also utilized for describing station positions (e.g. in connection with GNSS or the International Terrestrial Reference Frame – ITRF). However, today spherical harmonic expansions based on satellite data could also be employed (cf. [134]).

The linearization process leads to the disturbing (or anomalous) potential T defined as

$$T_P = W_P - U_P , \tag{44}$$

where U is the normal gravity potential associated with the level ellipsoid. Accordingly, the gravity vector and other gravity field observables are approximated by corresponding reference quantities based on the level ellipsoid, leading to gravity anomalies Δg, height anomalies ζ, geoid undulations N, etc. The main advantage of the linearization process is that the residual quantities (with respect to the known ellipsoidal reference field) are in general four to five orders of magnitude smaller than the original ones, and in addition they are less position dependent.

Hence, the disturbing potential T takes over the role of W as the new fundamental target quantity, to which all other gravity field quantities of interest are related. Accordingly, the gravity anomaly is given by

$$\Delta g_P = g_P - \gamma_Q = -\frac{\partial T}{\partial h} + \frac{1}{\gamma}\frac{\partial \gamma}{\partial h}T - \frac{\partial \gamma}{\partial h}\left(W_0^{(i)} - U_0\right), \tag{45}$$

where g_P is the gravity acceleration at the observation point P (at the Earth's surface or above), γ_Q is the normal gravity acceleration at a known linearization point Q (telluroid, Q is located on the same ellipsoidal normal as P at a distance H^N above the ellipsoid, or equivalently $U_Q = W_P$; for further details, see [134]), the partial derivatives are with respect to the ellipsoidal height h, and $\delta W_0^{(i)} = W_0^{(i)} - U_0$ is the potential difference between the zero level height reference surface ($W_0^{(i)}$) and the normal gravity potential U_0 at the surface of the level ellipsoid. Equation (45) is also denoted as the fundamental equation of physical geodesy; it represents a boundary condition that has to be fulfilled by solutions of the Laplace equation for the disturbing potential T, sought within the framework of GBVPs. Moreover, the subscripts P and Q are dropped on the right side of Eq. (45), noting that it must be evaluated at the known telluroid point (boundary surface).

In a similar way, Bruns's formula gives the height anomaly or quasigeoid height as a function of T in the form

$$\zeta^{(i)} = h - H^{N(i)} = \frac{T}{\gamma} - \frac{W_0^{(i)} - U_0}{\gamma} = \frac{T}{\gamma} - \frac{\delta W_0^{(i)}}{\gamma} = \zeta + \zeta_0^{(i)}, \tag{46}$$

implying that $\zeta^{(i)}$ and ζ are associated with the corresponding zero level surfaces $W = W_0^{(i)}$ and $W = U_0$, respectively. The $\delta W_0^{(i)}$ term is also denoted as height system bias and is frequently omitted in the literature, implicitly assuming that $W_0^{(i)}$ equals U_0. However, when aiming at a consistent derivation of absolute potential values, the $\delta W_0^{(i)}$ term has to be taken into consideration.

Hence, all linearized gravity field observables are linked to the disturbing potential T, which has the important property of being harmonic outside the Earth's surface and regular (zero) at infinity. Consequently, solutions of T are developed in the framework of potential theory and GBVPs, i.e. solutions of the Laplace equation are sought that fulfil certain boundary conditions. Now, the first option to compute T is based on the well-known spherical harmonic expansion, using coefficients derived from satellite data alone or in combination with terrestrial data (e.g., EGM2008; EGM – Earth Gravitational Model [144]), yielding

$$T(\theta, \lambda, r) = \sum_{n=0}^{n_{\max}} \left(\frac{a}{r}\right)^{n+1} \sum_{m=-n}^{n} \overline{T}_{nm} \overline{Y}_{nm}(\theta, \lambda) \tag{47}$$

with

$$\overline{Y}_{nm}(\theta, \lambda) = \overline{P}_{n|m|}(\cos\theta) \begin{cases} \cos m\lambda \text{ for } m \geq 0 \\ \sin |m|\lambda \text{ for } m < 0 \end{cases}, \tag{48}$$

$$\overline{T}_{nm} = \frac{GM}{a} \begin{cases} \Delta\overline{C}_{nm} \text{ for } m \geq 0 \\ \Delta\overline{S}_{nm} \text{ for } m < 0 \end{cases}, \tag{49}$$

where (θ, λ, r) are spherical coordinates, n, m are integers denoting the degree and order, GM is the geocentric gravitational constant (gravitational constant G times the mass of the Earth M), a is in the first instance an arbitrary constant, but is typically set equal to the semimajor axis of a reference ellipsoid, $\overline{P}_{nm}(\cos\theta)$ are the fully normalized associated Legendre functions of the first kind, and $\Delta\overline{C}_{nm}$, $\Delta\overline{S}_{nm}$ are the (fully normalized) spherical harmonic coefficients (also denoted as Stokes's constants), representing the difference in the gravitational potential between the real Earth and the level ellipsoid.

Regarding the uncertainty of a gravity field quantity computed from a global spherical harmonic model up to some fixed degree $n_{\max}$, the coefficient uncertainties lead to the so-called commission error based on the law of error propagation, and the omitted coefficients above degree $n_{\max}$, which are not available in the model, lead to the corresponding omission error. With dedicated satellite gravity field missions such as GRACE (Gravity Recovery and Climate Experiment) and GOCE (Gravity field and steady-state Ocean Circulation Explorer), the long wavelength geoid and quasigeoid can today be determined with low uncertainty, e.g., about 1 mm at 200 km resolution ($n = 95$) and 1 cm at 150 km resolution ($n = 135$) from GRACE (e.g. [145]), and 1.5 cm at about 110 km resolution ($n = 185$) from the GOCE mission (e.g. [146, 147]). However the corresponding omission error at these wavelengths is still quite significant with values at the level of several decimetres, e.g., 0.94 m for $n = 90$, 0.42 m for $n = 200$, and 0.23 m for $n = 360$. For the ultra-high degree geopotential model EGM2008 [144], which combines satellite and terrestrial data and is complete up to degree and order 2159, the omission error is 0.023 m, while the commission error is about 5 to 20 cm, depending on the region and the corresponding data quality. The above uncertainty estimates are based on the published potential coefficient standard deviations as well as a statistical model for the estimation of corresponding omission errors, but do not include the uncertainty contribution of GM (zero degree term in Eq. (47)); hence, the latter term, contributing about 3 mm in terms of the height anomaly (corresponds to about 0.5 ppb; see [148, 149]), has to be added in quadrature to the figures given above. Further details on the uncertainty estimates can be found in [134].

Based on these considerations it is clear that satellite measurements alone will never be able to supply the complete geopotential field with sufficient accuracy, which is due to the signal attenuation with height and the required satellite altitudes of a few 100 km. Only a combination of the highly accurate and homogeneous (long wavelength) satellite gravity fields with high-resolution terrestrial data (mainly gravity and topography data with a resolution down to 1–2 km and below) can cope with this task. In this respect, the satellite and terrestrial data complement each other in an ideal way, with the satellite data accurately providing the long wavelength field structures, while the terrestrial data sets, which have potential weaknesses in large-scale accuracy and coverage, mainly contribute the short wavelength features. However, in the future, also height anomalies derived from common GNSS and clock points may contribute to regional gravity field modelling (see Sect. 6).

Consequently, regional solutions for the disturbing potential and other gravity field parameters have to be developed, which typically have a higher resolution (down to

1–2 km) than global spherical harmonic models. Based on the developments of Molodensky (e.g., [143]), the disturbing potential T can be derived from a series of surface integrals, involving gravity anomalies and heights over the entire Earth's surface, which in the first instance can be symbolically written as

$$T = \mathbf{M}(\Delta g) \, , \tag{50}$$

where $\mathbf{M}$ is the Molodensky operator and Δg are the gravity anomalies over the entire Earth's surface.

Further details on regional gravity field modelling are given in [81, 134], including the solution of Molodensky's problem, the remove-compute-restore (RCR) procedure, the spectral combination approach, data requirements, and uncertainty estimates for the disturbing potential and quasigeoid heights. These investigations show that quasigeoid heights can be obtained today with an estimated uncertainty of 1.9 cm, where the major contributions come from the spectral band below spherical harmonic degree 360. Furthermore, this uncertainty estimate represents an optimistic scenario and is only valid for the case that a state-of-the-art global satellite model (e.g. a 5th generation GOCE model [147]) is employed and sufficient high-resolution and high-quality terrestrial gravity and terrain data sets (especially gravity measurements with a spacing of a few kilometers and an uncertainty lower than 1 mGal) are available around the point of interest (e.g. within a distance of 50–100 km), see also [150, 151]. Fortunately, such a data situation exists for most of the metrology institutes with optical clock laboratories – at least in Europe. Furthermore, the perspective exists to improve the uncertainty of the calculated quasigeoid heights [81].

Now, once the disturbing potential values T are computed, either from a global geopotential model by Eq. (47), or from a regional solution by Eq. (50) based on Molodensky's theory, the gravity potential W, needed for the relativistic redshift corrections, can be computed most straightforwardly from Eq. (44) as

$$W_P = U_P + T_P \, , \tag{51}$$

where the basic requirement is that the position of the given point P in space must be known accurately (e.g. from GNSS observations), as the normal potential U is strongly height-dependent, while T is only weakly height dependent with a maximum vertical gradient of a few parts in 10^{-3} $\mathrm{m^2 s^{-2}}$ per metre. The above equation also makes clear that the predicted potential values W_P are in the end independent of the choice of W_0 and U_0 used for the linearization. Furthermore, by combining equation (51) with (46), and representing U as a function of U_0 and the ellipsoidal height h, the following alternative expressions for W (at point P) can be derived as

$$W_P = U_0 - \bar{\gamma}(h - \zeta) = U_0 - \bar{\gamma}\left(h - \zeta^{(i)}\right) + \delta W_0^{(i)} \, , \tag{52}$$

which demonstrates that ellipsoidal heights (e.g. from GNSS) and the results from gravity field modelling in the form of the quasigeoid heights (height anomalies) ζ or the disturbing potential T are required, whereby a similar equation can be derived for the geoid undulations N. Consequently, the above approach (Eqs. (51) and (52)) is denoted here somewhat loosely as the GNSS/geoid approach, which is also known in the literature as the GNSS/GBVP approach (the geodetic boundary value problem is the basis for computing the disturbing potential T; see, e.g., [152, 153]).

The GNSS/geoid approach depends strongly on precise gravity field modelling (disturbing potential T, metric height anomalies ζ or geoid undulations N) and precise GNSS positions (ellipsoidal heights h) for the points of interest, with the advantage that it delivers the absolute gravity potential W, which is not directly observable and is therefore always based on the assumption that the gravitational potential is regular (zero) at infinity (see above). In addition, the GNSS/geoid approach allows the derivation of the height system bias term $\delta W_0^{(i)}$ based on Eq. (46) together with at least one (better several) common GNSS and levelling stations in combination with the gravimetrically determined disturbing potential T.

4.4 Uncertainty Considerations

The following uncertainty considerations are based on heights, but corresponding potential values can easily be obtained by multiplying the meter values with an average gravity value (e.g. 9.81 ms^{-2} or roughly 10 ms^{-2}). Regarding the geometric levelling and the GNSS/geoid approach, the most direct and accurate way to derive potential differences over short distances is the geometric levelling technique, as standard deviations of 0.2–1.0 mm can be attained for a 1 km double-run levelling with appropriate technical equipment [137]. However, the uncertainty of geometric levelling depends on many factors, with some of the levelling errors behaving in a random manner and propagating with the square root of the number of individual set-ups or the distance, respectively, while other errors of systematic type may propagate with distance in a less favourable way. Consequently, it is important to keep in mind that geometric levelling is a differential technique and hence may be susceptible to systematic errors; examples include the differences between the second and third geodetic levelling in Great Britain (about 0.2 m in the north–south direction over about 1000 km distance [154]), corresponding differences between an old and new levelling in France (about 0.25 m from the Mediterranean Sea to the North Sea, also mainly in north–south direction, distance about 900 km [155]), and inconsistencies of more than ± 1 m across Canada and the USA (differences between different levellings and with respect to an accurate geoid [156–158]). In addition, a further complication with geometric levelling in different countries is that the results are usually based on different tide gauges with offsets between the corresponding zero level surfaces, which, for example, reach more than 0.5 m across Europe. Furthermore, in some countries the levelling observations are about 100 years old and thus may not represent the actual situation due to possibly occurring recent vertical crustal movements.

With respect to the GNSS/geoid approach, the uncertainty of the GNSS positions is today more or less independent of the interstation distance. For instance, the station coordinates provided by the International GNSS Service (IGS) or the IERS (e.g. ITRF2008) reach vertical accuracies of about 5–10 mm (cf. [159–161]). The uncertainty of the quasigeoid heights (height anomalies) is discussed in the previous subsection, showing that a standard deviation of 1.9 cm is possible in a best-case scenario. Moreover, the values are nearly uncorrelated over longer distances, with a correlation of less than 10% beyond a distance of about 180 km [81]. Aiming at the determination of the absolute gravity potential W according to Eq. (51) or (52), which is the main advantage of the GNSS/geoid technique over the geometric levelling approach, both the uncertainties of GNSS and the quasigeoid have to be considered. Assuming a standard deviation of 1.9 cm for the quasigeoid heights and 1 cm for the GNSS ellipsoidal heights without correlations between both quantities, a standard deviation of 2.2 cm is finally obtained (in terms of heights) for the absolute potential values based on the GNSS/geoid approach. Thus, for contributions of optical clocks to international timescales, which require the absolute potential W_P relative to a conventional zero potential W_0 (see Sect. 3.5), the relativistic redshift correction can be computed with an uncertainty of about 2×10^{-18}. This is the case more or less everywhere in the world where high-resolution regional gravity field models have been developed on the basis of a state-of-the-art global satellite model in combination with sufficient terrestrial gravity field data. On the other hand, for potential differences over larger distances of a few 100 km (i.e. typical distances between different metrology institutes), the statistical correlations of the quasigeoid values virtually vanish, which then leads to a standard deviation for the potential difference of 3.2 cm in terms of height, i.e. $\sqrt{2}$ times the figure given above for the absolute potential (according to the law of error propagation), which again has to be considered as a best-case scenario. This would also hold for intercontinental connections between metrology institutes, provided again that sufficient regional high-resolution terrestrial data exist around these places. Furthermore, in view of future refined satellite and terrestrial data, the perspective exists to improve the uncertainty of the relativistic redshift corrections from the level of a few parts in 10^{18} to one part in 10^{18} or below. According to this, over long distances across national borders, the GNSS/geoid approach should be a better approach than geometric levelling (see also [81]).

5 Relativistic Redshift Corrections for the Realization of TT from Geodetic Methods

An atomic clock, in order to contribute to the realization of Terrestrial Time (TT), needs to be corrected for the relativistic redshift (see Sect. 3.5, Eq. (31)). In this section we present some results from the ITOC (International Timescales with Optical Clocks) project, in particular those linked to the determination of unified relativistic redshift corrections for several European metrology institutes.

5.1 The ITOC Project

The ITOC project ([60]; see also http://projects.npl.co.uk/itoc/) was a 3 years (2013–2016) EURAMET joint research project funded by the European Community's Seventh Framework Programme, ERA-NET Plus. This project was done in the context of a future optical redefinition of the SI second (see e.g. [162–165]). An extensive programme of comparisons between high accuracy European optical atomic clocks has been performed, verifying the estimated uncertainty budgets of the optical clocks. Relativistic effects influencing clock comparisons have been evaluated at an improved level of accuracy, and the potential benefits that optical clocks could bring to the field of geodesy have been demonstrated.

Several optical frequency ratio measurements as well as independent absolute frequency measurements of optical lattice clocks have been made locally at the following NMIs (National Metrology Institutes): INRIM (Istituto Nazionale di Ricerca Metrologica, Torino, Italy), LNE-SYRTE (Laboratoire national de métrologie et d'essais – Système de Références Temps-Espace, Paris, France), NPL (National Physical Laboratory, Teddington, UK), and PTB (Physikalisch-Technische Bundesanstalt, Braunschweig, Germany), all of whom operate one or more than one type of optical clock, as well as Caesium primary frequency standards (see e.g. [78, 106, 107, 166]). Distant comparisons have also been performed between the same laboratories with a broadband version of two-way satellite time and frequency transfer (TWSTFT).

A proof-of-principle experiment has been realized to show that the relativistic redshift of optical clocks can be exploited to measure gravity potential differences over medium–long baselines. A transportable ^{87}Sr optical lattice clock has been developed at PTB [105]. It has been transported to the Laboratoire Souterrain de Modane (LSM) in the Fréjus road tunnel through the Alps between France and Italy. There it was compared, using a transportable frequency comb from NPL, to the caesium fountain primary frequency standard at INRIM, via a coherent fibre link and a second optical frequency comb operated by INRIM. A physical model has been formulated to describe the relativistic effects relevant to time and frequency transfer over optical fibre links, and has been used to evaluate the relativistic corrections for the fibre links now in place between NPL, LNE-SYRTE and PTB, as well as to provide guidelines on the importance of exact fibre routing for time and frequency transfer via optical fibre links (see [122] and Sect. 3.3.2).

Within the ITOC project, the gravity potential has been determined by IfE/LUH (Institut für Erdmessung, Leibniz Universität Hannover) with significantly improved accuracy at the sites participating in optical clock comparisons within the project (INRIM, LNE-SYRTE, LSM, NPL and PTB). Levelling measurements and gravity surveys have been performed at INRIM, LSM, OBSPARIS, NPL and PTB, the latter including at least one absolute gravity observation at each site. These measurements have been integrated into the existing European gravity database and used for the computation of a new version of the European Gravimetric (Quasi) Geoid, EGG2015 (see [167] and Sect. 5.3). Time-variable gravity potential signals induced by tides and non-tidal mass redistributions have also been calculated for the optical clock comparison sites [129]. Finally, the potential contributions of combined GNSS and optical clock measurements for determining the gravity potential at high spatial resolution have been studied theoretically, which will be presented in Sect. 6.

5.2 The GNSS and Levelling Campaigns

Within the ITOC project, GNSS and levelling observations were performed at the NMIs INRIM, LNE-SYRTE, NPL, and PTB, as well as the collaborator LSM (not an NMI) to calculate the relativistic redshift corrections. First of all, some general recommendations were developed for carrying out the measurements to ensure accuracies in the millimetre range for the levelling results and better than one centimetre for the GNSS (ellipsoidal) heights (see [81]). In general, it is recommended to install fixed markers in all local laboratories close to the clock tables to allow an easy height transfer to the clocks (e.g. with a simple spirit level used for building construction), and to connect these markers by geometric levelling with millimetre uncertainty to the existing national levelling networks and at least two (better several) GNSS stations. This is to support local clock comparisons at the highest level, and to apply the GNSS/geoid approach to obtain also the absolute potential values for remote clock comparisons and contributions to international timescales, while at the same time improving the redundancy and allowing a mutual control of GNSS, levelling, and (quasi)geoid data.

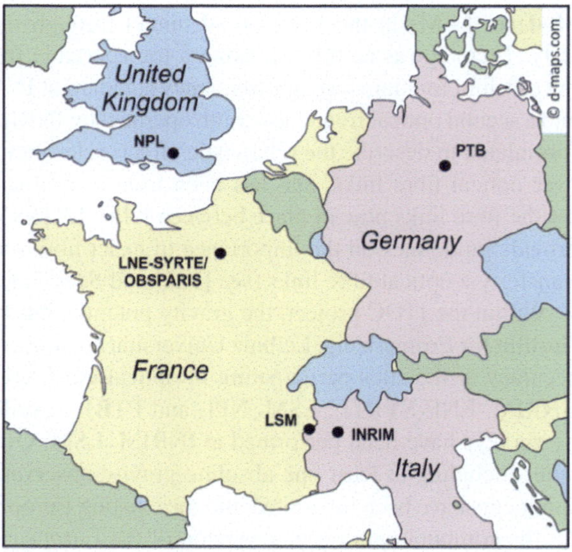

Fig. 8 Map showing the locations of the INRIM, LNE-SYRTE, NPL, PTB, and LSM sites

The actual levelling and GNSS measurements were mainly taken by local surveyors on behalf of the respective NMIs, and the NMIs provided all results to Leibniz Universität Hannover (LUH; Institut für Erdmessung) for further processing and homogenisation. The locations of the above mentioned ITOC clock sites are shown in Fig. 8.

The coordinates of all GNSS stations were referred to the ITRF2008 at its associated standard reference epoch 2005.0, with three-dimensional Cartesian and ellipsoidal coordinates being available. The geometric levelling results were based in the first instance on the corresponding national vertical reference networks, which are the following:

- DHHN92 is the official German height reference system; it is based on the Amsterdam tide gauge and consists of normal heights.
- NGF-IGN69 is the official French height reference system; it is based on the Marseille tide gauge and also consists of normal heights. In addition, selected levelling lines were re-observed since 2000, which lead to the so-called (NGF)–NIREF network, differing from the old network mainly by a south-to-north trend of 31.0 mm per degree latitude [155].
- ODN (Ordnance Datum Newlyn, established by Ordnance Survey) is the height reference system for mainland Great Britain; it is based on the Newlyn tide gauge and consists of orthometric heights.
- IGM is the Italian height reference system (established by Istituto Geografico Militare); it is based on the Genova tide gauge and consists of orthometric heights.

The different zero level surfaces of the above national height systems (datum) were taken into account by transforming all national heights into the unified European Vertical Reference System (EVRS) using its latest realization EVRF2007 (European Vertical Reference Frame 2007). The EVRF2007 is based on a common adjustment of all available European levelling networks in terms of geopotential numbers, which are finally transformed into normal heights. The measurements within the UELN (United European Levelling Network) originate from very different epochs, but reductions for vertical crustal movements were only applied for the (still ongoing) post-glacial isostatic adjustment (GIA) in northern Europe; for further details on EVRF2007, see [168]. However, as GIA hardly affects the aforementioned clock sites, while other sources of vertical crustal movements are not known, the EVRF2007 heights are considered as stable in time in the following.

For the conversion of the national heights into the vertical reference frame EVRF2007, nearby common points with heights in both systems were utilized; this information was kindly provided by Bundesamt für Kartographie und Geodäsie (BKG) in Germany (M. Sacher, personal communication, 9 October 2015). If such information is not available, the CRS-EU webpage (Coordinate Reference Systems in Europe; http://www.crs-geo.eu, also operated by BKG) can be used, which gives, besides a description of all the national and international coordinate and height reference systems for the participating European countries, up to three transformation parameters (height bias and two tilt parameters) for the transformation of the national heights into EVRF2007 and a statement on the quality of this transformation.

The EVRF2007 heights were computed as

$$H^{N(\mathrm{EVRF2007})} = H^{(\mathrm{national})} + \Delta H^{(\mathrm{national})} , \tag{53}$$

where $\Delta H^{(\mathrm{national})}$ is a constant shift for each NMI site. The following offsets $\Delta H^{(\mathrm{national})}$ were employed:

- PTB: +0.006 m (DHHN92),
- LNE-SYRTE: −0.479 m (NGF-IGN69),
- NPL: −0.010 m (ODN, based on the official EVRF2007 results), −0.144 m (ODN, own estimate, see below)
- INRIM: −0.307 m (IGM),
- LSM: −0.307 m (IGM).

The accuracy of the above transformation depends on the accuracy of the input heights as well as the number of identical points, giving RMS residuals of the transformation between 2 mm (Germany) and 35 mm (Italy). A further note is necessary for the computation of the NPL offset. The offset of −0.010 m is based on the official EVRF2007 heights, which rely on hydrodynamic levelling (see [169]), but do not include the 1994 channel tunnel levelling. Therefore, a first attempt was made to consider the new channel tunnel levelling as well as the new levelling measurements in France (NIREF, see above); this was done by starting with an offset of −0.479 m for LNE-SYRTE, plus a correction for the NGF-IGN69 tilt between LNE-SYRTE and Coquelle (channel tunnel entrance in France) of −0.065 m (south-north slope = −31.0 mm per degree latitude, latitude difference = 3.105° [155]), plus an offset of +0.400 m for the difference between ODN and NGF-IGN69 from the channel tunnel levelling [170], resulting in an offset of −0.144 m for NPL. In Sect. 5.4 it is shown that the new offset leads to a better agreement between the geometric levelling and the GNSS/geoid approach.

Further details on the local levelling results and corresponding GNSS observations at some of the aforementioned clock sites can be found in [81]. In general, the uncertainty of the local levellings is at the few millimeters level, and the uncertainty of the GNSS ellipsoidal heights is estimated to be better than 10 mm. Moreover, care has to be taken in the handling of the permanent parts of the tidal corrections (for details, see, e.g., [132–134]. The IAG has recommended to use the so-called "zero tide system" (resolutions no. 9 and 16 from the year 1983; cf.[135]), which is implemented in the European height reference frame EVRF2007 and the European gravity field modelling performed at LUH (e.g. EGG2015, see below). On the other hand, most GNSS coordinates (including the ITRF and IGS results) refer to the "non-tidal (or tide-free) system". Hence, for consistency with the IAG recommendations and the other quantities involved (EVRF2007 heights, quasigeoid), the ellipsoidal heights from GNSS were converted from the non-tidal to the zero-tide system based on the following formula from [133] with

$$h_{zt} = h_{nt} + 60.34 - 179.01 \sin^2 \phi - 1.82 \sin^4 \phi \ \ [\mathrm{mm}] , \tag{54}$$

where ϕ is the ellipsoidal latitude, and h_{nt} and h_{zt} are the non-tidal and zero-tide ellipsoidal heights, respectively. Hence, the zero-tide heights over Europe are about 3–5 cm smaller than the corresponding non-tidal heights.

5.3 The European Gravimetric Quasigeoid Model EGG2015

The latest European gravimetric quasigeoid model EGG2015 [167] was employed to determine absolute potential values based on Eqs. (51) and (52), as needed for the derivation of the relativistic redshift corrections in the context of international timescales. The major differences between EGG2015 and the previous EGG2008 model [134] are the inclusion of additional gravity measurements carried out recently around the aforementioned ITOC clock sites [60] and the use of a newer geopotential model based on the GOCE satellite mission instead of EGM2008. The new gravity measurements around the clock sites were carried out by LUH, taking at least one absolute gravity observation (with the LUH FG5X-220 instrument) plus additional relative gravity observations (relative to the established absolute points) around all ITOC sites. The total number of new gravity points is 36 for INRIM, 100 for LNE-SYRTE, 123 for LSM, 66 for NPL, and 46 for PTB, where most of the measurements were taken around LSM due to the high mountains and corresponding strong gravity field variations. Overall, the purpose of the new gravity measurements was threefold, namely to perform spot checks of the largely historic gravity data base (consistency check), to add new observations in areas void of gravity data so far (coverage improvement), and to serve for future geodynamic and meteorological purposes (infrastructure improvement), with the ultimate goal of improving the reliability and accuracy of the computed quasigeoid model.

EGG2015 was computed from surface gravity data in combination with topographic information and the geopotential model GOCO05S [146] based on the RCR technique. The estimated uncertainty (standard deviation) of the absolute quasigeoid values is 1.9 cm; further details including correlation information can be found in [81, 134].

5.4 Gravity Potential Determination

First, a consistency check between the GNSS and levelling heights at each clock site was performed by evaluating the differences between the GNSS ellipsoidal heights and the normal heights from levelling, computed as $\zeta_{\mathrm{GNSS}} = h_{zt} - H^{N(\mathrm{EVRF2007})}$, also denoted as GNSS/levelling quasigeoid heights (h_{zt} is referring to ITRF2008, epoch 2005.0, zero-tide system; $H^{N(\mathrm{EVRF2007})}$ is based on EVRF2007; see Sect. 5.2). As the distances between the GNSS stations at each NMI site are typically only a few 100 m, the quasigeoid at each site can be approximated in the first instance by a horizontal plane, but a more general and better way (especially for larger interstation distances) is the comparison with a high-resolution gravimetric quasigeoid model, such as EGG2015. After computing the differences ($\zeta_{\mathrm{GNSS}} - \zeta_{\mathrm{EGG2015}}$), the main quantities of interest are the residuals about the mean difference, giving RMS values of 11 mm (max. 16 mm) for INRIM, 5 mm (max. 5 mm) for LNE-SYRTE, 6 mm (max. 8 mm) for NPL, and 4 mm (max. 6 mm) for PTB, while for LSM only a control through two RTK (Real Time Kinematics - a differential GNSS technique) positions exists, giving a RMS difference of 17 mm (max. 29 mm); this proves an excellent consistency of the GNSS and levelling results at all clock sites. Although initial results were worse for the PTB and LNE-SYRTE sites, the problem was traced to an incorrect identification of the corresponding antenna reference points (ARPs); at the PTB site, an error of 16 mm was found for station PTBB, and at LNE-SYRTE, there was a difference of 29 mm between the ARP and the levelling benchmark and an additional error in the ARP height of 8 mm at station OPMT. It should be noted that, due to the high consistency of the GNSS and levelling data at all sites, even quite small problems in the ARP heights (below 1 cm) could be detected and corrected after on-site inspections and additional verification measurements. This also strongly supports the recommendation to have sufficient redundancy in the GNSS and levelling stations.

Now, in order to apply the GNSS/geoid approach according to Eqs. (51) and (52), ellipsoidal heights are required for all stations of interest. However, initially GNSS coordinates are only available for a few selected points at each NMI site, while for most of the other laboratory points near the clocks, only levelled heights exist. Therefore, based on Eq. (46), a quantity $\delta\zeta$ is defined as

$$\delta\zeta = \left(h - H^{N(i)}\right) - \zeta^{(i)} , \tag{55}$$

which should be zero in theory, but is not in practice due to the uncertainties in the quantities involved (GNSS, levelling, quasigeoid). However, if a high-resolution quasigeoid model is employed (such as EGG2015), the term $\delta\zeta$ should be small and represent only long-wavelength features, mainly due to systematic levelling errors over large distances as well as long-wavelength quasigeoid errors. In this case, an average (constant) value $\overline{\delta\zeta}$ (based on the common GNSS and levelling benchmarks) can be used at each NMI site to convert all levelled heights into ellipsoidal heights by using

$$h^{(\mathrm{adj})} = H^{N(i)} + \zeta^{(i)} + \overline{\delta\zeta} = H^{N(i)} + \left(\zeta + \zeta_0^{(i)}\right) + \overline{\delta\zeta} , \tag{56}$$

which is based on Eq. (42). This has the advantage that locally (at each NMI) the consistency is kept between the levelling results, on the one hand, and the GNSS/quasigeoid results on the other hand. Consequently, the final potential differences between stations at each NMI are identical for the GNSS/geoid and geometric levelling approach, which is reasonable, as locally the uncertainty of levelling is usually lower than that of the GNSS/quasigeoid results.

Based on the ellipsoidal heights (according to Eq. (56)) and the EVRF2007 normal heights (based on Eq. (53)), the gravity potential values can finally be derived for all relevant stations, using both the geometric levelling approach (Eq. (43)) and the GNSS/geoid approach (Eq. (51) or (52)). The results from both approaches are provided in the form of geopotential numbers according to Eq. (39) with

$$C^{(\mathrm{IAU})} = W_0^{(\mathrm{IAU})} - W_P \tag{57}$$

where the conventional value $W_0^{(\mathrm{IAU})} = c^2 L_G \approx 62,636,856.00\,\mathrm{m}^2\,\mathrm{s}^2$ is used, following the IERS2010 conventions and the IAU resolutions for the definition of TT (see Sect. 3.5). The geopotential numbers C are more convenient than the absolute potential values W_P due to their smaller numerical values and direct usability for the derivation of the (static) relativistic redshift corrections according to Eq. (32). The geopotential numbers C derived from Eq. (57) are typically given in the geopotential unit (gpu; 1 gpu $= 10\mathrm{m}^2\,\mathrm{s}^{-2}$), resulting in numerical values of C that are about 2% smaller than the numerical height values. Regarding the geometric levelling approach, the value $W_0^{(\mathrm{EVRF2007})} = 62,636,857.86\,\mathrm{m}^2\,\mathrm{s}^{-2}$ based on the European EUVN_DA GNSS/levelling data set from [171] is utilized in Eq. (43), giving $C^{(\mathrm{lev})}$. For the GNSS/geoid approach according to Eq. (51) or (52), the disturbing potential T or the corresponding height anomaly values ζ are taken from the EGG2015 model, and the normal potential $U_0 = 62,636,860.850\,\mathrm{m}^2\,\mathrm{s}^{-2}$, associated with the surface of the underlying GRS80 (Geodetic Reference System 1980; see [172]) level ellipsoid, is used, resulting in $C^{(\mathrm{GNSS/geoid})}$. Furthermore, the mean normal gravity values $\bar{\gamma}$ are also based on the GRS80 level ellipsoid; for further details, see [81]. Taking all this into account, leads to the following discrepancies between the geopotential numbers from the GNSS/geoid and the geometric levelling approach, defined in the sense $\Delta C = C^{(\mathrm{GNSS/geoid})} - C^{(\mathrm{lev})}$:

- PTB: −0.017 gpu,
- LNE-SYRTE: −0.109 gpu,
- NPL: −0.275 gpu (with ODN offset based on official EVRF2007 results),
  −0.144 gpu (with ODN offset based on own estimate, see above),
- INRIM: +0.019 gpu,
- LSM: −0.087 gpu.

The above results show first of all that the two approaches differ at the few decimetre level over Europe, that the consideration of the new French and channel tunnel levelling leads to a better agreement, and that the implementation of the national height system offsets was done correctly, recalling, e.g., that the difference between the French and German zero level surfaces is about half a metre. However, as the above differences ΔC are directly depending on the chosen reference potential $W_0^{(\mathrm{EVRF2007})}$ for EVRF2007, potential differences between two stations and the corresponding discrepancies between the GNSS/geoid and the geometric levelling approach are discussed as well in the following. Regarding potential differences, the discrepancies between both approaches amount to −0.106 gpu for the connection INRIM/LSM, −0.036 gpu for INRIM/PTB, −0.092 gpu for PTB/LNE-SYRTE, −0.166 gpu for LNE-SYRTE/NPL (−0.035 gpu based on own ODN offset, see above), as well as −0.258 gpu for PTB/NPL (−0.127 gpu based on own ODN offset, see above), respectively.

Regarding the significance of the aforementioned discrepancies in the potential differences between both geodetic approaches (levelling, GNSS/geoid), these have to be discussed in relation to the corresponding uncertainties of levelling, GNSS, and the quasigeoid model. Denker et al. [81] discuss the uncertainties (standard deviation) from single line levelling connections and the EVRF2007 network adjustment, indicating a factor 2.5 improvement due to the network adjustment. The EVRF2007 network adjustment gives a standard deviation of about 20 mm for the height connection PTB/LNE-SYRTE, while the corresponding standard deviations for the connections PTB/NPL and LNE-SYRTE/NPL are both about 80 mm (M. Sacher, BKG, Leipzig, Germany, personal communication, 10 May 2017), the latter being dominated by the uncertainty of the hydrodynamic levelling across the English Channel. However, these internal uncertainty estimates from the network adjustment do not consider any systematic levelling error contributions. On the other hand, the GNSS ellipsoidal heights have uncertainties below 10 mm, the uncertainty of EGG2015 has been discussed above, yielding a standard deviation of 19 mm for the absolute values and about 27 mm for corresponding differences over longer distances, and therefore some of the larger discrepancies between the two geodetic approaches (levelling versus GNSS/geoid) have to be considered as statistically significant. Hence, as systematic errors in levelling at the decimetre level exist over larger distances in the order of 1000km (e.g. in France, UK, and North America; see above), it is hypothesized that the largest uncertainty contribution to the discrepancies between both geodetic approaches comes from geometric levelling (see also [81]). Consequently, geometric levelling is recommended mainly for shorter distances of up to several ten kilometres, where it can give millimetre uncertainties, while over long distances, the GNSS/geoid approach should be a better approach than geometric levelling, and it can also deliver absolute potential values needed for contributions to international timescales.

5.5 *Unified Relativistic Redshift Corrections*

The results from the gravity potential determination from both the geometric levelling and the GNSS/geoid approach are given in Table 1 for the two ITOC sites PTB and LNE-SYRTE as typical examples; further results for the other ITOC sites are foreseen for a separate publication, and corresponding results for further sites in Germany are documented in [81]. Based on the discussion in the preceding section as well as Sect. 4.4, Table 1 gives the relativistic redshift corrections only for the GNSS/geoid approach, which can be considered as the recommended values. The redshift corrections are based on the conventional value $W_0^{(\mathrm{IAU})}$, following the IERS2010 conventions and the IAU resolutions for the definition of TT (see Sect. 3.5), using equation (32). The uncertainty of the given relativistic redshift corrections based on the GNSS/geoid approach amounts to about 2×10^{-18} (see above). All operations from the measurements to the final values of the unified relativistic redshift corrections are summarized in the flowchart given in Fig. 9.

Finally, as the results from the geometric levelling approach and the GNSS/geoid approach are presently inconsistent at the decimetre level across Europe, the more or less direct observation of gravity potential differences through optical clock comparisons (with targeted fractional accuracies of 10^{-18}, corresponding to 1cm in height) is eagerly awaited as a means for resolving the existing discrepancies between different geodetic techniques and remedying the geodetic height determination problem over large distances. A first attempt in this direction was the comparison of two strontium optical clocks between PTB and LNE-SYRTE via a fibre link, showing an uncertainty and agreement with the geodetic results of about 5×10^{-17} [75]. This was mainly limited by the uncertainty and instability of the participating clocks, which is likely to improve in the near future.

Furthermore, for clocks with performance at the 10^{-17} level and below, time-variable effects in the gravity potential, especially solid Earth and ocean tides, have to be considered and can also serve as a method of evaluating the performance of the optical clocks (i.e. a detectability test). Recent analysis of optical clock comparisons already included temporal variations [130, 131]. Then, after further improvements in the optical clock performance, conclusive geodetic results can be anticipated in the future, and clock networks may also contribute to the establishment of the International Height Reference System (IHRS).

6 Contribution of Chronometric Geodesy to the Determination of the Geoid

In geodesy, geoid determination is understood as the determination of the shape and size of the geoid with respect to a well-defined coordinate reference system, which usually means the determination of the height of the geoid (geoid height) above a given reference ellipsoid. The problem is solved within the framework of potential theory and GBVPs, where the task is to find a harmonic function (i.e. the disturbing potential T) everywhere outside the Earth's masses (possibly after mass displacements and reductions), which fulfills certain boundary conditions. In principle, all measurements that can be mathematically linked to the disturbing potential T (e.g. gravity anomalies, vertical deflections, gradiometer observations, and pointwise disturbing potential values itself), can contribute to the solution, but in practice gravity measurements usually play the main role in combination with topographic and global satellite gravity information (also denoted as the gravimetric method, see above). A very flexible approach, with the capability to combine all the aforementioned (inhomogeneous) measurements of different kinds and the option to predict (output) heterogeneous quantities related to T, is the least-squares collocation (LSC) method [173].

Table 1 Ellipsoidal coordinates (latitude, longitude, height; ϕ, λ, $h^{(adj.)}$) referring to ITRF2008 reference frame (epoch 2005.0; GRS80 ellipsoid; zero-tide system), normal heights $H^{N(\mathrm{EVRF2007})}$ based on EVRF2007, geopotential numbers based on the geometric levelling ($C^{(\mathrm{lev})}$) and GNSS/geoid approach ($C^{(\mathrm{GNSS/geoid})}$) relative to the IAU2000 conventional reference potential $W_0^{(\mathrm{IAU})}$ and differences ΔC thereof, as well as the relativistic redshift correction based on the GNSS/geoid approach

Station	ϕ			λ			$h^{(adj.)}$	$H^{N(i)}$	$C^{(\mathrm{lev})}$	$C^{(\mathrm{GNSS/geoid})}$	ΔC	Redshift
	[°]	[']	['']	[°]	[']	['']	[m]	[m]	[10 m^2s^{-2}]	[10 m^2s^{-2}]	[10 m^2s^{-2}]	[10^{-16}]
PTB, Braunschweig, Germany												
PTBB	52	17	46.28177	10	27	35.08676	130.201	87.442	85.617	85.600	−0.017	−95.243
LB03	52	17	49.94834	10	27	37.63590	143.514	100.758	98.684	98.667	−0.017	−109.782
AF02	52	17	30.90851	10	27	28.21874	123.716	80.945	79.242	79.225	−0.017	−88.150
MB02	52	17	47.22270	10	27	50.49262	144.932	102.173	100.072	100.055	−0.017	−111.326
KB01	52	17	45.2	10	27	33.1	119.627	76.867	75.241	75.224	−0.017	−83.698
KB02	52	17	46.3	10	27	35.1	119.708	76.949	75.321	75.304	−0.017	−83.787
LNE-SYRTE, Paris, France												
100	48	50	7.99682	2	20	8.38896	105.652	61.394	60.039	59.930	−0.109	−76.845
A	48	50	10.90277	2	20	10.55555	130.964	86.706	84.868	84.759	−0.109	−76.851
OPMT	48	50	9.31198	2	20	5.77891	122.546	78.288	76.611	76.502	−0.109	−81.712

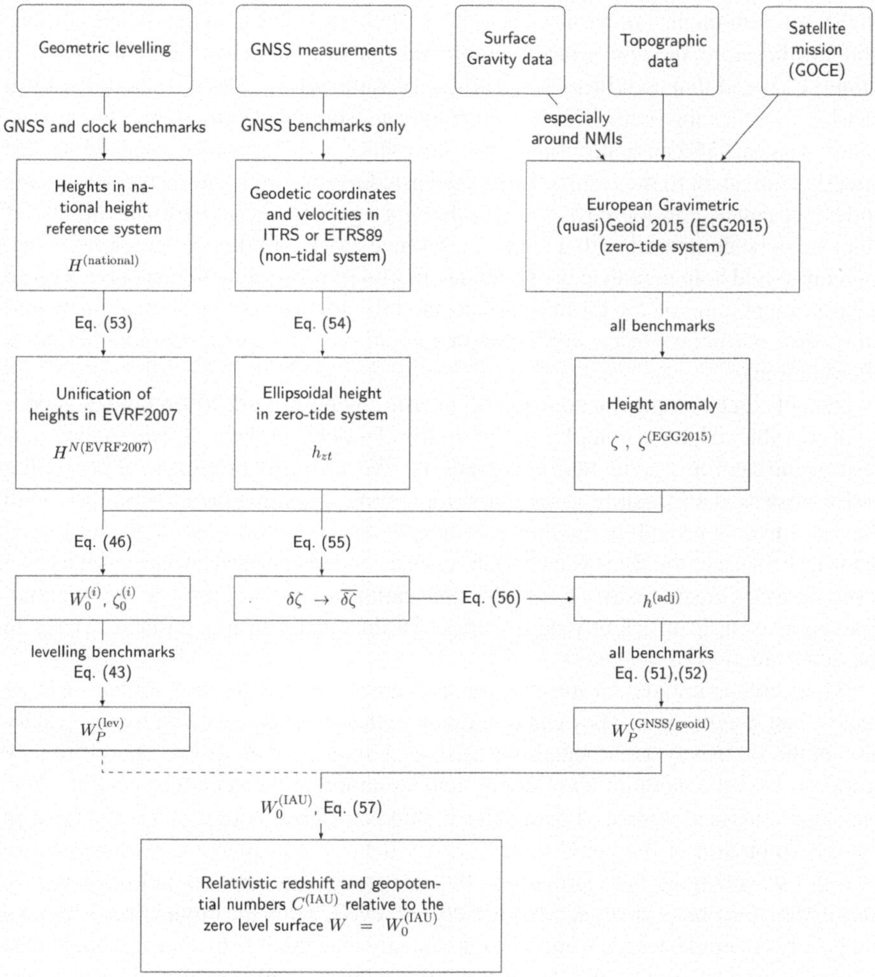

Fig. 9 Flowchart: from measurements to the determination of a unified relativistic redshift clock correction at European scale. It is possible to extend this chart to the worldwide scale wherever a high quality gravimetric model of the geoid exists

Regarding the use of clocks for gravity field modelling and geoid determination, this always implies that also precise positions of the clock points with respect to a well-defined reference system are required. This concerns mainly the ellipsoidal height, which should be available with the same (or lower) uncertainty than the clock-based physical heights or potential values, such that gravity field related quantities $N = h - H$ or $\zeta = h - H_N$ (cf. Eq. (42)) can be obtained, establishing a direct link to the disturbing potential T (e.g. through Eq. (46)); this is exactly the same situation as a combination of GNSS and geometric levelling (so-called GNSS/levelling), as employed since many years (e.g. [174, 175]). Consequently, always clock plus

GNSS measurements are required for gravity field modelling and geoid determination. Furthermore, in view of further improved clocks at (or below) the 10^{-18} level, it should be noted that an ellipsoidal height uncertainty of 5–10 mm is about the limit of what is achievable with GNSS today, requiring static and sufficiently long observation sessions and an appropriate post-processing. Clock measurements alone are directly equivalent to the results from geometric levelling and gravity measurements and hence can be considered as a height (but not a geoid) determination technique; if clocks can be compared with a (space) reference clock with known potential value, then this could help to realize the geoid, i.e. to find its position with respect to a given measurement point on the Earth's surface, but this still does not mean that one would know the coordinates of the corresponding geoid point (i.e. its ellipsoidal height or geoid height).

Distant clock comparisons and GNSS measurements provide a new kind of geodetic observable, which is complementary to the classical geodetic measurements (terrestrial and satellite gravity field observations). We have seen in Sect. 4.3 that satellite and terrestrial data (mainly gravity and topography) complement each other, with the satellite data providing the long wavelength field structures, while the terrestrial data contributes to the short wavelength features. Indeed, terrestrial data (gravity and topography) is most sensitive to small-scale spatial variations of the gravity potential. For this reason, insufficiently dense terrestrial data can lead to significant errors in the determination of the geoid.

By nature, potential data are smoother and more sensitive to mass sources at large scales than gravity data. They can complement the information given by the gravity data in the same way as the satellite data does, but on smaller scales. Therefore they could provide the medium wavelength field structure, in between the spectral information of classical terrestrial data and satellite data. They could reduce the error in the determination of the geoid where gravity data are too sparse to reconstruct the medium wavelengths field structures. Indeed, gravity data are sometimes sparsely distributed: the plains are generally densely surveyed, while the mountainous regions are poorly covered because some areas are mostly inaccessible by conventional gravity surveys. Clock and GNSS data nearby these inaccessible areas could reduce the error in the determination of the geoid.

To illustrate the potential benefits of clocks and GNSS in geodesy, the determination of the geopotential at high spatial resolution, about 10 km, was investigated in [65]. The tested region is the Massif Central in France. It is interesting because it is characterized by smooth, moderate altitude mountains and volcanic plateaus, leading to variations of the gravitational field over a range of spatial scales. In such type of region, the scarcity of gravity data is an important limitation in deriving accurate high resolution geopotential models.

6.1 Methodology

The simulations are based on synthetic data (gravity and potential/clock data) and consist in comparing the quality of the geopotential reconstruction solutions from the gravity data, with or without taking into account clock data. In the following, "clock data" is considered as disturbing potential T values derived from clock and GNSS measurements as outlined above (on the basis of Eq. (46)). The synthetic gravity and potential data are sampled by using a state of the art geopotential model [176, EIGEN-6C4] up to degree and order 2000 (i.e. 10 km resolution), and some spatial distribution of points. The solutions are estimated thanks to an inversion method, requiring a covariance model to interpolate the data, and they are compared to a reference model. In more details, the numerical process is presented below and sketched up in Fig. 10:

1. Step 1: Generation of the reference model of the disturbing potential $\overline{T}$ with program GEOPOT [177], which allows to compute the gravity field related quantities at given locations by using mainly a geopotential model. The long wavelengths of the gravity field covered by the satellites and longer than the extent of the local area are removed, providing centered or close to zero data for the determination of a local covariance function. The terrain effects are removed with the help of the topographic potential model dV_ELL_RET2012 [178];
2. Step 2: Generation of the synthetic data δg and T from a realistic spatial distribution. A white noise is then added to δg and T, with a standard deviation of $0.1\,\mathrm{m}^2\,\mathrm{s}^{-2}$ (i.e. 1 cm on the geoid) for clocks and 1 mGal for gravimetric measurements;
3. Step 3: Estimation of the disturbing potential $\widetilde{T}$ from the synthetic data δg only and then in combination with the synthetic data T on the 10-km grid using the Least-Squares Collocation (LSC) method. In this step, a logarithmic 3D covariance function is employed [179]. This model has the advantage to provide the auto-covariances (ACF) and cross-covariances (CCF) of the potential T and its derivatives in closed-form expressions. Parameters of this model are adjusted to the empirical ACF of δg with the program GPFIT [180]. Note that low frequencies are included in this covariance function, which were not removed as done in step 1.
4. Final step : Evaluation of the potential recovery quality for selected data situations by comparing the statistics of the residuals δ between the estimated values $\widetilde{T}$ and the reference model $\overline{T}$.

Let us underline that in this work, we use synthetic potential data while a network of clocks would give access to potential differences between the clocks. We indeed assume that the clock-based potential differences have been connected to one or a few reference points, without introducing additional biases larger than the assumed clock uncertainties. In order to have more realistic simulations, we should add the noise due to uncertainty of the geometric coordinates of the clock, especially the vertical component. This is a work in progress. However, if this error is below

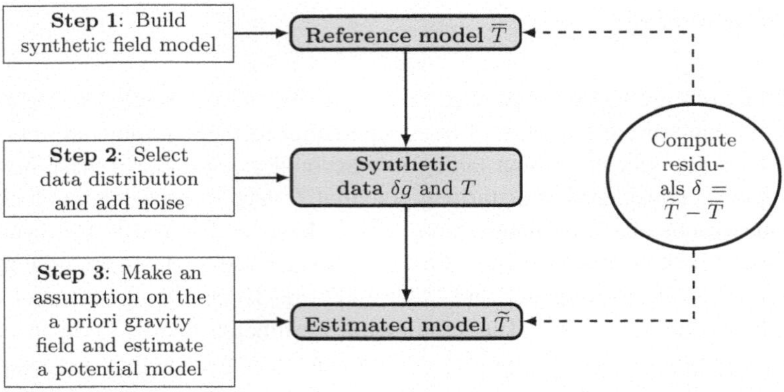

Fig. 10 Scheme of the numerical approach used to evaluate the contribution of atomic clocks

the accuracy of the clock, i.e. $0.1\ \text{m}^2\ \text{s}^{-2}$ (1 cm on the geoid), it will not change the main conclusions of this work.

6.2 Data Set

The locations of the gravimetric data are chosen to reproduce a realistic distribution of measurements. Their spatial distribution can be obtained from the BGI (International Gravimetric Bureau) database, then under-sampled by using a data reduction process, as plotted in blue in Fig. 11. For this test case, the clock measurements (red markers) are put only where existing land gravity data are located and in areas where the gravity data coverage is poor. Moreover, in order to avoid clock points to be too close to each other, a minimal distance is defined between them.

6.3 Contribution of Clocks

In Fig. 12, it is shown that adding the clock-based potential values to the existing gravimetric data set can notably improve the reconstruction of the potential T. In Fig. 12a, the 4374 gravimetric data are used as input and the disturbing potential is estimated with a bias $\mu_T \approx 0.041\ \text{m}^2\ \text{s}^{-2}$ (4.1 mm) and a rms $\sigma_T \approx 0.25\ \text{m}^2\ \text{s}^{-2}$ (2.5 cm). By combining the gravimetric measurements and the 33 potential measurements, see Fig. 12a, the bias is improved by one order of magnitude ($\mu_T \approx -0.002\ \text{m}^2\ \text{s}^{-2}$ or -0.2 mm) and the standard deviation by a factor 3 ($\sigma_T \approx 0.07\ \text{m}^2\ \text{s}^{-2}$ or 7 mm). From the comparison of Fig. 12a, b it is clear that the pure gravimetric solution exhibits a significant trend, which may be related to

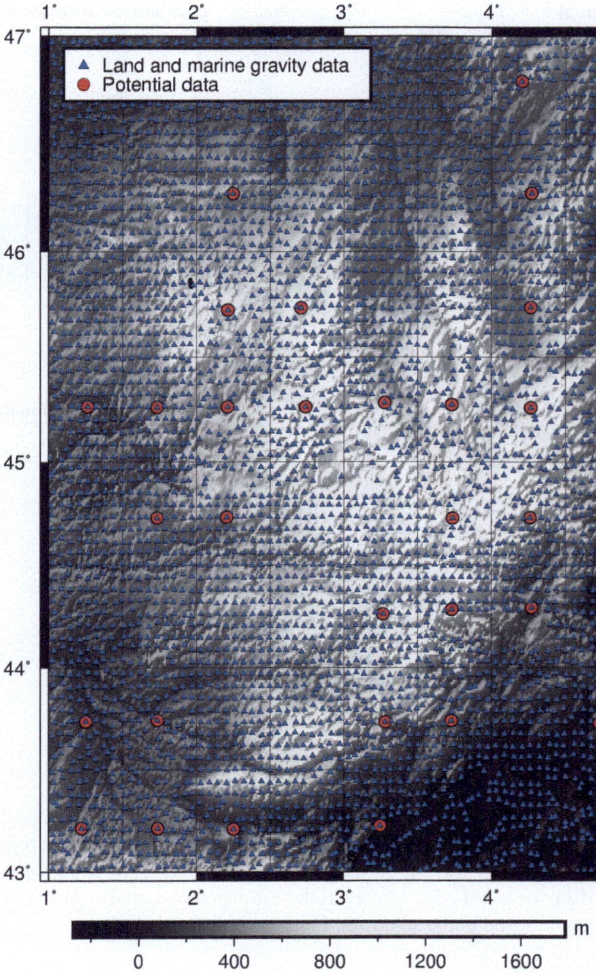

Fig. 11 Spatial distribution of 4374 gravity data and 33 clock data used in the synthetic tests

the data collection area and covariance function used, while the additional potential data effectively remove this trend.

Another important conclusion stemming from our simulations is that for solving the problem of gravity field recovery, it is not required to have a dense clock network. As shown in [65], only a very few percent of clock measurements compared to the number of needed gravity data is sufficient. A more detailed study discussing the role of different parameters, such as the noise level in the data, effects of the resolution of gravity measurements and modeling errors can be found in [65].

As a result of this work, ways to optimize clock location points have begun in order to answer to a practical question: where to put the geopotential measurements to minimize the residuals and improve further the determination of the gravity field?

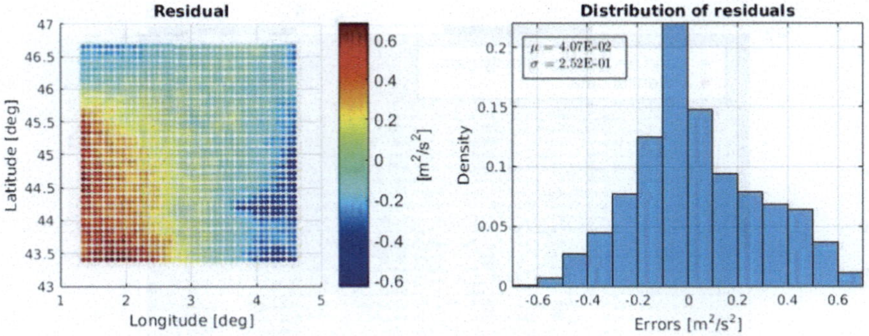

(a) Without clock data.

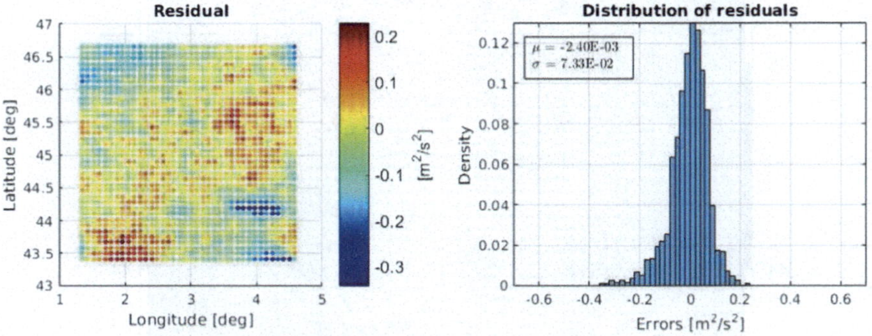

(b) With clock data.

Fig. 12 Accuracy of the disturbing potential T reconstruction on a regular 10-km grid in Massif Central, obtained by comparing the reference model and the reconstructed one. In figure (**a**), the estimation is realized from the 4374 gravimetric data δg only, and in figure (**b**) by adding 33 potential data T to the gravity data. To avoid edge effects in the estimated potential recovery, a grid edge cutoff of 30 km has been removed in the solutions. Figures published in [65].

This is important when the gravimetric measurements can be tarnished by correlated errors. For this, we have implemented the optimization of a spatial distribution of clocks completing a pre-existing gravimetric network, by using the genetic algorithm ϵ-MOEA (Multi Objective Evolutionary Algorithm, see [181]).

7 Conclusions

We presented in this chapter what is chronometric geodesy, introducing notions and methods, both theoretical and experimental. The interest in this rather new topic is raised by the tremendous ameliorations of atomic clocks in the last decade; it is at the crossroads of general relativity and physical geodesy. On the one hand, physical geodesy is essential in order to model the relativistic redshift in distant clock comparisons, as well as to establish global timescales such as the TAI. On the other hand, when the limitations of physical geodesy are reached in terms of method inaccuracies, then the clock comparison observables, which give directly gravity potential differences, could bring something new for physical geodesy.

The question whether these ideas will emerge one day as operational methods depends to a large part on technological challenges. The development of sufficiently accurate and transportable optical clocks is not a barrier, and several projects go in this direction [105, 182, 183]. The frequency transfer method is more challenging, especially on global scales (see Sect. 3.3). Optical fibre transfers fully meet the expectations of current and future optical clocks, but are limited to continental scales and are available only along predefined paths. Phase coherent free space optical links are being developed, but are currently limited by the effect of atmospheric turbulence. This method would be more adapted to global scales, especially if we think about some islands in the middle of the ocean, which are unlikely to be linked with an optical fibre.

Finally, we have to speak about the stability and integration of time of optical clocks. Some recent techniques such as three-dimensional optical lattice clocks [104] allow to greatly improve the integration times necessary to attain some specified accuracy for the clock. This would permit, in a distant comparison, to obtain the variations of the gravity potential with a good time resolution, and could lead to new ideas for the study of geophysical phenomenon.

Acknowledgements The authors would like to thank Jérôme Lodewyck (SYRTE/Paris Observatory) for providing Fig. 1, and Martina Sacher (Bundesamt für Kartographie und Geodäsie, BKG, Leipzig, Germany) for providing information on the EVRF2007 heights and uncertainties, the associated height transformations, and a new UELN adjustment in progress.
This research was supported by the European Metrology Research Programme (EMRP) within the Joint Research Project "International Timescales with Optical Clocks" (SIB55 ITOC), as well as the Deutsche Forschungsgemeinschaft (DFG) within the Collaborative Research Centre 1128 "Relativistic Geodesy and Gravimetry with Quantum Sensors (geo-Q)", project C04. The EMRP is jointly funded by the EMRP participating countries within EURAMET and the European Union. We gratefully acknowledge financial support from Labex FIRST-TF and ERC AdOC (Grant No. 617553).

References

1. R.V. Pound, G.A. Rebka, Resonant absorption of the 14.4-kev γ ray from 0.10-μsec fe^{57}. Phys. Rev. Lett. **3**(12), 554–556 (1959)
2. R.V. Pound, G.A. Rebka, Gravitational red-shift in nuclear resonance. Phys. Rev. Lett. **3**(9), 439–441 (1959)
3. R.V. Pound, G.A. Rebka, Apparent weight of photons. Phys. Rev. Lett. **4**(7), 337–341 (1960)
4. R.V. Pound, J.L. Snider, Effect of gravity on gamma radiation. Phys. Rev. **140**(3B), B788–B803 (1965)
5. Norman F. Ramsey, History of early atomic clocks. Metrologia **42**(3), S1 (2005)
6. Sigfrido Leschiutta, The definition of the 'atomic' second. Metrologia **42**(3), S10 (2005)

7. J. Terrien, News from the international bureau of weights and measures. Metrologia **4**(1), 41 (1968)
8. Leonard S. Cutler, Fifty years of commercial caesium clocks. Metrologia **42**(3), S90 (2005)
9. B. Guinot, E.F. Arias, Atomic time-keeping from 1955 to the present. Metrologia **42**(3), S20 (2005)
10. J.C. Hafele, R.E. Keating, Around-the-world atomic clocks: predicted relativistic time gains. Science **177**(4044), 166–168 (1972)
11. J.C. Hafele, R.E. Keating, Around-the-world atomic clocks: observed relativistic time gains. Science **177**, 168–170 (1972)
12. L. Briatore, S. Leschiutta, Evidence for the Earth gravitational shift by direct atomic-time-scale comparison. Nuovo Cim. B **37**(2), 219–231 (1977)
13. Jean-Marc Lévy-Leblond, One more derivation of the Lorentz transformation. Am. J. Phys. **44**(3), 271–277 (1976)
14. C.W. Misner, K.S. Thorne, J.A. Wheeler, *Gravitation* (W. H. Freeman, San Francisco, 1973)
15. Jeffrey M. Cohen, Harry E. Moses, New test of the synchronization procedure in noninertial systems. Phys. Rev. Lett. **39**(26), 1641–1643 (1977)
16. J.M. Cohen, H.E. Moses, A. Rosenblum, Clock-transport synchronization in noninertial frames and gravitational fields. Phys. Rev. Lett. **51**(17), 1501–1502 (1983)
17. J.M. Cohen, H.E. Moses, A. Rosenblum, Electromagnetic synchronisation of clocks with finite separation in a rotating system. Class. Quantum Grav. **1**(6), L57 (1984)
18. M.F. Podlaha, Note on the Cohen, Moses and Rosenblum letter about the slow-clock transport synchronization in noninertial reference systems. Lett. Nuovo Cim. **40**(7), 223–224 (1984)
19. N. Ashby, D.W. Allan, Coordinate time on and near the Earth. Phys. Rev. Lett. **53**(19), 1858–1858 (1984)
20. N. Ashby, D.W. Allan, Coordinate time on and near the Earth (erratum). Phys. Rev. Lett. **54**(3), 254–254 (1985)
21. M. Born, *Einstein's Theory of Relativity* (Dover Publications, New York, 1962)
22. C. Møller, *Theory of Relativity*, 2nd edn. (Oxford University Press, Oxford, 1976)
23. N. Ashby, D.W. Allan, Practical implications of relativity for a global coordinate time scale. Radio Sci. **14**(4), 649–669 (1979)
24. D.W. Allan, N. Ashby, *Coordinate Time in the Vicinity of the Earth*, vol. 114 (1986), pp. 299–312
25. A.J. Skalafuris, Current theoretical attempts toward synchronization of a global satellite network. Radio Sci. **20**(6), 1529–1536 (1985)
26. N. Ashby, Relativity in the global positioning system. Living Rev. Relativ. **6**, 1 (2003)
27. B. Coll, *A Principal Positioning System for the Earth*, vol. 14 (2003), pp. 34–38, arXiv:gr-qc/0306043
28. Carlo Rovelli, GPS observables in general relativity. Phys. Rev. D **65**(4), 044017 (2002)
29. Marc Lachièze-Rey, The covariance of GPS coordinates and frames. Class. Quantum Grav. **23**(10), 3531 (2006)
30. B. Coll, J.A. Morales, Symmetric frames on Lorentzian spaces. J. Math. Phys. **32**(9), 2450–2455 (1991)
31. M. Blagojević, J. Garecki, F.W. Hehl, Y.N. Obukhov, Real null coframes in general relativity and GPS type coordinates. Phys. Rev. D **65**(4), 044018 (2002)
32. P. Delva, U. Kostić, A. Čadež, Numerical modeling of a Global Navigation Satellite System in a general relativistic framework. Adv. Space Res. **47**(2), 370–379 (2011)
33. D. Bini, A. Geralico, M.L. Ruggiero, A. Tartaglia, Emission versus Fermi coordinates: applications to relativistic positioning systems. Class. Quantum Grav. **25**(20), 205011 (2008)
34. A. Tartaglia, Emission coordinates for the navigation in space. Acta Astronaut. **67**(5), 539–545 (2010)
35. N. Puchades, D. Sáez, Relativistic positioning: four-dimensional numerical approach in Minkowski space-time. Astrophys. Space Sci. **341**(2), 631–643 (2012)
36. D. Bunandar, S.A. Caveny, R.A. Matzner, Measuring emission coordinates in a pulsar-based relativistic positioning system. Phys. Rev. D **84**(10), 104005 (2011)

37. V.A. Brumberg, General discussion, in *Relativity in celestial mechanics and astrometry*, proceedings of the IAU symposium No.114, ed. by J. Kovalesky and V.A. Brumberg (D. Reidel publishing company, 1986)
38. B. Guinot, Is the International Atomic Time TAI a coordinate time or a proper time? Celest. Mech. **38**, 155–161 (1986)
39. B. Guinot, P.K. Seidelmann, Time scales - their history, definition and interpretation. Astron. Astrophys. **194**, 304–308 (1988)
40. T.-Y. Huang, B.-X. Xu, J. Zhu, H. Zhang, The concepts of International Atomic Time (TAI) and Terrestrial Dynamic Time (TDT). Astron. Astrophys. **220**, 329–334 (1989)
41. V.A. Brumberg, S.M. Kopeikin, Relativistic time scales in the solar system. Celest. Mech. Dyn. Astron. **48**, 23–44 (1990)
42. S.A. Klioner, The problem of clock synchronization: a relativistic approach. Celest. Mech. Dyn. Astron. **53**(1), 81–109 (1992)
43. N. Ashby, B. Bertotti, Relativistic perturbations of an earth satellite. Phys. Rev. Lett. **52**(7), 485–488 (1984)
44. T. Fukushima, The Fermi coordinate system in the post-Newtonian framework. Celest. Mech. **44**, 61–75 (1988)
45. N. Ashby, B. Bertotti, Relativistic effects in local inertial frames. Phys. Rev. D **34**(8), 2246–2259 (1986)
46. S.M. Kopejkin, Celestial coordinate reference systems in curved space-time. Celest. Mech. **44**, 87–115 (1988)
47. M.H. Soffel, *Relativity in Astrometry, Celestial Mechanics, and Geodesy* (Springer, Berlin, 1989)
48. V.A. Brumberg, S.M. Kopejkin, Relativistic theory of celestial reference frames, in *Reference Frames, Astrophysics and Space Science Library*, vol. 154, ed. by J. Kovalevsky, I.I. Mueller, B. Kolaczek (Springer, Netherlands, 1989), pp. 115–141
49. V.A. Brumberg, S.M. Kopejkin, Relativistic reference systems and motion of test bodies in the vicinity of the earth. Nuovo Cim. **103**(1), 63–98 (1989)
50. T. Damour, M. Soffel, C. Xu, General-relativistic celestial mechanics. I. Method and definition of reference systems. Phys. Rev. D **43**(10), 3273–3307 (1991)
51. S.M. Kopeikin, M. Efroimsky, G. Kaplan, *Relativistic Celestial Mechanics of the Solar System* (Wiley, New York, 2011)
52. U. Kostić, M. Horvat, A. Gomboc, Relativistic positioning system in perturbed spacetime. Class. Quantum Grav. **32**(21), 215004 (2015)
53. D. Bini, B. Mashhoon, Relativistic gravity gradiometry. Phys. Rev. D **94**(12), 124009 (2016)
54. M. Soffel, S.A. Klioner, G. Petit, P. Wolf, S.M. Kopeikin, P. Bretagnon, V.A. Brumberg, N. Capitaine, T. Damour, T. Fukushima, B. Guinot, T.-Y. Huang, L. Lindegren, C. Ma, K. Nordtvedt, J.C. Ries, P.K. Seidelmann, D. Vokrouhlický, C.M. Will, C. Xu, The IAU 2000 resolutions for astrometry, celestial mechanics, and metrology in the relativistic framework: explanatory supplement. Astron. J. **126**(6), 2687 (2003)
55. S.A. Klioner, N. Capitaine, W.M. Folkner, B. Guinot, T.-Y. Huang, S.M. Kopeikin, E.V. Pitjeva, P.K. Seidelmann, M.H. Soffel, Units of relativistic time scales and associated quantities, in *Relativity in Fundamental Astronomy: Dynamics, Reference Frames, and Data Analysis, Proceedings of the International Astronomical Union*, vol. 5 (2009), pp. 79–84
56. E.F. Arias, G. Panfilo, G. Petit, Timescales at the BIPM. Metrologia **48**(4), S145 (2011)
57. A. Bjerhammar, Discrete approaches to the solution of the boundary value problem in physical geodesy. Bolletino di geodesia e scienze affini **2**, 185–241 (1975)
58. M. Vermeer, Chronometric levelling, Technical report, Finnish Geodetic Institute, Helsinki (1983)
59. A. Bjerhammar, On a relativistic geodesy. Bull. Geodesique **59**(3), 207–220 (1985)
60. H.S. Margolis, R.M. Godun, P. Gill, L.A.M. Johnson, S.L. Shemar, P.B. Whibberley, D. Calonico, F. Levi, L. Lorini, M. Pizzocaro, P. Delva, S. Bize, J. Achkar, H. Denker, L. Timmen, C. Voigt, S. Falke, D. Piester, C. Lisdat, U. Sterr, S. Vogt, S. Weyers, J. Gersl, T. Lindvall, M. Merimaa, International timescales with optical clocks (ITOC), in *European Frequency and*

Time Forum International Frequency Control Symposium (EFTF/IFC), 2013 Joint (2013), pp. 908–911
61. J. Grotti, S. Koller, S. Vogt, S. Häfner, U. Sterr, C. Lisdat, H. Denker, C. Voigt, L. Timmen, A. Rolland, F.N. Baynes, H.S. Margolis, M. Zampaolo, P. Thoumany, M. Pizzocaro, B. Rauf, F. Bregolin, A. Tampellini, P. Barbieri, M. Zucco, G.A. Costanzo, C. Clivati, F. Levi, D. Calonico, Geodesy and metrology with a transportable optical. Nat. Phys. **14**(5), 437 (2018)
62. V.A. Brumberg, E. Groten, On determination of heights by using terrestrial clocks and GPS signals. J. Geod. **76**(1), 49–54 (2002)
63. R. Bondarescu, M. Bondarescu, G. Hetényi, L. Boschi, P. Jetzer, J. Balakrishna, Geophysical applicability of atomic clocks: direct continental geoid mapping. Geophys. J. Int. **191**(1), 78–82 (2012)
64. C.W. Chou, D.B. Hume, T. Rosenband, D.J. Wineland, Optical clocks and relativity. Science **329**(5999), 1630–1633 (2010)
65. G. Lion, I. Panet, P. Wolf, C. Guerlin, S. Bize, P. Delva, Determination of a high spatial resolution geopotential model using atomic clock comparisons. J. Geod. (2017), pp. 1–15
66. A. Bjerhammar, Relativistic geodesy. Technical Report NON118 NGS36, NOAA Technical Report (1986)
67. M. Soffel, H. Herold, H. Ruder, M. Schneider, Relativistic theory of gravimetric measurements and definition of the geoid. Manuscr. Geod. **13**, 143–146 (1988)
68. B. Hofmann-Wellenhof, H. Moritz, *Physical Geodesy* (Springer Science & Business Media, 2006)
69. S.M. Kopejkin, Relativistic manifestations of gravitational fields in gravimetry and geodesy. Manuscr. Geod. **16**, 301–312 (1991)
70. J. Müller, M. Soffel, S.A. Klioner, Geodesy and relativity. J. Geod. **82**(3), 133–145 (2007)
71. S.M. Kopeikin, E.M. Mazurova, A.P. Karpik, Towards an exact relativistic theory of Earth's geoid undulation. Phys. Lett. A **379**(26–27), 1555–1562 (2015)
72. S.M. Kopeikin, W. Han, E. Mazurova, Post-Newtonian reference ellipsoid for relativistic geodesy. Phys. Rev. D **93**(4), 044069 (2016)
73. S.M. Kopeikin, Reference ellipsoid and geoid in chronometric geodesy. Front. Astron. Space Sci. 3 (2016)
74. J. Guéna, S. Weyers, M. Abgrall, C. Grebing, V. Gerginov, P. Rosenbusch, S. Bize, B. Lipphardt, H. Denker, N. Quintin, S.M.F. Raupach, D. Nicolodi, F. Stefani, N. Chiodo, S. Koke, A. Kuhl, F. Wiotte, F. Meynadier, E. Camisard, C. Chardonnet, Y. Le Coq, M. Lours, G. Santarelli, A. Amy-Klein, R. Le Targat, O. Lopez, P.E. Pottie, G. Grosche, First international comparison of fountain primary frequency standards via a long distance optical fiber link. Metrologia **54**(3), 348 (2017)
75. C. Lisdat, G. Grosche, N. Quintin, C. Shi, S.M.F. Raupach, C. Grebing, D. Nicolodi, F. Stefani, A. Al-Masoudi, S. Dörscher, S. Häfner, J.-L. Robyr, N. Chiodo, S. Bilicki, E. Bookjans, A. Koczwara, S. Koke, A. Kuhl, F. Wiotte, F. Meynadier, E. Camisard, M. Abgrall, M. Lours, T. Legero, H. Schnatz, U. Sterr, H. Denker, C. Chardonnet, Y. Le Coq, G. Santarelli, A. Amy-Klein, R. Le Targat, J. Lodewyck, O. Lopez, P.-E. Pottie, A clock network for geodesy and fundamental science. Nat. Commun. **7**, 12443 (2016)
76. M. Schioppo, R.C. Brown, W.F. McGrew, N. Hinkley, R.J. Fasano, K. Beloy, T.H. Yoon, G. Milani, D. Nicolodi, J.A. Sherman, N.B. Phillips, C.W. Oates, A.D. Ludlow, Ultrastable optical clock with two cold-atom ensembles. Nat. Photonics **11**(1), 48–52 (2017)
77. N. Huntemann, C. Sanner, B. Lipphardt, Chr. Tamm, E. Peik, Single-ion atomic clock with 3×10^{-18} systematic uncertainty. Phys. Rev. Lett. **116**(6), 063001 (2016)
78. H.S. Margolis, P. Gill, Least-squares analysis of clock frequency comparison data to deduce optimized frequency and frequency ratio values. Metrologia **52**(5), 628 (2015)
79. H.S. Margolis, P. Gill, Determination of optimized frequency and frequency ratio values from over-determined sets of clock comparison data. J. Phys.: Conf. Ser. **723**(1), 012060 (2016)
80. P. Wolf, G. Petit, Relativistic theory for clock syntonization and the realization of geocentric coordinate times. Astron. Astrophys. **304**, 653 (1995)

81. H. Denker, L. Timmen, C. Voigt, S. Weyers, E. Peik, H.S. Margolis, P. Delva, P. Wolf, G. Petit, Geodetic methods to determine the relativistic redshift at the level of 10^{-18} in the context of international timescales – a review and practical results. J. Geod. **92**(5), 487–516 (2018)
82. W. Torge, *Geodesy*, 2nd edn. (Berlin; New York, W. de Gruyter, 1991)
83. A. Bauch, Time and frequency comparisons using radiofrequency signals from satellites. Comptes Rendus Phys. **16**(5), 471–479 (2015)
84. E. Samain, Clock comparison based on laser ranging technologies. Int. J. Mod. Phys. D **24**(08), 1530021 (2015)
85. G. Petit, A. Kanj, S. Loyer, J. Delporte, F. Mercier, F. Perosanz, 1×10^{-16} frequency transfer by GPS PPP with integer ambiguity resolutio. Metrologia **52**(2), 301 (2015)
86. S. Droste, C. Grebing, J. Leute, S.M.F. Raupach, A. Matveev, T.W. Hänsch, A. Bauch, R. Holzwarth, G. Grosche, Characterization of a 450 km baseline GPS carrier-phase link using an optical fiber link. New J. Phys. **17**(8), 083044 (2015)
87. J. Leute, N. Huntemann, B. Lipphardt, C. Tamm, P.B.R. Nisbet-Jones, S.A. King, R.M. Godun, J.M. Jones, H.S. Margolis, P.B. Whibberley, A. Wallin, M. Merimaa, P. Gill, E. Peik, Frequency comparison of $^{171}Yb^{+}$ ion optical clocks at ptb and npl via GPS PPP. IEEE Trans. Ultrason. Ferroelectr. Freq. Control **63**(7), 981–985 (2016)
88. P. Dubé, J.E. Bernard, M. Gertsvolf, Absolute frequency measurement of the 88 Sr + clock transition using a GPS link to the SI second. Metrologia **54**(3), 290 (2017)
89. C.F.A. Baynham, R.M. Godun, J.M. Jones, S.A. King, P.B.R. Nisbet-Jones, F. Baynes, A. Rolland, P.E.G. Baird, K. Bongs, P. Gill, H.S. Margolis, Absolute frequency measurement of the optical clock transition in with an uncertainty of using a frequency link to international atomic time. J. Mod. Opt. **65**(2), 221–227 (2018)
90. D. Kirchner, Two-way time transfer via communication satellites. Proc. IEEE **79**(7), 983–990 (1991)
91. D. Piester, A. Bauch, L. Breakiron, D. Matsakis, B. Blanzano, O. Koudelka, Time transfer with nanosecond accuracy for the realization of International atomic time. Metrologia **45**(2), 185 (2008)
92. M. Fujieda, T. Gotoh, J. Amagai, Advanced two-way satellite frequency transfer by carrier-phase and carrier-frequency measurements. J. Phys.: Conf. Ser. **723**(1), 012036 (2016)
93. H. Hachisu, M. Fujieda, S. Nagano, T. Gotoh, A. Nogami, T. Ido, St Falke, N. Huntemann, C. Grebing, B. Lipphardt, Ch. Lisdat, D. Piester, Direct comparison of optical lattice clocks with an intercontinental baseline of 9000 km. Opt. Lett. **39**(14), 4072–4075 (2014)
94. F. Meynadier, P. Delva, C. le Poncin-Lafitte, C. Guerlin, P. Wolf, Atomic clock ensemble in space (ACES) data analysis. Class. Quantum Grav. **35**(3), 035018 (2018)
95. L. Cacciapuoti, Ch. Salomon, Space clocks and fundamental tests: The ACES experiment. Eur. Phys. J. Spec. Top. **172**(1), 57–68 (2009)
96. Ph. Laurent, D. Massonnet, L. Cacciapuoti, C. Salomon, The ACES/PHARAO space mission. Comptes rendus de l'Académie des sciences. Physique **16**(5), 540–552 (2015)
97. P. Exertier, E. Samain, C. Courde, M. Aimar, J.M. Torre, G.D. Rovera, M. Abgrall, P. Uhrich, R. Sherwood, G. Herold, U. Schreiber, P. Guillemot, Sub-ns time transfer consistency: a direct comparison between GPS CV and T2L2. Metrologia **53**(6), 1395 (2016)
98. G.D. Rovera, M. Abgrall, C. Courde, P. Exertier, P. Fridelance, Ph. Guillemot, M. Laas-Bourez, N. Martin, E. Samain, R. Sherwood, J.-M. Torre, P. Uhrich, A direct comparison between two independently calibrated time transfer techniques: T2L2 and GPS Common-Views. J. Phys.: Conf. Ser. **723**(1), 012037 (2016)
99. E. Samain, P. Vrancken, P. Guillemot, P. Fridelance, P. Exertier, Time transfer by laser link (T2L2): characterization and calibration of the flight instrument. Metrologia **51**(5), 503 (2014)
100. E. Samain, P. Exertier, C. Courde, P. Fridelance, P. Guillemot, M. Laas-Bourez, J.-M. Torre, Time transfer by laser link: a complete analysis of the uncertainty budget. Metrologia **52**(2), 423 (2015)
101. P. Exertier, E. Samain, N. Martin, C. Courde, M. Laas-Bourez, C. Foussard, Ph. Guillemot, Time transfer by laser link: data analysis and validation to the ps level. Adv. Space Res. **54**(11), 2371–2385 (2014)

102. J. Kodet, M. Vacek, P. Fort, I. Prochazka, J. Blazej, *Photon Counting Receiver for the Laser Time Transfer, Optical Design, and Construction*, vol. 8072 (2011), pp. 80720A (International Society for Optics and Photonics, 2011)
103. I. Prochazka, J. Kodet, J. Blazej, Note: space qualified photon counting detector for laser time transfer with picosecond precision and stability. Rev. Sci. Instrum. **87**(5), 056102 (2016)
104. S.L. Campbell, R.B. Hutson, G.E. Marti, A. Goban, N. Darkwah Oppong, R.L. McNally, L. Sonderhouse, J.M. Robinson, W. Zhang, B.J. Bloom, J. Ye, A fermi-degenerate three-dimensional optical lattice clock. Science **358**(6359), 90–94 (2017)
105. S.B. Koller, J. Grotti, St. Vogt, A. Al-Masoudi, S. Dörscher, S. Häfner, U. Sterr, Ch. Lisdat, Transportable optical lattice clock with 7×10^{-17} uncertainty. Phys. Rev. Lett. **118**(7), 073601 (2017)
106. R. Tyumenev, M. Favier, S. Bilicki, E. Bookjans, R. Le Targat, J. Lodewyck, D. Nicolodi, Y. Le Coq, M. Abgrall, J. Guéna, L. De Sarlo, S. Bize, Comparing a mercury optical lattice clock with microwave and optical frequency standards. New J. Phys. **18**(11), 113002 (2016)
107. J. Lodewyck, S. Bilicki, E. Bookjans, J.-L. Robyr, C. Shi, G. Vallet, R. Le Targat, D. Nicolodi, Y. Le Coq, J. Guéna, M. Abgrall, P. Rosenbusch, S. Bize, Optical to microwave clock frequency ratios with a nearly continuous strontium optical lattice clock. Metrologia **53**(4), 1123 (2016)
108. K. Predehl, G. Grosche, S.M.F. Raupach, S. Droste, O. Terra, J. Alnis, Th Legero, T.W. Hänsch, Th Udem, R. Holzwarth, H. Schnatz, A 920-kilometer optical fiber link for frequency metrology at the 19th decimal place. Science **336**(6080), 441–444 (2012)
109. O. Lopez, A. Haboucha, B. Chanteau, C. Chardonnet, A. Amy-Klein, G. Santarelli, Ultra-stable long distance optical frequency distribution using the Internet fiber network. Opt. Express **20**(21), 23518–23526 (2012)
110. N. Chiodo, K. Djerroud, O. Acef, A. Clairon, P. Wolf, Lasers for coherent optical satellite links with large dynamics. Appl. Opt. **52**(30), 7342–7351 (2013)
111. K. Djerroud, O. Acef, A. Clairon, P. Lemonde, C.N. Man, E. Samain, P. Wolf, Coherent optical link through the turbulent atmosphere. Opt. Lett. **35**(9), 1479–1481 (2010)
112. F.R. Giorgetta, W.C. Swann, L.C. Sinclair, E. Baumann, I. Coddington, N.R. Newbury, Optical two-way time and frequency transfer over free space. Nat. Photonics **7**(6), 434–438 (2013)
113. J.-D. Deschênes, L.C. Sinclair, F.R. Giorgetta, W.C. Swann, E. Baumann, H. Bergeron, M. Cermak, I. Coddington, N.R. Newbury, Synchronization of distant optical clocks at the femtosecond level. Phys. Rev. X **6**(2), 021016 (2016)
114. L.C. Sinclair, F.R. Giorgetta, W.C. Swann, E. Baumann, I. Coddington, N.R. Newbury, Optical phase noise from atmospheric fluctuations and its impact on optical time-frequency transfer. Phys. Rev. A **89**(2), 023805 (2014)
115. L.C. Sinclair, W.C. Swann, H. Bergeron, E. Baumann, M. Cermak, I. Coddington, J.-D. Deschênes, F.R. Giorgetta, J.C. Juarez, I. Khader, K.G. Petrillo, K.T. Souza, M.L. Dennis, N.R. Newbury, Synchronization of clocks through 12 km of strongly turbulent air over a city. Appl. Phys. Lett. **109**(15), 151104 (2016)
116. C. Robert, J.-M. Conan, P. Wolf, Impact of turbulence on high-precision ground-satellite frequency transfer with two-way coherent optical links. Phys. Rev. A **93**(3), 033860 (2016)
117. L. Blanchet, C. Salomon, P. Teyssandier, P. Wolf, Relativistic theory for time and frequency transfer to order. Astron. Astrophys. **370**(1), 10 (2001)
118. C. Le Poncin-Lafitte, B. Linet, P. Teyssandier, World function and time transfer: general post-Minkowskian expansions. Class. Quantum Grav. **21**(18), 4463 (2004)
119. P. Teyssandier, C. Le Poncin-Lafitte, General post-Minkowskian expansion of time transfer functions. Class. Quantum Grav. **25**(14), 145020 (2008)
120. A. Hees, S. Bertone, C. Le Poncin-Lafitte, Frequency shift up to the 2-PM approximation, in *SF2A-2012: Proceedings of the annual meeting of the French Society of Astronomy and Astrophysics* (2012), pp. 145–148, arXiv:1210.2577
121. J.L. Synge, *Relativity: The General Theory*, 1st edn. (North-Holland Publishing Company, Amsterdam, 1960)
122. J. Geršl, P. Delva, P. Wolf, Relativistic corrections for time and frequency transfer in optical fibres. Metrologia **52**(4), 552 (2015)

123. A. Rülke, G. Liebsch, M. Sacher, U. Schäfer, U. Schirmer, J. Ihde, Unification of European height system realizations. J. Geod. Sci. **2**(4), 343–354 (2013)
124. G. Petit, P. Wolf, P. Delva, Atomic time, clocks, and clock comparisons in relativistic spacetime: a review, in *Frontiers in Relativistic Celestial Mechanics - Volume 2: Applications and Experiments*, ed. by S.M. Kopeikin. De Gruyter Studies in Mathematical Physics (De Gruyter, 2014), pp. 249–279
125. L. Sánchez, Towards a vertical datum standardisation under the umbrella of Global Geodetic Observing System. J. Geod. Sci. **2**(4), 325–342 (2012)
126. M. Burša, S. Kenyon, J. Kouba, Z. Šíma, V. Vatrt, V. Vtek, M. Vojtíšková, The geopotential value W_0 for specifying the relativistic atomic time scale and a global vertical reference system. J. Geod. **81**(2), 103–110 (2006)
127. N. Dayoub, S.J. Edwards, P. Moore, The Gauss-Listing geopotential value W_0 and its rate from altimetric mean sea level and GRACE. J. Geod. **86**(9), 681–694 (2012)
128. S. Jevrejeva, J.C. Moore, A. Grinsted, Sea level projections to AD2500 with a new generation of climate change scenarios. Glob. Planet. Chang. **80–81**, 14–20 (2012)
129. C. Voigt, H. Denker, L. Timmen, Time-variable gravity potential components for optical clock comparisons and the definition of international time scales. Metrologia **53**(6), 1365 (2016)
130. T. Takano, M. Takamoto, I. Ushijima, N. Ohmae, T. Akatsuka, A. Yamaguchi, Y. Kuroishi, H. Munekane, B. Miyahara, H. Katori, Real-time geopotentiometry with synchronously linked optical lattice clocks. Nat. Photonics **10**(10), 662–666 (2016)
131. P. Delva, J. Lodewyck, S. Bilicki, E. Bookjans, G. Vallet, R. Le Targat, P.-E. Pottie, C. Guerlin, F. Meynadier, C. Le Poncin-Lafitte, others, Test of special relativity using a fiber network of optical clocks. Phys. Rev. Lett. **118**(22), 221102 (2017)
132. J. Mäkinen, J. Ihde, The permanent tide in height systems, in *Observing our Changing Earth, International Association of Geodesy Symposia* (Springer, Berlin, 2009), pp. 81–87
133. J. Ihde, J. Mäkinen, M. Sacher, Conventions for the definition and realization of a European Vertical Reference System (EVRS) – EVRS Conventions 2007. EVRS Conventions V5.1, Bundesamt für Kartographie and Geodäsie, Finnish Geodetic Institute, 2008-12-17. Technical report, 2008
134. H. Denker, Regional gravity field modeling: theory and practical results, in *Sciences of Geodesy*, vol. II, ed. by G. Xu (Springer, Berlin, 2013), pp. 185–291
135. Bull. Geodesique **58**(3), 309–323 (1984)
136. W.A. Heiskanen, H. Moritz, *Physical Geodesy* (W.H. Freeman and Company, San Francisco, London, 1967)
137. W. Torge, J. Müller, *Geodesy*, 4th edn. (De Gruyter, Berlin, Boston, 2012)
138. F. Condi, C. Wunsch, Gravity field variability, the geoid, and ocean dynamics, in *V Hotine-Marussi Symposium on Mathematical Geodesy, International Association of Geodesy Symposia*, vol. 127 (Springer, Berlin, 2004), pp. 285–292
139. H. Drewes, F. Kuglitsch, J. Adám, S. Rózsa, The Geodesist's handbook 2016. J. Geod. **90**(10), 907–1205 (2016)
140. J. Ihde, R. Barzaghi, U. Marti, L. Sànchez, M. Sideris, H. Drewes, C. Foerste, T. Gruber, G. Liebsch, G. Pail, Report of the ad-hoc group on an international height reference system (IHRS), in *Reports 2011-2015, Number 39 in IAG Travaux* (2015), pp. 549–557
141. M. Burša, J. Kouba, M. Kumar, A. Müller, K. Raděj, S.A. True, V. Vatrt, M. Vojtíšková, Geoidal geopotential and world height system. Stud. Geophys. Geod. **43**(4), 327–337 (1999)
142. L. Sánchez, R. Čunderlk, N. Dayoub, K. Mikula, Z. Minarechová, Z. Šíma, V. Vatrt, M. Vojtíšková, A conventional value for the geoid reference potential W_0. J. Geod. **90**(9), 815–835 (2016)
143. M.S. Molodenskii, V.F. Eremeev, M.I. Yurkina, *Methods for Study of the External Gravitational Field and Figure of the Earth* (Israel Program for Scientific Translations, Jerusalem, 1962)
144. N.K. Pavlis, S.A. Holmes, S.C. Kenyon, J.K. Factor, The development and evaluation of the Earth Gravitational Model 2008 (EGM2008). J. Geophys. Res. **117**(B4), B04406 (2012)

145. T. Mayer-Gürr, N. Zehentner, B. Klinger, A. Kvas, ITSG-Grace2014: a new GRACE gravity field release computed in Graz, 2014. GRACE Science Team Meeting (GSTM), Potsdam, 29 Sept.-01 Oct. 2014
146. T. Mayer-Gürr, G. Team, The combined satellite gravity field model GOCO05s, 2015. EGU General Assembly 2015, Vienna (2015)
147. J.M. Brockmann, N. Zehentner, E. Höck, R. Pail, I. Loth, T. Mayer-Gürr, W.-D. Schuh, EGM_TIM_RL05: an independent geoid with centimeter accuracy purely based on the GOCE mission. Geophys. Res. Lett. **41**(22), 8089–8099 (2014)
148. D.E. Smith, R. Kolenkiewicz, P.J. Dunn, M.H. Torrence, Earth scale below a part per billion from Satellite Laser Ranging, in *Geodesy Beyond 2000, International Association of Geodesy Symposia*, vol. 121 (Springer, Berlin, 2000), pp. 3–12
149. J.C. Ries, The scale of the terrestrial reference frame from VLBI and SLR, 2014. IERS Unified Analysis Workshop, Pasadena, CA, 27–28 June 2014
150. R. Forsberg, Modelling the fine-structure of the geoid: methods, data requirements and some results. Surv. Geophys. **14**(4–5), 403–418 (1993)
151. C. Jekeli, O. Error, Data requirements, and the fractal dimension of the geoid, in *VII Hotine-Marussi Symposium on Mathematical Geodesy, International Association of Geodesy Symposia*, vol. 137 (Springer, Berlin, 2012), pp. 181–187
152. R. Rummel, P. Teunissen, Height datum definition, height datum connection and the role of the geodetic boundary value problem. Bull. Geodesique **62**(4), 477–498 (1988)
153. B. Heck, R. Rummel, Strategies for solving the vertical datum problem using terrestrial and satellite geodetic data, in *Sea Surface Topography and the Geoid, International Association of Geodesy Symposia*, vol. 104 (Springer, New York, 1990), pp. 116–128
154. J. Kelsey, D.A. Gray, Geodetic aspects concerning possible subsidence in southeastern England. Phil. Trans. R. Soc. Lond. A **272**(1221), 141–149 (1972)
155. P. Rebischung, H. Duquenne, F. Duquenne, The new French zero-order levelling network – first global results and possible consequences for UELN, 2008, *EUREF Symposium, Brussels, June 18-21, 2008* (2008)
156. M. Véronneau, R. Duvai, J. Huang, A gravimetric geoid model as a vertical datum in Canada. Geomatica **60**, 165–172 (2006)
157. D. Smith, M. Véronneau, D.R. Roman, J. Huang, Y. Wang, M. Sideris, Towards the unification of the vertical datums over the North American continent, 2010. *IAG Comm. 1 Symposium 2010, Reference Frames for Applications in Geosciences (REFAG2010), Marne-La-Vallée, France, 4–8 Oct. 2010* (2010)
158. D.A. Smith, S.A. Holmes, X. Li, S. Guillaume, Y.M. Wang, B. Bürki, D.R. Roman, T.M. Damiani, Confirming regional 1 cm differential geoid accuracy from airborne gravimetry: the Geoid slope validation survey of 2011. J. Geod. **87**(10–12), 885–907 (2013)
159. Z. Altamimi, X. Collilieux, L. Métivier, ITRF2008: an improved solution of the international terrestrial reference frame. J. Geod. **85**(8), 457–473 (2011)
160. Z. Altamimi, P. Rebischung, L. Métivier, X. Collilieux, ITRF2014: a new release of the International Terrestrial Reference Frame modeling nonlinear station motions. J. Geophys. Res. Solid Earth **121**, 6109–6131 (2016)
161. M. Seitz, D. Angermann, H. Drewes, Accuracy assessment of the ITRS 2008 realization of DGFI: DTRF2008, in *Reference Frames for Applications in Geosciences, International Association of Geodesy Symposia*, vol. 138 (Springer, Berlin, 2013), pp. 87–93
162. H. Margolis, Timekeepers of the future (2014), https://www.nature.com/articles/nphys2834
163. F. Riehle, Towards a redefinition of the second based on optical atomic clocks. Comptes Rendus Phys. **16**(5), 506–515 (2015)
164. P. Gill, Is the time right for a redefinition of the second by optical atomic clocks? J. Phys.: Conf. Ser. **723**(1), 012053 (2016)
165. F. Riehle, Optical clock networks. Nat. Photonics **11**(1), 25–31 (2017)
166. S. Falke, N. Lemke, C. Grebing, B. Lipphardt, S. Weyers, G. Vladislav, N. Huntemann, C. Hagemann, A. Al-Masoudi, S. Häfner, S. Vogt, S. Uwe, C. Lisdat, A strontium lattice clock with 3×10^{-17} inaccuracy and its frequency. New J. Phys. **16**(7), 073023 (2014)

167. H. Denker, A new European Gravimetric (Quasi)Geoid EGG2015, 2015. XXVI General Assembly of the International Union of Geodesy and Geophysics (IUGG), Earth and Environmental Sciences for Future Generations, Prague, Czech Republic, 22 June–02 July 2015 (Poster)
168. M. Sacher, J. Ihde, G. Liebsch, J. Mäkinen, EVRF2007 as realization of the European vertical reference system, 2008, *EUREF Symposium, Brussels, Belgium, 18–21 June 2008* (2008)
169. D.E. Cartwright, J. Crease, A comparison of the geodetic reference levels of England and France by means of the sea surface. Proc. R. Soc. Lond. A **273**(1355), 558–580 (1963)
170. M. Greaves, R. Hipkin, C. Calvert, C. Fane, P. Rebischung, F. Duquenne, A. Harmel, A. Coulomb, H. Duquenne, Connection of British and French levelling networks – Applications to UELN, 2007, *EUREF Symposium, London, June 6–9, 2007* (2007)
171. A. Kenyeres, M. Sacher, J. Ihde, H. Denker, U. Marti, EUVN densification action – final report. Technical report (2010), https://evrs.bkg.bund.de/SharedDocs/Downloads/EVRS/EN/Publications/EUVN-DA_FinalReport.pdf?__blob=publicationFile&v=1
172. H. Moritz, Geodetic reference system 1980. J. Geod. **74**(1), 128–133 (2000)
173. H. Moritz, *Advanced Physical Geodesy* (Wichmann, Karlsruhe, 1980)
174. H. Denker, Hochauflösende regionale Schwerefeldbestimmung mit gravimetrischen und topographischen Daten. PhD thesis, Wiss. Arb. d. Fachr. Verm.wesen d. Univ. Hannover, Nr. 156, Hannover, 1988
175. H. Denker, Evaluation and improvement of the EGG97 quasigeoid model for Europe by GPS and leveling data, in *Second Continental Workshop on the Geoid in Europe, Proceed., Report of Finnish Geodetic Institute, Masala*, vol. 98:4, ed. by M. Vermeer, J. Ádám (1998), pp. 53–61
176. C. Förste, S.L. Bruinsma, O. Abrikosov, J.-M. Lemoine, J.C. Marty, F. Flechtner, G. Balmino, F. Barthelmes, R. Biancale, EIGEN-6C4 The latest combined global gravity field model including GOCE data up to degree and order 2190 of GFZ Potsdam and GRGS Toulouse (2014)
177. D.A. Smith, There is no such thing as "The" EGM96 geoid: subtle points on the use of a global geopotential model. Technical report, IGeS Bulletin No. 8, International Geoid Service, Milan, Italy, 1998
178. S.J. Claessens, C. Hirt, Ellipsoidal topographic potential: New solutions for spectral forward gravity modeling of topography with respect to a reference ellipsoid. J. Geophys. Res. Solid Earth **118**(11), 5991–6002 (2013)
179. R. Forsberg, A new covariance model for inertial gravimetry and gradiometry. J. Geophys. Res. **92**(B2), 1305–1310 (1987)
180. R. Forsberg. An overview manual for the GRAVSOFT – Geodetic Gravity Field Modelling Programs. Technical report, DNSC – Danios National Space Center, 2003
181. D. Coulot, P. Rebischung, A. Pollet, L. Grondin, G. Collot, Global optimization of GNSS station reference networks. GPS Solut. **19**(4), 569–577 (2015)
182. J. Cao, P. Zhang, J. Shang, K. Cui, J. Yuan, S. Chao, S. Wang, H. Shu, X. Huang, A compact, transportable single-ion optical clock with 7.8×10^{17} systematic uncertainty. Appl. Phys. B **123**(4), 112 (2017)
183. M. Yasuda, T. Tanabe, T. Kobayashi, D. Akamatsu, T. Sato, A. Hatakeyama, Laser-controlled cold ytterbium atom source for transportable optical clocks. J. Phys. Soc. Jpn. **86**(12), 125001 (2017)

Measuring the Gravitational Field in General Relativity: From Deviation Equations and the Gravitational Compass to Relativistic Clock Gradiometry

Yuri N. Obukhov and Dirk Puetzfeld

Abstract How does one measure the gravitational field? We give explicit answers to this fundamental question and show how all components of the curvature tensor, which represents the gravitational field in Einstein's theory of General Relativity, can be obtained by means of two different methods. The first method relies on the measuring the accelerations of a suitably prepared set of test bodies relative to the observer. The second method utilizes a set of suitably prepared clocks. The methods discussed here form the basis of relativistic (clock) gradiometry and are of direct operational relevance for applications in geodesy.

1 Introduction

The measurement of the gravitational field lies at the heart of gravitational physics and geodesy. Here we provide the relativistic foundation and present two methods for the operational determination of the gravitational field.

In Einstein's theory of gravitation, i.e. General Relativity (GR), the gravitational field manifests itself in the form of the Riemannian curvature tensor $R_{abc}{}^{d}$ [1, 2]. This 4th-rank tensor can be defined as a measure of the noncommutativity of the parallel transport process of the underlying spacetime manifold M. In terms of the covariant derivative ∇_a, and for a mixed tensor $T^c{}_d$, it is introduced via

$$(\nabla_a \nabla_b - \nabla_b \nabla_a)\, T^c{}_d = R_{abe}{}^{c}\, T^e{}_d - R_{abd}{}^{e}\, T^c{}_e. \tag{1}$$

Y. N. Obukhov
Theoretical Physics Laboratory, Nuclear Safety Institute, Russian Academy of Sciences, B. Tulskaya 52, 115191 Moscow, Russia
e-mail: obukhov@ibrae.ac.ru

D. Puetzfeld (✉)
Center of Applied Space Technology and Microgravity (ZARM), University of Bremen, 28359 Bremen, Germany
e-mail: dirk.puetzfeld@zarm.uni-bremen.de
URL: http://puetzfeld.org

D. Puetzfeld and C. Lämmerzahl (eds.), *Relativistic Geodesy*,
Fundamental Theories of Physics 196, https://doi.org/10.1007/978-3-030-11500-5_3

Note that our curvature conventions differ from those in [2, 3], see also Tables 1 and 2 in Appendix for a directory of symbols used throughout the article. General Relativity is formulated on a four-dimensional (pseudo) Riemannian spacetime with the metric g_{ab} of the signature $(+1, -1, -1, -1)$ which is compatible with the connection in the sense of $\nabla_c g_{ab} = 0$. Therefore the curvature tensor in Einstein's theory has twenty (20) independent components for the most general field configurations produced by nontrivial matter sources, whereas in vacuum the number of independent components reduces to ten (10). As compared to Newton's theory, the gravitational field thus has more degrees of freedom in the relativistic framework.

The smooth tensor field $g_{ab}(x^c)$ introduces the metricity relations on the spacetime manifold M: an interval ("distance") between any two close points $x \in M$ and $x + dx \in M$ is defined by

$$ds^2 = g_{ab}\, dx^a dx^b. \tag{2}$$

The metric and connection (g, ∇) underlie the formalism of Synge's world function [2] which plays a crucial role in the methods of measurement of the gravitational field in GR and in its natural extensions.

A central question in General Relativity, and consequently in relativistic geodesy, is how these components of the gravitational field can be determined in an operational way.

1.1 Method 1: Measuring the Gravitational Field by Means of Test Bodies

Method 1 utilizes a suitably prepared set of test bodies in order to determine all components of the curvature of spacetime and thereby the gravitational field. This method relies on the measurement of the acceleration between the test-bodies and the observer. Historically, Felix Pirani [1] was the first to point out that one could determine the full Riemann tensor with the help of a (sufficiently large) number of test bodies in the vicinity of observer's world line. Pirani's suggestion to measure the curvature was based on the equation which describes the dynamics of a vector connecting two adjacent geodesics in spacetime. In the literature this equation is known as a Jacobi equation, or a geodesic deviation equation; its early derivations in a Riemannian context can be found in [4–6].

A modern derivation and extension of the deviation equation, based on [7], is presented in the next section. In particular, it is explicitly shown, how a suitably prepared set of test bodies can be used to determine all components of the curvature of spacetime (and thereby to measure the gravitational field) with the help of an exact solution for the components of the Riemann tensor in terms of the mutual accelerations between the constituents of a cloud of test bodies and the observer. This can be viewed as an explicit realization of Szekeres' "gravitational compass" [8], or Synge's "curvature detector" [2]. In geodetic terms, such a solution represents a realization of a relativistic gradiometer or tensor gradiometer, which has a direct

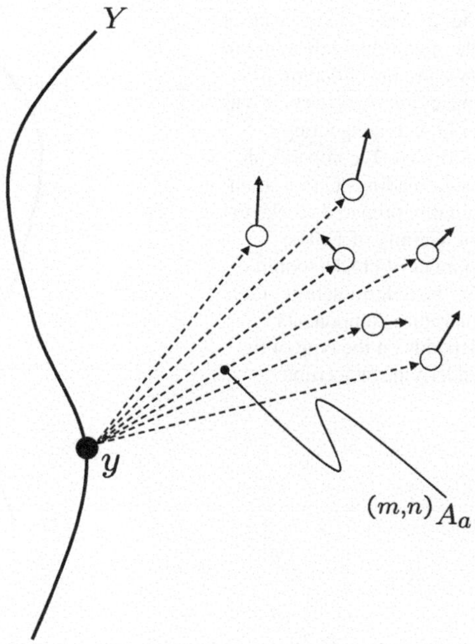

Fig. 1 Method 1: Sketch of the operational procedure to measure the curvature of spacetime. An observer moving along a world line Y monitors the accelerations $^{(m,n)}A_a$ to a set of suitably prepared test bodies (hollow circles). The number of test bodies required for the determination of all curvature components depends on the type of the underlying spacetime

operational relevance and forms the basis of relativistic gradiometry. The operational procedure, see Fig. 1, is to monitor the accelerations of a set of test bodies w.r.t. to an observer moving along a reference world line Y. A mechanical analogue would be to measure the forces between the test bodies and the reference body via a spring connecting them.

Method 1 relies on the standard geodesic deviation equation. A modern covariant derivation of this equation, as well as its generalization to higher orders will be provided to make the presentation self-contained. Furthermore, we provide an explicit exact solution for the curvature components in terms of the mutual accelerations between the constituents of a cloud of test bodies and the observer. Our presentation is mainly based on [7].

1.2 Method 2: Measuring the Gravitational Field by Means of Clocks

Method 2 utilizes a suitably prepared set of clocks to determine all components of the gravitational field in General Relativity, see Fig. 2. In contrast to the gravitational compass, the method relies on the frequency comparison between the clocks from the ensemble and the one carried by the observer. We call such an experimental setup a clock compass, in analogy to the usual gravitational compass, or in geodetic language a clock gradiometer.

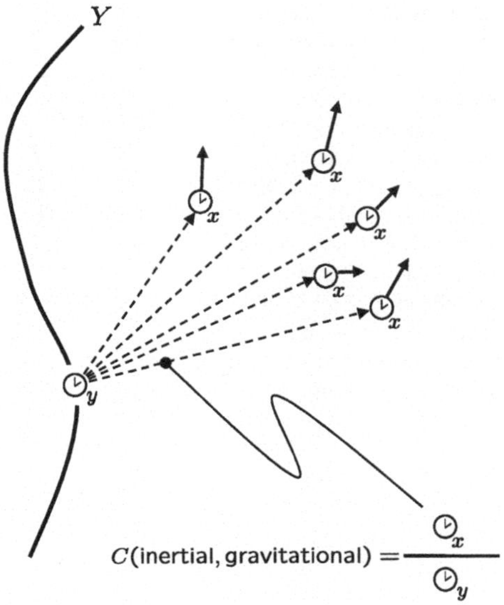

Fig. 2 Method 2: Sketch of the operational procedure to measure the curvature of spacetime. An observer with a clock moving along a world line Y compares his clock readings C to a set of suitably prepared clocks in the vicinity of Y. The number of clocks required for the determination of all curvature components depends on the type of the underlying spacetime

We base our review on [9] and pay particular attention to the construction of the underlying reference frame. As in the case of the gravitational compass, our results are of direct operational relevance for the setup of networks of clocks, for example in the context of relativistic geodesy.

2 Theoretical Foundations

In this section we present the theoretical foundations for method 1 and method 2.

For method 1 we start by comparing two general curves in an arbitrary spacetime manifold and work out an equation for the generalized deviation vector between those two curves in Sect. 2.1.

For method 2 we first show in Sect. 2.2 how the metric along an arbitrary world line can be expressed in terms of geometrical and kinematical parameters. This result is then used in Sect. 2.3 to derive the frequency ratio of two clocks moving on two general curves, again within an arbitrary spacetime manifold.

2.1 Method 1: Comparison of Two General Curves

Consider two curves $Y(t)$ and $X(\tilde{t})$ with general parameters t and $\tilde{t}$, i.e. are not necessarily the proper time on the given curves. Now we connect two points $x \in X$

and $y \in Y$ on the two curves by the geodesic joining the two points (we assume that this geodesic is unique).

For the geodesic connecting the two general curves $Y(t)$ and $X(\tilde{t})$ we have the world function introduced as an integral

$$\sigma(x, y) := \frac{\epsilon}{2}\left(\int_x^y d\tau\right)^2 \tag{3}$$

over the geodesic curve γ connecting the spacetime points x and y. Here $d\tau = \sqrt{g_{ab}u^a u^b}\, d\lambda$ is the differential of the proper time along the geodesic which is defined as the curve $\gamma = \{x^a(\lambda)\}$, such that the tangent vector $u^a = dx^a/d\lambda$ is parallely transported $Du^a/d\lambda = 0$, and $\epsilon = \pm 1$ for timelike/spacelike curves. Along with the world function $\sigma(x, y)$, another important bitensor is the parallel propagator $g^y{}_x(x, y)$ that allows for the parallel transportation of objects along the unique geodesic that links the points x and y. For example, given a vector V^x at x, the corresponding vector at y is obtained by means of the parallel transport along the geodesic curve as $V^y = g^y{}_x(x, y)V^x$. For more details see, e.g., [2, 10] or Sect. 5 in [3]. A compact summary of useful formulas in the context of the bitensor formalism can also be found in the appendices A and B of [11]. Note that we will use the condensed notation when the spacetime point to which an index of a bitensor belongs can be directly read from an index itself. Indices attached to the world-function always denote covariant derivatives, at the given point, i.e. $\sigma_y := \nabla_y \sigma$, hence we do not make explicit use of the semicolon in case of the world-function.

Conceptually, the closest object to the connecting vector between the two points is the covariant derivative of the world function: σ^y. Note though that σ^y is just tangent at that point (its length being the the geodesic length between y and x), only in flat spacetime it coincides with the connecting vector. Keeping in mind such an interpretation, let us now work out a propagation equation for this "generalized" connecting vector along the reference curve, cf. Fig. 3. Following our conventions the reference curve will be $Y(t)$ and we define the generalized connecting vector to be:

$$\eta^y := -\sigma^y. \tag{4}$$

Taking its covariant total derivative, we have

$$\begin{aligned} \frac{D}{dt}\eta^{y_1} &= -\frac{D}{dt}\sigma^{y_1}\left(Y(t), X(\tilde{t})\right) \\ &= -\sigma^{y_1}{}_{y_2}\frac{\partial Y^{y_2}}{\partial t} - \sigma^{y_1}{}_{x_2}\frac{\partial X^{x_2}}{\partial \tilde{t}}\frac{d\tilde{t}}{dt} \\ &= -\sigma^{y_1}{}_{y_2}u^{y_2} - \sigma^{y_1}{}_{x_2}\tilde{u}^{x_2}\frac{d\tilde{t}}{dt}, \end{aligned} \tag{5}$$

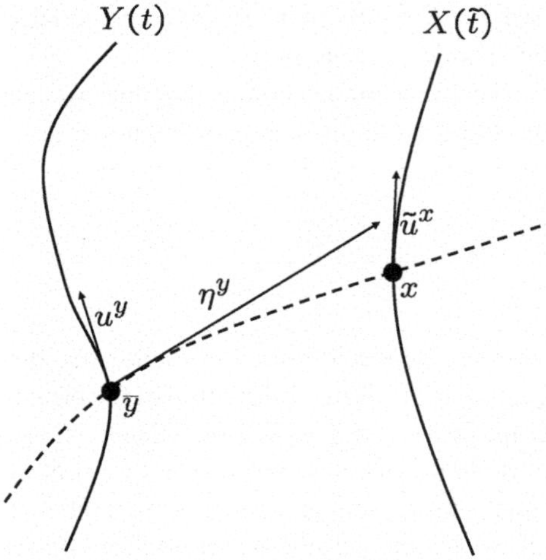

Fig. 3 Sketch of the two arbitrarily parametrized world lines $Y(t)$ and $X(\tilde{t})$, and the geodesic connecting two points on these world line. The (generalized) deviation vector along the reference world line Y is denoted by η^y

where in the last line we defined the velocities along the two curves Y and X. As usual, $\sigma^y{}_{x_1 \dots y_2 \dots} := \nabla_{x_1} \dots \nabla_{y_2} \dots (\sigma^y)$ denote the higher order covariant derivatives of the world function. We continue by taking the second derivative of (5), which yields

$$\frac{D^2}{dt^2}\eta^{y_1} = -\sigma^{y_1}{}_{y_2 y_3} u^{y_2} u^{y_3} - 2\sigma^{y_1}{}_{y_2 x_3} u^{y_2} \tilde{u}^{x_3} \frac{d\tilde{t}}{dt}$$
$$-\sigma^{y_1}{}_{y_2} a^{y_2} - \sigma^{y_1}{}_{x_2 x_3} \tilde{u}^{x_2} \tilde{u}^{x_3} \left(\frac{d\tilde{t}}{dt}\right)^2$$
$$-\sigma^{y_1}{}_{x_2} \tilde{a}^{x_2} \left(\frac{d\tilde{t}}{dt}\right)^2 - \sigma^{y_1}{}_{x_2} \tilde{u}^{x_2} \frac{d^2\tilde{t}}{dt^2}, \qquad (6)$$

here we introduced the accelerations $a^y := Du^y/dt$, and $\tilde{a}^x := D\tilde{u}^x/d\tilde{t}$. Equation (6) is already the generalized deviation equation, but the goal is to have all the quantities therein defined along the reference wordline Y.

We now derive some auxiliary formulas, by introducing the inverse of the second derivative of the world function via the following equations:

$$\overset{-1}{\sigma}{}^{y_1}{}_x \sigma^x{}_{y_2} = \delta^{y_1}{}_{y_2}, \qquad \overset{-1}{\sigma}{}^{x_1}{}_y \sigma^y{}_{x_2} = \delta^{x_1}{}_{x_2}. \qquad (7)$$

Multiplication of (5) by $\overset{-1}{\sigma}{}^{x_3}{}_{y_1}$ then yields

$$\tilde{u}^{x_3} \frac{d\tilde{t}}{dt} = -\overset{-1}{\sigma}{}^{x_3}{}_{y_1} \sigma^{y_1}{}_{y_2} u^{y_2} + \overset{-1}{\sigma}{}^{x_3}{}_{y_1} \frac{D\sigma^{y_1}}{dt}$$
$$= K^{x_3}{}_{y_2} u^{y_2} - H^{x_3}{}_{y_1} \frac{D\sigma^{y_1}}{dt}. \qquad (8)$$

In the last line we defined two auxiliary quantities $K^x{}_y$ and $H^x{}_y$ – the notation follows the terminology of Dixon. Equation (8) allows us to formally express the the velocity along the curve X in terms of the quantities which are defined at Y and then "propagated" by $K^x{}_y$ and $H^x{}_y$. Using (8) in (6) we arrive at:

$$\begin{aligned}\frac{D^2}{dt^2}\eta^{y_1} &= -\sigma^{y_1}{}_{y_2 y_3} u^{y_2} u^{y_3} - \sigma^{y_1}{}_{y_2} a^{y_2} \\ &\quad -2\sigma^{y_1}{}_{y_2 x_3} u^{y_2} \left(K^{x_3}{}_{y_4} u^{y_4} - H^{x_3}{}_{y_4} \frac{D\sigma^{y_4}}{dt} \right) \\ &\quad -\sigma^{y_1}{}_{x_2 x_3} \left(K^{x_2}{}_{y_4} u^{y_4} - H^{x_2}{}_{y_4} \frac{D\sigma^{y_4}}{dt} \right) \\ &\quad \times \left(K^{x_3}{}_{y_5} u^{y_5} - H^{x_3}{}_{y_5} \frac{D\sigma^{y_5}}{dt} \right) \\ &\quad -\sigma^{y_1}{}_{x_2} \frac{D}{dt} \left(K^{x_2}{}_{y_3} u^{y_3} - H^{x_2}{}_{y_3} \frac{D\sigma^{y_3}}{dt} \right). \end{aligned} \tag{9}$$

We may derive an alternative version of this equation – by using (8) multiplied by $dt/d\tilde{t}$ – which yields

$$\tilde{u}^{x_3} = K^{x_3}{}_{y_2} u^{y_2} \frac{dt}{d\tilde{t}} - H^{x_3}{}_{y_1} \frac{D\sigma^{y_1}}{dt} \frac{dt}{d\tilde{t}}, \tag{10}$$

and inserted into (6):

$$\frac{D^2}{dt^2}\eta^{y_1} = -\sigma^{y_1}{}_{y_2 y_3} u^{y_2} u^{y_3} - \sigma^{y_1}{}_{y_2} a^{y_2} - \sigma^{y_1}{}_{x_2} \tilde{a}^{x_2} \left(\frac{d\tilde{t}}{dt} \right)^2 \tag{11}$$

$$\begin{aligned} &-2\sigma^{y_1}{}_{y_2 x_3} u^{y_2} \left(K^{x_3}{}_{y_4} u^{y_4} - H^{x_3}{}_{y_4} \frac{D\sigma^{y_4}}{dt} \right) \\ &-\sigma^{y_1}{}_{x_2 x_3} \left(K^{x_2}{}_{y_4} u^{y_4} - H^{x_2}{}_{y_4} \frac{D\sigma^{y_4}}{dt} \right) \\ &\times \left(K^{x_3}{}_{y_5} u^{y_5} - H^{x_3}{}_{y_5} \frac{D\sigma^{y_5}}{dt} \right) \\ &-\sigma^{y_1}{}_{x_2} \frac{dt}{d\tilde{t}} \frac{d^2\tilde{t}}{dt^2} \left(K^{x_2}{}_{y_3} u^{y_3} - H^{x_2}{}_{y_3} \frac{D\sigma^{y_3}}{dt} \right). \end{aligned} \tag{12}$$

Note that we may determine the factor $d\tilde{t}/dt$ by requiring that the velocity along the curve X is normalized, i.e. $\tilde{u}^x \tilde{u}_x = 1$, in which case (8) yields

$$\frac{d\tilde{t}}{dt} = \tilde{u}_{x_1} K^{x_1}{}_{y_2} u^{y_2} - \tilde{u}_{x_1} H^{x_1}{}_{y_2} \frac{D\sigma^{y_2}}{dt}. \tag{13}$$

Expansion of quantities on the reference world line The generalized (exact) deviation Eqs. (9) and (12) contain quantities which are not defined along the reference curve, in particular the covariant derivatives of the world function. We now make use of the covariant expansions of these quantities, which read (for details, see [12]):

$$\sigma^{y_0}{}_{x_1} = g^{y'}{}_{x_1}\left(-\delta^{y_0}{}_{y'} + \sum_{k=2}^{\infty}\frac{1}{k!}\alpha^{y_0}{}_{y' y_2 \ldots y_{k+1}}\sigma^{y_2}\cdots\sigma^{y_{k+1}}\right), \tag{14}$$

$$\sigma^{y_0}{}_{y_1} = \delta^{y_0}{}_{y_1} - \sum_{k=2}^{\infty}\frac{1}{k!}\beta^{y_0}{}_{y_1 y_2 \ldots y_{k+1}}\sigma^{y_2}\cdots\sigma^{y_{k+1}}, \tag{15}$$

$$g^{y_0}{}_{x_1;x_2} = g^{y'}{}_{x_1}g^{y''}{}_{x_2}\left(\frac{1}{2}R^{y_0}{}_{y'y''y_3}\sigma^{y_3} + \sum_{k=2}^{\infty}\frac{1}{k!}\gamma^{y_0}{}_{y'y''y_3\ldots y_{k+2}}\sigma^{y_3}\cdots\sigma^{y_{k+2}}\right), \tag{16}$$

$$g^{y_0}{}_{x_1;y_2} = g^{y'}{}_{x_1}\left(\frac{1}{2}R^{y_0}{}_{y'y_2y_3}\sigma^{y_3} + \sum_{k=2}^{\infty}\frac{1}{k!}\gamma^{y_0}{}_{y'y_2y_3\ldots y_{k+2}}\sigma^{y_3}\cdots\sigma^{y_{k+2}}\right). \tag{17}$$

The coefficients α, β, γ in these expansions are polynomials constructed from the Riemann curvature tensor and its covariant derivatives. The first coefficients read (as one can also check using computer algebra [13]):

$$\alpha^{y_0}{}_{y_1y_2y_3} = -\frac{1}{3}R^{y_0}{}_{(y_2y_3)y_1}, \tag{18}$$

$$\beta^{y_0}{}_{y_1y_2y_3} = \frac{2}{3}R^{y_0}{}_{(y_2y_3)y_1}, \tag{19}$$

$$\alpha^{y_0}{}_{y_1y_2y_3y_4} = \frac{1}{2}\nabla_{(y_2}R^{y_0}{}_{y_3y_4)y_1}, \tag{20}$$

$$\beta^{y_0}{}_{y_1y_2y_3y_4} = -\frac{1}{2}\nabla_{(y_2}R^{y_0}{}_{y_3y_4)y_1}, \tag{21}$$

$$\alpha^{y_0}{}_{y_1y_2y_3y_4y_5} = -\frac{7}{15}R^{y_0}{}_{(y_2y_3|y'|}R^{y'}{}_{y_4y_5)y_1} - \frac{3}{5}\nabla_{(y_5}\nabla_{y_4}R^{y_0}{}_{y_2y_3)y_1}, \tag{22}$$

$$\beta^{y_0}{}_{y_1y_2y_3y_4y_5} = \frac{8}{15}R^{y_0}{}_{(y_2y_3|y'|}R^{y'}{}_{y_4y_5)y_1} + \frac{2}{5}\nabla_{(y_5}\nabla_{y_4}R^{y_0}{}_{y_2y_3)y_1}, \tag{23}$$

$$\gamma^{y_0}{}_{y_1y_2y_3y_4} = \frac{1}{3}\nabla_{(y_3}R^{y_0}{}_{|y_1|y_4)y_2}, \tag{24}$$

$$\gamma^{y_0}{}_{y_1y_2y_3y_4y_5} = \frac{1}{4}R^{y_0}{}_{y_1y'(y_3}R^{y'}{}_{y_4y_5)y_2} + \frac{1}{4}\nabla_{(y_5}\nabla_{y_4}R^{y_0}{}_{|y_1y_2|y_3)}. \tag{25}$$

These results allow us to derive the third derivatives of the world function appearing in (9) and (12), i.e. we have up to the second order in the deviation vector:

$$\sigma^{y_0}{}_{y_1 y_2} = -\frac{2}{3}R^{y_0}{}_{(y_2 y_3) y_1}\sigma^{y_3} - \frac{1}{2}\left(\frac{1}{2}\nabla_{y_2}R^{y_0}{}_{(y_3 y_4) y_1} - \frac{1}{3}\nabla_{y_3}R^{y_0}{}_{(y_2 y_4) y_1}\right)\sigma^{y_3}\sigma^{y_4} - \frac{1}{6}\lambda^{y_0}{}_{y_1 y_2 y_3 y_4 y_5}\sigma^{y_3}\sigma^{y_4}\sigma^{y_5} + \mathcal{O}(\sigma^4), \tag{26}$$

$$\sigma^{y_0}{}_{y_1 x_2} = g^{y'}{}_{x_2}\left(\frac{2}{3}R^{y_0}{}_{(y' y_3) y_1}\sigma^{y_3} - \frac{1}{4}\nabla_{(y'}R^{y_0}{}_{y_3 y_4) y_1}\sigma^{y_3}\sigma^{y_4} + \frac{1}{6}\mu^{y_0}{}_{y_1 y' y_3 y_4 y_5}\sigma^{y_3}\sigma^{y_4}\sigma^{y_5}\right) + \mathcal{O}(\sigma^4), \tag{27}$$

$$\sigma^{y_0}{}_{x_1 x_2} = -g^{y'}{}_{x_1}g^{y''}{}_{x_2}\left[\left(\frac{1}{2}R^{y_0}{}_{y' y'' y_3} - \frac{1}{3}R^{y_0}{}_{(y'' y_3) y'}\right)\sigma^{y_3} + \left(\frac{1}{6}\nabla_{(y_3}R^{y_0}{}_{|y'|y_4) y''} + \frac{1}{4}\nabla_{(y''}R^{y_0}{}_{y_3 y_4) y'}\right)\sigma^{y_3}\sigma^{y_4} + \frac{1}{6}\nu^{y_0}{}_{y' y'' y_3 y_4 y_5}\sigma^{y_3}\sigma^{y_4}\sigma^{y_5}\right] + \mathcal{O}(\sigma^4). \tag{28}$$

Here we introduced a compact notation for the combinations of the second covariant derivatives of the curvature and the quadratic polynomial of the curvature tensor (in symbolic form, "$\nabla\nabla R + R\cdot R$"):

$$\lambda^{y_0}{}_{y_1 y_2 y_3 y_4 y_5} = \beta^{y_0}{}_{y_1 y_3 y_4 y_5 ; y_2} + \beta^{y_0}{}_{y_1 y_2 y_3 y_4 y_5} - 3\beta^{y_0}{}_{y_1 y' (y_3}\beta^{y'}{}_{|y_2|y_4 y_5)}, \tag{29}$$

$$\mu^{y_0}{}_{y_1 y_2 y_3 y_4 y_5} = \beta^{y_0}{}_{y_1 y_2 y_3 y_4 y_5} - 3\beta^{y_0}{}_{y_1 y' (y_3}\alpha^{y'}{}_{|y_2|y_4 y_5)}, \tag{30}$$

$$\nu^{y_0}{}_{y_1 y_2 y_3 y_4 y_5} = \gamma^{y_0}{}_{y_1 y_2 y_3 y_4 y_5} + \alpha^{y_0}{}_{y_1 y_2 y_3 y_4 y_5} - 3\alpha^{y_0}{}_{y_1 y' (y_3}\alpha^{y'}{}_{|y_2|y_4 y_5)} - \frac{1}{4}R^{y'}{}_{y_1 y_2 (y_3}\alpha^{y_0}{}_{|y'|y_4 y_5)}. \tag{31}$$

Substituting the coefficients of the expansions (14)–(16) we obtain the explicit (complicated) expressions which we do not display here.

For the symmetrized versions of (26) and (28) we obtain

$$\sigma^{y_0}{}_{(y_1 y_2)} = \frac{1}{3}R^{y_0}{}_{(y_1 y_2) y_3}\sigma^{y_3} - \frac{1}{4}\left(\nabla_{(y_1}R^{y_0}{}_{|y_3 y_4|y_2)} + \frac{1}{3}\nabla_{y_3}R^{y_0}{}_{(y_1 y_2) y_4}\right)\sigma^{y_3}\sigma^{y_4} - \frac{1}{6}\lambda^{y_0}{}_{(y_1 y_2) y_3 y_4 y_5}\sigma^{y_3}\sigma^{y_4}\sigma^{y_5} + \mathcal{O}(\sigma^4), \tag{32}$$

$$\sigma^{y_0}{}_{(x_1 x_2)} = g^{y'}{}_{(x_1}g^{y''}{}_{x_2)}\Big[-\frac{2}{3}R^{y_0}{}_{(y' y'') y_3}\sigma^{y_3}$$

$$+\frac{1}{4}\left(\nabla_{y_3} R^{y_0}{}_{(y'y'')y_4} - \frac{1}{3}\nabla_{(y'} R^{y_0}{}_{|y_3 y_4|y'')}\right)\sigma^{y_3}\sigma^{y_4}$$
$$\left. - \frac{1}{6}\nu^{y_0}{}_{(y'y'')y_3 y_4 y_5}\sigma^{y_3}\sigma^{y_4}\sigma^{y_5}\right] + \mathcal{O}(\sigma^4), \tag{33}$$

Furthermore we need the expansions of $K^x{}_y$ and $\overset{-1}{\sigma}{}^x{}_y = -H^x{}_y$:

$$\overset{-1}{\sigma}{}^{x_1}{}_{y_2} = -g^{x_1}{}_{y'}\left(\delta^{y'}{}_{y_2} - \frac{1}{6}R^{y'}{}_{(y_3 y_4)y_2}\sigma^{y_3}\sigma^{y_4}\right.$$
$$\left. + \frac{1}{12}\nabla_{(y_3} R^{y'}{}_{y_4 y_5)y_2}\sigma^{y_3}\sigma^{y_4}\sigma^{y_5}\right) + \mathcal{O}(\sigma^4), \tag{34}$$

$$K^{x_1}{}_{y_2} = g^{x_1}{}_{y'}\left(\delta^{y'}{}_{y_2} - \frac{1}{2}R^{y'}{}_{(y_3 y_4)y_2}\sigma^{y_3}\sigma^{y_4}\right.$$
$$\left. + \frac{1}{6}\nabla_{(y_3} R^{y'}{}_{y_4 y_5)y_2}\sigma^{y_3}\sigma^{y_4}\sigma^{y_5}\right) + \mathcal{O}(\sigma^4). \tag{35}$$

From this one can derive the recurring term in (12) up to the needed order, i.e.

$$\left(K^{x_1}{}_{y_2}u^{y_2} - H^{x_1}{}_{y_2}\frac{D\sigma^{y_2}}{dt}\right) = g^{x_1}{}_{y'}\left[u^{y'} - \frac{D\sigma^{y'}}{dt}\right.$$
$$- \frac{1}{2}R^{y'}{}_{(y_3 y_4)y_2}\sigma^{y_3}\sigma^{y_4}\left(u^{y_2} - \frac{1}{3}\frac{D\sigma^{y_2}}{dt}\right)$$
$$\left. + \frac{1}{6}\nabla_{(y_3} R^{y'}{}_{y_4 y_5)y_2}u^{y_2}\sigma^{y_3}\sigma^{y_4}\sigma^{y_5}\right] + \mathcal{O}(\sigma^4). \tag{36}$$

With these expansions at hand we are finally able to develop the deviation Eq. (12) up to the third order.

Denote $\tilde{a}^{y_1} = g^{y_1}{}_{x_2}\tilde{a}^{x_2}$ in accordance with the definition of the parallel propagator, and introduce

$$\phi^{y_1}{}_{y_2 y_3 y_4 y_5 y_6} = \lambda^{y_1}{}_{y_2 y_3 y_4 y_5 y_6} - 2\mu^{y_1}{}_{y_2 y_3 y_4 y_5 y_6} + \nu^{y_1}{}_{y_2 y_3 y_4 y_5 y_6}. \tag{37}$$

The deviation equation up to the third order reads

$$\frac{D^2}{dt^2}\eta^{y_1} = \tilde{a}^{y_1}\left(\frac{d\tilde{t}}{dt}\right)^2 - a^{y_1} + \frac{dt}{d\tilde{t}}\frac{d^2\tilde{t}}{dt^2}u^{y_1} + \frac{D\eta^{y_1}}{dt}\frac{dt}{d\tilde{t}}\frac{d^2\tilde{t}}{dt^2}$$
$$-\eta^{y_4}R^{y_1}{}_{y_2 y_3 y_4}\left(u^{y_2}u^{y_3} + 2u^{y_3}\frac{D\eta^{y_2}}{dt}\right)$$
$$+\eta^{y_4}\eta^{y_5}\left\{u^{y_2}u^{y_3}\left(\frac{1}{2}\nabla_{y_2}R^{y_1}{}_{y_4 y_5 y_3} - \frac{1}{3}\nabla_{y_4}R^{y_1}{}_{y_2 y_3 y_5}\right)\right.$$

$$
\begin{aligned}
&+\frac{1}{3}R^{y_1}{}_{y_4y_5y_2}\left[a^{y_2}+\frac{1}{2}\tilde{a}^{y_2}\left(\frac{d\tilde{t}}{dt}\right)^2-u^{y_2}\frac{dt}{d\tilde{t}}\frac{d^2\tilde{t}}{dt^2}\right]\Bigg\} \\
&-\frac{1}{6}\eta^{y_4}\eta^{y_5}\eta^{y_6}\Bigg\{\phi^{y_1}{}_{y_2y_3y_4y_5y_6}u^{y_2}u^{y_3} \\
&-\frac{1}{2}\nabla_{(y_4}R^{y_1}{}_{y_5y_6)y_2}\left[a^{y_2}+\tilde{a}^{y_2}\left(\frac{d\tilde{t}}{dt}\right)^2-u^{y_2}\frac{dt}{d\tilde{t}}\frac{d^2\tilde{t}}{dt^2}\right]\Bigg\} \\
&-\frac{1}{2}u^{y'}\frac{D\eta^{y''}}{dt}\eta^{y_2}\eta^{y_3}\Bigg(-\nabla_{(y''}R^{y_1}{}_{y_2y_3)y_1}+\nabla_{y_2}R^{y_1}{}_{(y'y'')y_3} \\
&-\frac{1}{3}\nabla_{(y'}R^{y_1}{}_{|y_2y_3|y'')}\Bigg)-\frac{2}{3}\frac{D\eta^{y_2}}{dt}\frac{D\eta^{y_3}}{dt}\eta^{y_4}R^{y_1}{}_{y_2y_3y_4}+\mathcal{O}(\sigma^4). \qquad (38)
\end{aligned}
$$

We would like to stress that the generalized deviation Eq. (38) is completely general. In particular, it allows for a comparison of two general, i.e. not necessarily geodetic, world lines in spacetime. Various special cases of (38) qualitatively reproduce all the previous results in the literature, see in particular [14–20].

2.2 *Method 2: Reference Frame (Inertial and Gravitational Effects)*

The above discussion of the deviation equation made clear that a suitable choice of coordinates is crucial for the successful determination of the gravitational field. In particular, the operational realization of the coordinates is of importance when it comes to actual measurements.

From an experimentalists perspective so-called (generalized) Fermi coordinates appear to be realizable operationally. There have been several suggestions for such coordinates in the literature in different contexts [2, 21–46]. In the following we are going to derive the line element in the vicinity of a world line, representing an observer in an arbitrary state of motion, in generalized Fermi coordinates.

Fermi normal coordinates Following [24] we start by taking successive derivatives of the usual geodesic equation. This generates a set of equations of the form (for $n \geq 2$)

$$
\frac{d^n x^a}{ds^n} = -\Gamma_{b_1\ldots b_n}{}^a \frac{dx^{b_1}}{ds}\cdots\frac{dx^{b_n}}{ds}, \qquad (39)
$$

where the Γ objects with $n \geq 3$ lower indices are defined by the recurrent relation

$$
\Gamma_{b_1\ldots b_n}{}^a := \partial_{(b_1}\Gamma_{b_2\ldots b_n)}{}^a - (n-1)\,\Gamma_{c(b_1\ldots b_{n-2}}{}^a\,\Gamma_{b_{n-1}b_n)}{}^c \qquad (40)
$$

from the components of the linear connection $\Gamma_{bc}{}^{a}$. A solution $x^a = x^a(s)$ of the geodesic equation may then be expressed as a series

$$\begin{aligned} x^a &= x^a\big|_0 + s\,\frac{dx^a}{ds}\bigg|_0 + \frac{s^2}{2}\,\frac{d^2x^a}{ds^2}\bigg|_0 + \frac{s^3}{6}\,\frac{d^3x^a}{ds^3}\bigg|_0 + \cdots \\ &= q^a + s v^a - \frac{s^2}{2}\,\overset{0}{\Gamma}_{bc}{}^{a}\,v^b v^c - \frac{s^3}{6}\,\overset{0}{\Gamma}_{bcd}{}^{a}\,v^b v^c v^d - \cdots, \end{aligned} \tag{41}$$

where in the last line we used $q^a := x^a|_0$, $v^a := \frac{dx^a}{ds}\big|_0$, and $\overset{0}{\Gamma}_{\dots}{}^{a} := \Gamma_{\dots}{}^{a}|_0$ for constant quantities at the point around which the series development is performed.

Now let us setup coordinates centered on the reference curve Y to describe an adjacent point X. For this we consider a unique geodesic connecting Y and X. We define our coordinates in the vicinity of a point on $Y(s)$, with proper time s, by using a tetrad $\lambda_b{}^{(\alpha)}$ which is Fermi transported along Y, i.e.

$$X^0 = s, \qquad X^\alpha = \tau \xi^b \lambda_b{}^{(\alpha)}. \tag{42}$$

Here $\alpha = 1, \dots, 3$, and τ is the proper time along the (spacelike) geodesic connecting $Y(s)$ and X. The ξ^b are constants, and it is important to notice that the tetrads are functions of the proper time s along the reference curve Y, but independent of τ. See Fig. 4 for further explanations. By means of this linear ansatz (42) for the coordinates in the vicinity of Y, we obtain for the derivatives w.r.t. τ along the connecting geodesic ($n \geq 1$):

$$\frac{d^n X^0}{d\tau^n} = 0, \qquad \frac{dX^\alpha}{d\tau} = \xi^b \lambda_b{}^{(\alpha)}, \qquad \frac{d^{n+1} X^\alpha}{d\tau^{n+1}} = 0. \tag{43}$$

In other words, in the chosen coordinates (42), along the geodesic connecting Y and X, one obtains for the derivatives ($n \geq 2$)

$$\Gamma_{b_1 \dots b_n}{}^{a}\,\frac{dX^{b_1}}{d\tau} \cdots \frac{dX^{b_n}}{d\tau} = 0. \tag{44}$$

This immediately yields

$$\Gamma_{\beta_1 \dots \beta_n}{}^{a} = 0, \tag{45}$$

along the connecting curve, in the region covered by the linear coordinates as defined above.

The Fermi normal coordinate system cannot cover the whole spacetime manifold. By construction, it is a good way to describe the physical phenomena in a small region around the world line of an observer. The smallness of the corresponding domain depends on the motion of the latter, in particular, on the magnitudes of acceleration $|a|$ and angular velocity $|\omega|$ of the observer which set the two characteristic lengths:

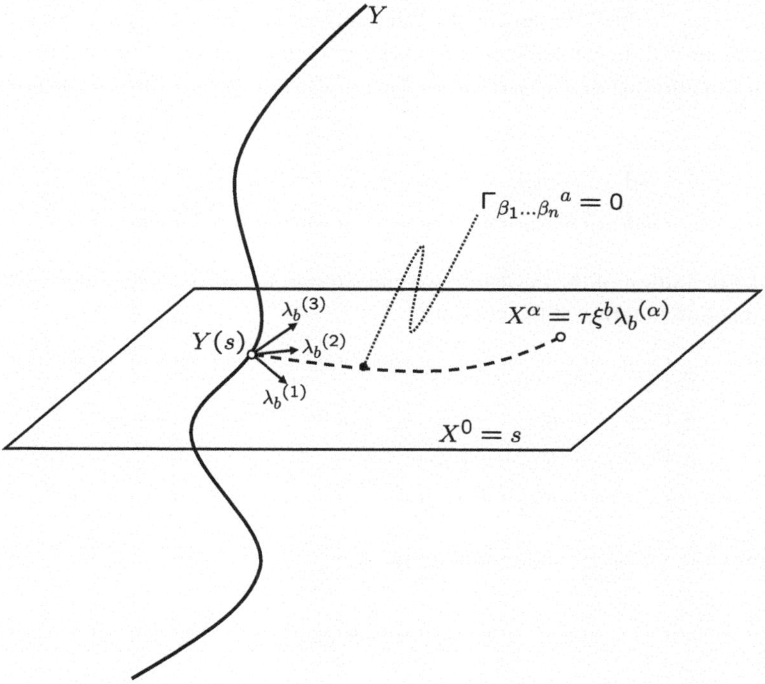

Fig. 4 Construction of the coordinate system around the reference curve Y. Coordinates of a point X in the vicinity of $Y(s)$ – s representing the proper time along Y – are constructed by means of a tetrad $\lambda_b{}^{(\alpha)}$. Here τ is the proper time along the (spacelike) geodesic connecting Y and X. By choosing a linear ansatz for the coordinates the derivatives of the connection vanish along the geodesic connecting Y and X

$\ell_{\rm tr} = c^2/|a|$ and $\ell_{\rm rot} = c/|\omega|$. The Fermi coordinate system X^α provides a good description for the region $|X|/\ell \ll 1$. For example, this condition is with a high accuracy valid in terrestrial laboratories since $\ell_{\rm tr} = c^2/|g_\oplus| \approx 10^{16}$ m (one light year), and $\ell_{\rm rot} = c/|\Omega_\oplus| \approx 4 \times 10^{12}$ m (27 astronomical units). Note, however, that for a particle accelerated in a storage ring $\ell \approx 10^{-6}$ m. Furthermore, the region of validity of the Fermi coordinate system is restricted by the strength of the gravitational field in the region close to the reference curve, $\ell_{\rm grav} = \min\{|R_{abcd}|^{-1/2}, |R_{abcd}|/|R_{abcd,e}|\}$, so that the curvature should have not yet caused geodesics to cross. We always assume that there is a unique geodesic connecting Y and X.

Explicit form of the connection At the lowest order, in flat spacetime, the connection of a noninertial system that is accelerating with a^α and rotating with angular velocity ω^α at the origin of the coordinate system is

$$\Gamma_{00}{}^{0} = \Gamma_{\alpha\beta}{}^{c} = 0, \quad \Gamma_{00}{}^{\alpha} = a^\alpha, \quad \Gamma_{0\alpha}{}^{0} = a_\alpha, \quad \Gamma_{0\beta}{}^{\alpha} = -\varepsilon^\alpha{}_{\beta\gamma}\omega^\gamma . \tag{46}$$

Hereafter $\varepsilon_{\alpha\beta\gamma}$ is the 3-dimensional totally antisymmetric Levi-Civita symbol, and the Euclidean 3-dimensional metric $\delta_{\alpha\beta}$ is used to raise and lower the spatial (Greek) indices, in particular $a_\alpha = \delta_{\alpha\beta}a^\beta$ and $\varepsilon^\alpha{}_{\beta\gamma} = \delta^{\alpha\delta}\varepsilon_{\delta\beta\gamma}$. For the time derivatives we have

$$\begin{aligned}
&\partial_0\Gamma_{00}{}^0 = \partial_0\Gamma_{\alpha\beta}{}^c = 0, \quad \partial_0\Gamma_{00}{}^\alpha = \partial_0 a^\alpha =: b^\alpha,\\
&\partial_0\Gamma_{0\alpha}{}^0 = b_\alpha, \ \partial_0\Gamma_{0\beta}{}^\alpha = -\varepsilon^\alpha{}_{\beta\gamma}\partial_0\omega^\gamma =: -\varepsilon^\alpha{}_{\beta\gamma}\eta^\gamma.
\end{aligned} \tag{47}$$

From the definition of the curvature we can express the next order of derivatives of the connection in terms of the curvature:

$$\begin{aligned}
\partial_\alpha\Gamma_{00}{}^0 &= b_\alpha - a_\beta\varepsilon^\beta{}_{\alpha\gamma}\omega^\gamma,\\
\partial_\alpha\Gamma_{00}{}^\beta &= -R_{0\alpha 0}{}^\beta - \varepsilon^\beta{}_{\alpha\gamma}\eta^\gamma + a_\alpha a^\beta - \delta^\beta_\alpha\omega_\gamma\omega^\gamma + \omega_\alpha\omega^\beta,\\
\partial_\alpha\Gamma_{0\beta}{}^0 &= -R_{0\alpha\beta}{}^0 - a_\alpha a_\beta,\\
\partial_\alpha\Gamma_{0\beta}{}^\gamma &= -R_{0\alpha\beta}{}^\gamma + \varepsilon^\gamma{}_{\alpha\delta}\omega^\delta a_\beta.
\end{aligned} \tag{48}$$

Using (45), we derive the spatial derivatives

$$\partial_\alpha\Gamma_{\beta\gamma}{}^d = \frac{2}{3}R_{\alpha(\beta\gamma)}{}^d, \tag{49}$$

see also the general solution given in the Appendix B of [7].

Explicit form of the metric In order to determine, in the vicinity of the reference curve Y, the form of the metric at the point X in coordinates y^a centered on Y, we start again with an expansion of the metric around the reference curve

$$g_{ab}|_X = g_{ab}|_Y + g_{ab,c}\big|_Y\, y^c + \frac{1}{2}\, g_{ab,cd}\big|_Y\, y^c y^d + \cdots. \tag{50}$$

Of course in normal coordinates we have $g_{ab}|_Y = \eta_{ab}$, whereas the derivatives of the metric have to be calculated, and the result actually depends on which type of coordinates we want to use. The derivatives of the metric may be expressed just by successive differentiation of the metricity condition $\nabla_c g_{ab} = 0$:

$$\begin{aligned}
g_{ab,c} &= 2\, g_{d(a}\Gamma_{b)c}{}^d,\\
g_{ab,cd} &= 2\left(\partial_d g_{e(a}\Gamma_{b)c}{}^e + \partial_d\Gamma_{c(a}{}^e g_{b)e}\right),\\
&\ \ \vdots
\end{aligned} \tag{51}$$

In other words, we can iteratively determine the metric by plugging in the explicit form of the connection and its derivatives from above.

In combination with (46) one finds:

$$g_{00,0} = g_{0\alpha,0} = g_{\alpha\beta,0} = g_{\alpha\beta,\gamma} = 0,$$
$$g_{00,\alpha} = 2a_\alpha, \quad g_{0\alpha,\beta} = \varepsilon_{\alpha\beta\gamma}\omega^\gamma. \tag{52}$$

For the second order derivatives of metric we obtain, again using (51) in combination with (47), (48), and (52):

$$g_{00,00} = g_{0\alpha,00} = g_{\alpha\beta,00} = g_{\alpha\beta,\gamma 0} = 0, \quad g_{00,\alpha 0} = 2b_\alpha,$$
$$g_{0\alpha,\beta 0} = -\varepsilon^\gamma{}_{\beta\delta}\eta^\delta g_{\alpha\gamma} = \varepsilon_{\alpha\beta\gamma}\eta^\gamma,$$
$$g_{00,\alpha\beta} = -2R_{0\beta\alpha}{}^0 + 2a_\alpha a_\beta - 2\delta_{\alpha\beta}\omega_\gamma\omega^\gamma + 2\omega_\alpha\omega_\beta,$$
$$g_{0\alpha,\beta\gamma} = -\frac{4}{3}R_{\alpha(\beta\gamma)}{}^0, \quad g_{\alpha\beta,\gamma\delta} = \frac{2}{3}R_{\gamma(\alpha\beta)\delta}. \tag{53}$$

Note that $R_{0\beta\alpha}{}^0 + R_{\alpha 0\beta}{}^0 + R_{\beta\alpha 0}{}^0 \equiv 0$, in view of the Ricci identity. Since $R_{\beta\alpha 0}{}^0 = 0$, we thus find $R_{0\beta\alpha}{}^0 = R_{0(\beta\alpha)}{}^0$.

As a result, we derive the line element in the Fermi coordinates (up to the second order):

$$\begin{aligned} ds^2|_X(y^0, y^\alpha) = {} & (dy^0)^2\big[1 + 2a_\alpha y^\alpha + 2b_\alpha y^\alpha y^0 \\ & + (a_\alpha a_\beta - \delta_{\alpha\beta}\omega_\gamma\omega^\gamma + \omega_\alpha\omega_\beta - R_{0\alpha\beta 0})y^\alpha y^\beta\big] \\ & + 2dy^0 dy^\alpha\Big[\varepsilon_{\alpha\beta\gamma}\omega^\gamma y^\beta + \varepsilon_{\alpha\beta\gamma}\eta^\gamma y^\beta y^0 - \frac{2}{3}R_{\alpha\beta\gamma 0}y^\beta y^\gamma\Big] \\ & - dy^\alpha dy^\beta\left[\delta_{\alpha\beta} - \frac{1}{3}R_{\gamma\alpha\beta\delta}y^\gamma y^\delta\right] + \mathcal{O}(3). \end{aligned} \tag{54}$$

It is worthwhile to notice that we can recast this result into

$$\begin{aligned} ds^2|_X(y^0, y^\alpha) = {} & (dy^0)^2\big[1 + 2\overline{a}_\alpha y^\alpha + (\overline{a}_\alpha\overline{a}_\beta - \delta_{\alpha\beta}\overline{\omega}_\gamma\overline{\omega}^\gamma + \overline{\omega}_\alpha\overline{\omega}_\beta \\ & - R_{0\alpha\beta 0})y^\alpha y^\beta\big] + 2dy^0 dy^\alpha\Big[\varepsilon_{\alpha\beta\gamma}\overline{\omega}^\gamma y^\beta - \frac{2}{3}R_{\alpha\beta\gamma 0}y^\beta y^\gamma\Big] \\ & - dy^\alpha dy^\beta\left[\delta_{\alpha\beta} - \frac{1}{3}R_{\gamma\alpha\beta\delta}y^\gamma y^\delta\right] + \mathcal{O}(3), \end{aligned} \tag{55}$$

by introducing $\overline{a}_\alpha = a_\alpha + y^0\partial_0 a_\alpha = a_\alpha + y^0 b_\alpha$ and $\overline{\omega}_\alpha = \omega_\alpha + y^0\partial_0\omega_\alpha = \omega_\alpha + y^0$ η_α which represent the power expansion of the time dependent acceleration and angular velocity.

2.3 Method 2: Apparent Behavior of Clocks

The results from the last section may now be used to describe the behavior of clocks in the vicinity of the reference world line, around which the coordinates were constructed.

There is one interesting peculiarity about writing the metric like in (54), i.e. one obtains clock effects which depend on the acceleration of the clock (just integrate along a curve in those coordinates and the terms with a and ω will of course contribute to the proper time along the curve). This behavior of clocks is of course due to the choice of the noninertial observer, and they are *only* present along curves which do not coincide with the observers world line. Recall that, by construction, one has Minkowski's metric along the world line of the observer, which is also the center of the coordinate system in which (54) is written – all inertial effects vanish at the origin of the coordinate system.

Flat case We start with the flat spacetime and switch to a quantity which is directly measurable, i.e. the proper time quotient of two clocks located at Y and X. It is worthwhile to note that for a flat spacetime, $R_{ijk}{}^{l} = 0$, the interval (54) reduces to the Hehl-Ni [47] line element of a noninertial (rotating and accelerating) system:

$$ds^2|_X(y^0, y^\alpha) = (1 + \overline{a}_\alpha y^\alpha)^2 (dy^0)^2 - \delta_{\alpha\beta}(dy^\alpha + \varepsilon^\alpha{}_{\mu\nu}\overline{\omega}^\mu y^\nu\, dy^0) \times (dy^\beta + \varepsilon^\beta{}_{\rho\sigma}\overline{\omega}^\rho y^\sigma\, dy^0) + \mathcal{O}(3), \tag{56}$$

From (54) we derive

$$\left(\frac{ds|_X}{ds|_Y}\right)^2 = \left(\frac{dy^0}{ds|_Y}\right)^2 \Big[1 - \delta_{\alpha\beta}v^\alpha v^\beta + 2a_\alpha y^\alpha + 2b_\alpha y^\alpha y^0 + y^\alpha y^\beta\left(a_\alpha a_\beta - \delta_{\alpha\beta}\omega_\gamma\omega^\gamma + \omega_\alpha\omega_\beta\right) + 2v^\alpha \varepsilon_{\alpha\beta\gamma}\left(y^\beta\omega^\gamma + y^0 y^\beta \eta^\gamma\right)\Big] + \mathcal{O}(3) \tag{57}$$

$$= 1 + \frac{1}{1 - \delta_{\alpha\beta}v^\alpha v^\beta}\Big[2a_\alpha y^\alpha + 2b_\alpha y^\alpha y^0 + y^\alpha y^\beta\left(a_\alpha a_\beta - \delta_{\alpha\beta}\omega_\gamma\omega^\gamma + \omega_\alpha\omega_\beta\right) + 2v^\alpha \varepsilon_{\alpha\beta\gamma}\left(y^\beta\omega^\gamma + y^0 y^\beta \eta^\gamma\right)\Big] + \mathcal{O}(3). \tag{58}$$

Here we introduced the velocity $v^\alpha = dy^\alpha/dy^0$. Defining

$$V^\alpha := v^\alpha + \varepsilon^\alpha{}_{\beta\gamma}\overline{\omega}^\beta y^\gamma, \tag{59}$$

we can rewrite the above relation more elegantly as

$$\left(\frac{ds|_X}{ds|_Y}\right)^2 = \left(\frac{dy^0}{ds|_Y}\right)^2 \left[(1 + \overline{a}_\alpha y^\alpha)^2 - \delta_{\alpha\beta}V^\alpha V^\beta\right] + \mathcal{O}(3). \tag{60}$$

Equation (58) is reminiscent of the situation which we encountered in case of the gravitational compass, i.e. we may look at this measurable quantity depending on how we prepare the

$$C\left(y^\alpha, y^0, v^\alpha, a^\alpha, \omega^\alpha, b^\alpha, \eta^\alpha\right) := \left(\frac{ds|_X}{ds|_Y}\right)^2. \tag{61}$$

Curved case Now let us investigate the curved spacetime, after all we are interested in measuring the gravitational field by means of clock comparison. The frequency ratio becomes:

$$\begin{aligned}\left(\frac{ds|_X}{ds|_Y}\right)^2 = 1 + \frac{1}{1-\delta_{\alpha\beta}v^\alpha v^\beta}\Big[& 2a_\alpha y^\alpha + 2b_\alpha y^\alpha y^0 \\ & + y^\alpha y^\beta\left(a_\alpha a_\beta - R_{0\alpha\beta 0} - \delta_{\alpha\beta}\omega_\gamma\omega^\gamma + \omega_\alpha\omega_\beta\right) \\ & + 2v^\alpha \varepsilon_{\alpha\beta\gamma}\left(y^\beta\omega^\gamma + y^0 y^\beta \eta^\gamma\right) - \frac{4}{3}v^\alpha y^\beta y^\gamma R_{\alpha\beta\gamma 0} \\ & + \frac{1}{3}v^\alpha v^\beta y^\gamma y^\delta R_{\gamma\alpha\beta\delta}\Big] + \mathcal{O}(3).\end{aligned} \tag{62}$$

Analogously to the flat case in (61), we introduce a shortcut for the measurable frequency ratio in a curved background, denoting its dependence on different quantities as $C\left(y^\alpha, y^0, v^\alpha, a^\alpha, \omega^\alpha, b^\alpha, \eta^\alpha, R_{\alpha\beta\gamma\delta}\right)$.

Note that in the flat, as well as in the curved case, the frequency ratio becomes independent of b^α and η^α on the three-dimensional slice with fixed y^0 (since we can always choose our coordinate time parameter $y^0 = 0$), i.e. we have $C\left(y^\alpha, v^\alpha, a^\alpha, \omega^\alpha\right)$ and $C\left(y^\alpha, v^\alpha, a^\alpha, \omega^\alpha, R_{\alpha\beta\gamma\delta}\right)$ respectively.

3 Operational Determination of the Gravitational Field

3.1 *Method 1: Relativistic Gradiometry/Gravitational Compass*

The determination of the curvature of spacetime in the context of deviation equations has been discussed in [2, 8, 48]. In particular, Szekeres coined in [8] the notion of a "gravitational compass." From now on we will adopt this notion for a set of suitably prepared test bodies which allow for the measurement of the curvature and, thereby, the gravitational field.

The operational procedure is to monitor the accelerations of a set of test bodies w.r.t. to an observer moving on the reference world line Y. A mechanical analogue would be to measure the forces between the test bodies and the reference body via a spring connecting them.

Rewriting the deviation equation We now describe the configurations of test bodies which allow for a complete determination of all curvature components in a Riemannian background spacetime. For concreteness, our analysis will be based on the standard geodesic deviation equation, as well as one of its generalizations. Our starting point is the standard geodesic deviation equation, i.e.

$$\frac{D^2}{ds^2}\eta^a = R^a{}_{bcd}u^b\eta^c u^d. \tag{63}$$

Since we want to express the curvature in terms of measured quantities, i.e. the velocities and the accelerations, we rewrite this equation in terms of the standard (non-covariant) derivative w.r.t. the proper time.

In order to simplify the resulting equation we employ normal coordinates, i.e. we have on the world line of the reference test body

$$\Gamma_{ab}{}^c|_Y = 0, \qquad \partial_a\Gamma_{bc}{}^d|_Y = \frac{2}{3}R_{a(bc)}{}^d. \tag{64}$$

In terms of the standard total derivative w.r.t. to the proper time s, the deviation Eq. (63) takes the form:

$$\frac{d^2}{ds^2}\eta^a \stackrel{|_Y}{=} \frac{2}{3}R^a{}_{bcd}u^b\eta^c u^d. \tag{65}$$

However, what actually seems to be measured by a compass at the reference point Y is the lower components of the relative acceleration. For the lower index position, in terms of the ordinary derivative in normal coordinates, the deviation Eq. (63) takes the form

$$\frac{d^2}{ds^2}\eta_a \stackrel{|_Y}{=} \frac{4}{3}R_{abcd}u^b\eta^c u^d. \tag{66}$$

Explicit compass setup Let us consider a general 6-point compass. In addition to the reference test body on the world line we will use the following geometrical setup of the 5 remaining test bodies:

$$\begin{aligned} {}^{(1)}\eta^a &= \begin{pmatrix}0\\1\\0\\0\end{pmatrix}, \quad {}^{(2)}\eta^a = \begin{pmatrix}0\\0\\1\\0\end{pmatrix}, \quad {}^{(3)}\eta^a = \begin{pmatrix}0\\0\\0\\1\end{pmatrix}, \\ {}^{(4)}\eta^a &= \begin{pmatrix}0\\1\\1\\0\end{pmatrix}, \quad {}^{(5)}\eta^a = \begin{pmatrix}0\\0\\1\\1\end{pmatrix}. \end{aligned} \tag{67}$$

In addition to the positions of the compass constituents, we have to make a choice for the velocity of the reference test body/observer. In the following we will use (m) different compasses, each of these compasses will have a different velocity (associated) with the reference test body. In other words, we consider (m) different compasses or reference test bodies, all of which are located at the world line reference point Y (at the same time), and all these (m) observers measure the relative accelerations to all five test bodies placed at the positions given in (67). The left-hand sides of (66) are the measured accelerations and in the following we refer to them by ${}^{(m,n)}A_a$. Furthermore, we also introduced the compass index ${}^{(m)}u^a$ for the velocities. In other words, for (m) compasses and (n) bodies in one compass, we have the following set of equations:

$$
{}^{(m,n)}A_a \overset{|_Y}{=} \frac{4}{3} R_{abcd}{}^{(m)}u^b\,{}^{(n)}\eta^c\,{}^{(m)}u^d. \tag{68}
$$

What remains to be chosen, apart from the $(n = 1 \ldots 5)$ positions of bodies in one compass, is the number (m) and the actual directions in which each compass/observer shall move. Of course in the end we want to minimize both numbers, i.e. (m) and (n), which are needed to determine all curvature components.

$$
\begin{aligned}
{}^{(1)}u^a &= \begin{pmatrix} c_{10} \\ 0 \\ 0 \\ 0 \end{pmatrix}, \; {}^{(2)}u^a = \begin{pmatrix} c_{20} \\ c_{21} \\ 0 \\ 0 \end{pmatrix}, \; {}^{(3)}u^a = \begin{pmatrix} c_{30} \\ 0 \\ c_{32} \\ 0 \end{pmatrix}, \\
{}^{(4)}u^a &= \begin{pmatrix} c_{40} \\ 0 \\ 0 \\ c_{43} \end{pmatrix}, \; {}^{(5)}u^a = \begin{pmatrix} c_{50} \\ c_{51} \\ c_{52} \\ 0 \end{pmatrix}, \; {}^{(6)}u^a = \begin{pmatrix} c_{60} \\ 0 \\ c_{62} \\ c_{63} \end{pmatrix}.
\end{aligned} \tag{69}
$$

The $c_{(m)a}$ here are just constants, chosen appropriately to ensure the normalization of the 4-velocity of each compass.

In summary, we are going to consider $(m) = 1 \ldots 6$ compasses, each of them with 6-points, where the five reference points are always the $(n) = 1 \ldots 5$ from (67).

Explicit curvature components The 20 independent components of the curvature tensor can be explicitly determined in terms of the accelerations ${}^{(m,n)}A_a$ and velocities ${}^{(m)}u^a$ by making use of the deviation Eq. (68) with the help of the compass configuration given in (67) and (69). The result reads as follows:

$$
01: R_{1010} = \frac{3}{4}{}^{(1,1)}A_1 c_{10}^{-2}, \tag{70}
$$

$$
02: R_{2010} = \frac{3}{4}{}^{(1,1)}A_2 c_{10}^{-2}, \tag{71}
$$

$$
03: R_{3010} = \frac{3}{4}{}^{(1,1)}A_3 c_{10}^{-2}, \tag{72}
$$

$$04: R_{2020} = \frac{3}{4}{}^{(1,2)}A_2 c_{10}^{-2}, \tag{73}$$

$$05: R_{3020} = \frac{3}{4}{}^{(1,2)}A_3 c_{10}^{-2}, \tag{74}$$

$$06: R_{3030} = \frac{3}{4}{}^{(1,3)}A_3 c_{10}^{-2}, \tag{75}$$

$$07: R_{2110} = \frac{3}{4}{}^{(2,1)}A_2 c_{21}^{-1} c_{20}^{-1} - R_{2010} c_{21}^{-1} c_{20}, \tag{76}$$

$$08: R_{3110} = \frac{3}{4}{}^{(2,1)}A_3 c_{21}^{-1} c_{20}^{-1} - R_{3010} c_{21}^{-1} c_{20}, \tag{77}$$

$$09: R_{0212} = \frac{3}{4}{}^{(3,1)}A_0 c_{32}^{-2} + R_{2010} c_{32}^{-1} c_{30}, \tag{78}$$

$$10: R_{1212} = \frac{3}{4}{}^{(2,2)}A_2 c_{21}^{-2} - R_{2020} c_{20}^{2} c_{21}^{-2} - 2R_{0212} c_{21}^{-1} c_{20}, \tag{79}$$

$$11: R_{3220} = \frac{3}{4}{}^{(3,2)}A_3 c_{32}^{-1} c_{30}^{-1} - R_{3020} c_{32}^{-1} c_{30}, \tag{80}$$

$$12: R_{0313} = \frac{3}{4}{}^{(4,1)}A_0 c_{43}^{-2} + R_{3010} c_{43}^{-1} c_{40}, \tag{81}$$

$$13: R_{1313} = \frac{3}{4}{}^{(2,3)}A_3 c_{21}^{-2} - R_{3030} c_{20}^{2} c_{21}^{-2} - 2R_{0313} c_{21}^{-1} c_{20}, \tag{82}$$

$$14: R_{0323} = \frac{3}{4}{}^{(4,2)}A_0 c_{43}^{-2} + R_{3020} c_{43}^{-1} c_{40}, \tag{83}$$

$$15: R_{2323} = \frac{3}{4}{}^{(4,2)}A_2 c_{43}^{-2} - R_{2020} c_{43}^{-2} c_{40}^{2} + 2R_{3220} c_{43}^{-1} c_{40}, \tag{84}$$

$$\begin{aligned} 16: R_{3132} = {} & \frac{3}{8}{}^{(5,3)}A_3 c_{52}^{-1} c_{51}^{-1} - \frac{1}{2} R_{3030} c_{52}^{-1} c_{51}^{-1} c_{50}^{2} \\ & - R_{0313} c_{52}^{-1} c_{50} - R_{0323} c_{51}^{-1} c_{50} - \frac{1}{2} R_{1313} c_{52}^{-1} c_{51} \\ & - \frac{1}{2} R_{2323} c_{52} c_{51}^{-1}, \end{aligned} \tag{85}$$

$$\begin{aligned} 17: R_{1213} = {} & \frac{3}{8}{}^{(6,1)}A_1 c_{63}^{-1} c_{62}^{-1} - \frac{1}{2} R_{1010} c_{63}^{-1} c_{62}^{-1} c_{60}^{2} \\ & + R_{2110} c_{63}^{-1} c_{60} + R_{3110} c_{62}^{-1} c_{60} - \frac{1}{2} R_{1212} c_{63}^{-1} c_{62} \\ & - \frac{1}{2} R_{1313} c_{63} c_{62}^{-1}, \end{aligned} \tag{86}$$

There are still 3 components of the curvature tensor missing. To determine them, we notice that the following relation between the remaining equations is at our disposal:

$$R_{0312} - R_{0231} = \frac{3}{4}{}^{(2,2)}A_3 c_{20}^{-1} c_{21}^{-1} - R_{3020} c_{20} c_{21}^{-1} - R_{3121} c_{21} c_{20}^{-1}, \tag{87}$$

$$R_{0231} - R_{0123} = \frac{3}{4}{}^{(4,1)}A_2 c_{40}^{-1} c_{43}^{-1} - R_{2010} c_{40} c_{43}^{-1} - R_{2313} c_{43} c_{40}^{-1}. \tag{88}$$

Subtracting (87) from (88) and using the Ricci identity we find:

$$\begin{aligned} 18: R_{0231} = &\frac{1}{4}{}^{(4,1)}A_2 c_{40}^{-1} c_{43}^{-1} - \frac{1}{4}{}^{(2,2)}A_3 c_{20}^{-1} c_{21}^{-1} \\ &+ \frac{1}{3}\big(R_{3020} c_{20} c_{21}^{-1} + R_{3121} c_{21} c_{20}^{-1} \\ &- R_{2010} c_{40} c_{43}^{-1} - R_{2313} c_{43} c_{40}^{-1}\big), \end{aligned} \tag{89}$$

$$\begin{aligned} 19: R_{0312} = &\frac{1}{4}{}^{(4,1)}A_2 c_{40}^{-1} c_{42}^{-1} + \frac{1}{2}{}^{(2,2)}A_3 c_{20}^{-1} c_{21}^{-1} \\ &- \frac{1}{3}\big(2R_{3020} c_{20} c_{21}^{-1} + 2R_{3121} c_{21} c_{20}^{-1} \\ &+ R_{2010} c_{40} c_{43}^{-1} + R_{2313} c_{43} c_{40}^{-1}\big). \end{aligned} \tag{90}$$

Finally, by reinsertion of (87) in one of the remaining compass equations, one obtains:

$$\begin{aligned} 20: R_{3212} = &\frac{3}{4}{}^{(4,1)}A_3 c_{20}^{-1} c_{21}^{-1} c_{50} c_{52}^{-1} - \frac{3}{4}{}^{(5,2)}A_3 c_{51}^{-1} c_{52}^{-1} \\ &+ R_{3121} c_{52}^{-1}\left(c_{51} - c_{50} c_{21} c_{20}^{-1}\right) + R_{3220} c_{50} c_{51}^{-1} \\ &+ R_{3020} c_{50} c_{52}^{-1}\left(c_{50} c_{51}^{-1} - c_{20} c_{21}^{-1}\right). \end{aligned} \tag{91}$$

By examination of the components given in (70)–(91), we conclude that for a full determination of the curvature one needs 13 test bodies, see Fig. 5 for a sketch of the solution.

Vacuum spacetime In vacuum the number of independent components of the curvature is reduced to the 10 components of the Weyl tensor C_{abcd}. Replacing R_{abcd} in the compass solution (70)–(91), and taking into account the symmetries of Weyl we may use a reduced compass setup to completely determine the gravitational field, i.e.

$$01: C_{1010} = \frac{3}{4}{}^{(1,1)}A_1 c_{10}^{-2}, \tag{92}$$

$$02: C_{2010} = \frac{3}{4}{}^{(1,1)}A_2 c_{10}^{-2}, \tag{93}$$

$$03: C_{3010} = \frac{3}{4}{}^{(1,1)}A_3 c_{10}^{-2}, \tag{94}$$

$$04: C_{2020} = \frac{3}{4}{}^{(1,2)}A_2 c_{10}^{-2}, \tag{95}$$

$$05: C_{3020} = \frac{3}{4}{}^{(1,2)}A_3 c_{10}^{-2}, \tag{96}$$

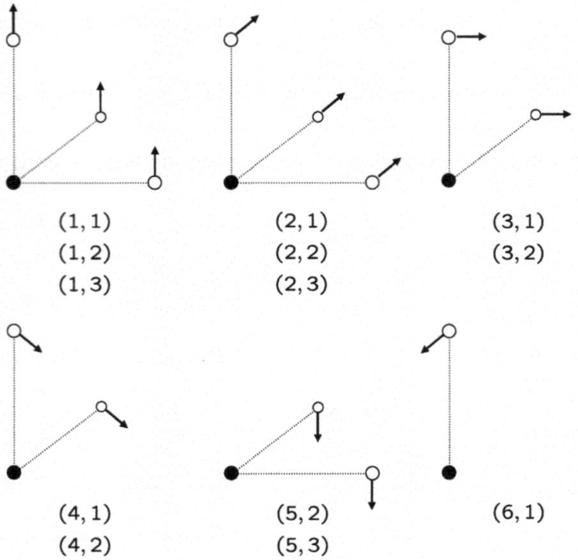

Fig. 5 Symbolical sketch of the explicit compass solution in (70)–(91). In total 13 suitably prepared test bodies (hollow circles) are needed to determine all curvature components. The reference body is denoted by the black circle. Note that with the standard deviation equation all ${}^{(1\ldots6)}u^a$, but only ${}^{(1\ldots3)}\eta^a$ are needed in the solution

$$06: C_{2110} = \frac{3}{4}{}^{(2,1)}A_2 c_{21}^{-1} c_{20}^{-1} - C_{2010} c_{21}^{-1} c_{20}, \tag{97}$$

$$07: C_{3110} = \frac{3}{4}{}^{(2,1)}A_3 c_{21}^{-1} c_{20}^{-1} - C_{3010} c_{21}^{-1} c_{20}, \tag{98}$$

$$08: C_{0212} = \frac{3}{4}{}^{(3,1)}A_0 c_{32}^{-2} + C_{2010} c_{32}^{-1} c_{30}, \tag{99}$$

$$\begin{aligned} 09: C_{0231} = {} & \frac{1}{4}{}^{(4,1)}A_2 c_{40}^{-1} c_{43}^{-1} - \frac{1}{4}{}^{(2,2)}A_3 c_{20}^{-1} c_{21}^{-1} \\ & + \frac{1}{3} C_{3020}\left(c_{20} c_{21}^{-1} + c_{21} c_{20}^{-1}\right) \\ & - \frac{1}{3} C_{2010}\left(c_{40} c_{43}^{-1} + c_{43} c_{40}^{-1}\right), \end{aligned} \tag{100}$$

$$\begin{aligned} 10: C_{0312} = {} & \frac{1}{4}{}^{(4,1)}A_2 c_{40}^{-1} c_{42}^{-1} + \frac{1}{2}{}^{(2,2)}A_3 c_{20}^{-1} c_{21}^{-1} \\ & - \frac{2}{3} C_{3020}\left(c_{20} c_{21}^{-1} + c_{21} c_{20}^{-1}\right) \\ & + \frac{1}{3} C_{2010}\left(c_{40} c_{43}^{-1} + c_{43} c_{40}^{-1}\right). \end{aligned} \tag{101}$$

All the other components of the Weyl tensor are obtained from the above by making use of the double-self-duality property $C_{abcd} = -\frac{1}{4}\epsilon_{abef}\epsilon_{cdgh}C^{efgh}$, where ϵ_{abcd} is the totally antisymmetric Levi-Civita tensor with $\epsilon_{0123} = 1$, and the Ricci identity. See Fig. 6 for a sketch of the solution.

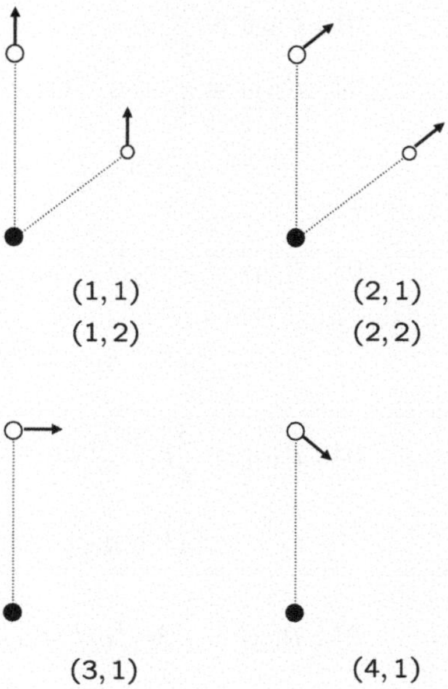

Fig. 6 Symbolical sketch of the explicit compass solution in (92)–(101) for the vacuum case. In total 6 suitably prepared test bodies (hollow circles) are needed to determine all components of the Weyl tensor. The reference body is denoted by the black circle. Note that in vacuum, with the standard deviation equation, all ${}^{(1...4)}u^a$, but only ${}^{(1...2)}\eta^a$ are needed in the solution

3.2 *Method 2: Relativistic Clock Gradiometry/Gravitational Clock Compass*

Now we turn to the determination of the curvature in a general spacetime by means of clocks. We consider the non-vacuum case first, when one needs to measure 20 independent components of the Riemann curvature tensor $R_{abc}{}^d$.

Again we start by rearranging the system (61):

$$\begin{aligned} &{}^{(n)}y^{\alpha\,(n)}y^{\beta}\left(-R_{0\alpha\beta0} - \frac{4}{3}R_{\gamma\alpha\beta0}\,{}^{(m)}v^{\gamma} + \frac{1}{3}R_{\alpha\gamma\delta\beta}\,{}^{(m)}v^{\gamma\,(m)}v^{\delta}\right) \\ &= B({}^{(n)}y^{\alpha},\,{}^{(m)}v^{\alpha},\,{}^{(p)}a^{\alpha},\,{}^{(q)}\omega^{\alpha}), \end{aligned} \tag{102}$$

where

$$\begin{aligned} B(y^{\alpha}, v^{\alpha}, a^{\alpha}, \omega^{\alpha}) := (1 - v^2)(C - 1) - 2a_{\alpha}y^{\alpha} - y^{\alpha}y^{\beta}\Big(a_{\alpha}a_{\beta} \\ -\delta_{\alpha\beta}\omega_{\gamma}\omega^{\gamma} + \omega_{\alpha}\omega_{\beta}\Big) - 2v^{\alpha}\varepsilon_{\alpha\beta\gamma}y^{\beta}\omega^{\gamma}. \end{aligned} \tag{103}$$

Analogously to our analysis of the gravitational compass [7], we may now consider different setups of clocks to measure as many curvature components as possible. The system in (102) yields (please note that only the position and the velocity indices are indicated):

$$01 : R_{1010} = {}^{(1,1)}B, \tag{104}$$

$$02 : R_{2110} = \frac{3}{4}c_{22}^{-1}c_{42}^{-1}\,(c_{22} - c_{42})^{-1}\Big({}^{(1,1)}Bc_{22}^2 - {}^{(1,1)}Bc_{42}^2 + {}^{(1,2)}Bc_{42}^2 - {}^{(1,4)}Bc_{22}^2\Big), \tag{105}$$

$$03 : R_{1212} = -3c_{22}^{-1}c_{42}^{-1}\,(c_{22} - c_{42})^{-1}\Big({}^{(1,1)}Bc_{22} - {}^{(1,1)}Bc_{42} + {}^{(1,2)}Bc_{42} - {}^{(1,4)}Bc_{22}\Big), \tag{106}$$

$$04 : R_{3110} = \frac{3}{4}c_{33}^{-1}c_{63}^{-1}\,(c_{33} - c_{63})^{-1}\Big({}^{(1,1)}Bc_{33}^2 - {}^{(1,1)}Bc_{63}^2 + {}^{(1,3)}Bc_{63}^2 - {}^{(1,6)}Bc_{33}^2\Big), \tag{107}$$

$$05 : R_{1313} = -3c_{33}^{-1}c_{63}^{-1}\,(c_{33} - c_{63})^{-1}\Big({}^{(1,1)}Bc_{33} - {}^{(1,1)}Bc_{63} + {}^{(1,3)}Bc_{63} - {}^{(1,6)}Bc_{33}\Big), \tag{108}$$

$$06 : R_{1213} = \frac{3}{2}c_{52}^{-1}c_{53}^{-1}\Big(- {}^{(1,5)}B + R_{1010} - \frac{4}{3}R_{2110}c_{52} - \frac{4}{3}R_{3110}c_{53} - \frac{1}{3}R_{1212}c_{52}^2 - \frac{1}{3}R_{1313}c_{53}^2\Big), \tag{109}$$

$$07 : R_{2020} = {}^{(2,2)}B, \tag{110}$$

$$08 : R_{0212} = \frac{3}{4}c_{11}^{-1}\Big({}^{(2,1)}B - R_{2020} + \frac{1}{3}R_{1212}c_{11}^2\Big), \tag{111}$$

$$09 : R_{3220} = \frac{3}{4}c_{33}^{-1}c_{53}^{-1}\,(c_{33} - c_{53})^{-1}\Big({}^{(2,2)}Bc_{33}^2 - {}^{(2,2)}Bc_{53}^2 + {}^{(2,3)}Bc_{53}^2 - {}^{(2,5)}Bc_{33}^2\Big), \tag{112}$$

$$10 : R_{2323} = -3c_{33}^{-1}c_{53}^{-1}\,(c_{33} - c_{53})^{-1}\Big({}^{(2,2)}Bc_{33} - {}^{(2,2)}Bc_{53} + {}^{(2,3)}Bc_{53} - {}^{(5,2)}Bc_{33}\Big), \tag{113}$$

$$11 : R_{3212} = \frac{3}{2}c_{61}^{-1}c_{63}^{-1}\Big(- {}^{(2,6)}B + R_{2020} + \frac{4}{3}R_{0212}c_{61} - \frac{4}{3}R_{3220}c_{63} - \frac{1}{3}R_{1212}c_{61}^2 - \frac{1}{3}R_{2323}c_{63}^2\Big), \tag{114}$$

$$12 : R_{3030} = {}^{(3,3)}B, \tag{115}$$

$$13 : R_{0313} = \frac{3}{4}c_{11}^{-1}\left({}^{(3,1)}B - R_{3030} + \frac{1}{3}R_{1313}c_{11}^2\right), \tag{116}$$

$$14 : R_{0323} = \frac{3}{4}c_{22}^{-1}\left({}^{(3,2)}B - R_{3030} + \frac{1}{3}R_{2323}c_{22}^2\right), \tag{117}$$

$$15 : R_{3132} = \frac{3}{2}c_{41}^{-1}c_{42}^{-1}\bigg(- {}^{(3,4)}B + R_{3030} + \frac{4}{3}R_{0313}c_{41} + \frac{4}{3}R_{0323}c_{42} \\ -\frac{1}{3}R_{1313}c_{41}^2 - \frac{1}{3}R_{2323}c_{42}^2\bigg), \tag{118}$$

$$16 : R_{2010} = \frac{1}{2}\bigg({}^{(4,1)}B - R_{1010} - R_{2020} - \frac{4}{3}R_{0212}c_{11} \\ -\frac{4}{3}R_{2110}c_{11} + \frac{1}{3}R_{1212}c_{11}^2\bigg), \tag{119}$$

$$17 : R_{3020} = \frac{1}{2}\bigg({}^{(5,2)}B - R_{2020} - R_{3030} - \frac{4}{3}R_{0323}c_{22} \\ -\frac{4}{3}R_{3220}c_{22} + \frac{1}{3}R_{2323}c_{22}^2\bigg), \tag{120}$$

$$18 : R_{3010} = \frac{1}{2}\bigg({}^{(6,1)}B - R_{1010} - R_{3030} - \frac{4}{3}R_{0313}c_{11} \\ -\frac{4}{3}R_{3110}c_{11} + \frac{1}{3}R_{1313}c_{11}^2\bigg). \tag{121}$$

Introducing abbreviations

$$K_1 := \frac{3}{4}c_{33}^{-1}\bigg[- {}^{(4,3)}B + R_{1010} + 2R_{2010} + R_{2020} \\ -\frac{4}{3}(R_{3110} + R_{3220})c_{33} - \frac{1}{3}(R_{1313} + 2R_{3132} + R_{2323})c_{33}^2\bigg], \tag{122}$$

$$K_2 := \frac{3}{4}c_{11}^{-1}\bigg[- {}^{(5,1)}B + R_{2020} + 2R_{3020} + R_{3030} \\ +\frac{4}{3}(R_{0212} + R_{0313})c_{11} - \frac{1}{3}(R_{1212} + 2R_{1213} + R_{1313})c_{11}^2\bigg], \tag{123}$$

$$K_3 := \frac{3}{4}c_{22}^{-1}\bigg[- {}^{(6,2)}B + R_{1010} + 2R_{3010} + R_{3030} \\ -\frac{4}{3}(R_{2110} + R_{0323})c_{22} - \frac{1}{3}(R_{1212} + 2R_{3212} + R_{2323})c_{22}^2\bigg], \tag{124}$$

we find the remaining three curvature components

$$19 : R_{1023} = \frac{1}{3}(K_3 - K_1)\,, \tag{125}$$

$$20 : R_{2013} = \frac{1}{3}(K_2 - K_1)\,, \tag{126}$$

$$21 : R_{3021} = \frac{1}{3}(K_3 - K_2)\,. \tag{127}$$

See Fig. 7 for a symbolical sketch of the solution. The B's in these equations can be explicitly resolved in terms of the C's

$${}^{(1,1)}B = \left(1 - c_{11}^2\right)\left({}^{(1,1)}C - 1\right), \tag{128}$$

$${}^{(1,2)}B = \left(1 - c_{22}^2\right)\left({}^{(1,2)}C - 1\right), \tag{129}$$

$${}^{(1,3)}B = \left(1 - c_{33}^2\right)\left({}^{(1,3)}C - 1\right), \tag{130}$$

$${}^{(1,4)}B = \left(1 - c_{41}^2 - c_{42}^2\right)\left({}^{(1,4)}C - 1\right), \tag{131}$$

$${}^{(1,5)}B = \left(1 - c_{52}^2 - c_{53}^2\right)\left({}^{(1,5)}C - 1\right), \tag{132}$$

$${}^{(1,6)}B = \left(1 - c_{61}^2 - c_{63}^2\right)\left({}^{(1,6)}C - 1\right), \tag{133}$$

$${}^{(2,1)}B = \left(1 - c_{11}^2\right)\left({}^{(2,1)}C - 1\right), \tag{134}$$

$${}^{(2,2)}B = \left(1 - c_{22}^2\right)\left({}^{(2,2)}C - 1\right), \tag{135}$$

$${}^{(2,3)}B = \left(1 - c_{33}^2\right)\left({}^{(2,3)}C - 1\right), \tag{136}$$

$${}^{(2,5)}B = \left(1 - c_{52}^2 - c_{53}^2\right)\left({}^{(2,5)}C - 1\right), \tag{137}$$

$${}^{(2,6)}B = \left(1 - c_{61}^2 - c_{63}^2\right)\left({}^{(2,6)}C - 1\right), \tag{138}$$

$${}^{(3,1)}B = \left(1 - c_{11}^2\right)\left({}^{(3,1)}C - 1\right), \tag{139}$$

$${}^{(3,2)}B = \left(1 - c_{22}^2\right)\left({}^{(3,2)}C - 1\right), \tag{140}$$

$${}^{(3,3)}B = \left(1 - c_{33}^2\right)\left({}^{(3,3)}C - 1\right), \tag{141}$$

$${}^{(3,4)}B = \left(1 - c_{41}^2 - c_{42}^2\right)\left({}^{(3,4)}C - 1\right), \tag{142}$$

$${}^{(4,1)}B = \left(1 - c_{11}^2\right)\left({}^{(4,1)}C - 1\right), \tag{143}$$

$${}^{(4,3)}B = \left(1 - c_{33}^2\right)\left({}^{(4,1)}C - 1\right), \tag{144}$$

$${}^{(5,1)}B = \left(1 - c_{11}^2\right)\left({}^{(5,1)}C - 1\right), \tag{145}$$

$${}^{(5,2)}B = \left(1 - c_{22}^2\right)\left({}^{(5,2)}C - 1\right), \tag{146}$$

$${}^{(6,1)}B = \left(1 - c_{11}^2\right)\left({}^{(6,1)}C - 1\right), \tag{147}$$

$${}^{(6,2)}B = \left(1 - c_{22}^2\right)\left({}^{(6,2)}C - 1\right). \tag{148}$$

Vacuum spacetime In vacuum the number of independent components of the curvature is reduced to the 10 components of the Weyl tensor C_{abcd}. Replacing R_{abcd} in the compass solution (104)–(121), and taking into account the symmetries of the Weyl

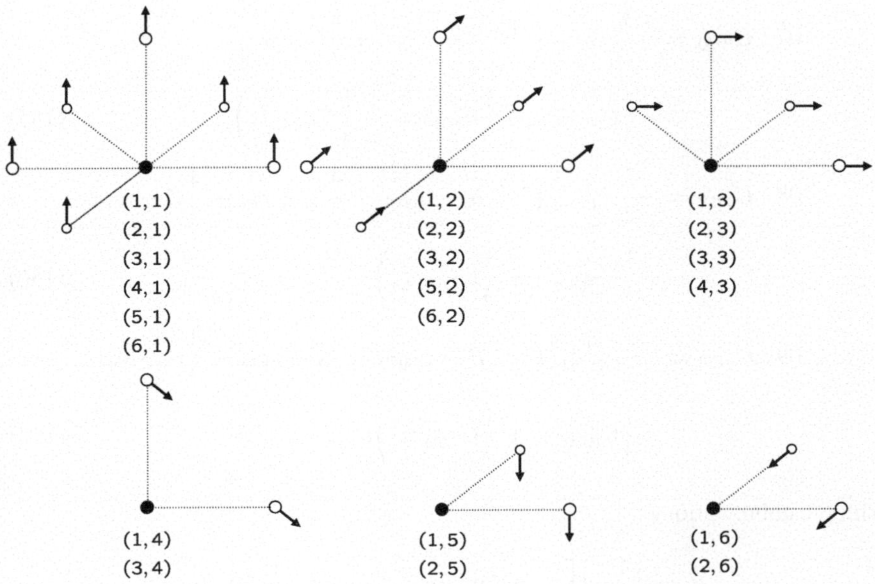

Fig. 7 Symbolical sketch of the explicit solution for the curvature (104)–(127). In total 21 suitably prepared clocks (hollow circles) are needed to determine all curvature components. The observer is denoted by the black circle. Note that all ${}^{(1\ldots6)}v^a$, but only ${}^{(1\ldots3)}y^a$ are needed in the solution

tensor, we may use a reduced clock setup to completely determine the gravitational field. All other components may be obtained from the double self-duality property $C_{abcd} = -\frac{1}{4}\varepsilon_{abef}\varepsilon_{cdgh}C^{efgh}$.

$$01: C_{2323} = -{}^{(1,1)}B, \tag{149}$$

$$\begin{aligned} 02: C_{0323} = {} & \frac{3}{4}c_{22}^{-1}c_{42}^{-1}\left(c_{22}-c_{42}\right)^{-1}\Big({}^{(1,1)}Bc_{22}^2 - {}^{(1,1)}Bc_{42}^2 \\ & + {}^{(1,2)}Bc_{42}^2 - {}^{(1,4)}Bc_{22}^2\Big), \end{aligned} \tag{150}$$

$$\begin{aligned} 03: C_{3030} = {} & 3c_{22}^{-1}c_{42}^{-1}\left(c_{22}-c_{42}\right)^{-1}\Big({}^{(1,1)}Bc_{22} - {}^{(1,1)}Bc_{42} \\ & + {}^{(1,2)}Bc_{42} - {}^{(1,4)}Bc_{22}\Big), \end{aligned} \tag{151}$$

$$04: C_{2020} = {}^{(2,2)}B, \tag{152}$$

$$05: C_{3220} = \frac{3}{4}c_{33}^{-1}\left({}^{(1,3)}B + C_{2323} - \frac{1}{3}C_{2020}c_{33}^2\right), \tag{153}$$

$$06: C_{0313} = -\frac{3}{4}c_{11}^{-1}\left({}^{(2,1)}B - C_{2020} - \frac{1}{3}C_{3030}c_{11}^2\right), \tag{154}$$

$$07: C_{3020} = -\frac{3}{2}c_{52}^{-1}c_{53}^{-1}\Big({}^{(1,5)}B + C_{2323} + \frac{4}{3}C_{0323}c_{52} - \frac{4}{3}C_{3220}c_{53} - \frac{1}{3}C_{3030}c_{52}^2 - \frac{1}{3}C_{2020}c_{53}^2\Big), \tag{155}$$

$$08: C_{3212} = -\frac{3}{2}c_{61}^{-1}c_{63}^{-1}\Big({}^{(2,6)}B - C_{2020} + \frac{4}{3}C_{0313}c_{61} + \frac{4}{3}C_{3220}c_{63} - \frac{1}{3}C_{3030}c_{61}^2 + \frac{1}{3}C_{2323}c_{63}^2\Big), \tag{156}$$

$$09: C_{3132} = -\frac{3}{2}c_{41}^{-1}c_{42}^{-1}\Big({}^{(3,4)}B - C_{3030} - \frac{4}{3}C_{0313}c_{41} - \frac{4}{3}C_{0323}c_{42} - \frac{1}{3}C_{2020}c_{41}^2 + \frac{1}{3}C_{2323}c_{42}^2\Big). \tag{157}$$

With the abbreviations

$$K_1 := \frac{3}{4}c_{33}^{-1}\Big[-{}^{(4,3)}B - C_{2323} + 2C_{3132} + C_{2020} + \frac{1}{3}(C_{2020} - 2C_{3132} - C_{2323})c_{33}^2\Big], \tag{158}$$

$$K_2 := \frac{3}{4}c_{11}^{-1}\Big[-{}^{(5,1)}B + C_{2020} + 2C_{3020} + C_{3030} + \frac{1}{3}(C_{3030} - 2C_{3020} + C_{2020})c_{11}^2\Big], \tag{159}$$

$$K_3 := -\frac{3}{4}c_{22}^{-1}\Big[{}^{(6,2)}B + C_{2323} - 2C_{3212} - C_{3030} - \frac{1}{3}(C_{3030} - 2C_{3212} - C_{2323})c_{22}^2\Big], \tag{160}$$

the remaining three curvature components read

$$10: C_{1023} = \frac{1}{3}(K_3 - K_1), \tag{161}$$

$$11: C_{2013} = \frac{1}{3}(K_2 - K_1), \tag{162}$$

$$12: C_{3021} = \frac{1}{3}(K_3 - K_2). \tag{163}$$

A symbolical sketch of the solution is given in Fig. 8.

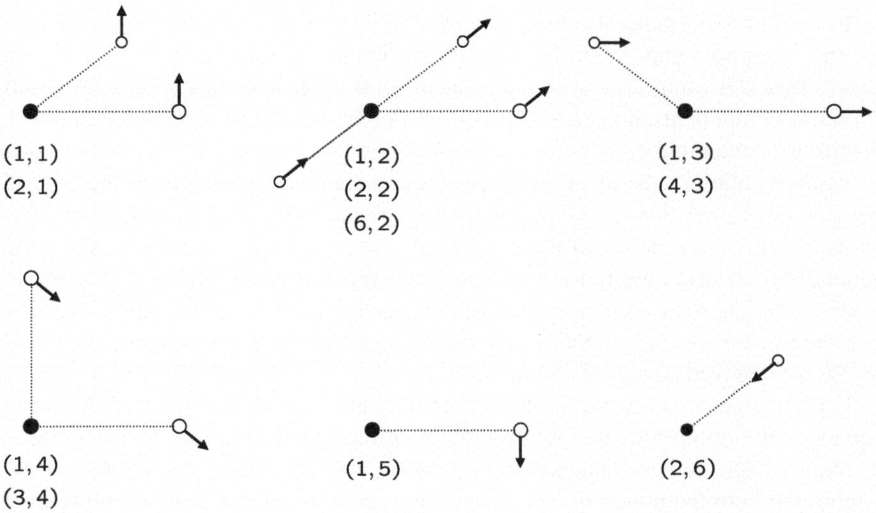

Fig. 8 Symbolical sketch of the explicit vacuum solution for the curvature (149)–(163). In total 11 suitably prepared clocks (hollow circles) are needed to determine all curvature components. The observer is denoted by the black circle. Note that all $^{(1\ldots6)}v^a$, but only $^{(1\ldots3)}y^a$ are needed in the solution

4 Summary

4.1 Method 1: Summary

In the framework of Synge's world function approach, we have derived a generalized covariant deviation Eq. (12) which is valid for arbitrary world lines and in general background spacetimes. Making use of systematic expansions of the exact deviation equation up to the third order in the world function, we obtain the final result (38) which can be viewed as a generalization of the well-known geodesic deviation equation. Furthermore, our results encompass several suggestions for a generalized deviation equation from the literature as special cases, and therefore may serve a unified framework for further studies.

In Sect. 3.1 we have shown how deviation equations can be used to determine the curvature of spacetime. For this we extended the notion of a gravitational compass [8] and worked out compass setups for general as well as for vacuum spacetimes. One setup is based on the standard geodesic deviation equation, and another is based on the next order generalization given in which goes beyond the linearized case. For both cases we provided the explicit compass solution which allows for a full determination of the curvature.

In contrast to the general considerations in [2, 8] we give an explicit exact solution for the compass setup. With the standard deviation equation, as well as with the generalized deviation equation, we need at least 13 test bodies to determine all curvature components in a general spacetime. For the standard deviation we therefore obtain the same number of bodies as in [48], however it is worthwhile to note that no explicit solution was given in [48] for a non-vacuum spacetime. In the case of a generalized deviation equation our findings are at odds with the results in [48]. However, this discrepancy in the generalized case comes as no surprise since the generalized equation used in [48] – which was previously derived in [49] – differs from our equation. In vacuum spacetimes, we have explicitly shown that the number of required test bodies is reduced to 6, for the standard deviation equation, and to 5, for the generalized deviation equation.

Furthermore, it is interesting to note that in the case of the standard deviation equation, the opinion of the authors [2, 8] differs when it comes to the number of required test bodies. This seems to be related to the counting scheme and the interpretation of the notion of a compass. Since no explicit compass solutions were given in [2, 8], one cannot make a comparison to our results. In the case of [48], we were not able to verify that the given solution does fulfill the compass equations derived in that work. However, the agreement on the number of required bodies in combination with the standard deviation is reassuring.

Our results are of direct operational relevance and form the basis for many experiments. Important applications range from the description of gravitational wave detectors to the study of satellite configurations for gravitational field mapping in relativistic geodesy.

4.2 Method 2: Summary

Section 3.2 describes an experimental setup which we call a clock compass, in analogy to the usual gravitational compass [7, 8]. We have shown that a suitably prepared set of clocks can be used to determine all components of the gravitational field, i.e. the curvature, in General Relativity, as well as to describe the state of motion of a noninertial observer.

Working out explicit clock compass setups in different situations, we have demonstrated that in general 6 clocks are needed to determine the linear acceleration as well as the rotational velocity, while 4 clocks will suffice in case of the velocity. Furthermore, we prove that one needs 21 and 11 clocks, respectively, to determine all curvature components in a general curved spacetime and in vacuum. In view of possible future experimental realizations it is interesting to note that restrictions regarding the choice of clock velocities in a setup lead to restrictions regarding the number of determinable curvature components.

Our results are of direct operational relevance for the setup of networks of clocks, especially in the context of relativistic geodesy. In geodetic terms, the given clock configurations may be thought of as a clock gradiometers. Taking into account the steadily increasing accuracy of clocks [50], these results should be combined with those from a gradiometric context, for example in the form of a hybrid gravitational compass – which combines acceleration as well as clock measurements in one setup. Another possible application is the detection of gravitational waves by means of clock as well as standard interferometric techniques. An interesting question is a possible reduction of the number of measurements by a combination of different techniques.

5 Outlook: Operational Determination of the Gravitational Field in Theories Beyond GR

In the previous sections, we have shown how the deviation equation as well as an ensemble of clocks can be used to measure the gravitational field in GR. However, the results were limited to theories in a Riemannian background. While such theories are justified in many physical situations, several modern gravitational theories [51–53] reach significantly beyond the Riemannian geometrical framework. In particular it is already well-known [12, 54, 55], that in the description of test bodies with intrinsic degrees of freedom – like spin – there is a natural coupling to the post-Riemannian features of spacetime. Therefore, in view of possible tests of gravitational theories by means of structured test bodies, a further extension of the deviation equation to post-Riemannian geometries is needed.

In the following we present a generalized deviation equation in a Riemann–Cartan background, allowing for spacetimes endowed with torsion, the presentation is based on [56]. This equation describes the dynamics of the connecting vector which links events on two general (adjacent) world lines. Our results are valid for any theory in a Riemann–Cartan background, in particular they apply to Einstein–Cartan theory [57] as well as to Poincaré gauge theory [58, 59]. Interestingly, Synge was apparently the first who derived the deviation equation for the Riemann–Cartan geometry [60].

5.1 World Function and Deviation Equation

Let us briefly recapitulate the relevant steps which lead to the generalized deviation equation: We want to compare two general curves $Y(t)$ and $X(\tilde{t})$ in an arbitrary spacetime manifold. Here t and $\tilde{t}$ are general parameters, i.e. not necessarily the proper time on the given curves. In contrast to the Riemannian case, see Sect. 2.1, we now connect two points $x \in X$ and $y \in Y$ on the two curves by the autoparallel joining the two points (we assume that this autoparallel is unique). An autoparallel is a curve along which the velocity vector is transported parallel to itself with respect to the connection on the spacetime manifold. In a Riemannian space autoparallel curves coincide with geodesic lines. Along the autoparallel we have the world function σ,

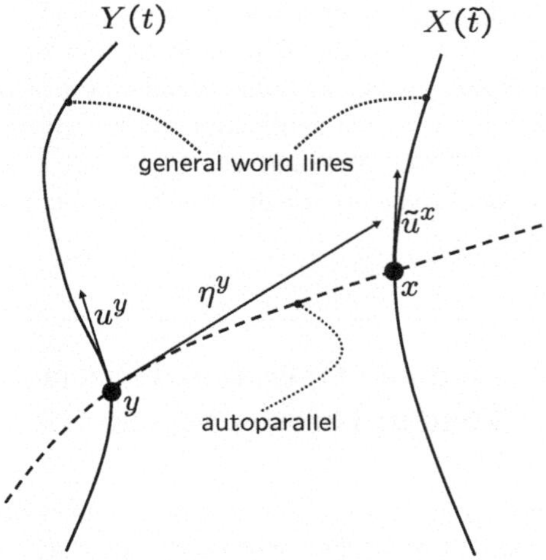

Fig. 9 Sketch of the two arbitrarily parametrized world lines $Y(t)$ and $X(\tilde{t})$, and the (dashed) autoparallel connecting two points on these world line. The generalized deviation vector along the reference world line Y is denoted by η^y

and conceptually the closest object to the connecting vector between the two points is the covariant derivative of the world function, denoted at the point y by σ^y, cf. Fig. 9.

At this point, the generality of our derivation of the deviation equation from Sect. 2.1 pays of, i.e. the exact deviation equation given in Eq. 12 can be directly used in the case in which the connecting curve is an autoparallel therefore, to recapitulate, we have:

$$\begin{aligned}\frac{D^2}{dt^2}\eta^{y_1} &= -\sigma^{y_1}{}_{y_2y_3}u^{y_2}u^{y_3} - \sigma^{y_1}{}_{y_2}a^{y_2} - \sigma^{y_1}{}_{x_2}\tilde{a}^{x_2}\left(\frac{d\tilde{t}}{dt}\right)^2 \\ &\quad -2\sigma^{y_1}{}_{y_2x_3}u^{y_2}\left(K^{x_3}{}_{y_4}u^{y_4} - H^{x_3}{}_{y_4}\frac{D\sigma^{y_4}}{dt}\right) \\ &\quad -\sigma^{y_1}{}_{x_2x_3}\left(K^{x_2}{}_{y_4}u^{y_4} - H^{x_2}{}_{y_4}\frac{D\sigma^{y_4}}{dt}\right) \\ &\quad \times\left(K^{x_3}{}_{y_5}u^{y_5} - H^{x_3}{}_{y_5}\frac{D\sigma^{y_5}}{dt}\right) \\ &\quad -\sigma^{y_1}{}_{x_2}\frac{dt}{d\tilde{t}}\frac{d^2\tilde{t}}{dt^2}\left(K^{x_2}{}_{y_3}u^{y_3} - H^{x_2}{}_{y_3}\frac{D\sigma^{y_3}}{dt}\right). \end{aligned} \tag{164}$$

Again the factor $d\tilde{t}/dt$ by requiring that the velocity along the curve X is normalized, i.e. $\tilde{u}^x\tilde{u}_x = 1$, in which case we have

$$\frac{d\tilde{t}}{dt} = \tilde{u}_{x_1}K^{x_1}{}_{y_2}u^{y_2} - \tilde{u}_{x_1}H^{x_1}{}_{y_2}\frac{D\sigma^{y_2}}{dt}. \tag{165}$$

Equation (164) is the exact generalized deviation equation, it is completely general and can be viewed as the extension of the standard geodesic deviation (Jacobi) equation to any order.

5.2 World Function in Riemann–Cartan Spacetime

In order to arrive at an expanded approximate version of the deviation equation, we need to work out the properties of a world function based on autoparallels in a Riemann–Cartan background. In contrast to a Riemannian spacetime a Riemann–Cartan spacetime is endowed with an asymmetric connection $\Gamma_{ab}{}^{c}$, and there will be differences when it comes to the basic properties of a world function σ based on autoparallels. We base our presentation on [56], other relevant references which contain some results in a Riemann–Cartan context are [61–68].

For a world function σ based on autoparallels, we have the following basic relations in the case of spacetimes with asymmetric connections:

$$\sigma^{x}\sigma_{x} = \sigma^{y}\sigma_{y} = 2\sigma, \tag{166}$$

$$\sigma^{x_2}\sigma_{x_2}{}^{x_1} = \sigma^{x_1}, \tag{167}$$

$$\sigma_{x_1x_2} - \sigma_{x_2x_1} = T_{x_1x_2}{}^{x_3}\partial_{x_3}\sigma. \tag{168}$$

Note in particular the change in (168) due to the presence of the spacetime torsion $T_{x_1x_2}{}^{x_3}$, which leads to $\sigma_{x_1x_2} \neq \sigma_{x_2x_1}$, in contrast to the symmetric Riemannian case, in which $\overline{\sigma}_{x_1x_2} \overset{\mathrm{s}}{=} \overline{\sigma}_{x_2x_1}$ holds.[1]

In many calculations the limiting behavior of a bitensor $B_{...}(x, y)$ as x approaches the references point y is required. This so-called coincidence limit of a bitensor $B_{...}(x, y)$ is a tensor

$$[B_{...}] = \lim_{x\to y} B_{...}(x, y), \tag{169}$$

at y and will be denoted by square brackets. In particular, for a bitensor B with arbitrary indices at different points (here just denoted by dots), we have the rule [2]

$$[B_{...}]_{;y} = \left[B_{...;y}\right] + \left[B_{...;x}\right]. \tag{170}$$

[1] We use "s" to indicate relations which only hold for symmetric connections and denote Riemannian objects by the overbar.

We collect the following useful identities for the world function σ:

$$[\sigma] = [\sigma_x] = [\sigma_y] = 0, \tag{171}$$
$$[\sigma_{x_1 x_2}] = [\sigma_{y_1 y_2}] = g_{y_1 y_2}, \tag{172}$$
$$[\sigma_{x_1 y_2}] = [\sigma_{y_1 x_2}] = -g_{y_1 y_2}, \tag{173}$$
$$[\sigma_{x_3 x_1 x_2}] + [\sigma_{x_2 x_1 x_3}] = 0. \tag{174}$$

Note that up to the second covariant derivative the coincidence limits of the world function match those in spacetimes with symmetric connections. However, at the next (third) order the presence of the torsion leads to

$$[\sigma_{x_1 x_2 x_3}] = \frac{1}{2}\left(T_{y_1 y_3 y_2} + T_{y_2 y_3 y_1} + T_{y_1 y_2 y_3}\right) = K_{y_2 y_1 y_3}, \tag{175}$$

where in the last line we made use of the contortion[2] $K_{ab}{}^c = \overline{\Gamma}_{ab}{}^c - \Gamma_{ab}{}^c$. With the help of (170) we obtain for the other combinations with three indices:

$$[\sigma_{y_1 x_2 x_3}] = -[\sigma_{y_1 y_2 x_3}] = [\sigma_{y_1 y_2 y_3}] = K_{y_2 y_1 y_3}. \tag{176}$$

The non-vanishing of these limits leads to added complexity in subsequent calculations compared to the Riemannian case.

At the fourth order we have

$$K_{y_1}{}^{y}{}_{y_2} K_{y_3 y y_4} + K_{y_1}{}^{y}{}_{y_3} K_{y_2 y y_4} + K_{y_1}{}^{y}{}_{y_4} K_{y_2 y y_4}$$
$$+[\sigma_{x_4 x_1 x_2 x_3}] + [\sigma_{x_3 x_1 x_2 x_4}] + [\sigma_{x_2 x_1 x_3 x_4}] = 0, \tag{177}$$

and in particular

$$[\sigma_{x_1 x_2 x_3 x_4}] = \frac{1}{3}\nabla_{y_1}\left(K_{y_3 y_2 y_4} + K_{y_4 y_2 y_3}\right) + \frac{1}{3}\nabla_{y_3}\left(3K_{y_2 y_1 y_4} - K_{y_1 y_2 y_4}\right)$$
$$+\frac{1}{3}\nabla_{y_4}\left(3K_{y_2 y_1 y_3} - K_{y_1 y_2 y_3}\right) + \pi_{y_1 y_2 y_3 y_4}, \tag{178}$$
$$[\sigma_{x_1 x_2 x_3 y_4}] = -\frac{1}{3}\nabla_{y_1}\left(K_{y_3 y_2 y_4} + K_{y_4 y_2 y_3}\right) - \frac{1}{3}\nabla_{y_3}\left(3K_{y_2 y_1 y_4} - K_{y_1 y_2 y_4}\right)$$
$$+\frac{1}{3}\nabla_{y_4} K_{y_1 y_2 y_3} - \pi_{y_1 y_2 y_3 y_4}, \tag{179}$$
$$[\sigma_{x_1 x_2 y_3 y_4}] = \frac{1}{3}\nabla_{y_1}\left(K_{y_4 y_2 y_3} + K_{y_3 y_2 y_4}\right) - \frac{1}{3}\nabla_{y_4} K_{y_1 y_2 y_3}$$
$$-\frac{1}{3}\nabla_{y_3} K_{y_1 y_2 y_4} + \pi_{y_1 y_2 y_4 y_3}, \tag{180}$$

[2]The contortion $K_{y_2 y_1 y_3}$ should not be confused with the Jacobi propagator $K^x{}_y$.

$$
\begin{aligned}
[\sigma_{x_1 y_2 y_3 y_4}] &= -\frac{1}{3}\nabla_{y_1}\left(K_{y_3 y_4 y_2} + K_{y_2 y_4 y_3}\right) + \frac{1}{3}\nabla_{y_3} K_{y_1 y_4 y_2} + \frac{1}{3}\nabla_{y_2} K_{y_1 y_4 y_3} \\
&\quad + \nabla_{y_4} K_{y_3 y_1 y_2} - \pi_{y_1 y_4 y_3 y_2}, \qquad (181) \\
[\sigma_{y_1 y_2 y_3 y_4}] &= \frac{1}{3}\nabla_{y_4}\left(-2K_{y_2 y_3 y_1} + K_{y_1 y_3 y_2}\right) - \frac{1}{3}\nabla_{y_2} K_{y_4 y_3 y_1} - \frac{1}{3}\nabla_{y_1} K_{y_4 y_3 y_2} \\
&\quad - \nabla_{y_3} K_{y_2 y_4 y_1} + \pi_{y_4 y_3 y_2 y_1}, \qquad (182) \\
\pi_{y_1 y_2 y_3 y_4} &:= \frac{1}{3}\Big[K_{y_1 y_2}{}^{y}\left(K_{y_3 y_4 y} + K_{y_4 y_3 y}\right) - K_{y_1 y_3}{}^{y}\left(K_{y_4 y_2 y} + K_{y y_2 y_4}\right) \\
&\quad - K_{y_1 y_4}{}^{y}\left(K_{y_3 y_2 y} + K_{y y_2 y_3}\right) - 3K_{y_2 y_1}{}^{y} K_{y_3 y_4 y} + K_{y_3 y_1}{}^{y} K_{y y_2 y_4} \\
&\quad + K_{y_4 y_1}{}^{y} K_{y y_2 y_3} + R_{y_1 y_3 y_2 y_4} + R_{y_1 y_4 y_2 y_3}\Big]. \qquad (183)
\end{aligned}
$$

Again, we note the added complexity compared to the Riemannian case, in which we have $[\sigma_{x_1 x_2 x_3 x_4}] \stackrel{\$}{=} \frac{1}{3}\left(\overline{R}_{y_2 y_4 y_1 y_3} + \overline{R}_{y_3 y_2 y_1 y_4}\right)$ at the fourth order. In particular, we observe the occurrence of derivatives of the contortion in (178)–(182).

Finally, let us collect the basic properties of the so-called parallel propagator $g^{y}{}_{x} := e^{y}_{(a)} e^{(a)}_{x}$, defined in terms of a parallelly propagated tetrad $e^{y}_{(a)}$, which in turn allows for the transport of objects, i.e. $V^{y} = g^{y}{}_{x} V^{x}, \quad V^{y_1 y_2} = g^{y_1}{}_{x_1} g^{y_2}{}_{x_2} V^{x_1 x_2}$, etc., along an autoparallel:

$$
\begin{aligned}
g^{y_1}{}_{x} g^{x}{}_{y_2} &= \delta^{y_1}{}_{y_2}, \quad g^{x_1}{}_{y} g^{y}{}_{x_2} = \delta^{x_1}{}_{x_2}, \qquad (184) \\
\sigma^{x}\nabla_{x} g^{x_1}{}_{y_1} &= \sigma^{y}\nabla_{y} g^{x_1}{}_{y_1} = 0, \\
\sigma^{x}\nabla_{x} g^{y_1}{}_{x_1} &= \sigma^{y}\nabla_{y} g^{y_1}{}_{x_1} = 0, \qquad (185) \\
\sigma_{x} = -g^{y}{}_{x}\sigma_{y}, &\quad \sigma_{y} = -g^{x}{}_{y}\sigma_{x}. \qquad (186)
\end{aligned}
$$

Note in particular the coincidence limits of its derivatives

$$
\begin{aligned}
\left[g^{x_0}{}_{y_1}\right] &= \delta^{y_0}{}_{y_1}, \qquad (187) \\
\left[g^{x_0}{}_{y_1;x_2}\right] &= \left[g^{x_0}{}_{y_1;y_2}\right] = 0, \qquad (188) \\
\left[g^{x_0}{}_{y_1;x_2 x_3}\right] &= -\left[g^{x_0}{}_{y_1;x_2 y_3}\right] = \left[g^{x_0}{}_{y_1;x_2 x_3}\right] \\
&= -\left[g^{x_0}{}_{y_1;y_2 y_3}\right] = \frac{1}{2} R^{y_0}{}_{y_1 y_2 y_3}. \qquad (189)
\end{aligned}
$$

In the next section we will derive an expanded approximate version of the deviation equation. For this we first work out the expanded version of quantities around the reference world line Y. In particular, we make use of the covariant expansion technique [2, 3] on the basis of the autoparallel world function.

Expanded Riemann–Cartan deviation equation For a general bitensor $B_{\dots}$ with a given index structure, we have the following general expansion, up to the third order (in powers of σ^y):

$$\begin{aligned} B_{y_1\dots y_n} &= A_{y_1\dots y_n} + A_{y_1\dots y_{n+1}}\sigma^{y_{n+1}} \\ &\quad + \frac{1}{2}A_{y_1\dots y_{n+1}y_{n+2}}\sigma^{y_{n+1}}\sigma^{y_{n+2}} + \mathcal{O}\left(\sigma^3\right), \end{aligned} \tag{190}$$

$$A_{y_1\dots y_n} := \left[B_{y_1\dots y_n}\right], \tag{191}$$

$$A_{y_1\dots y_{n+1}} := \left[B_{y_1\dots y_n;y_{n+1}}\right] - A_{y_1\dots y_n;y_{n+1}}, \tag{192}$$

$$\begin{aligned} A_{y_1\dots y_{n+2}} &:= \left[B_{y_1\dots y_n;y_{n+1}y_{n+2}}\right] - A_{y_1\dots y_n y_0}\left[\sigma^{y_0}{}_{y_{n+1}y_{n+2}}\right] \\ &\quad - A_{y_1\dots y_n;y_{n+1}y_{n+2}} - 2A_{y_1\dots y_n(y_{n+1};y_{n+2})}. \end{aligned} \tag{193}$$

With the help of (190) we are able to iteratively expand any bitensor to any order, provided the coincidence limits entering the expansion coefficients can be calculated. The expansion for bitensors with mixed index structure can be obtained from transporting the indices in (190) by means of the parallel propagator.

In order to develop an approximate form of the generalized deviation Eq. (164) up to the second order, we need the following expansions of the derivatives of the world function:

$$\sigma_{y_1y_2} = g_{y_1y_2} + K_{y_2y_1y_3}\sigma^{y_3} + \mathcal{O}\left(\sigma^2\right), \tag{194}$$

$$\sigma_{y_1x_2} = -g_{y_1x_2} + g_{x_2}{}^{y}K_{y_3yy_1}\sigma^{y_3} + \mathcal{O}\left(\sigma^2\right), \tag{195}$$

$$\begin{aligned} \sigma_{y_1y_2y_3} &= K_{y_2y_1y_3} + \frac{1}{3}\Big[\nabla_{y_4}\left(K_{y_2y_3y_1} + K_{y_1y_3y_2}\right) \\ &\quad - \nabla_{y_2}K_{y_4y_3y_1} - \nabla_{y_1}K_{y_4y_3y_2} - 3\nabla_{y_3}K_{y_2y_4y_1} \\ &\quad + 3\pi_{y_4y_3y_2y_1}\Big]\sigma^{y_4} + \mathcal{O}\left(\sigma^2\right), \end{aligned} \tag{196}$$

$$\begin{aligned} \sigma_{y_1y_2x_3} &= g_{x_3}{}^{y_3}\Big\{K_{y_2y_3y_1} - \frac{1}{3}\Big[\nabla_{y_3}\left(K_{y_2y_4y_1} + K_{y_1y_4y_2}\right) \\ &\quad - \nabla_{y_2}K_{y_3y_4y_1} - \nabla_{y_1}K_{y_3y_4y_2} \\ &\quad + 3\pi_{y_3y_4y_2y_1}\Big]\sigma^{y_4}\Big\} + \mathcal{O}\left(\sigma^2\right), \end{aligned} \tag{197}$$

$$\begin{aligned} \sigma_{y_1x_2x_3} &= g_{x_2}{}^{y_2}g_{x_3}{}^{y_3}\Big\{K_{y_3y_1y_2} \\ &\quad + \frac{1}{3}\Big[\nabla_{y_2}\left(K_{y_4y_3y_1} + K_{y_1y_3y_4}\right) \\ &\quad - \nabla_{y_4}\left(K_{y_2y_3y_1} + 3K_{y_3y_1y_2}\right) \\ &\quad - \nabla_{y_1}K_{y_2y_3y_4} + 3\pi_{y_2y_3y_4y_1}\Big]\sigma^{y_4}\Big\} + \mathcal{O}\left(\sigma^2\right). \end{aligned} \tag{198}$$

The Jacobi propagators are approximated as follows

$$H^{x_1}{}_{y_2} = g^{x_1}{}_{y_2} + K_{y_3 y_2}{}^{x_1}\sigma^{y_3} + \mathcal{O}\left(\sigma^2\right), \tag{199}$$

$$K^{x_1}{}_{y_2} = g^{x_1}{}_{y_2} + \left(K_{y_2}{}^{x_1}{}_{y_3} + K_{y_3 y_2}{}^{x_1}\right)\sigma^{y_3} + \mathcal{O}\left(\sigma^2\right), \tag{200}$$

which in turn allows for an expansion of the recurring term entering (164):

$$K^{x_1}{}_{y_2}u^{y_2} - H^{x_1}{}_{y_2}\frac{D\sigma^{y_2}}{dt} = g^{x_1}{}_{y'}\left(u^{y'} - \frac{D\sigma^{y'}}{dt} + T_{y_2 y_3}{}^{y'}u^{y_2}\sigma^{y_3}\right) + \mathcal{O}\left(\sigma^2\right). \tag{201}$$

Synchronous parametrization Before writing down the expanded version of the generalized deviation equation, we will simplify the latter by choosing a proper parametrization of the neighboring curves. The factors with the derivatives of the parameters t and $\tilde{t}$ appear in (164) due to the non-synchronous parametrization of the two curves. It is possible to make things simpler by introducing the synchronization of parametrization. Namely, we start by rewriting the velocity as

$$u^y = \frac{dY^y}{dt} = \frac{d\tilde{t}}{dt}\frac{dY^y}{d\tilde{t}}. \tag{202}$$

That is, we now parametrize the position on the first curve by the same variable $\tilde{t}$ that is used on the second curve. Accordingly, we denote

$$\underset{\sim}{u}^y = \frac{dY^y}{d\tilde{t}}. \tag{203}$$

By differentiation, we then derive

$$a^y = \frac{d^2\tilde{t}}{dt^2}\underset{\sim}{u}^y + \left(\frac{d\tilde{t}}{dt}\right)^2 \underset{\sim}{a}^y, \tag{204}$$

where

$$\underset{\sim}{a}^y = \frac{D}{d\tilde{t}}\underset{\sim}{u}^y = \frac{D^2 Y^y}{d\tilde{t}^2}. \tag{205}$$

Analogously, we derive for the derivative of the deviation vector

$$\frac{D^2\eta^y}{dt^2} = \frac{d^2\tilde{t}}{dt^2}\frac{D\eta^y}{d\tilde{t}} + \left(\frac{d\tilde{t}}{dt}\right)^2\frac{D^2\eta^y}{d\tilde{t}^2}. \tag{206}$$

Now everything is synchronous in the sense that both curves are parametrized by $\tilde{t}$.

As a result, the exact deviation Eq. (164) is recast into a simpler form

$$\frac{D^2}{d\tilde{t}^2}\eta^{y_1} = -\sigma^{y_1}{}_{y_2}\underset{\sim}{a}^{y_2} - \sigma^{y_1}{}_{x_2}\tilde{a}^{x_2} - \sigma^{y_1}{}_{y_2 y_3}\underset{\sim}{u}^{y_2}\underset{\sim}{u}^{y_3}$$
$$-2\sigma^{y_1}{}_{y_2 x_3}\underset{\sim}{u}^{y_2}\left(K^{x_3}{}_{y_4}\underset{\sim}{u}^{y_4} - H^{x_3}{}_{y_4}\frac{D\sigma^{y_4}}{d\tilde{t}}\right)$$
$$-\sigma^{y_1}{}_{x_2 x_3}\left(K^{x_2}{}_{y_4}\underset{\sim}{u}^{y_4} - H^{x_2}{}_{y_4}\frac{D\sigma^{y_4}}{d\tilde{t}}\right)$$
$$\times\left(K^{x_3}{}_{y_5}\underset{\sim}{u}^{y_5} - H^{x_3}{}_{y_5}\frac{D\sigma^{y_5}}{d\tilde{t}}\right). \tag{207}$$

Explicit expansion of the deviation equation Substituting the expansions (194)–(201) into (207), we obtain the final result

$$\frac{D^2}{d\tilde{t}^2}\eta^{y_1} = \tilde{a}^{y_1} - \underset{\sim}{a}^{y_1} + T_{y_2 y_3}{}^{y_1}\underset{\sim}{u}^{y_2}\frac{D\eta^{y_3}}{d\tilde{t}} - \left(\Delta^{y_1}{}_{y_2 y_3 y_4}\underset{\sim}{u}^{y_2}\underset{\sim}{u}^{y_3}\right.$$
$$\left. + K_{y_2 y_4}{}^{y_1}\underset{\sim}{a}^{y_2} - K_{y_4 y_2}{}^{y_1}\tilde{a}^{y_2}\right)\eta^{y_4} + \mathcal{O}\left(\sigma^2\right), \tag{208}$$

where we introduced the abbreviation

$$\Delta_{y_1 y_2 y_3 y_4} := 2\pi_{y_3 y_4 y_2 y_1} - \pi_{y_4 y_3 y_2 y_1} - \pi_{y_2 y_3 y_4 y_1} + T_{y' y_2 y_1}T_{y_4 y_3}{}^{y'}$$
$$-2\nabla_{y_2}K_{(y_1 y_3) y_4} + \nabla_{y_1}K_{y_2 y_3 y_4} - \nabla_{y_4}K_{y_2 y_3 y_1}. \tag{209}$$

It should be understood that the last expression is contracted with $\underset{\sim}{u}^{y_2}\underset{\sim}{u}^{y_3}$ and hence the symmetrization is naturally imposed on the indices $(y_2 y_3)$.

Equation (208) allows for the comparison of two general world lines in Riemann–Cartan spacetime, which are not necessarily geodetic or autoparallel. It therefore represents the generalization of the deviation equation derived in Eq. (38).

Riemannian case A great simplification is achieved in a Riemannian background, when

$$\overline{\Delta}_{y_1 y_2 y_3 y_4} = 2\overline{\pi}_{y_3 y_4 y_2 y_1} - \overline{\pi}_{y_4 y_3 y_2 y_1} - \overline{\pi}_{y_2 y_3 y_4 y_1}$$
$$= \overline{R}_{y_1 y_3 y_2 y_4}, \tag{210}$$

and (208) is reduced to

$$\frac{D^2}{d\tilde{t}^2}\eta^{y_1} \stackrel{\text{s}}{=} \tilde{a}^{y_1} - \underset{\sim}{a}^{y_1} - \overline{R}^{y_1}{}_{y_2 y_3 y_4}\underset{\sim}{u}^{y_2}\underset{\sim}{u}^{y_3}\eta^{y_4} + \mathcal{O}\left(\sigma^2\right). \tag{211}$$

Along geodesic curves, this equation is further reduced to the well-known geodesic deviation (Jacobi) equation.

Choice of coordinates In order to utilize the deviation equation for measurements or in a gravitational compass setup [2, 7, 8, 48], the occurring covariant total derivatives need to be rewritten and an appropriate coordinate choice needs to be made. The left-hand side of the deviation equation takes the form:

$$\frac{D^2\eta_a}{dt^2} = \dot{u}^b\nabla_b\eta_a + \overset{\circ\circ}{\eta}_a - 2u^b\Gamma_{ba}{}^d\,\overset{\circ}{\eta}_d - u^b u^c\Gamma_{cb}{}^d\partial_d\eta_a$$
$$-u^b u^c\eta_e\left(\partial_c\Gamma_{ba}{}^e - \Gamma_{cb}{}^d\Gamma_{da}{}^e - \Gamma_{ca}{}^d\Gamma_{bd}{}^e\right). \tag{212}$$

Here we used $\overset{\circ}{\eta}^a := d\eta^a/dt$ for the standard total derivative.

Observe that the first term on the right-hand side vanishes in the case of autoparallel curves ($\dot{u}^a := Du^a/dt = 0$). Also note the symmetrization of the connection imposed by the velocities in some terms.

Rewriting the connection in terms of the contortion and switching to normal coordinates [23, 24, 69–73] along the world line, which we assume to be an autoparallel, yields

$$\frac{D^2\eta_a}{dt^2} \overset{|_Y}{=} \overset{\circ\circ}{\eta}_a + 2u^b K_{ba}{}^d\,\overset{\circ}{\eta}_d + u^b u^c K_{cb}{}^d\partial_d\eta_a$$
$$+u^b u^c\eta_e\left(\partial_c K_{ba}{}^e - \frac{2}{3}\overline{R}_{c(ba)}{}^e + K_{cb}{}^d K_{da}{}^e - K_{ca}{}^d K_{bd}{}^e\right). \tag{213}$$

Note the appearance of a term containing the partial (not ordinary total) derivative of the deviation vector, in contrast to the Riemannian case.

The first term in the second line may be rewritten as an ordinary total derivative, i.e. $u^b u^c\eta_e\partial_c K_{ba}{}^e = u^b\eta_e\overset{\circ}{K}_{ba}{}^e$, but this is still inconvenient when recalling the compass equation, which will contain terms with covariant derivatives of the contortion.

5.3 *Operational Interpretation*

Some thoughts about the operational interpretation of the coordinate choice are in order. In particular, it should be stressed that we did not specify any *physical* theory in which the deviation Eq. (207) should be applied. Or, stated the other way round, the derived deviation equation is of completely *geometrical* nature, i.e. it describes the change of the deviation vector between points on two general curves in Riemann–Cartan spacetime. From the mathematical perspective, the choice of coordinates should be solely guided by the simplicity of the resulting equation. In this sense, our previous choice of normal coordinates appears to be appropriate. But what about the physical interpretation, or better, the operational realization of such coordinates?

Let us recall the coordinate choice in General Relativity in a Riemannian background. In this case normal coordinates also have a clear operational meaning, which is related to the motion of structureless test bodies in General Relativity. As is well known, such test bodies move along the geodesic equation. In other words, we could – at least in principle – identify a normal coordinate system by the local observation of test bodies. If other external forces are absent, normal coordinates will locally

– where "locally" refers to the observers laboratory on the reference world line – lead to straight line motion of test bodies. In this sense, there is a clear operational procedure for the realization of normal coordinates.

Here we are in a more general situation, since we have not yet specified which gravitational theory we are considering in the geometrical Riemann–Cartan background. The physical choice of a gravity theory will be crucial for the operational realization of the coordinates. Recall the form of the equations of motion for a very large class [12, 55] of gravitational theories, which also allow for additional internal degrees of freedom, in particular for spin. In this case the equations of motion are no longer given by the geodesic equation or, as it is sometimes erroneously postulated in the literature, by the autoparallel equation. In such theories, test bodies exhibit an additional spin-curvature coupling, which leads to non-geodesic motion, even locally.

In the context of gravitational theories beyond GR, one should therefore be aware of the fact, that for the experimental realization of the normal coordinates, one now has to make sure to use the correct equation of motion and, consequently, the correct type of test body. Taking the example of a theory with spin-curvature coupling, like Einstein–Cartan theory, this would eventually lead to the usage of test bodies with vanishing spin – since those still move on standard geodesics, and therefore lead to an identical procedure as in the general relativistic case, i.e. one adopts coordinates in which the motion of those test bodies becomes rectilinear.

5.4 Summary

This concludes or outlook and the generalization of the deviation equation to a Riemann–Cartan geometry. The generalization should serve as a foundation for the test of gravitational theories which make use of post-Riemannian geometrical structures. As we have discussed in detail, the operational usability of the Riemann–Cartan deviation equation differs from the one in a general relativistic context, which was also noticed quite early in [51]. In contrast to the Riemannian case, an algebraic realization of a gravitational compass [2, 7, 8, 48] on the basis of the deviation equation is out of the question due to the appearance of derivatives of the torsion even at the lowest orders. It remains to be shown which additional concepts and assumptions are needed in order to fully realize a gravitational compass in a Riemann–Cartan background.

Acknowledgements This work was supported by the Deutsche Forschungsgemeinschaft (DFG) through the grant PU 461/1-1 (D.P.). The work of Y.N.O. was partially supported by PIER ("Partnership for Innovation, Education and Research" between DESY and Universität Hamburg) and by the Russian Foundation for Basic Research (Grant No. 16-02-00844-A).

Appendix

A Directory of Symbols

Table 1 Directory of symbols

Symbol	Explanation
Geometrical quantities	
g_{ab}	Metric
$\sqrt{-g}$	Determinant of the metric
δ^a_b	Kronecker symbol
ε_{abcd}, $\varepsilon_{\alpha\beta\gamma}$	(4D, 3D) Levi-Civita symbol
s, τ	Proper time
x^a, y^a	Coordinates
$\lambda_b{}^{(\alpha)}$	(Fermi propagated) tetrad
$Y(s)$, $X(\tau)$	(Reference) world line
ξ^a	Constants in spatial Fermi coordinates
$\overline{\Gamma}_{ab}{}^c$, $\Gamma_{ab}{}^c$	(Levi-Civita) connection
$T_{ab}{}^c$, $K_{ab}{}^c$	Torsion, contortion
$^*\Gamma_{ab\ldots}{}^c$	Derivative of connection (normal coordinates)
$R_{abc}{}^d$, $C_{abc}{}^d$	Riemann, Weyl curvature
σ	World function
η^y	Deviation vector
$g^{y_0}{}_{x_0}$	Parallel propagator
$K^x{}_y$, $H^x{}_y$	Jacobi propagators
Misc	
u^a, a^b	4-velocity, 4-acceleration
v^α, ω^α, V^α	(Linear, rotational, combined) 3-velocity
b^α, η^α	Derivative of (linear, rotational) acceleration
Operators	
∂_i, ","	Partial derivative
∇_i, ";"	Covariant derivative
$\frac{D}{ds} =$ "$\cdot$"	Total covariant derivative
$\frac{d}{ds} =$ "$\circ$"	Total derivative
"$[\ldots]$"	Coincidence limit
"$\overline{}$"	Riemannian object

Table 2 Directory of symbols (continued)

Symbol	Explanation
Auxiliary quantities (Method 1)	
${}^{(m,n)}A_a$, ${}^{(m,n,p)}A_a$	Accelerations of compass constituents
$\alpha^{y_0}{}_{y_1\dots y_n}$, $\beta^{y_0}{}_{y_1\dots y_n}$, $\gamma^{y_0}{}_{y_1\dots y_n}$	Expansion coefficients
$c_{(m)a}$, $d_{(m)a}$	Constants
$\phi^{y_1}{}_{y_2\dots}$, $\lambda^{y_1}{}_{y_2\dots}$, $\mu^{y_1}{}_{y_2\dots}$,	Abbreviations
$\Delta_{i_1\dots}$, $\Xi_{i_1\dots}$	
Auxiliary quantities (Method 2)	
C	Frequency ratio
A, B, $K_{1,2,3}$	Abbreviations
Auxiliary quantities (Outlook)	
$A_{y_1\dots y_n}$	Expansion coefficient
$\pi_{y_1 y_2 y_3 y_4}$	Abbreviation

References

1. F.A.E. Pirani, On the physical significance of the Riemann tensor. Acta Phys. Pol. **15**, 389 (1956)
2. J.L. Synge, *Relativity: The General Theory* (North-Holland, Amsterdam, 1960)
3. E. Poisson, A. Pound, I. Vega, The motion of point particles in curved spacetime. Living Rev. Relativ. **14**(1), 7 (2011)
4. T. Levi-Civita, Sur l'écart géodésique. Math. Ann. **97**, 291 (1927)
5. J.L. Synge, The first and second variations of the length integral in Riemannian space. Proc. Lond. Math. Soc. **25**, 247 (1926)
6. J.L. Synge, On the geometry of dynamics. Phil. Trans. R. Soc. Lond. A **226**, 31 (1927)
7. D. Puetzfeld, Y.N. Obukhov, Generalized deviation equation and determination of the curvature in general relativity. Phys. Rev. D **93**, 044073 (2016)
8. P. Szekeres, The gravitational compass. J. Math. Phys. **6**, 1387 (1965)
9. D. Puetzfeld, Y.N. Obukhov, C. Lämmerzahl, Gravitational clock compass in general relativity. Phys. Rev. D **98**, 024032 (2018)
10. B.S. DeWitt, R.W. Brehme, Radiation damping in a gravitational field. Ann. Phys. (N.Y.) **9**, 220 (1960)
11. D. Puetzfeld, Y.N. Obukhov, Covariant equations of motion for test bodies in gravitational theories with general nonminimal coupling. Phys. Rev. D **87**, 044045 (2013)
12. D. Puetzfeld, Y.N. Obukhov, Equations of motion in metric-affine gravity: a covariant unified framework. Phys. Rev. D **90**, 084034 (2014)
13. A.C. Ottewill, B. Wardell, Transport equation approach to calculations of Hadamard Green functions and non-coincident DeWitt coefficients. Phys. Rev. D **84**, 104039 (2011)
14. D.E. Hodgkinson, A modified theory of geodesic deviation. Gen. Relativ. Gravit. **3**, 351 (1972)
15. S.L. Bażański, Kinematics of relative motion of test particles in general relativity. Ann. H. Poin. A **27**, 115 (1977)
16. A.N. Aleksandrov, K.A. Piragas, Geodesic structure: I. Relative dynamics of geodesics. Theor. Math. Phys. **38**, 48 (1978)
17. B. Schutz, On generalized equations of geodesic deviation, in *Galaxies, Axisymmetric Systems, and Relativity*, vol 17, ed. by M.A.H. MacCallum (Cambridge University Press, Cambridge, 1985), p. 237

18. C. Chicone, B. Mashhoon, The generalized Jacobi equation. Class. Quantum Gravity **19**, 4231 (2002)
19. T. Mullari, R. Tammelo, On the relativistic tidal effects in the second approximation. Class. Quantum Gravity **23**, 4047 (2006)
20. J. Vines, Geodesic deviation at higher orders via covariant bitensors. Gen. Relativ. Gravit. **47**, 59 (2015)
21. E. Fermi, Sopra i fenomeni che avvengono in vicinanza di una linea oraria. Atti. Accad. Naz. Lincei Cl, Sci. Fis. Mat. Nat. Rend. **31**, 21, 51, 101 (1922)
22. E. Fermi, *Collected Papers*, vol 1, ed. by E. Amaldi, E. Persico, F. Rasetti, E. Segrè (University of Chicago Press, Chicago, 1962)
23. O. Veblen, Normal coordinates for the geometry of paths. Proc. Natl. Acad. Sci. (USA) **8**, 192 (1922)
24. O. Veblen, T.Y. Thomas, The geometry of paths. Trans. Am. Math. Soc. **25**, 551 (1923)
25. J.L. Synge, A characteristic function in Riemannian space and its application to the solution of geodesic triangles. Proc. Lond. Math. Soc. **32**, 241 (1931)
26. A.G. Walker, Relative coordinates. Proc. R. Soc. Edinb. **52**, 345 (1932)
27. F.K. Manasse, C.W. Misner, Fermi normal coordinates and some basic concepts in differential geometry. J. Math. Phys. **4**, 735 (1963)
28. C.W. Misner, K.S. Thorne, J.A. Wheeler, *Gravitation* (Freeman, San Francisco, 1973)
29. W.-T. Ni, On the proper reference frame and local coordinates of an accelerated observer in special relativity. Chin. J. Phys. **15**, 51 (1977)
30. B. Mashhoon, Tidal radiation. Astrophys. J. **216**, 591 (1977)
31. W.-T. Ni, M. Zimmermann, Inertial and gravitational effects in the proper reference frame of an accelerated, rotating observer. Phys. Rev. D **17**, 1473 (1978)
32. W.-Q. Li, W.-T. Ni, On an accelerated observer with rotating tetrad in special relativity. Chin. J. Phys. **16**, 214 (1978)
33. W.-T. Ni, Geodesic triangles and expansion of the metrics in normal coordinates. Chin. J. Phys. **16**, 223 (1978)
34. W.-Q. Li, W.-T. Ni, Coupled inertial and gravitational effects in the proper reference frame of an accelerated, rotating observer. J. Math. Phys. **20**, 1473 (1979)
35. W.-Q. Li, W.-T. Ni, Expansions of the affinity, metric and geodesic equations in Fermi normal coordinates about a geodesic. J. Math. Phys. **20**, 1925 (1979)
36. N. Ashby, B. Bertotti, Relativistic effects in local inertial frames. Phys. Rev. D **34**, 2246 (1986)
37. A.M. Eisele, On the behaviour of an accelerated clock. Helv. Phys. Acta **60**, 1024 (1987)
38. T. Fukushima, The Fermi coordinate system in the post-Newtonian framework. Celest. Mech. **44**, 1024 (1988)
39. O. Semerák, Stationary frames in the Kerr field. Gen. Relativ. Gravit. **25**, 1041 (1993)
40. K.-P. Marzlin, Fermi coordinates for weak gravitational fields. Phys. Rev. D **50**, 888 (1994)
41. D. Bini, A. Geralico, R.T. Jantzen, Kerr metric, static observers and Fermi coordinates. J. Math. Phys. **22**, 4729 (2005)
42. C. Chicone, B. Mashhoon, Explicit Fermi coordinates and tidal dynamics in de Sitter and Gödel spacetime. Phys. Rev. D **74**, 064019 (2006)
43. D. Klein, P. Collas, General transformation formulas for Fermi-Walker coordinates. Class. Quantum Gravity **25**, 145019 (2008)
44. D. Klein, P. Collas, Exact Fermi coordinates for a class of space-times. J. Math. Phys. **51**, 022501 (2010)
45. P. Delva, M.-C. Angonin, Extended Fermi coordinates. Gen. Relativ. Gravit. **44**, 1 (2012)
46. S.G. Turyshev, O.L. Minazzoli, V.T. Toth, Accelerating relativistic reference frames in Minkowski space-time. J. Math. Phys. **53**, 032501 (2012)
47. F.W. Hehl, W.-T. Ni, Inertial effects of a Dirac particle. Phys. Rev. D **42**, 2045 (1990)
48. I. Ciufolini, M. Demianski, How to measure the curvature of space-time. Phys. Rev. D **34**, 1018 (1986)
49. I. Ciufolini, Generalized geodesic deviation equation. Phys. Rev. D **34**, 1014 (1986)

50. R. Rodrigo, V. Dehant, L. Gurvits, M. Kramer, R. Park, P. Wolf, J. Zarnecki (eds.), *High Performance Clocks with Special Emphasis on Geodesy and Geophysics and Applications to Other Bodies of the Solar System*, vol. 63, Space Sciences Series of ISSI (Springer, Netherlands, 2018)
51. F.W. Hehl, P. von der Heyde, G.D. Kerlick, J.M. Nester, General relativity with spin and torsion: foundations and prospects. Rev. Mod. Phys. **48**, 393 (1976)
52. M. Blagojević, F.W. Hehl, *Gauge Theories of Gravitation: A Reader with Commentaries* (Imperial College Press, London, 2013)
53. V.N. Ponomarev, A.O. Barvinsky, Y.N. Obukhov, *Gauge Approach and Quantization Methods in Gravity Theory* (Nauka, Moscow, 2017)
54. F.W. Hehl, Y.N. Obukhov, D. Puetzfeld, On Poincaré gauge theory of gravity, its equations of motion, and gravity probe B. Phys. Lett. A **377**, 1775 (2013)
55. Y.N. Obukhov, D. Puetzfeld, Multipolar test body equations of motion in generalized gravity theories, in *Equations of Motion in Relativistic Gravity*, vol. 179, Fundamental Theories of Physics, ed. by D. Puetzfeld, et al. (Springer, Cham, 2015), p. 67
56. D. Puetzfeld, Y.N. Obukhov, Deviation equation in Riemann-Cartan spacetime. Phys. Rev. D **97**, 104069 (2018)
57. A. Trautman, Einstein-Cartan theory, in *Encyclopedia of Mathematical Physics*, vol. 2, ed. by J.-P. Francoise, G.L. Naber, S.T. Tsou (Elsevier, Oxford, 2006), p. 189
58. Y.N. Obukhov, Poincaré gauge gravity: selected topics. Int. J. Geom. Methods Mod. Phys. **03**, 95 (2006)
59. Y.N. Obukhov, Poincaré gauge gravity: an overview. Int. J. Geom. Methods Mod. Phys. **15**, Supp. 1 (2018) 1840005
60. J.L. Synge, Geodesics in non-holonomic geometry. Math. Ann. **99**, 738 (1928)
61. W.H. Goldthorpe, Spectral geometry and $SO(4)$ gravity in a Riemann-Cartan spacetime. Nucl. Phys. B **170**, 307 (1980)
62. H.T. Nieh, M.L. Yan, Quantized Dirac field in curved Riemann-Cartan background: I. Symmetry properties, Green's function. Ann. Phys. (N.Y.) **138**, 237 (1982)
63. N.H. Barth, Heat kernel expansion coefficient: I. An extension. J. Phys. A Math. Gen. **20**, 857 (1987)
64. S. Yajima, Evaluation of the heat kernel in Riemann-Cartan space. Class. Quantum Gravity **13**, 2423 (1996)
65. S.S. Manoff, Auto-parallel equation as Euler-Lagrange's equation in spaces with affine connections and metrics. Gen. Relativ. Gravit. **32**, 1559 (2000)
66. S.S. Manoff, Deviation equations of Synge and Schild over spaces with affine connections and metrics. Int. J. Mod. Phys. A **16**, 1109 (2001)
67. B.Z. Iliev. Deviation equations in spaces with a transport along paths. JINR Commun. E2-94-40, Dubna, 1994 (2003)
68. R.J. van den Hoogen, Towards a covariant smoothing procedure for gravitational theories. J. Math. Phys. **58**, 122501 (2017)
69. T.Y. Thomas, *The Differential Invariants of Generalized Spaces* (Cambridge University Press, Cambridge, 1934)
70. J.A. Schouten, *Ricci-Calculus. An Introduction to Tensor Analysis and its Geometric Applications*, 2nd edn. (Springer, Berlin, 1954)
71. I.G. Avramidi, A covariant technique for the calculation of the one-loop effective action. Nucl. Phys. B **355**, 712 (1991)
72. I.G. Avramidi, Covariant methods for the calculation of the effective action in quantum field theory and investigation of higher-derivative quantum gravity. Ph.D. thesis, Moscow State University (1986), English version arXiv:hep-th/9510140
73. A.Z. Petrov, *Einstein Spaces* (Pergamon, Oxford, 1969)

A Snapshot of J. L. Synge

Peter A. Hogan

Abstract A brief description is given of the life and influence on relativity theory of Professor J. L. Synge accompanied by some technical examples to illustrate his style of work.

1 Introduction

When I was a postdoctoral fellow working with Professor Synge in the School of Theoretical Physics of the Dublin Institute for Advanced Studies he was fifty–one years older than me and he remained research active for another twenty years. John Lighton Synge FRS was born in Dublin on 23rd. March, 1897 and died in Dublin on 30th. March, 1995. As well as his emphasis on, and mastery of, the geometry of space–time he had a unique delivery, both verbal and written, which I will try to convey in the course of this short article. But first the basic facts of his academic life are as follows: He was educated in St. Andrew's College, Dublin and entered Trinity College, University of Dublin in 1915. He graduated B.A. (1919), M.A. (1922) and Sc.D. (1926). He was Assistant Professor of Mathematics in the University of Toronto (1920–1925), subsequently returning to Trinity College Dublin as Professor of Natural Philosophy (1925–1930) and then left for the University of Toronto again to take up the position of Professor of Applied Mathematics (1930–1943). From there he went to Ohio State University as chairman of the Mathematics Department (1943–1946) followed by Head, Mathematics Department at Carnegie Institute of Technology, Pittsburgh (1946–1948) before returning to Dublin to establish his school of relativity in the Dublin Institute for Advanced Studies. He officially retired when he was seventy–five years old.

Synge was prolific, publishing 250 papers and 11 books. In 1986 he wrote, but did not publish, some informal autobiographical notes [1], which he described as being for his family and descendants and to aid obituary writers, and which are deposited in

P. A. Hogan (✉)
School of Physics, University College Dublin, Belfield, Dublin 4, Ireland
e-mail: peter.hogan@ucd.ie

D. Puetzfeld and C. Lämmerzahl (eds.), *Relativistic Geodesy*,
Fundamental Theories of Physics 196, https://doi.org/10.1007/978-3-030-11500-5_4

the library of the School of Theoretical Physics of the Dublin Institute for Advanced Studies.

Early in his career he published his first important paper: "On the Geometry of Dynamics", *Phil. Trans. Roy. Soc.* A **226** (1926), 31–106. Of this work he said [1]; "I sent a copy to T. Levi-Cività, and in return he sent me a copy of a paper by him, just appearing. Our papers had in common the equation of geodesic deviation, now familiar to relativists, but he had done it using an indefinite line element, appropriate to relativity, whereas my line element was positive definite."

2 A Scheme of Approximation

Synge placed great emphasis on working things out for oneself, writing that [2] "the lust for calculation must be tempered by periods of inaction, in which the mechanism is completely unscrewed and then put together again. It is the decarbonisation of the mind." As an illustration of this activity I give a weak field approximation scheme published in 1970 by Synge [3] which has the advantage that it can be described without reference to an example. This is a topic which, by 1970, had become a standard entry in textbooks on general relativity and one might be forgiven for thinking that by then the last word had been said on it.

We first need some basic objects and notation. In Minkowskian space-time Synge liked to use imaginary time (which some people find maddening!) and to write the position 4–vector in rectangular Cartesians and time as

$$x_a = (x, y, z, it) \text{ with } a = 1, 2, 3, 4 \text{ and } i = \sqrt{-1},$$

with the index in the covariant or lower position. The Minkowskian metric tensor in these coordinates has components δ_{ab} (the Kronecker delta). If the metric tensor of a space–time has components of the form

$$g_{ab} = \delta_{ab} + \gamma_{ab} ,$$

then he defined the "truncated Einstein tensor" $\hat{G}^{ab}$ via

$$G^{ab} = L_{ab} + \hat{G}^{ab} ,$$

where G^{ab} is the Einstein tensor calculated with the metric g_{ab} and

$$L_{ab} = \frac{1}{2}(\gamma_{ab,cc} + \gamma_{cc,ab} - \gamma_{ac,cb} - \gamma_{bc,ca}) - \frac{1}{2}\delta_{ab}(\gamma_{cc,dd} - \gamma_{cd,cd}) .$$

The energy–momentum–stress tensor of matter giving rise to a gravitational field has components T^{ab}. With these preliminaries Synge's strategy is as follows:

(1) Given T^{ab}, generate a sequence of metrics $\underset{M}{g_{ab}} = \delta_{ab} + \underset{M}{\gamma_{ab}}$ $(M = 0, 1, 2, \ldots, N)$;
(2) Approximations are introduced by expressing the components T^{ab} in terms of a small parameter;
(3) Integrability conditions, equivalent to the equations of motion, are imposed to terminate the sequence at a term which satisfies Einstein's field equations with a predicted order of approximation in terms of the small parameter.

The sequence is constructed as follows: with

$$\underset{M}{\overset{*}{\gamma}_{ab}} = \underset{M}{\gamma_{ab}} - \frac{1}{2}\delta_{ab}\,\underset{M}{\gamma_{cc}} \quad (M = 0, 1, 2, \ldots\ N)\,,$$

and

$$\underset{M}{H^{ab}} = T^{ab} + (8\,\pi)^{-1}\underset{M}{\hat{G}^{ab}}(\underset{M}{\gamma}) \quad (M = 0, 1, 2, \ldots\ N)\,,$$

define the sequence $\{\underset{M}{\gamma_{ab}}\}$ by

$$\underset{0}{\gamma_{ab}} = 0 \quad \text{and} \quad \underset{M}{\overset{*}{\gamma}_{ab}} = 16\,\pi\, K^{ab}_{rs}\,\underset{M-1}{H^{rs}} \quad (M = 1, 2, 3, \ldots\ N)\,.$$

Here K^{ab}_{rs} is an operator defined by

$$K^{ab}_{rs} = -\delta_{ar}\,\delta_{bs}\,J + J(\delta_{ar}\,D_{bs} + \delta_{bs}\,D_{ar} - \delta_{ab}\,D_{rs})J\,,$$

with $D_a = \partial/\partial x_a$, $D_{ab} = \partial^2/\partial x_a \partial x_b$. The operator J is the inverse d'Alembertian:

$$Jf(\vec{x}, t) = -\frac{1}{4\pi}\int \frac{f(\vec{x}', t - |\vec{x} - \vec{x}'|)}{|\vec{x} - \vec{x}'|} d_3x'\,.$$

Synge proved that the integrals involved in the implementation of the operator K converge if the physical system is stationary ($T^{ab}{}_{,4} = 0$) for some period in the past. He called this property J–convergence.

Approximations are introduced as follows: all $\{\underset{M}{\gamma_{ab}}\}$ defined above satisfy the coordinate conditions

$$\underset{M}{\overset{*}{\gamma}_{ab,b}} = 0 \quad (M = 0, 1, 2, \ldots\ N)\,.$$

Introduce approximations by assuming $T^{ab} = O(k)$ for some dimensionless parameter k then

$$\underset{M}{\gamma_{ab}} - \underset{M-1}{\gamma_{ab}} = O(k^M) \quad (M = 1, 2, 3, \ldots, N)\,,$$

from the definition of γ_{ab}, and

$$\underset{M}{\hat{G}}^{ab} - \underset{M-1}{\hat{G}}^{ab} = O(k^{M+1})\,,$$

from the quadratic nature of $\hat{G}^{ab}$. To obtain a solution of Einstein's field equations in the Nth. approximation, terminate the sequence $\{\underset{M}{\gamma_{ab}}\}$ at the Nth. term by imposing the *Integrability Conditions/Equations of Motion* in the Nth. approximation:

$$\underset{N-1}{H}{}^{ab}{}_{,b} \equiv T^{ab}{}_{,b} + (8\,\pi)^{-1}\,\underset{N-1}{\hat{G}}{}^{ab}{}_{,b} = 0\,.$$

Now

$$\underset{N}{\gamma^*_{ab}} = -16\,\pi\,J\,\underset{N-1}{H}{}^{ab}\,,$$

and

$$\underset{N}{G}{}^{ab} + 8\,\pi\,T^{ab} = O(k^{N+1})\,,$$

showing that Einstein's field equations are approximately satisfied in this sense. This scheme was subsequently utilised for the study of equations of motion in general relativity [4–7].

In an amusing spin-off Synge [8] constructed the following divergence–free pseudo–tensor: first write the vanishing covariant divergence of the energy–momentum–stress tensor in the equivalent forms

$$T^{ab}{}_{|b} = 0 \;\Leftrightarrow\; T^{ab}{}_{,b} + K_a = 0\,.$$

Here $K_a = \Gamma^a_{cb}\,T^{cb} + \Gamma^b_{cb}\,T^{ac}$ is not a tensor (so the position of the index a is not significant; Γ^a_{bc} are the components of the Riemannian connection calculated with the metric tensor g_{ab}). Then define the pseudo–vector

$$Q_a = J\,K_a \;\Rightarrow\; \Box Q_a = K_a\,,$$

(with J the operator introduced above and $\Box$ the Minkowskian d'Alembertian operator) and define the pseudo–tensor

$$\varphi_{ab} = Q_{a,b} + Q_{b,a} - \delta_{ab}\,Q_{c,c}\,.$$

It thus follows that

$$\varphi_{ab,b} = \Box Q_a + Q_{b,ab} - Q_{c,ca} = \Box Q_a = K_a = -T^{ab}{}_{,b}\,.$$

Hence

$$\tau^{ab} = T^{ab} + \varphi_{ab} = \tau^{ba} ,$$

is *a pseudo–tensor with vanishing divergence* ($\tau^{ab}{}_{,b} = 0$). However Synge offered, in his characteristic style, these words of warning: "I refrain from attaching the words momentum and energy to this pseudo–tensor or to integrals formed from it, because I believe that we are barking up the wrong tree if we attach such important physical terms to mathematical constructs which lack the essential invariance property fundamental in general relativity."

3 Lorentz Transformations

Synge gave a succinct description of his early education when he wrote [1]: "Although there are great gaps in my scientific equipment - like Hadamard, I could never get my teeth into group theory - I think I have ranged more widely than most. I might easily have stuck to classical subjects in which I was well trained as an undergraduate (dynamics, hydrodynamics, elasticity), but I wanted to take part in the new subjects, and in due course I mastered relativity but not quantum theory." True to this background, when considering Lorentz transformations, Synge thought of the analogy with "the kinematics of a rigid body with a fixed point" (in [9]) and thus the construction of a general rotation in three dimensional Euclidean space in terms of the Euler angles. For Lorentz transformations the analogy requires six transformations of an orthonormal tetrad to another orthonormal tetrad, involving three pseudo angles (the arguments of hyperbolic functions) and three Euclidean angles. While this perspective is interesting the resulting formalism is not well suited to discussing the detailed effect of Lorentz transformations on the null cone. In the second edition of his text on special relativity Synge thanked I. Robinson and A. Taub "for pointing out an error in Chap. IV of this book as first published (1955): singular Lorentz transformations were overlooked."

Taub was using spinors but Robinson had encountered the singular case in a novel way [10–12]: Robinson was interested in the Schwarzschild solution in the limit $m \to +\infty$. Starting with the Eddington–Finkelstein form

$$ds^2 = -r^2 \frac{(dx^2 + dy^2)}{\left\{1 + \frac{1}{4}(x^2 + y^2)\right\}^2} + 2\,du\,dr + \left(1 - \frac{2\,m}{r}\right) du^2$$

and, using a clever coordinate transformation, Robinson wrote this in the form

$$ds^2 = -\frac{r^2}{\cosh^2 \lambda\xi}(d\xi^2 + d\eta^2) + 2\,du\,dr + \left(\lambda^2 - \frac{2}{r}\right) du^2 , \quad \lambda = m^{-1/3}$$

Taking the limit $\lambda \to 0$ ($\Leftrightarrow m \to +\infty$) this becomes

$$ds^2 = -r^2(d\xi^2 + d\eta^2) + 2\,du\,dr - \frac{2}{r}du^2$$

This is another (different from Schwarzschild) Robinson–Trautman [13] type D vacuum space–time. The metric tensor has one term singular at $r = 0$. This line element can be written in the form

$$ds^2 = -T^{4/3}(dX^2 + dY^2) - T^{-2/3}dZ^2 + dT^2$$

which is a Kasner [14] solution of Einstein's vacuum field equations. If we remove the term singular at $r = 0$ above we have a line element

$$ds^2 = -r^2(d\xi^2 + d\eta^2) + 2\,du\,dr$$

This is flat space–time and $r = 0$ is a null geodesic. Hence

$$\xi \to \xi + a\ ,\quad \eta \to \eta + b\ ,\quad u \to u\ ,\quad r \to r$$

where a, b are real constants, constitutes *a Lorentz transformation leaving only the null direction* $r = 0$ *invariant*. This is a singular Lorentz transformation (or null rotation) and the example moreover shows that such transformations exist and constitute a two–parameter Abelian subgroup of the Lorentz group.

4 Synge on an Observation of E. T. Whittaker

I mentioned at the outset that Professor Synge remained research active well into old age. To demonstrate this I want to give an example of some work carried out when he was eighty–eight years old. For several years, starting in the early 1980s, he and I found it convenient to correspond via letter. This allowed easy exchange of the results of calculations before the age of email. He typed his letters, including equations, on an ancient machine which he had used for years. The example I want to give involves an observation due to E. T. Whittaker and to do it justice I must first give a fairly extensive introduction.

Whittaker (in [15, 16]) was concerned with the Liénard–Wiechert electromagnetic field of a moving charge e so we will need some notation which we can briefly summarise as follows:

(1) Line element: $ds^2 = \eta_{ij}dX^i\,dX^j = -dX^2 - dY^2 - dZ^2 + dT^2$.
(2) World line of charge: $X^i = w^i(u)$; $v^i(u) = dw^i/du$ with $v^i\,v_i = +1$ ($\Rightarrow$ v^i = 4–velocity, u = arc length or proper time); $a^i = dv^i/du$ = 4–acceleration $\Rightarrow$ $a^i\,v_i = 0$)
(3) Retarded distance of X^i from $X^i = w^i(u)$:

$$r = \eta_{ij}(X^i - w^i(u))v^j \geq 0\,;\; \eta_{ij}(X^i - w^i(u))(X^j - w^j(u)) = 0\,.$$

Let $X^i - w^i(u) = r\,k^i$ then $k^i\,k_i = 0$ and $k^i\,v_i = +1$. Parametrise the *direction* of k^i by x, y such that

$$k^i = P_0^{-1}\left(-x, -y, -1 + \frac{1}{4}(x^2 + y^2), 1 + \frac{1}{4}(x^2 + y^2)\right),$$

and then the normalisation $k^i\,v_i = +1$ implies

$$P_0 = x\,v^1(u) + y\,v^2(u) + \left\{1 - \frac{1}{4}(x^2 + y^2)\right\} v^3(u) + \left\{1 + \frac{1}{4}(x^2 + y^2)\right\} v^4(u)\,.$$

Whittaker observed that the Liénard–Wiechert 4–potential

$$A^i = \frac{e\,v^i}{r} \quad \Rightarrow \quad A^i{}_{,i} = 0 = \Box A^i\,,$$

could be written, modulo a gauge transformation, in the form

$$A^i = \frac{e\,v^i}{r} = K^{ij}\,F_{,j} + {}^*K^{ij}\,G_{,j}\,,$$

where $K^{ij} = -K^{ji}$ is a constant real bivector, with ${}^*K_{ij} = \frac{1}{2}\epsilon_{ijkl}\,K^{kl}$ its dual, and F, G are real–valued functions each satisfying the Minkowskian wave equation

$$\Box F = 0 \text{ and } \Box G = 0\,.$$

To establish this in coordinates x, y, r, u we need

$$\frac{\partial}{\partial X^i} = -\frac{P_0^2}{r}\left(\frac{\partial k_i}{\partial x}\frac{\partial}{\partial x} + \frac{\partial k_i}{\partial y}\frac{\partial}{\partial y}\right) + v_i\frac{\partial}{\partial r} + k_i\left\{\frac{\partial}{\partial u} - (1 - r\,a_i\,k^i)\frac{\partial}{\partial r}\right\},$$

and

$$\Box = -\frac{P_0^2}{r^2}\left(\frac{\partial^2}{\partial x^2} + \frac{\partial^2}{\partial y^2}\right) - (1 - 2\,a_i\,k^i\,r)\left(\frac{\partial^2}{\partial r^2} + \frac{2}{r}\frac{\partial}{\partial r}\right) + 2\,\frac{\partial^2}{\partial u\partial r} + \frac{2}{r}\frac{\partial}{\partial u}\,.$$

In coordinates x, y, r, u Whittaker's two wave functions are [17]

$$F = -\frac{e}{2}\log(x^2 + y^2) \text{ and } G = -e\,\tan^{-1}\frac{y}{x}\,,$$

(two *harmonic* functions) and thus

$$\frac{\partial F}{\partial X^i} = \frac{e\,P_0^2}{r\,(x^2+y^2)}\left(x\,\frac{\partial k_i}{\partial x} + y\,\frac{\partial k_i}{\partial y}\right) ,$$

and

$$\frac{\partial G}{\partial X^i} = \frac{e\,P_0^2}{r\,(x^2+y^2)}\left(x\,\frac{\partial k_i}{\partial y} - y\,\frac{\partial k_i}{\partial x}\right) .$$

Define

$$K^{ij} = \delta^i_3\,\delta^j_4 - \delta^i_4\,\delta^j_3 \quad \text{and} \quad L^{ij} = \delta^i_1\,\delta^j_2 - \delta^i_2\,\delta^j_1 = {}^*K^{ij} ,$$

then

$$A^i = K^{ij}\,F_{,j} + {}^*K^{ij}\,G_{,j} = \frac{e\,v^i}{r} + \eta^{ij}\,\Psi_{,j} ,$$

with

$$\Psi = e\,\log\{r\,P_0^{-1}\sqrt{x^2+y^2}\} .$$

Whittaker pointed out that this decomposition is analogous to the splitting of a plane light wave into two plane polarised components. A notable fact is that *almost every vacuum Maxwell field can be resolved into two parts in this way*. The presentation of Whittaker's observation in coordinates x, y, r, u facilitates the derivation of the explicit decomposition (see [17]) for the Goldberg–Kerr electromagnetic field [18]. The second, and final, part of the introduction, to enable us to appreciate Synge's contribution, involves a simple proof of this decomposition of a vacuum Maxwell field in general.

We are working in Minkowskian space–time and we shall write the line element as given above in rectangular Cartesian coordinates and time $X^i = (X, Y, Z, T)$ with $i = 1, 2, 3, 4$. In addition we shall make use of the following basis vector fields:

$$k^i\frac{\partial}{\partial X^i} = \frac{\partial}{\partial Z} + \frac{\partial}{\partial T} ,\quad l^i\frac{\partial}{\partial X^i} = -\frac{\partial}{\partial Z} + \frac{\partial}{\partial T} ,$$
$$m^i\frac{\partial}{\partial X^i} = \frac{\partial}{\partial X} + i\frac{\partial}{\partial Y} ,\quad \bar{m}^i\frac{\partial}{\partial X^i} = \frac{\partial}{\partial X} - i\frac{\partial}{\partial Y} .$$

All scalar products (with respect to the Minkowskian metric) of the pairs of these vectors vanish except $k^i\,l_i = +2$ and $m^i\,\bar{m}_i = -2$. In what follows a complex self–dual bivector satisfies: $A_{ij} = -A_{ji}$ and ${}^*A_{ij} = i\,A_{ij}$ and a complex anti–self–dual bivector satisfies: $B_{ij} = -B_{ji}$ and ${}^*B_{ij} = -i\,B_{ij}$, with the star denoting the Hodge dual. A basis of complex *anti–self–dual* bivectors is given by

$$m_{ij} = m_i\,k_j - m_j\,k_i\ ,\ n_{ij} = \bar{m}_i\,l_j - \bar{m}_j\,l_i\ ,$$

and

$$l_{ij} = m_i\,\bar{m}_j - \bar{m}_i\,m_j + l_i\,k_j - l_j\,k_i\;.$$

Let $F_{ij} = -F_{ji}$ be a candidate for a real Maxwell bivector. Since $F_{ij} + i^* F_{ij}$ is an anti–self–dual complex bivector it can be expanded on the basis above as

$$F_{ij} + i^* F_{ij} = \phi_0\,n_{ij} + \phi_1\,l_{ij} + \phi_2\,m_{ij}\;,$$

where ϕ_0, ϕ_1, ϕ_2 are complex–valued functions of X^i. Maxwell's Equations

$$(F^{ij} + i^* F^{ij})_{,j} = 0\;,$$

imply integrability conditions for the existence of a complex–valued function $Q(X^i)$ such that:

(a) Q is a wave function: $\Box Q = 0 \;\Leftrightarrow\; \bar{m}^i\,m^j\,Q_{,ij} = k^i\,l^j\,Q_{,ij}$;
(b) $\phi_0 = \frac{1}{4}k^i\,m^j\,Q_{,ij}\;,\;\phi_1 = \frac{1}{4}k^i\,l^j\,Q_{,ij}\;,\;\phi_2 = -\frac{1}{4}l^i\,\bar{m}^j\,Q_{,ij}\;.$

Let $\bar{l}_{ij}$ denote the complex conjugate of l_{ij}, then $\bar{l}_{ij}$ is *self–dual*. Define

$$W_{ij} = \frac{1}{4}\bar{l}_i{}^p\,Q_{,pj} - \frac{1}{4}\bar{l}_j{}^p\,Q_{,pi} = -W_{ji}\;.$$

Since Q is a wave function it follows that W_{ij} is *anti–self–dual*. Expressing W_{ij} on the anti–self–dual bivector basis, and using (b) above, results in

$$W_{ij} = F_{ij} + i^* F_{ij}\;.$$

Hence with $\frac{1}{4}\bar{l}_{ij} = K_{ij} - i^* K_{ij}$ and $Q = U + iV$, we can write

$$F_{ij} = A_{i,j} - A_{j,i} \quad \text{with} \quad A_i = K_i{}^j\,U_{,j} + {}^*K_i{}^j\,V_{,j}\;.$$

Thus in general an analytic solution of Maxwell's vacuum field equations on Minkowskian space–time can be constructed from a pair of real wave functions U, V and a constant real bivector $K_{ij} = -K_{ji}$. The classic paper on this type of result for zero rest mass, spin s fields is that of Penrose [19] (see also Stewart [20]).

When I wrote out this proof (incorporated into [21]) and sent it to Synge his response was characteristic. He worked it all out for himself and sent me the following proof in December, 1985:

Synge's proof begins with

Lemma: With $X^i = (X, Y, Z, T), \eta_{ij} = \text{diag}(-1, -1, -1, +1), F_{ij} = -F_{ji}$ a Maxwell field so that $F^{ij}{}_{,j} = 0\;;\; F_{j,k} + F_{ki,j} + F_{jk,i} = 0$ then

$$F_{ij} = 0 \text{ at } T = 0 \;\Rightarrow\; F_{ij} = 0 \text{ for all } T\;.$$

"You cannot make energy out of nothing" (Synge).

Corollary: If F_{ij} and H_{ij} are Maxwell fields then

$$F_{ij} = H_{ij} \text{ at } T = 0 \Rightarrow F_{ij} = H_{ij} \text{ for all } T .$$

With these preliminaries Synge stated the following:

Theorem: Given a Maxwell field F_{ij} and

$$H_{ij} = K_i{}^l\, U_{,lj} + {}^*K_i{}^l\, V_{,lj} - K_j{}^l\, U_{,li} - {}^*K_j{}^l\, V_{,li} ,$$

with $K_{ij} = -K_{ji}$ = constants and U, V wave functions, then H_{ij} is a Maxwell field and there exists K_{ij}, U, V such that

$$H_{ij} = F_{ij} \text{ at } T = 0 .$$

Comment: Clearly H_{ij} is a solution of Maxwell's equations. The choice of K_{ij}, U, V is not unique. The theorem demands only their existence.

Proof Choose $K^{ij} = \delta^i_3\,\delta^j_4 - \delta^i_3\,\delta^j_4$ then ${}^*K^{ij} = \delta^i_1\,\delta^j_2 - \delta^i_2\,\delta^j_1$ and writing out $H_{ij} = F_{ij}$ at $T = 0$ we find the following pairs of equations for the Cauchy data U, V, $U_{,4}$, $V_{,4}$ for the wave functions at $T = 0$: (all equations evaluated at $T = 0$)
(A): $(U_{,4})_{,1} = F_{13} + V_{,23}$ and $(U_{,4})_{,2} = F_{23} - V_{,13}$;
(B): $(V_{,4})_{,1} = F_{24} - U_{,23}$ and $(V_{,4})_{,2} = -F_{14} + U_{,13}$;
(C): $U_{,11} + U_{,22} = -F_{34}$ and $V_{,11} + V_{,22} = -F_{12}$.

If the equations (A) are consistent and if the equations (B) are consistent then (A), (B) and (C) can in principle be solved for the Cauchy data. The consistency follows from the assumption that F_{ij} is a Maxwell field since then (A) implies that

$$(U_{,4})_{,12} - (U_{,4})_{,21} = F_{13,2} - F_{23,1} + V_{,232} + V_{,131} = F_{13,2} + F_{32,1} + F_{21,3} = 0 ,$$

and (B) implies that

$$(V_{,4})_{,12} - (V_{,4})_{,21} = F_{24,2} + F_{14,1} - U_{,232} - U_{,131} = F_{24,2} + F_{14,1} + F_{34,3} = 0 ,$$

and the theorem is established.

5 Epilogue

When visitors came to the Center for Relativity in the University of Texas at Austin, Alfred Schild, the founder of the Center and one of Synge's former collaborators [22] would enthusiastically point out to them that this was where Roy Kerr found his solution. This raises the question: what were the stand–out works produced in Professor Synge's school of relativity in Dublin? I discussed this with George Ellis

Fig. 1 J. L. Synge
1897–1995

some time ago and we concluded that Felix Pirani's study of the physical significance of the Riemann tensor [23] and Werner Israel's proof of the uniqueness of the static black hole (uncharged [24] and charged [25]) are *arguably* the most profound products of Synge's school.

When Synge turned ninety years of age a small conference was organised in his honour. His status within Ireland was reflected in the report in a national newspaper which stated: "President Hillery [Head of State] attended a special event in the Dublin Institute for Advanced Studies yesterday to wish a happy 90th. birthday to Professor Emeritus J. L. Synge, Ireland's most distinguished mathematician of the present century. Although he has been retired for fifteen years, the professor, a nephew of the playwright J. M. Synge, published three papers last year and has two more at present in the course of publication" [*Irish Times*, 23rd. March, 1987].

My photograph of Professor Synge (Fig. 1) was taken in July, 1987 in my back garden. Also present were two of Synge's former students, Dermott Mc Crea (see [4, 5, 7] for example) and Stephen O'Brien (of the O'Brien–Synge junction conditions [26]) together with Bill Bonnor who was visiting from the University of London.

References

1. J.L. Synge, Autobiography. Dublin Institute for Advanced Studies, unpublished (1986)
2. J.L. Synge, *Relativity: The General Theory* (North-Holland Publishing Company, Amsterdam, 1966)
3. J.L. Synge, Proc. R. Ir. Acad. **A59**, 11 (1970)
4. P.A. Hogan, J.D. McCrea, GRG J. **5**, 77 (1974)
5. J.D. McCrea, G.M. O'Brien, GRG J. **9**, 1101 (1977)
6. G.M. O'Brien, GRG J. **10**, 129 (1979)
7. J.D. McCrea, GRG J. **13**, 397 (1981)

8. J.L. Synge, Nature **215**, 102 (1967)
9. J.L. Synge, *Relativity: The Special Theory* (North-Holland Publishing Company, Amsterdam, 1965)
10. Private communication to the author from I. Robinson
11. W. Rindler, A. Trautman, *Gravitation and Geometry, Bibliopolis, Naples* (1987), p. 13
12. C. Barrabès, P.A. Hogan, *Advanced General Relativity: Gravity Waves, Spinning Particles, and Black Holes* (Oxford University Press, Oxford, 2013)
13. I. Robinson, A. Trautman, Proc. R. Soc. **A265**, 463 (1962)
14. E. Kasner, Trans. Am. Math. Soc. **27**, 155 (1925)
15. E.T. Whittaker, Proc. Lond. Math. Soc. **1**, 367 (1903)
16. E.T. Whittaker, *A History of the Theories of Aether and Electricity* (Nelson, London, 1958), p. 410
17. G.F.R. Ellis, P.A. Hogan, Ann. Phys. (N.Y.) **210**, 178 (1991)
18. J.N. Goldberg, R.P. Kerr, J. Math. Phys. **5**, 172 (1964)
19. R. Penrose, Proc. R. Soc. **A284**, 159 (1965)
20. J.M. Stewart, Proc. R. Soc. **A367**, 527 (1979)
21. P.A. Hogan, J. Math. Phys. **28**, 2087 (1987)
22. J.L. Synge, A. Schild, *Tensor Calculus* (University of Toronto Press, Toronto, 1949)
23. F.A.E. Pirani, Acta Phys. Pol. **15**, 389 (1956)
24. W. Israel, Phys. Rev. **164**, 1776 (1967)
25. W. Israel, Commun. Math. Phys. **8**, 245 (1968)
26. J.L. Synge, S. O'Brien. Commun. Dubl. Inst. Adv. Stud. **A9** (1952)

General Relativistic Gravity Gradiometry

Bahram Mashhoon

Abstract Gravity gradiometry within the framework of the general theory of relativity involves the measurement of the elements of the relativistic tidal matrix, which is theoretically obtained via the projection of the spacetime curvature tensor upon the nonrotating orthonormal tetrad frame of a geodesic observer. The behavior of the measured components of the curvature tensor under Lorentz boosts is briefly described in connection with the existence of certain special tidal directions. Relativistic gravity gradiometry in the exterior gravitational field of a rotating mass is discussed and a gravitomagnetic beat effect along an inclined spherical geodesic orbit is elucidated.

1 Newtonian Gravity Gradiometry

Consider a distribution of matter of density $\rho(t, \mathbf{x})$ and the corresponding Newtonian gravitational potential $\Phi(t, \mathbf{x})$ in an inertial frame of reference. In a source-free region of space, we imagine two nearby test masses m_a and m_b that fall freely in the potential Φ along trajectories $\mathbf{x}_a(t)$ and $\mathbf{x}_b(t)$, respectively. Choosing one of these as the reference trajectory, we are interested in the *relative* motion of these test particles. With $\mathbf{x}_b(t)$ as the fiducial path, let us define $\boldsymbol{\xi}(t) = \mathbf{x}_a(t) - \mathbf{x}_b(t)$. Newton's second law of motion implies that the instantaneous deviation vector $\boldsymbol{\xi}(t)$ between the neighboring paths satisfies the tidal equation

$$\frac{d^2\xi^i}{dt^2} + \kappa^i{}_j\,\xi^j + O(|\boldsymbol{\xi}|^2) = 0\,, \tag{1}$$

B. Mashhoon (✉)
Department of Physics and Astronomy, University of Missouri,
Columbia, MO 65211, USA
e-mail: MashhoonB@missouri.edu

B. Mashhoon
School of Astronomy, Institute for Research in Fundamental Sciences (IPM),
P. O. Box 19395-5531, Tehran, Iran

D. Puetzfeld and C. Lämmerzahl (eds.), *Relativistic Geodesy*,
Fundamental Theories of Physics 196, https://doi.org/10.1007/978-3-030-11500-5_5

where $-\kappa^i{}_j\, \xi^j$ is the first-order tidal acceleration and

$$\kappa_{ij}(t, \mathbf{x}) = \frac{\partial^2 \Phi}{\partial x^i \partial x^j} \tag{2}$$

is the symmetric tidal matrix evaluated along the reference trajectory. In the Newtonian theory of gravitation, gravity gradiometry involves the measurement of $\kappa_{ij}(t, \mathbf{x})$, which is the gradient of the acceleration of gravity and can be determined, in principle, by means of Eq. (1).

The tidal matrix in Eq. (2) is independent of the test masses m_a and m_b as a consequence of the principle of equivalence. The principle of equivalence of gravitational and inertial masses ensures the universality of the gravitational interaction. The modern history of the science of gravity gradiometry can be traced back to the pioneering efforts of L. Eötvös, who employed a torsion-balance method to test the principle of equivalence (1889–1922).

In Eq. (2), Poisson's equation for Φ, $\nabla^2\Phi = 4\pi G\, \rho$, reduces to Laplace's equation, $\nabla^2\Phi = 0$, in the source-free region under consideration. In this case, $\nabla^2\kappa_{ij} = 0$ and hence each element of the Newtonian tidal matrix is a *harmonic* function. Moreover, tidal matrix (2) is *traceless*; therefore, the shape of a tidally deformed test body would generally tend to either a cigar-like or a pancake-like configuration when tides are dominant, since the symmetric and traceless tidal matrix can in general have either two positive and one negative or one positive and two negative eigenvalues, respectively.

In recent years, gravity gradiometers of high sensitivity have been developed; indeed, the Paik gravity gradiometer employs superconducting quantum interference devices [1–3]. Furthermore, gravity gradients can now be measured via atom interferometry as well [4, 5]. Gravity gradiometry has many important practical applications. The magnitude of a gravity gradient is usually expressed in units of Eötvös, $1\,\mathrm{E} = 10^{-9}\,\mathrm{s}^{-2}$.

To extend the treatment of gravity gradiometry to the relativistic domain, it is necessary to introduce the quasi-inertial Fermi normal coordinate system that can provide a physically meaningful interpretation of the measurement of relative motion within the framework of general relativity (GR). In GR, masses m_a and m_b follow geodesics and a hypothetical observer comoving with the fiducial test mass m_b would set up in the neighborhood of the reference trajectory a laboratory where the motion of m_a could be monitored. Such a quasi-inertial frame is represented by the Fermi normal coordinate system [6–8]. In our treatment of Fermi coordinates in the next section, we employ an extended framework [9–12], since in practice nongravitational accelerations and rotations may be present.

2 Fermi Coordinates

To develop the relativistic analogs of Eqs. (1) and (2), we consider a congruence of future-directed timelike paths representing the world lines of test masses in a gravitational field. Next, we choose a reference path in the congruence and establish a local quasi-inertial Fermi system of geodesic coordinates in its neighborhood. This is necessary in order to provide a physically meaningful interpretation of the measurement of relative motion from the standpoint of the observer comoving with the reference test mass along the fiducial world line $\bar{x}^{\mu}(\tau)$. The observer has proper time τ and carries an orthonormal tetrad frame $\lambda^{\mu}{}_{\hat{\alpha}}(\tau)$ along $\bar{x}^{\mu}$; that is, $g_{\mu\nu}\,\lambda^{\mu}{}_{\hat{\alpha}}\,\lambda^{\nu}{}_{\hat{\beta}} = \eta_{\hat{\alpha}\hat{\beta}}$, where $g_{\mu\nu}$ is the spacetime metric and $\eta_{\hat{\alpha}\hat{\beta}}$ is the Minkowski metric given by diag$(-1, 1, 1, 1)$ in our convention. Here, $\lambda^{\mu}{}_{\hat{0}}(\tau) = d\bar{x}^{\mu}/d\tau$ is the observer's temporal axis and its local frame is carried along its path according to

$$\frac{D\lambda^{\mu}{}_{\hat{\alpha}}}{d\tau} = \phi_{\hat{\alpha}}{}^{\hat{\beta}}\,\lambda^{\mu}{}_{\hat{\beta}}\,, \tag{3}$$

where $\phi_{\hat{\alpha}\hat{\beta}}$ is the observer's antisymmetric acceleration tensor. Greek indices run from 0 to 3, while Latin indices run from 1 to 3. The signature of the spacetime metric is +2 and units are chosen such that $c = G = 1$, unless specified otherwise.

In close analogy with the electromagnetic field tensor, we can decompose the acceleration tensor into its "electric" and "magnetic" components, namely, $\phi_{\hat{\alpha}\hat{\beta}} \to (-\mathbf{A}, \mathbf{\Omega})$, where $\mathbf{A}(\tau)$ is a spacetime scalar that represents the translational acceleration of the fiducial observer and $\mathbf{\Omega}(\tau)$ is a spacetime scalar that represents its rotational acceleration. More precisely, the reference observer in general follows an accelerated world line with

$$\frac{d^2\bar{x}^{\mu}}{d\tau^2} + \Gamma^{\mu}_{\nu\sigma}\,\frac{d\bar{x}^{\nu}}{d\tau}\,\frac{d\bar{x}^{\sigma}}{d\tau} = \mathcal{A}^{\mu}\,, \tag{4}$$

where

$$\mathcal{A}^{\mu} = A^{\hat{i}}\,\lambda^{\mu}{}_{\hat{i}} \tag{5}$$

and $\mathbf{\Omega}$ is the angular velocity of the rotation of the observer's spatial frame with respect to a locally nonrotating (i.e. Fermi–Walker transported) frame.

At each event $\bar{x}^{\mu}(\tau)$ along the reference world line, we imagine all spacelike geodesic curves that start out from this event and are normal to the reference world line. These generate a local hypersurface. Let x^{μ} be an event on this hypersurface sufficiently close to the reference world line such that there is a *unique* spacelike geodesic of proper length σ that connects $\bar{x}^{\mu}(\tau)$ to x^{μ}. We define ξ^{μ} to be a unit spacelike vector that is tangent to the unique spacelike geodesic at $\bar{x}^{\mu}(\tau)$, so that $\xi_{\mu}(\tau)\,\lambda^{\mu}{}_{\hat{0}}(\tau) = 0$. Then, to event x^{μ} one assigns Fermi coordinates $X^{\hat{\mu}}$, where

$$X^{\hat{0}} := \tau\,, \qquad X^{\hat{i}} := \sigma\,\xi^{\mu}(\tau)\,\lambda_{\mu}{}^{\hat{i}}(\tau)\,. \tag{6}$$

The reference observer has Fermi coordinates $X^{\hat{\mu}} = (\tau, 0, 0, 0)$ and is thus permanently fixed at the spatial origin of the Fermi coordinate system. Henceforth, we find it convenient to express Fermi coordinates as $X^{\hat{\mu}} = (T, \mathbf{X})$, where $|\mathbf{X}| = \sigma$. When $\sigma \neq 0$, $X^{\hat{i}}/\sigma = \xi^{\mu}(\tau)\,\lambda_{\mu}{}^{\hat{i}}(\tau)$, for $i = 1, 2, 3$, are the corresponding direction cosines at proper time τ along $\bar{x}^{\mu}$.

The Fermi coordinate system is admissible in a cylindrical domain along $\bar{x}^{\mu}$ of radius $|\mathbf{X}| \sim \mathfrak{R}$, where $\mathfrak{R}$ is a certain minimal radius of curvature of spacetime along the reference world line.

The spacetime metric in Fermi coordinates is given by

$$ds^2 = g_{\hat{\mu}\hat{\nu}}(T, \mathbf{X})\, dX^{\hat{\mu}}\, dX^{\hat{\nu}}\,, \tag{7}$$

where

$$g_{\hat{0}\hat{0}} = -P^2 + Q^2 - R_{\hat{0}\hat{i}\hat{0}\hat{j}}\, X^{\hat{i}}\, X^{\hat{j}} + O(|\mathbf{X}|^3)\,, \tag{8}$$

$$g_{\hat{0}\hat{i}} = Q_{\hat{i}} - \frac{2}{3}\, R_{\hat{0}\hat{j}\hat{i}\hat{k}}\, X^{\hat{j}}\, X^{\hat{k}} + O(|\mathbf{X}|^3)\,, \tag{9}$$

$$g_{\hat{i}\hat{j}} = \delta_{\hat{i}\hat{j}} - \frac{1}{3}\, R_{\hat{i}\hat{k}\hat{j}\hat{l}}\, X^{\hat{k}}\, X^{\hat{l}} + O(|\mathbf{X}|^3)\,. \tag{10}$$

Here, P and $\mathbf{Q}$,

$$P := 1 + \mathbf{A}(T) \cdot \mathbf{X}\,, \qquad \mathbf{Q} := \boldsymbol{\Omega}(T) \times \mathbf{X}\,, \tag{11}$$

are related to the local translational and rotational accelerations of the reference observer, respectively, and

$$R_{\hat{\alpha}\hat{\beta}\hat{\gamma}\hat{\delta}}(T) := R_{\mu\nu\rho\sigma}\, \lambda^{\mu}{}_{\hat{\alpha}}\, \lambda^{\nu}{}_{\hat{\beta}}\, \lambda^{\rho}{}_{\hat{\gamma}}\, \lambda^{\sigma}{}_{\hat{\delta}} \tag{12}$$

is the projection of the Riemann curvature tensor along $\bar{x}^{\mu}$ on the tetrad frame of the reference observer.

Fermi coordinates are invariantly defined and can have advantages over other physically motivated coordinate systems such as radar coordinates [13]; therefore, they have been applied in many different contexts. For instance, Fermi coordinates have been employed to elucidate dynamics of astrophysical jets [14–18].

3 Relativistic Gravity Gradiometry

In Einstein's GR, gravity gradiometry involves the measurement of the gravitational field, which is represented by the Riemannian curvature of spacetime. When an observer measures a gravitational field, the curvature tensor must be projected onto the tetrad frame of the observer.

It is now straightforward to express the equation of motion of any other test mass in the Fermi coordinate system and study the motion of the test mass relative to the fiducial test mass that follows world line $\bar{x}^\mu$. This general framework is necessary in practice, since the motion of the reference test mass may involve translational and rotational accelerations of nongravitational origin. These are absent, however, in the ideal case of purely tidal relative motion. To illustrate this ideal situation, let us assume that $\phi_{\hat{\alpha}\hat{\beta}} = 0$, so that the reference path $\bar{x}^\mu$ is a timelike geodesic and the orthonormal tetrad frame is parallel transported along the fiducial geodesic world line, i.e. $D\lambda^\mu{}_{\hat{\alpha}}/d\tau = 0$. The geodesic equation of motion of a free test particle in the corresponding Fermi coordinates relative to the reference test mass that is fixed at the spatial origin of Fermi coordinates can be expressed in terms of relative separation $\mathbf{X}$ as

$$\frac{d^2X^{\hat{i}}}{dT^2} + R_{\hat{0}\hat{i}\hat{0}\hat{j}}X^{\hat{j}} + 2\,R_{\hat{i}\hat{k}\hat{j}\hat{0}}V^{\hat{k}}X^{\hat{j}} + \frac{2}{3}\left(3R_{\hat{0}\hat{k}\hat{j}\hat{0}}V^{\hat{i}}V^{\hat{k}} + R_{\hat{i}\hat{k}\hat{j}\hat{l}}V^{\hat{k}}V^{\hat{l}} + R_{\hat{0}\hat{k}\hat{j}\hat{l}}V^{\hat{i}}V^{\hat{k}}V^{\hat{l}}\right)X^{\hat{j}} + O(|\mathbf{X}|^2) = 0\,. \tag{13}$$

This geodesic deviation equation is a generalized Jacobi equation [10] in which the rate of geodesic separation (i.e. the relative velocity of the test particle) $\mathbf{V} = d\mathbf{X}/dT$ is in general arbitrary; however, $|\mathbf{V}| < 1$ at $\mathbf{X} = 0$. It is clear from Eq. (13) that *all* of the curvature components in Eq. (12) can be measured from a careful study of the motion of the test masses in the congruence relative to the fiducial observer. Neglecting terms in the relative velocity $\mathbf{V}$, Eq. (13) reduces to the Jacobi equation,

$$\frac{d^2X^{\hat{i}}}{dT^2} + \mathcal{K}^{\hat{i}}{}_{\hat{j}}X^{\hat{j}} + O(|\mathbf{X}|^2) = 0\,, \tag{14}$$

which is the relativistic analog of the Newtonian tidal equation given by Eq. (1), and

$$\mathcal{K}_{\hat{i}\hat{j}} = R_{\hat{0}\hat{i}\hat{0}\hat{j}}\,. \tag{15}$$

This symmetric relativistic tidal matrix is traceless in Ricci-flat regions of spacetime and reduces in the nonrelativistic limit to the Newtonian tidal matrix (2).

The relativistic tidal matrix is thus determined by the projection of the Riemann curvature tensor upon the parallel-transported tetrad frame of the fiducial geodesic observer. The local spatial frame of the fiducial observer is unique up to a *constant* spatial rotation corresponding to the choice of the initial orthonormal triad in Eq. (3). The freedom in the choice of the initial local triad implies that the form of the tidal matrix is unique up to a constant spatial rotation.

The Jacobi equation can be used to study the influence of a gravitational field on the relative motion of nearby test masses in general relativity [19–22]. Einstein's field equations locally relate the energy-momentum tensor of matter to the Ricci tensor.

At any event in spacetime, the Riemann curvature tensor can be decomposed into a matter part and a part that is independent of matter; that is,

$$R_{\mu\nu\rho\sigma} = C_{\mu\nu\rho\sigma} + g_{\mu[\rho}\, R_{\sigma]\nu} - g_{\nu[\rho}\, R_{\sigma]\mu} - \frac{1}{6}\,(g_{\mu\rho}\, g_{\nu\sigma} - g_{\mu\sigma}\, g_{\nu\rho})\, R\,, \tag{16}$$

where $C_{\mu\nu\rho\sigma}$ is the traceless Weyl curvature tensor that represents the "free" gravitational field. At any point on the manifold, the Riemann tensor has in general 20 independent components, whereas the Ricci tensor has 10 independent components. Beyond any point on the spacetime manifold, the two parts of the curvature tensor are connected to each other via the Bianchi identity $R_{\mu\nu[\rho\sigma;\delta]} = 0$. Introducing decomposition (16) into the Jacobi equation and employing a canonical null tetrad frame, Szekeres has shown via the Petrov classification that the behavior of the free part of the gravitational field can be described in terms of the superposition of a transverse wave component, a longitudinal component and a Coulomb component [20]. The matter part has been treated in [22]. Some of the basic astrophysical applications of Eq. (14) have been studied in [9, 23, 24].

The Gravity Probe B ("GP-B") experiment has recently measured the exterior gravitomagnetic field of the Earth [25]. The gravitomagnetic field of a rotating mass contributes to the spacetime curvature and can thus influence the relative tidal motion of nearby test masses. In 1980, Braginsky and Polnarev [26] proposed an experiment to measure such an effect in a space platform in orbit around the Earth, since they claimed that such an approach could circumvent many of the difficulties associated with the GP-B experiment. However, in 1982, Mashhoon and Theiss [27, 28] showed that to measure the relativistic rotation-dependent tidal acceleration in a space platform, the local gyroscopes that would fix the local spatial frame carried by the space platform must satisfy the same performance criteria as in the GP-B experiment.

The achievements of the GP-B could possibly be integrated with Paik's superconducting gravity gradiometer [29] in future space experiments in order to measure the tidal influence of the gravitomagnetic field using an orbiting platform [30, 31]. We will consider the prediction of GR for the nature of the tidal matrix in such experiments in Sect. 5.

4 Special Tidal Directions

Let us return to the main focus of relativistic gravity gradiometry, namely, the determination of the Riemann curvature tensor projected on the tetrad frame of the fiducial observer as in Eq. (12). Taking advantage of the symmetries of the Riemann tensor, this quantity can be represented by a 6×6 matrix $\mathcal{R} = (\mathcal{R}_{IJ})$, where the indices I and J range over the set $(01, 02, 03, 23, 31, 12)$. Thus we can write

$$\mathcal{R} = \begin{bmatrix} \mathcal{E} & \mathcal{B} \\ \mathcal{B}^{\dagger} & \mathcal{S} \end{bmatrix}, \tag{17}$$

where $\mathcal{E}$ and $\mathcal{S}$ are symmetric 3×3 matrices and $\mathcal{B}$ is traceless. The tidal matrix $\mathcal{E}$ represents the "electric" components of the curvature tensor as measured by the fiducial observer, whereas $\mathcal{B}$ and $\mathcal{S}$ represent its "magnetic" and "spatial" components, respectively. Imagine next an observer that is boosted with speed β in a given direction with respect to the fiducial observer at the same event in spacetime. Let $\mathcal{R}'$ be the Riemann curvature tensor as measured by the boosted observer. It turns out that under the boost the elements of $\mathcal{E}$, $\mathcal{B}$ and $\mathcal{S}$ in the direction parallel to the direction of the boost are *not* affected, whereas those perpendicular to the direction of the boost are enhanced by γ^2, where $\gamma = (1 - \beta^2)^{-1/2}$ is the Lorentz factor; moreover, the mixed elements are enhanced by a factor of γ. This circumstance is reminiscent of the behavior of the electromagnetic field under a boost: The components of the electric field (E) and magnetic field (B) parallel to the direction of the boost remain the same as before, while those perpendicular to the direction of the boost are enhanced by a factor of γ.

In this way the strength of the gravitational field can be augmented by a factor of γ^2; alternatively, one can say that the radius of curvature of spacetime measured by the boosted observer is Lorentz contracted [32, 33]. In Ricci-flat regions of spacetime, Eq. (17) simplifies, since $\mathcal{S} = -\mathcal{E}$, $\mathcal{E}$ is traceless and $\mathcal{B}$ is symmetric. Hence, the Weyl curvature tensor with 10 independent components is completely determined by its "electric" and "magnetic" components that are symmetric and traceless 3×3 matrices.

These results imply that a gravity gradiometer would in general measure extremely strong tidal forces when it moves very fast ($\beta \to 1$). However, along certain exceptional directions in space, such as the radial direction in the exterior Schwarzschild spacetime, tidal forces remain finite as $\beta \to 1$ [32, 33]. Along such a *special tidal direction*, the corresponding world line of the boosted observer approaches a null direction in the local null cone as $\beta \to 1$. In this way, special tidal directions are associated with certain tidally *nondestructive null directions* in spacetime. The significance of these null directions can be further elucidated via the invariant Petrov classification of gravitational fields.

The Petrov classification involves the Weyl curvature tensor and provides an invariant characterization of a gravitational field. This can be accomplished, for instance, in terms of the *principal null directions* of the Weyl tensor. A vector k, $k_\alpha k^\alpha = 0$, which satisfies the condition

$$k_{[\alpha} C_{\mu]\nu\rho[\sigma} k_{\beta]} k^\nu k^\rho = 0 \tag{18}$$

is a principal null direction of the Weyl tensor. In a gravitational field, at least one and at most four such null vectors exist at each event in spacetime [34, 35].

The basic mathematical connection between the special tidal directions and the principal null directions of the Weyl tensor has been established by Beem and Parker [36] and Hall and Hossack [37]. It turns out that in general a nondestructive null direction at a point p in spacetime is a principal null direction of the Weyl tensor at p; moreover, it is a *repeated* principal null direction of the Weyl tensor at p

if and only if it is a Ricci eigendirection at p. A vector N^μ is a Ricci eigendirection at p if

$$R_{\mu\nu} N^\nu = \sigma N_\mu \tag{19}$$

for a real number σ at p. This means that in a Ricci-flat spacetime, or more generally when

$$R_{\mu\nu} = \Lambda g_{\mu\nu} \tag{20}$$

for a real number Λ, a special tidal direction at p corresponds to a *repeated principal null direction of the Weyl tensor* at p. A vector N^μ is a repeated principal null vector of the Weyl tensor at p if $N_\alpha N^\alpha = 0$ and

$$C_{\mu\nu\rho\sigma} N^\nu N^\sigma = \lambda N_\mu N_\rho \tag{21}$$

at p for a real number λ. There are at least zero and at most two such directions at each event in Ricci-flat spacetimes.

Let us assume that the Weyl tensor vanishes at p, then a nondestructive null direction at p exists if and only if it is a Ricci eigendirection at p. In this case, one can have $0, 1, 2$ or ∞ nondestructive null directions at p [37]. For instance, there are no special tidal directions in any of the standard Friedmann–Lemaître–Robertson–Walker cosmological models. However, every direction is a special tidal direction in a spacetime of constant nonzero curvature, namely, de Sitter (or anti-de Sitter) universe.

The behavior of the measured components of the Riemann curvature tensor under boosts along special tidal directions can be determined based on the results given in Ref. [33]. Let us consider, in particular, the Kerr gravitational field, which is of type D in the Petrov classification. The Weyl tensor at each point in this spacetime has two repeated principal null directions; therefore, there are two special tidal directions at each event. For example, along the axis of rotation, the outgoing and ingoing radial directions are the special tidal directions, see Ref. [14] for an extended treatment. In general, along the special tidal directions in Kerr spacetime, the curvature remains invariant under boosts ($\mathcal{R}' = \mathcal{R}$); in fact, the "electric" and "magnetic" components of the curvature can be made "parallel" such that the super-Poynting vector

$$\mathcal{P}_{\hat{i}} = -\epsilon_{\hat{i}\hat{j}\hat{k}} (\mathcal{E}\mathcal{B})_{\hat{j}\hat{k}} \tag{22}$$

vanishes. An analogous situation is encountered in the case of the electromagnetic field in an inertial frame in Minkowski spacetime. If the electromagnetic field is not *null*, so that the invariants $E^2 - B^2$ and $\mathbf{E} \cdot \mathbf{B}$ do not simultaneously vanish, then a boost with velocity v along the Poynting vector, i.e.

$$\frac{\mathbf{v}}{1 + v^2} = \frac{\mathbf{E} \times \mathbf{B}}{E^2 + B^2}, \tag{23}$$

renders the electric and magnetic fields parallel in the boosted frame. In the new inertial frame, the Poynting vector vanishes and any boost along the common direction of the fields leaves them invariant. The analogy between the electromagnetic field and algebraically special gravitational fields of types D and N has been treated in Ref. [33].

5 Tidal Matrix Around a Rotating Mass

To get some idea regarding the form of the relativistic tidal matrix, it is instructive to consider first the tidal field along stable circular orbits in the equatorial plane of the Kerr spacetime. The exterior Kerr metric can be expressed as [38]

$$ds^2 = -dt^2 + \frac{\Sigma}{\Delta} dr^2 + \Sigma\, d\theta^2 + (r^2 + a^2)\sin^2\theta\, d\varphi^2 + \frac{2Mr}{\Sigma}(dt - a\sin^2\theta\, d\varphi)^2\,, \tag{24}$$

where M is the mass of the gravitational source, $a = J/M$ is the specific angular momentum of the source, (t, r, θ, φ) are the standard Boyer–Lindquist coordinates and

$$\Sigma = r^2 + a^2\cos^2\theta\,, \qquad \Delta = r^2 - 2Mr + a^2\,. \tag{25}$$

The Kerr metric contains dimensionless gravitoelectric and gravitomagnetic potentials $\mathcal{U} = GM/(c^2 r)$ and $\mathcal{V} = GJ/(c^3 r^2)$, which correspond to the mass and angular momentum of the source, respectively. For instance, in the case of the Earth, we have $\mathcal{U}_\oplus \approx 6 \times 10^{-10}$ and $\mathcal{V}_\oplus \approx 4 \times 10^{-16}$.

We are interested in the tidal matrix along the circular equatorial trajectory of a fiducial test mass that follows a future-directed timelike geodesic world line about the Kerr source. The circular orbit has a fixed radial coordinate r_0 and orbital frequency [38]

$$\frac{d\varphi}{d\tau} = \frac{\omega_0}{(1 - 3\frac{M}{r_0} + 2a\omega_0)^{1/2}}\,, \tag{26}$$

where the Keplerian frequency ω_0 is given by

$$\omega_0^2 = \frac{M}{r_0^3}\,. \tag{27}$$

The circular geodesic orbit is such that at proper time $\tau = 0$, the azimuthal coordinate vanishes (i.e. $\varphi = 0$). Moreover, at this event, the initial directions of the orthonormal triad $\lambda^\mu{}_{\hat{i}}, i = 1, 2, 3$, point along the spherical polar coordinate directions. The spatial triad then undergoes parallel propagation along the circular orbit. The resulting radial and tangential components of the spatial frame, namely, $\lambda^\mu{}_{\hat{1}}$ and $\lambda^\mu{}_{\hat{3}}$, respectively, turn out to be periodic in τ with period $2\pi/\omega_0$. The difference between the orbital frequency (26) and the Keplerian frequency ω_0 leads to a combination of prograde geodetic and retrograde gravitomagnetic precessions of these frame components with respect to static inertial observers at spatial infinity in the asymptotically flat Kerr spacetime [39].

The tidal matrix is obtained as a certain symmetric and traceless projection of the Riemann curvature tensor evaluated along the orbit. The nonzero components of the tidal matrix consist of constant terms proportional to ω_0^2 plus terms that are periodic in τ with frequency $2\,\omega_0$ and can be expressed as [39]

$$\mathcal{K}_{\hat{1}\hat{1}} = \omega_0^2\,[1 - 3\gamma_0^2\,\cos^2(\omega_0\tau)]\,, \tag{28}$$

$$\mathcal{K}_{\hat{1}\hat{3}} = \mathcal{K}_{\hat{3}\hat{1}} = -\frac{3}{2}\,\omega_0^2\gamma_0^2\,\sin(2\,\omega_0\tau)\,, \tag{29}$$

$$\mathcal{K}_{\hat{2}\hat{2}} = \omega_0^2\,(3\gamma_0^2 - 2)\,, \tag{30}$$

$$\mathcal{K}_{\hat{3}\hat{3}} = \omega_0^2\,[1 - 3\gamma_0^2\,\sin^2(\omega_0\tau)]\,, \tag{31}$$

where γ_0 is given by

$$\gamma_0 = \left(\frac{r_0^2 - 2Mr_0 + a^2}{r_0^2 - 3Mr_0 + 2r_0^2 a\omega_0}\right)^{1/2}\,. \tag{32}$$

More generally, the tidal matrix for arbitrary timelike geodesics of Kerr spacetime has been calculated by Marck [40].

Let us next consider the tidal field along a tilted spherical orbit of fixed radial coordinate r_0 about a *slowly* rotating spherical mass. The exterior gravitational field is represented by the Kerr metric linearized in the angular momentum parameter a or, equivalently, the Schwarzschild metric plus the Thirring–Lense term. The symmetric and traceless tidal matrix can be obtained from [39]

$$\begin{aligned}
\mathcal{K}_{\hat{1}\hat{1}} &= \omega_0^2\,[1 - 3\,\Gamma^2\,\cos^2(\omega_0\tau)]\,,\\
\mathcal{K}_{\hat{1}\hat{2}} &= \mathcal{K}_{\hat{2}\hat{1}} = \omega_0^2\,\Xi\,\cos(\omega_0\tau)\,,\\
\mathcal{K}_{\hat{1}\hat{3}} &= \mathcal{K}_{\hat{3}\hat{1}} = -\frac{3}{2}\,\omega_0^2\,\Gamma^2\,\sin(2\,\omega_0\tau)\,,\\
\mathcal{K}_{\hat{2}\hat{2}} &= \omega_0^2\,(3\,\Gamma^2 - 2)\,,\\
\mathcal{K}_{\hat{2}\hat{3}} &= \mathcal{K}_{\hat{3}\hat{2}} = \omega_0^2\,\Xi\,\sin(\omega_0\tau)\,,\\
\mathcal{K}_{\hat{3}\hat{3}} &= \omega_0^2\,[1 - 3\,\Gamma^2\,\sin^2(\omega_0\tau)]\,,
\end{aligned} \tag{33}$$

where Γ and Ξ are given by

$$\Gamma := \left(\frac{1 - 2\frac{M}{r_0}}{1 - 3\frac{M}{r_0}}\right)^{1/2}\left(1 - \frac{a\,\omega_0\,\cos\alpha}{1 - 3\frac{M}{r_0}}\right) \tag{34}$$

and

$$\Xi := -3\,\frac{J}{M\,r_0^2\,\omega_0}\,\frac{\left(1-2\frac{M}{r_0}\right)^{1/2}(1+2\frac{M}{r_0})}{1-3\frac{M}{r_0}}\,\sin\alpha\,\sin\eta\,. \tag{35}$$

Here, the angle α denotes the inclination of the orbit with respect to the equatorial plane and η,

$$\eta := \omega\tau + \eta_0\,, \qquad \omega = \frac{\omega_0}{(1-3\frac{M}{r_0})^{1/2}}\,, \tag{36}$$

is the angular position of the reference test mass in the orbital plane measured from the line of the ascending node and η_0 is a constant angle. For $\alpha = 0$, the spherical orbit under consideration turns into the circular equatorial orbit, $\Xi = 0$, Γ reduces at the linear order in a to γ_0 and the tidal matrix agrees to first order in a with our previous results for the equatorial circular orbit in Kerr spacetime.

6 Beat Effect

The off-diagonal terms $\mathcal{K}_{\hat{1}\hat{2}} = \mathcal{K}_{\hat{2}\hat{1}}$ and $\mathcal{K}_{\hat{2}\hat{3}} = \mathcal{K}_{\hat{3}\hat{2}}$ in the tidal matrix (33) represent the *beat phenomenon* first pointed out in Ref. [27]. The beat effect involves frequencies ω and ω_0 with a beat frequency $\omega_F := \omega - \omega_0$. This is the frequency of the gravitoelectric (geodetic) Fokker precession of an ideal test gyro following a circular orbit about a spherical mass M. The tidal terms under consideration here that involve Ξ have dominant amplitudes that are proportional to the angular momentum J and are independent of the speed of light c.

In the work of Mashhoon and Theiss [27, 41–44], the resonance effect involving ω and ω_0 appeared in the calculation of the parallel-transported frame along the tilted spherical orbit about a rotating mass. It resulted in a small divisor phenomenon involving ω_F. For a near-Earth orbit, the Fokker period $2\pi/\omega_F$ is about 10^5 years; therefore, in practice the Mashhoon-Theiss effect shows up as a secular term in the corresponding off-diagonal elements of the tidal matrix with amplitude

$$\frac{9}{2}\,\frac{GJ}{c^2\,r_0^3}\,\omega_0^2\,\tau\,\sin\alpha\,, \tag{37}$$

which is consistent with the first post-Newtonian gravitomagnetic precession of the spatial frame [44, 45]. The possibility of measuring the Mashhoon-Theiss effect via neutron interferometry [46] has been discussed by Anandan [47].

In the first post-Newtonian approximation, the motion of an ideal test gyro of spin $\mathbf{S}$ in orbit about a central rotating mass can be written as [25]

$$\frac{d\mathbf{S}}{d\tau} = (\boldsymbol{\omega}_{ge} + \boldsymbol{\omega}_{gm}) \times \mathbf{S}\,, \tag{38}$$

where

$$\omega_{ge} = \frac{3}{2} \frac{GM}{c^2 r^3} \boldsymbol{\ell} , \qquad \omega_{gm} = \frac{G}{c^2 r^5}[3 (\mathbf{J} \cdot \mathbf{x}) \mathbf{x} - \mathbf{J} r^2] , \tag{39}$$

$|\mathbf{x}| = r$ and $\boldsymbol{\ell} = \mathbf{x} \times \mathbf{v}$ is the specific angular momentum of the gyro orbit. Here, $\boldsymbol{\omega}_{ge}$ is the (gravitoelectric) *geodetic* precession frequency of the gyroscope, while $\boldsymbol{\omega}_{gm}$ is its *gravitomagnetic* precession frequency. For the Earth, these precession frequencies have been directly measured via GP-B [25], which involved four superconducting gyroscopes and a telescope that were launched on 20 April 2004 into a polar Earth orbit of radius 642 km aboard a drag-free satellite.

During a satellite gradiometry experiment over a period of time τ, we expect that the spatial frame of the gradiometer would accumulate geodetic and gravitomagnetic precession angles that are of order $GM\, \omega_0 \tau/(c^2\, r_0)$ and $GJ\, \tau/(c^2\, r_0^3)$, respectively. From the comparison of these angles with Eq. (37), it is clear that only the post-Newtonian gravitomagnetic secular term survives in the calculation of the projection of the Riemann tensor onto the tetrad frame of the gradiometer for the case of the tilted spherical orbit. A recent detailed discussion of the beat effect is contained in Ref. [39], which should be consulted for a more complete treatment of relativistic gravity gradiometry in Kerr spacetime.

The results presented in the last two sections may be considered surprising and contrary to expectations. That is, it may appear on the basis of Eq. (15) that the main post-Newtonian terms in $(\mathcal{K}_{\hat{i}\hat{j}})$ can be obtained intuitively by combining Newtonian tides with the post-Newtonian motion of the spatial frame of the fiducial observer. However, in practice the projection of the Riemann tensor onto the frame of the fiducial observer involves detailed calculations in which the symmetries of the Riemann tensor need to be carefully taken into account.

Finally, the results presented here can be used to find the main relativistic effects in the motion of the Moon. Consider the nearly circular orbit of the Earth-Moon system about the Sun. In the Fermi normal coordinate system established along this orbit, the solar tidal acceleration $-\mathcal{K}_{\hat{i}\hat{j}}\, X^{\hat{j}}$ is a small perturbation on the dynamics of the Earth-Moon system. Here $\mathcal{K}_{\hat{i}\hat{j}}$ is essentially given by Eq. (33) and we recall that the ecliptic has a small inclination of $\alpha \approx 0.1$ with respect to the equatorial plane of the Sun. In this way, the main relativistic tidal effects in the motion of the Moon relative to the Earth caused by the gravitational field of the Sun have been determined [48–50].

7 Post-Schwarzschild Approximation

The exterior vacuum field of a spherically symmetry mass can be uniquely described by the Schwarzschild spacetime. Small deviations from spherical symmetry can then be treated in the post-Schwarzschild approximation scheme. This method was employed by Mashhoon and Theiss in their investigation of the relativistic tidal matrix for a gradiometer in orbit about a rotating mass [27, 41–44]. Thus in the first post-Schwarzschild approximation, the angular momentum of the central body is considered to first order while its mass is taken into account to all orders. The post-Schwarzschild method can be extended to include the quadrupole and higher moments of the central body. Indeed, the effect of oblateness, treated as a first-order static deformation of the source, has been investigated by Dietmar Theiss for a gravity gradiometer on a circular geodesic orbit of small inclination about a central oblate body [51].

8 Discussion

Gravity gradiometry in GR involves the measurement of a certain projection of the Riemannian curvature tensor of spacetime upon the orthonormal tetrad frame of an observer. In a satellite gravity gradiometry experiment in Earth orbit, the mass $M_\oplus$, angular momentum $J_\oplus$, quadrupole moment $Q_\oplus$ and higher moments of the Earth will all contribute to the result of the experiment. For an inclined spherical geodesic orbit about a slowly rotating mass, Eq. (33) gives the relativistic tidal matrix to all orders in the mass of the source M and to linear order in its angular momentum J. The result contains the beat phenomenon first pointed out by Mashhoon and Theiss [27, 41–44]. The corresponding influence of the quadrupole moment of the source has been studied by Theiss [51].

References

1. H.J. Paik, Superconducting tensor gravity gradiometer for satellite geodesy and inertial navigation. J. Astronaut. Sci. **29**, 1 (1981)
2. H.A. Chan, M.V. Moody, H.J. Paik, Null test of the gravitational inverse square law. Phys. Rev. Lett. **49**, 1745 (1982)
3. H.A. Chan, M.V. Moody, H.J. Paik, Superconducting gravity gradiometer for sensitive gravity measurements. II. Experiment. Phys. Rev. D **35**, 3572 (1987)
4. M.J. Snadden, J.M. McGuirk, P. Bouyer, K.G. Haritos, M.A. Kasevich, Measurement of the Earth's gravity gradient with an atom interferometer-based gravity gradiometer. Phys. Rev. Lett. **81**, 971 (1998)
5. J.M. McGuirk, G.T. Foster, J.B. Fixler, M.J. Snadden, M.A. Kasevich, Sensitive absolute-gravity gradiometry using atom interferometry. Phys. Rev. A **65**, 033608 (2002)
6. T. Levi-Civita, Sur l'écart géodésique. Math. Ann. **97**, 291 (1926)
7. J.L. Synge, On the geometry of dynamics. Phil. Trans. R. Soc. Lond. A **226**, 31–106 (1926)
8. J.L. Synge, *Relativity: The General Theory* (North-Holland, Amsterdam, 1960)
9. B. Mashhoon, Tidal radiation. Astrophys. J. **216**, 591–609 (1977)
10. C. Chicone, B. Mashhoon, The generalized Jacobi equation. Class. Quantum Gravity **19**, 4231–4248 (2002)
11. C. Chicone, B. Mashhoon, Ultrarelativistic motion: Inertial and tidal effects in Fermi coordinates. Class. Quantum Gravity **22**, 195–205 (2005)
12. C. Chicone, B. Mashhoon, Explicit Fermi coordinates and tidal dynamics in de Sitter and Gödel spacetimes. Phys. Rev. D **74**, 064019 (2006)

13. D. Bini, L. Lusanna, B. Mashhoon, Limitations of radar coordinates. Int. J. Mod. Phys. D **14**, 1413–1429 (2005)
14. B. Mashhoon, J.C. McClune, Relativistic tidal impulse. Mon. Not. R. Astron. Soc. **262**, 881–888 (1993)
15. C. Chicone, B. Mashhoon, B. Punsly, Relativistic motion of spinning particles in a gravitational field. Phys. Lett. A **343**, 1–7 (2005)
16. C. Chicone, B. Mashhoon, Tidal acceleration of ultrarelativistic particles. Astron. Astrophys. **437**, L39–L42 (2005)
17. C. Chicone, B. Mashhoon, B. Punsly, Dynamics of relativistic flows. Int. J. Mod. Phys. D **13**, 945–959 (2004)
18. D. Bini, C. Chicone, B. Mashhoon, Relativistic tidal acceleration of astrophysical jets. Phys. Rev. D **95**, 104029 (2017)
19. F.A.E. Pirani, On the physical significance of the Riemann tensor. Acta Phys. Pol. **15**, 389–405 (1956)
20. P. Szekeres, The gravitational compass. J. Math. Phys. (N.Y.) **6**, 1387 (1965)
21. J. Podolský, R. Švarc, Interpreting spacetimes of any dimension using geodesic deviation. Phys. Rev. D **85**, 044057 (2012)
22. R.F. Crade, G.S. Hall, The deviation of timelike geodesics in space-time. Phys. Lett. A **85**, 313–315 (1981)
23. B. Mashhoon, Tidal gravitational radiation. Astrophys. J. **185**, 83–86 (1973)
24. B. Mashhoon, On tidal phenomena in a strong gravitational field. Astrophys. J. **197**, 705–716 (1975)
25. C.W.F. Everitt et al., Gravity probe B: final results of a space experiment to test general relativity. Phys. Rev. Lett. **106**, 221101 (2011)
26. V.B. Braginsky, A.G. Polnarev, Relativistic spin-quadrupole gravitational effect. JETP Lett. **31**, 415–418 (1980)
27. B. Mashhoon, D.S. Theiss, Relativistic tidal forces and the possibility of measuring them. Phys. Rev. Lett. **49**, 1542 (1982)
28. B. Mashhoon, D.S. Theiss, Erratum: relativistic tidal forces and the possibility of measuring them. Phys. Rev. Lett. **49**, 1960 (1982)
29. H.J. Paik, Detection of the gravitomagnetic field using an orbiting superconducting gravity gradiometer: principles and experimental considerations. Gen. Relativ. Gravit. **40**, 907–919 (2008)
30. H.J. Paik, B. Mashhoon, C.M. Will, Detection of the gravitomagnetic field using an orbiting superconducting gravity gradiometer, in *Experimental Gravitational Physics*, ed. by P.F. Michelson, Hu. En-ke, G. Pizzella (World Scientific, Singapore, 1988), pp. 229–244
31. B. Mashhoon, H.J. Paik, C.M. Will, Detection of the gravitomagnetic field using an orbiting superconducting gravity gradiometer. Theoretical principles. Phys. Rev. D **39**, 2825 (1989)
32. B. Mashhoon, Wave propagation in a gravitational field. Phys. Lett. A **122**, 299–304 (1987)
33. B. Mashhoon, On the strength of a gravitational field. Phys. Lett. A **163**, 7–14 (1992)
34. H. Stephani, D. Kramer, M. MacCallum, C. Hoenselaers, E. Herlt, *Exact Solutions of Einstein's Field Equations*, 2nd edn. (Cambridge University Press, Cambridge, 2003)
35. J.B. Griffiths, J. Podolský, *Exact Space-Times in Einstein's General Relativity* (Cambridge University Press, Cambridge, 2009)
36. J.K. Beem, P.E. Parker, Sectional curvature and tidal accelerations. J. Math. Phys. (N.Y.) **31**, 819–827 (1990)
37. G.S. Hall, A.D. Hossack, Some remarks on sectional curvature and tidal accelerations. J. Math. Phys. (N.Y.) **34**, 5897–5899 (1993)
38. S. Chandrasekhar, *The Mathematical Theory of Black Holes* (Clarendon, Oxford, 1983)
39. D. Bini, B. Mashhoon, Relativistic gravity gradiometry. Phys. Rev. D **94**, 124009 (2016)
40. J.-A. Marck, Solution to the equations of parallel transport in Kerr geometry; tidal tensor. Proc. R. Soc. Lond. A **385**, 431 (1983)
41. B. Mashhoon, On a new gravitational effect of a rotating mass. Gen. Relativ. Gravit. **16**, 311 (1984)

42. D.S. Theiss, Ph.D. thesis, University of Cologne, Köln, Germany (1984)
43. D.S. Theiss, A general relativistic effect of a rotating spherical mass and the possibility of measuring it in a space experiment. Phys. Lett. A **109**, 19–22 (1985)
44. B. Mashhoon, Gravitational effects of rotating masses. Found. Phys. (Bergmann Festschrift), **15**, 497 (1985)
45. C.A. Blockley, G.E. Stedman, Gravitomagnetic effects along polar geodesics about a slowly rotating spherical mass in the PPN formalism. Phys. Lett. A **147**, 161–164 (1990)
46. H. Rauch, S.A. Werner, *Neutron Interferometry*, 2nd edn. (Oxford University Press, Oxford, 2015)
47. J. Anandan, Curvature effects in interferometry. Phys. Rev. D **30**, 1615–1624 (1984)
48. B. Mashhoon, D.S. Theiss, Gravitational influence of the rotation of the sun on the earth-moon system. Phys. Lett. A **115**, 333–337 (1986)
49. B. Mashhoon, D.S. Theiss, Relativistic lunar theory. Nuovo Cim. B **106**, 545–571 (1991)
50. B. Mashhoon, D.S. Theiss, Relativistic effects in the motion of the Moon. Lect. Notes Phys. **562**, 310–316 (2001)
51. D.S. Theiss, A new gravitational effect of a deformed mass. Phys. Lett. A **109**, 23–27 (1985)

Reference-Ellipsoid and Normal Gravity Field in Post-Newtonian Geodesy

Sergei Kopeikin

Abstract Modern geodesy is undergoing a crucial transformation from the Newtonian paradigm to the Einstein theory of general relativity. This is motivated by advances in developing quantum geodetic sensors including gravimeters and gradientometers, atomic clocks and fiber optics for making ultra-precise measurements of geoid and multipolar structure of Earth's gravitational field. At the same time, Very Long Baseline Interferometry, Satellite Laser Ranging and Global Navigation Satellite System have achieved an unprecedented level of accuracy in measuring spatial coordinates of reference points of the International Terrestrial Reference Frame and the world height system. The main geodetic reference standard to which gravimetric measurements of Earth's gravitational field are referred, is called normal gravity field which is represented in the Newtonian gravity by the field of a uniformly rotating, homogeneous Maclaurin ellipsoid having mass and quadrupole momentum equal to the total mass and (tide-free) quadrupole moment of the gravitational field of Earth. The present chapter extends the concept of the normal gravity field from the Newtonian theory to the realm of general relativity. We focus on the calculation of the post-Newtonian approximation of the normal field that would be sufficiently precise for near-future practical applications. We show that in general relativity the level surface of homogeneous and uniformly rotating fluid is no longer described by the Maclaurin ellipsoid in the most general case but represents an axisymmetric spheroid of the fourth order (PN spheroid) with respect to the geodetic Cartesian coordinates. At the same time, admitting post-Newtonian inhomogeneity of mass density in the form of concentric elliptical shells allows us to preserve the level surface of the fluid as an exact ellipsoid of rotation. We parametrize the mass density distribution and the level equipotential surface with two parameters which are intrinsically connected to the existence of the residual gauge freedom, and derive the post-Newtonian normal gravity field of the rotating spheroid both inside and outside of the rotating fluid body. The normal gravity field is given, similarly to the Newtonian gravity, in a closed form by a finite number of the ellipsoidal harmonics. We employ transformation from the

---

S. Kopeikin (✉)
Department of Physics and Astronomy, University of Missouri,
322 Physics Bldg, Columbia, MO 65211, USA
e-mail: kopeikins@missouri.edu
URL: https://physics.missouri.edu/people/kopeikin

D. Puetzfeld and C. Lämmerzahl (eds.), *Relativistic Geodesy*,
Fundamental Theories of Physics 196, https://doi.org/10.1007/978-3-030-11500-5_6

ellipsoidal to spherical coordinates to deduce a more conventional post-Newtonian multipolar expansion of scalar and vector gravitational potentials of the rotating spheroid. We compare these expansions with that of the normal gravity field generated by the Kerr metric and demonstrate that the Kerr metric has a fairly limited application in relativistic geodesy as it does not match the normal gravity field of the Maclaurin ellipsoid already in the Newtonian limit. We derive the post-Newtonian generalization of the *Somigliana* formula for the normal gravity field measured on the surface of the rotating PN spheroid and employed in practical work for measuring the Earth gravitational field anomalies. Finally, we discuss the possible choice of the gauge-dependent parameters of the normal gravity field model for practical applications and compare it with the existing EGM2008 model of gravitational field.

1 Introduction

This chapter reviews the results of our previous studies of the relativistic geoid, reference ellipsoid and equipotential surface of rotating fluid body which we conducted over decades and published in a number of articles [1–4] and in textbook [5].

1.1 Earth's Gravity Field in the Newtonian Theory

Gravitational field of the Earth has a complicated spatial structure that is also subject to short and long temporal variations [6–8]. Studying this structure and its time evolution is a primary goal of many scientific disciplines such as fundamental astronomy, celestial mechanics, geodesy, gravimetry, etc. The principal component of the Earth's gravity field is well approximated by radially-isotropic field that can be thought as being generated by either a point-like massive particle located at the geocenter or by a massive sphere (or a shell) having a spherically-symmetric distribution of mass inside it. According to the Newtonian gravity law the spheres (shells) of different size and/or of different spherically-symmetric stratifications of the mass density generate the same radially-isotropic gravitational field under condition that the masses of the spheres (shells) are equal. The same statement holds in general relativity where it is known under the name of Birkhoff's theorem [9]. The radially-isotropic component of the Earth's gravity field is often called a monopole as it is characterized by a single parameter - the Earth's mass, M. Generally speaking, the total mass M of the Earth is not constant because of the loss of hydrogen and helium from atmosphere, gradual cooling of the Earth's core and mantle, energy loss due to tidal friction, the dust accretion from an outer space, etc. Nonetheless, the temporal change of the Earth's overall mass is minuscule, $\dot{M}/M \leq 10^{-15}$ [https://en.wikipedia.org/wiki/Earth_mass], and can be neglected in most cases. Thus, in the present chapter we consider the Earth's mass, M, as constant. The time variability of mass does not affect the radial isotropy of the monopole field. It only changes its strength.

Monopole approximation is a good textbook example for discussion in undergraduate physics courses but it is insufficient for real scientific applications because Earths figure is not spherically-symmetric causing noticeable deviations from the radial isotropy of the gravity field. These deviations are taken into account by applying the next approximation in the multipolar expansion of the gravity field called ellipsoidal [6, 10]. This is achieved by modelling the distribution of Earth's matter as a rotating bi-axial ellipsoid with its center of mass located at the geocenter and the minor axis coinciding with the Earth's polar principal axis of inertia. Moreover, gravitational potential on the surface of the rotating ellipsoid is equated to the value of the gravitational potential, W_0, measured on geoid that is the equipotential surface which coincides with the undisturbed level of the world ocean [6, 8]. It is the ellipsoidal approximation which is called the normal gravity field and the corresponding ellipsoid of revolution is known as a (global) reference ellipsoid [8, chapter 4.2.1].

In the Newtonian theory of gravity the normal gravity field is uniquely specified in the exterior domain to ellipsoid by four parameters: the geocentric gravitational constant, GM, the semi-major axis of the reference ellipsoid, a, its flattening, f, and the nominal value of the Earth's rotational velocity, ω which are considered as fundamental geodetic constants [11, Table 1.1]. The normal field is used as a reference in description of the actual gravity field potential, W, of the Earth which can be represented as a linear superposition of the normal gravity field potential, U and a disturbing potential T, that is [8, 10]

$$W = U + T \,. \tag{1}$$

Notice that on the surface of rotating Earth there is also a centrifugal force besides the force of gravity. The potential Φ of the centrifugal force is considered as a well-defined quantity which can be easily calculated at each point of space. Therefore, although the potential Φ is small compared with U, it is not included to the perturbation T but considered as a part of the normal gravity field potential U that consists of the gravitational potential V of a non-rotating Earth, and the centrifugal potential Φ,

$$U = V + \Phi \,. \tag{2}$$

The disturbing potential, T, includes all high-order harmonics in the multipolar expansion of the gravitational field of the Earth associated with the, so-called, anomalies in the distribution of the mass density. The multipolar harmonics are functions of spatial coordinates (and time) that can be expressed in various mathematical forms. For example, the multipolar harmonics expressed in spherical coordinates are known as the gravity potential coefficients [6, 8] representing the, so-called, gravity disturbances or anomalies.

The gravity anomalies of the disturbing potential T have been consistently measured for a long time with a gradually growing accuracy. Originally, their measurements were limited to rather small, local regions of the Earth's surface and were conducted by means of gravimeters giving access to the absolute value of the gravity force and the deflection of vertical (plumb line) at the measuring point. The gravimetric ground-based measurements are indispensable for regional studies of the gravity field anomalies but they remain sparse and insufficient to build a comprehensive model of the global gravity field which is the primary task of geodesy. Advancement in constructing the global model of the gravity field was achieved with the help of the dedicated geodetic space missions like LAGEOS and satellite laser ranging techniques [12]. More recently, further progress have been spawned with the advent of space gravity gradiometers – GRACE and GOCE [13]. The overall set of measurements of the gravity field anomalies has been processed and summarized in 2008 in the form of the Earth Gravitational Model (EGM2008) that has been built and publicly released by the National Geospatial-Intelligence Agency (NGA) EGM Development Team in 2012 [14, 15]. The disturbing gravitational potential T of this model is complete up to all spherical harmonics of the degree and order 2159, and contains some additional potential coefficients up to degree 2190. Full access to the model's coefficients is provided on website of NGA [http://earth-info.nga.mil/GandG/wgs84/gravitymod/egm2008/index.html].

1.2 General Relativity in Geodesy. Why do We Need It?

Positions of reference points (geodetic stations) on the Earths surface can now be determined with precision at the level of few millimeters and their variation over time at the level of 1 mm/year, or even better [16]. Continuous geodetic observations become more and more fundamental for many Earth-science applications at the global and local levels like large scale and local Earth-crust deformation; global tectonic motion; redistribution of geophysical fluids on or near Earths surface including ocean, atmosphere, cryosphere, and the terrestrial hydrosphere; monitoring of the mean sea level and its variability for evaluation its impact on global warming, and many others [8]. All these important applications depend fundamentally on the availability and accuracy of the global International Terrestrial Reference System (ITRS). In addition to the above-mentioned geoscience applications, the ITRS – through its realization by an International Terrestrial Reference Frame (ITRF), is an indispensable reference needed to ensure the integrity of Global Navigation Satellite System (GNSS), such as GPS, GLONASS, Galileo, and clock's synchronization [11].

It is believed that the requirements of geoscience to measurement precision, including the most stringent one – the mean sea level variability, are to reach the availability of the reference frame that will be reliable, stable and accessible at the positional accuracy down to 1 mm, and stability of 0.1 mm/year [17]. It is crucial to reach this accuracy from both scientific and practical points of view as economy and safety of modern society is extremely vulnerable to even small changes in sea level [18]. Stability of the reference frame means that no discontinuity or drift should occur in its time evolution, especially for its defining physical parameters, namely the origin and the scale. Unfortunately, the current level of reference frame accuracy (based on the latest ITRF realization) is about ten times worse than the science requirement [19]. New technological methods and theoretical models are required to fill up this gap.

Relativistic effects of gravitational field of Earth have a fractional order of 1 ppb or about 1 cm on a geospatial scale. Albeit small, they depend globally on the geographic position of observer and produce a systematic bias in the height measurements, unless properly taken into account. It is, therefore, mandatory to switch from the Newtonian paradigm to general theory of relativity in order to thoroughly accommodate relativistic effects to geodesy. Nowadays, it is commonly accepted that a network of high precision clocks and their comparison will be able to significantly contribute to the solution of the problem of stability and accuracy of a new generation ITRF through very precise determination of height differences and relative velocities of clocks participating in the network [20]. In addition to the highly precise geometrical coordinates of the ITRF (such as ellipsoidal heights), clock measurements will help to consistently provide physical heights at the reference points of observatories [21–23]. Since clocks, according to general relativity, directly measure the difference of gravitational potential this opens up a fundamentally new conceptual basis for physical geodesy, that is, an unambiguous geoid determination and realization of a new global dynamic reference system.

For thorough theoretical description of clock's behaviour in gravitational field, one has to take into account all special and general relativistic effects like gravitational red shift, Doppler effect, gravitational (Shapiro) time delay, Sagnac effect, and even Lense-Thirring effect which appears as gravitomagnetic clock effect [24, 25]. All these effects depend on clocks relative motion and strength of gravitational field. Based on this we can solve the inverse problem to model the mass density and height variations affecting the clock measurement, e.g., related to the solid Earth tides. Gravitational effects associated with Earth's rotation and tides limit the metrological network of ground-based atomic clocks at fractional level 10^{-16} [26], and must be accurately calculated and subtracted from clock' readings. Further interesting details in developing theoretical and technological approaches to solving this problem can be found in the presentations of participants of ISSI-Bern workshop on spacetime metrology, clocks and relativistic geodesy [http://www.issibern.ch/teams/spacetimemetrology/], and in a review article [20].

This section would be incomplete without mentioning the other branches of modern geodesy which are tightly connected with the experimental gravitational physics and fully based on the mathematical apparatus of general relativity. This includes Very Long Baseline Interferometry (VLBI) that is used as a main tool of the International Earth Rotation Service (IERS) for monitoring precession, nutation and wobble (polar motion) as well as for producing the International Celestial and Terrestrial Reference Frames (ICRF and ITRF respectively). VLBI requires taking into account a stunning number of relativistic effects which are outlined in corresponding papers and recorded in IAU resolutions (see, e.g. [11, 27, 28]). Motion of geodetic satellites must take into account a significant number of relativistic effects as well, like geodetic precession, Lense-Thirring effect, relativistic quadrupole, relativistic tidal effects, etc. General-relativistic model of relativistic effects in the orbital motion of geodetic and navigation spacecraft has been worked out by Brumberg and Kopeikin

[29, 30] (c.f. [31]). It was numerically analysed in a number of recent papers [32–35] studying a feasibility of observing various relativistic effects. A particular attention has been recently paid to experimental measurement of the Lense-Thirring effect in the orbital motion of LAGEOS and LARES satellites [36–39]. This experimental study is especially important for relativistic astrophysics because the Lense-Thirring effect is considered as a main driving mechanism for the enormous release of energy in quasars and active galactic nuclei caused by accretion of matter on a central, supermassive Kerr black hole [40].

One of the important relativistic problem in geodesy is the description of normal gravity field of Earth represented as the international reference ellipsoid that is used as a reference for geodetic and gravimetric measurements. The goal of the present chapter is to provide the reader with a solution of this problem.

1.3 The Normal Gravity Field in Classic Geodesy and in General Relativity

In classic geodesy the normal gravity field of the Earth is generated by a rigidly rotating bi-axial ellipsoid which is made of a perfect (non-viscous) fluid of uniform density, ρ_c which value is determined from the known total mass and volume of the Earth. In the Newtonian theory this is the only possible distribution of mass density because any other mass distribution of rotating fluid yields the shape of the body being different from the bi-axial ellipsoid [10, 41]. Relativistic geodesy is an advanced branch of physical geodesy that is based on Einstein's general relativity which supersedes the Newtonian theory of gravity. General relativistic approach requires reconsidering the concept of the normal gravity field by taking into account the curvature of spacetime manifold and other post-Newtonian effects caused by Earth's mass.

General theory of relativity replaces a single gravitational potential, V, with ten potentials which are components of the metric tensor $g_{\alpha\beta}$, where, here and anywhere else, the Greek indexes $\alpha, \beta, \gamma, \ldots$ take values from the set $\{0, 1, 2, 3\}$. General relativity modifies gravitational field equation of the Newtonian theory correspondingly. More specifically, instead of a single Poisson equation for the scalar potential V, general relativity introduces ten partial differential equations of the second order for the metric tensor components. These equations are known as Einstein's equations [5]

$$R_{\alpha\beta} - \frac{1}{2} g_{\alpha\beta} R = \frac{8\pi G}{c^4} T_{\alpha\beta} \,, \tag{3}$$

where $R_{\alpha\beta}$ is the Ricci tensor, $R = g^{\alpha\beta} R_{\alpha\beta}$ is the Ricci scalar, $T_{\alpha\beta}$ is the tensor of energy-momentum of matter which is the source of gravitational field, G is the universal gravitational constant, and c is the fundamental speed of the Minkowski spacetime that is equal to the speed of light in vacuum or to the speed of propagation of weak gravitational waves.

The left part of (3) is called the Einstein tensor which is a hyperbolic differential operator of the second order in partial derivatives applied to the metric tensor $g_{\alpha\beta}$ [42]. The Einstein tensor is non-linear and, for this reason, Einstein's equations (3) cannot be solved exactly in the most physical situations of practical importance. In order to circumvent this difficulty researchers resort to iterative approximations to solve the Einstein equations. One of the most elaborated iterative schemes is called the post-Newtonian approximations (PNA) which basics are discussed in Sect. 2 in more detail. In many astrophysical applications (especially in gravitational-wave astronomy) one needs to make several post-Newtonian approximations for calculating observable effects [43]. For the purposes of relativistic geodesy and celestial mechanics of the solar system the first post-Newtonian approximation is usually sufficient [5] though there are indications that one may soon need a second PN approximation [44] and exact, axially-symmetric solutions of general relativity [2, 45, 46].

The problem of determination of a figure of rotating fluid body is formidably difficult already in the Newtonian theory [41]. It becomes even more complicated in general relativity because of non-linearity of the Einstein equations. Geophysics is interested in finding distribution of mass density inside the Earth to understand better its thermal behaviour and seismological response. The interior structure of the Earth is also important for the International Earth Rotation Service (IERS) to account properly for free core nutation (FCN) in calculation of polar wobble and tidal variations of the Earth's rotational velocity affected by the elasticity of the Earth's interior [11]. Geodesy does not require precise distribution of mass density inside the Earth as it basically needs to know the surface of equal geopotential (geoid) and the gravity anomalies in the domain being exterior to geoid. Geoid's reconstruction from the gravity anomalies utilizes the normal gravity field for solving the integral equations of the Stokes -Molodensky problem [8]. As a rule, the most simple, homogeneous distribution of mass density inside the Earth is used to model the normal gravity field. Attempts to operate with more realistic distributions of mass inside the Earth led to the models of the normal gravity field which turned out to be too complicated for practical computations and were abandoned.

We emphasize that the internal density distribution and the surface of the rotating fluid body taken for modelling the normal gravity field must be consistent with the laws of the theory of gravitation. In the Newtonian theory the surface of the uniformly rotating homogeneous fluid is a bi-axial ellipsoid of revolution - the Maclaurin ellipsoid [41]. More realistic, non-homogeneous distribution of mass of the rotating fluid does not allow it to be the ellipsoid of revolution yielding more complicated figure having a spheroidal surface [8, 10]. Such models have less practical significance in geodesy because of a more complicated structure of the normal gravity field.

One would think that modeling the normal gravity field in relativistic geodesy could be achieved by finding an exact solution of the Einstein equations which Newtonian limit corresponds to the homogeneous Maclaurin ellipsoid. Unfortunately, the exact solutions of general relativity describing gravity field of a single body consisting of homogeneous, incompressible fluid are currently known only for spherically-symmetric, non-rotating configurations [47–49]. There is a certain progress in understanding the general relativistic structure of rotating fluid configurations [50–52] but whether a rigidly rotating fluid body can be made of incompressible, homogeneous fluid, is not yet known. In the post-Newtonian approximation of general relativity it was found that a rigidly-rotating body consisting of a perfect fluid with homogeneous distribution of mass density inside it, can exist but it is not a bi-axial ellipsoid [53–56]. On the other hand, by assuming that the distribution of mass density has a post-Newtonian ellipsoidal component in addition to the constant density ρ_c, we can chose the parameters of the density distribution such that the figure of the rigidly rotating fluid will remain exactly ellipsoidal in the first (and higher-order) post-Newtonian approximations [3]. Thus, we have to make a decision what type of the post-Newtonian distribution of mass and the figure of the rotating fluid are to be used in relativistic geodesy. Depending on the choice we shall have slightly different relativistic descriptions of the normal gravitational field outside rotating spheroid. We will proceed by assuming that the surface of the rotating fluid has a small post-Newtonian spheroidal deviation and that the distribution of the fluid density is almost homogeneous with a small post-Newtonian correction taken in the form of a homeomorphic ellipsoidal distribution.

Thus, the normal gravity field in relativistic geodesy is a solution of the Einstein field equations with matter consisting of a uniformly rotating, perfect fluid of nearly constant density occupying a spheroidal volume. We notice that a number of researchers solved the Einstein equations to find out gravitational field of uniformly rotating bi-axial ellipsoid of constant density [57–59]. Their solutions are not directly applicable in relativistic geodesy for the shape of a uniformly rotating and incompressible homogeneous fluid is not an ellipsoid. Moreover, the authors of the papers [57–59] were mostly interested in astrophysical applications and never approached the problem from the geodetic point of view.

The chapter is organized as follows. Section 2 explains the post-Newtonian approximations in general relativity. We pay a special attention to various coordinate systems used in relativistic geodesy and transformations between them as well as to the Green functions used for solving the Einstein equations. Section 3 discusses the model of matter distribution used as a source of normal gravitational field and geometric shape of the post-Newtonian reference spheroid used for integration of the Newtonian and post-Newtonian gravitational potentials. Section 4 is devoted to a comprehensive calculation of the Newtonian gravitational potential. Section 5 presents details of the calculation of the post-Newtonian scalar and vector gravitational potentials. In Sect. 6 we discuss relativistic multipole expansion of gravitational field of rotating spheroid which includes both mass and spin multipole moments. Section 7 gives a full description of the normal gravity field of rotating spheroid in relativistic geodesy including definitions of equipotential surfaces, the gravity field potential, the figure of equilibrium of the rigidly rotating fluid, and the Somigliana formula for the normal gravity force on the surface of the rotating spheroid. Section 8 calculates the normal gravity field of the Kerr metric and compares it with that of a rigidly rotating spheroid made out of the ideal fluid. It proves that the Kerr metric is unsuitable for purposes of relativistic geodesy due to the peculiar structure of its multipole expansion.

1.4 Mathematical Symbols and Notations

The following notations are used throughout the present chapter:

- the spherical coordinates are denoted $\{R, \Theta, \Phi\}$,
- the ellipsoidal coordinates are denoted $\{\sigma, \theta, \phi\}$,
- the Greek indices $\alpha, \beta, \ldots$ run from 0 to 3,
- the Roman indices $i, j, \ldots$ run from 1 to 3,
- repeated Greek indices mean Einstein's summation from 0 to 3,
- repeated Roman indices mean Einstein's summation from 1 to 3,
- the unit matrix (also known as the Kronecker symbol) is denoted by $\delta_{ij} = \delta^{ij}$,
- the fully antisymmetric symbol Levi-Civita is denoted as $\varepsilon_{ijk} = \varepsilon^{ijk}$ with $\varepsilon_{123} = +1$,
- the bold letters $\boldsymbol{a} = a^i$, $\boldsymbol{b} = b^i$, etc., denote spatial 3-dimensional vectors,
- a dot between two spatial vectors, for example $\boldsymbol{a} \cdot \boldsymbol{b} = a^1 b^1 + a^2 b^2 + a^3 b^3 = \delta_{ij} a^i b^j$, means the Euclidean dot product,
- the cross between two vectors, for example $(\boldsymbol{a} \times \boldsymbol{b})^i \equiv \varepsilon^{ijk} a^j b^k$, means the Euclidean cross product,
- we use a shorthand notation for partial derivatives $\partial_\alpha = \partial / \partial x^\alpha$,
- covariant derivative with respect to a coordinate x^α is denoted as ∇_α,
- the Minkowski (flat) space-time metric $\eta_{\alpha\beta} = \mathrm{diag}(-1, +1, +1, +1)$,
- $g_{\alpha\beta}$ is the physical spacetime metric,
- the Greek indices are raised and lowered with the metric $\eta_{\alpha\beta}$,
- the Roman indices are raised and lowered with the Kronecker symbol δ^i_j,
- G is the universal gravitational constant,
- c is the fundamental speed of the Minkowski space,
- ω is a constant rotational velocity of rigidly rotating matter,
- ρ is a mass density distribution of matter,
- ρ_c is a constant central density of matter,
- a is a semi-major axis of the Maclaurin ellipsoid of revolution,
- b is a semi-minor axis of the Maclaurin ellipsoid of revolution,
- f is the geometric flattening: $f \equiv (a - b)/a$,
- ϵ is the *first* eccentricity of the Maclaurin ellipsoid: $\epsilon \equiv \sqrt{a^2 - b^2}/a = \sqrt{2f - f^2}$,
- $\varkappa$ is the *second* eccentricity of the Maclaurin ellipsoid: $\varkappa \equiv \sqrt{a^2 - b^2}/b = \epsilon/(1 - f)$,
- $\alpha \equiv a\epsilon = \sqrt{a^2 - b^2}$,
- $r \equiv R/\alpha$ is a dimensionless spherical radial coordinate,
- $\kappa \equiv \pi G \rho_c a^2 / c^2$ is a dimensionless parameter characterizing the strength of gravitational field on the surface of the field-generating body.

Other notations are explained in the text as they appear.

2 Post-Newtonian Approximations

2.1 *Harmonic Coordinates and the Metric Tensor*

Discussion of relativistic geodesy starts from the construction of the spacetime manifold for the case of a rigidly rotating fluid body having the same mass as the mass of the Earth. We shall employ Einstein's general relativity to build such a manifold though some other alternative theories of gravity discussed, for example in textbook [60], can be used as well. Einstein's gravitational field equations (3) represent a system of ten non-linear differential equations in partial derivatives for ten components of the (symmetric) metric tensor, $g_{\alpha\beta}$, which represents gravitational potentials generalizing the Newtonian gravitational potential V. Because the equations are difficult to solve exactly due to their non-linearity, we resort for their solution to the post-Newtonian approximations (PNA) [60, 61].

The PNA are the most effective in case of slowly-moving matter having a weak gravitational field. This is exactly the situation in the solar system which makes PNA highly appropriate for constructing relativistic theory of reference frames [27], and for relativistic celestial mechanics, astrometry and geodesy [5, 62, 63]. The PNA are based on the assumption that solution of the Einstein equations for the metric tensor can be presented in the form of a Taylor expansion of the metric tensor with respect to the inverse powers of the fundamental speed, c, that is equal to the speed of light in vacuum and the speed of weak gravitational waves in general relativity.

Exact mathematical formulation of a set of basic axioms required for doing the post-Newtonian expansion was given by Rendall [64]. Practically, it requires having several small parameters characterizing the source of gravity which is often is an isolated astronomical system comprised of extended bodies. The parameters are: $\varepsilon_i \sim v_i/c$, $\varepsilon_e \sim v_e/c$, and $\eta_i \sim U_i/c^2$, $\eta_e \sim U_e/c^2$, where v_i is a characteristic velocity of motion of matter inside the body, v_e is a characteristic velocity of the relative motion of the bodies with respect to each other, U_i is the internal gravitational potential of each body, and U_e is the external gravitational potential between the bodies. If one denotes a characteristic radius of a body as ℓ and a characteristic distance between the bodies as R, the estimates of the internal and external gravitational potentials will be, $U_i \simeq GM/\ell$ and $U_e \simeq GM/R$, where M is a characteristic mass of the body. Due to the virial theorem of the Newtonian gravity theory [53] the small parameters are not independent. Specifically, one has $\varepsilon_i^2 \sim \eta_i$ and $\varepsilon_e^2 \sim \eta_e$. Hence, parameters, ε_i and ε_e, characterizing the slow motion of matter, are sufficient in doing the iterative solution of the Einstein equations by the post-Newtonian approximations. Because within the solar system these parameters do not significantly differ from each other, we shall not distinguish between them. Quite often we shall use notation, $\kappa \equiv \pi G\rho_c a^2/c^2 \sim \eta_i$, to mark the powers of the fundamental speed c in the post-Newtonian terms.

We assume that physical spacetime within the solar system has the metric tensor denoted $g_{\alpha\beta}$. This spacetime is well-approximated in case of the slow-motion and weak-field post-Newtonian approximation, by a background manifold which is the Minkowski spacetime having the metric tensor denoted $\eta_{\alpha\beta} = \mathrm{diag}(-1, 1, 1, 1)$.

Einstein's equations admit a gauge freedom associated with the arbitrariness in choosing coordinate charts covering the spacetime manifold. The gauge freedom is used to simplify the structure of Einstein's equations. The most convenient choice is associated with the harmonic coordinates $x^\alpha = (x^0, x^i)$, where $x^0 = ct$, and t is the coordinate time. The class of the harmonic coordinates is used by the International Astronomical Union for description of the relativistic coordinates systems and for the data reduction [11, 27] as well as in relativistic geodesy [4, 65] The harmonic coordinates are defined by imposing the de Donder gauge condition on the metric tensor [66, 67],

$$\partial_\alpha \left(\sqrt{-g} g^{\alpha\beta}\right) = 0 \,. \tag{4}$$

Imposing the harmonic gauge greatly simplifies the Einstein equation (3) and allows us to solve them by the post-Newtonian iterations.

Because gravitational field of the solar system is weak and motion of matter is slow, we can solve Einstein's equations by post-Newtonian approximations. In fact, the first post-Newtonian approximation of general relativity is fully sufficient for the purposes of relativistic geodesy. We focus in this chapter on calculation of the normal gravitational field of the Earth generated by uniformly rotating ideal (perfect) fluid. Under these assumptions the spacetime interval has the following form [5]

$$ds^2 = g_{00}(t, \boldsymbol{x})c^2 dt^2 + 2g_{0i}(t, \boldsymbol{x})cdtdx^i + g_{ij}(t, \boldsymbol{x})dx^i dx^j \,, \tag{5}$$

where the post-Newtonian expressions for the metric tensor components read

$$g_{00}(t, \boldsymbol{x}) = -1 + \frac{2V(t, \boldsymbol{x})}{c^2} - \frac{2V^2(t, \boldsymbol{x})}{c^4} + \mathcal{O}\left(\frac{1}{c^6}\right) \,, \tag{6a}$$

$$g_{0i}(t, \boldsymbol{x}) = -\frac{4V^i(t, \boldsymbol{x})}{c^3} + \mathcal{O}\left(\frac{1}{c^5}\right) \,, \tag{6b}$$

$$g_{ij}(t, \boldsymbol{x}) = \delta_{ij}\left[1 + \frac{2V(t, \boldsymbol{x})}{c^2}\right] + \mathcal{O}\left(\frac{1}{c^4}\right) \,. \tag{6c}$$

Herein, the scalar potential $V = V(t, \boldsymbol{x})$ and a (gravitomagnetic) vector potential $V^i = V^i(t, \boldsymbol{x})$ are functions of time and spatial coordinates satisfying the Poisson equations,

$$\Delta V = -4\pi G\rho\left[1 + \frac{1}{c^2}\left(2v^2 + 2V + \Pi + \frac{3p}{\rho}\right)\right] \,, \tag{7}$$

$$\Delta V^i = -4\pi G\rho v^i \,, \tag{8}$$

with $\rho = \rho(t, \boldsymbol{x})$ being the mass density, $p = p(t, \boldsymbol{x})$ and $v^i = v^i(t, \boldsymbol{x})$ – pressure and velocity of matter respectively, and $\Pi = \Pi(t, \boldsymbol{x})$ is the specific internal energy of matter per unit mass. We emphasize that ρ is the local mass density of baryons per unit of invariant (3-dimensional) volume element $d\mathcal{V} = \sqrt{-g}u^0 d^3x$, where $u^0 = dt/d\tau$

is the time component of the 4-velocity of matter's particle, where $d\tau = \sqrt{-ds^2}/c$ is the proper time of the particle.[1] The local mass density, ρ, relates in the post-Newtonian approximation to the invariant mass density $\rho^* = \sqrt{-g}u^0\rho$, which post-Newtonian expression is given by [5]

$$\rho^* = \rho + \frac{\rho}{c^2}\left(\frac{1}{2}v^2 + 3V\right) + \mathcal{O}\left(\frac{1}{c^4}\right) . \tag{9}$$

The internal energy, Π, is related to pressure, p, and the local density, ρ, through the thermodynamic equation (the law of conservation of energy)

$$d\Pi + pd\left(\frac{1}{\rho}\right) = 0 , \tag{10}$$

and the equation of state, $p = p(\rho)$.

We shall further assume that the background matter rotates rigidly around fixed z axis with a constant angular velocity ω. This makes the background spacetime stationary with the background metric being independent of time. In the stationary spacetime, the mass density ρ^* obeys the *exact*, steady-state equation of continuity

$$\partial_i\left(\rho^* v^i\right) = 0 . \tag{11}$$

The velocity of the rigidly rotating fluid is a linear function of spatial coordinates,

$$v^i = \varepsilon^{ijk}\omega^j x^k , \tag{12}$$

where $\omega^i = (0, 0, \omega)$ is a constant angular velocity. Replacing velocity v^i in (11) with (12), and differentiating yield,

$$v^i \partial_i \rho = 0 , \tag{13}$$

which is equivalent to $d\rho/dt = 0$, and means that the linear velocity v^i of the fluid is tangent to the surfaces of constant density ρ.

2.2 Ellipsoidal and Spherical Coordinates

Equipotential surfaces of gravitational field produced by a rigidly rotating fluid body are closely approximated by biaxial ellipsoids. Therefore, it sounds reasonable to solve Einstein's equations in the oblate ellipsoidal coordinates. These coordinates are well known and widely used in geodesy [6, 8, 10]. In order to introduce these

[1] The minus sign in definition of the proper time appears because $ds^2 < 0$ due to the choice of the metric signature shown in (6a)–(6c).

coordinates let us consider a point $\mathcal{P}$ in space that is characterized by three Cartesian (harmonic) coordinates $\boldsymbol{x} = (x, y, z)$. We choose the origin of the coordinates at the center of mass of the rotating body with z-axis coinciding with the direction of the vector of the angular rotation, $\boldsymbol{\omega} = (\omega^i)$, and x and y axes lying in the equatorial plane.

In the Newtonian theory the surface of rotating, homogeneous fluid takes the shape of an oblate ellipsoid of revolution (Maclaurin ellipsoid) with a semi-major, a, and a semi-minor axis, $b = a\sqrt{1-\epsilon^2}$, where the constant parameter

$$\epsilon = \frac{\sqrt{a^2 - b^2}}{a} , \tag{14}$$

is called the *first eccentricity* [8], and $0 \le \epsilon \le 1$. The oblate ellipsoidal coordinates associated with the ellipsoid of revolution, are defined by a set of surfaces of confocal ellipsoids and hyperboloids being orthogonal to each other (see [https://en.wikipedia.org/wiki/Oblate_spheroidal_coordinates]). It means that the focal points of all the ellipsoids and hyperboloids coincide, and the distance of the focal points from the origin of the coordinates is given by the distance $\alpha = a\epsilon$.

In order to connect the Cartesian coordinates, (x, y, z), of the point $\mathcal{P}$ to the oblate ellipsoidal coordinates, (σ, θ, ϕ), we pass through $\mathcal{P}$ the surface of the ellipsoid which is confocal with the Maclaurin ellipsoid formed by the rotating homogeneous fluid. Geodetic definition of the transformation from the Cartesian to the ellipsoidal coordinates used in geodesy, is given, for example, in [10, eq. 1-103], and reads

$$x = \alpha\sqrt{1+\sigma^2}\sin\theta\cos\phi\ , \tag{15a}$$

$$y = \alpha\sqrt{1+\sigma^2}\sin\theta\sin\phi\ , \tag{15b}$$

$$z = \alpha\sigma\cos\theta\ , \tag{15c}$$

where the radial coordinate $\sigma \ge 0, 0 \le \theta \le \pi, 0 \le \phi \le 2\pi$, and the constant parameter $\alpha \equiv \sqrt{a^2 - b^2} = a\epsilon$. The interior domain, $\mathcal{V}$, of the ellipsoidal coordinate system is separated from the exterior domain, $(\mathcal{V}_{\rm ext})$, by the surface $\mathcal{S}$ of the Maclaurin ellipsoid. The interior domain is determined by conditions $0 \le \sigma \le 1/\varkappa$, and the exterior domain has $\sigma > 1/\varkappa$ respectively where the constant

$$\varkappa \equiv \frac{\epsilon}{\sqrt{1-\epsilon^2}} = \frac{\sqrt{a^2 - b^2}}{b} , \tag{16}$$

is called the *second eccentricity* [8], and we notice that $0 \le \varkappa \le \infty$. In terms of the second eccentricity, the focal parameter $\alpha = b\varkappa$.

It is worth noticing that Eq. (15) looks similar to that used for definition of the Boyer–Lindquist coordinates which have been used in astrophysical studies of the Kerr black hole that is an exact axisymmetric solution of vacuum Einstein's equation [68, chapter 17]. Nonetheless, the oblate ellipsoidal coordinates, $\{\sigma, \theta, \phi\}$ that we use in this chapter don't coincide with the Boyer–Lindquist coordinates which are connected to the original, non-harmonic, Kerr coordinates.

The volume of integration in the ellipsoidal coordinates is

$$d^3x = \alpha^3 \left(\sigma^2 + \cos^2\theta\right) d\sigma d\Omega\ , \tag{17}$$

where $d\Omega = \sin\theta d\theta d\phi$ is the infinitesimal element of the solid angle in the direction of the unit vector

$$\hat{\boldsymbol{x}} = \sin\theta \left(\hat{\boldsymbol{i}}\cos\phi + \hat{\boldsymbol{j}}\sin\phi\right) + \hat{\boldsymbol{k}}\cos\theta\ , \tag{18}$$

where $(\hat{\boldsymbol{i}}, \hat{\boldsymbol{j}}, \hat{\boldsymbol{k}})$ are the unit vectors along the axes of the Cartesian coordinates (x, y, z) respectively. Notice that the unit vector $\hat{\boldsymbol{x}}$ is different from the unit vector of the external normal $\hat{\boldsymbol{n}}$ to the surface $\mathcal{S}$ that is given by

$$\hat{\boldsymbol{n}} = \frac{\sqrt{1-\epsilon^2}\sin\theta\left(\hat{\boldsymbol{i}}\cos\phi + \hat{\boldsymbol{j}}\sin\phi\right) + \hat{\boldsymbol{k}}\cos\theta}{1-\epsilon^2\sin^2\theta}\ . \tag{19}$$

We also introduce the standard spherical coordinates, $\{R, \Theta, \Phi\}$ related to the harmonic coordinates, $x^\alpha = \{x, y, z\}$, by the relations

$$\begin{aligned} x &= R\sin\Theta\cos\Phi\ , \\ y &= R\sin\Theta\sin\Phi\ , \\ z &= R\cos\Theta\ . \end{aligned} \tag{20}$$

In what follows, it will be more convenient to use a dimensionless radial coordinate r by definition: $r \equiv R/\alpha$ so that $\alpha^2 r^2 = x^2 + y^2 + z^2$. The volume of integration in the spherical coordinates is

$$d^3x = \alpha^3 r^2 dr dO\ , \tag{21}$$

where $dO = \sin\Theta d\Theta d\Phi$ is the infinitesimal element of the solid angle in the spherical coordinates in the direction of the unit vector

$$\hat{\boldsymbol{X}} = \sin\Theta\left(\hat{\boldsymbol{i}}\cos\Phi + \hat{\boldsymbol{j}}\sin\Phi\right) + \hat{\boldsymbol{k}}\cos\Theta\ . \tag{22}$$

Comparing (15) and (20) we can find out a transformation between the oblate elliptical coordinates, (σ, θ, ϕ), and the spherical coordinates, (r, Θ, Φ), given by relations,

$$\sqrt{1+\sigma^2}\sin\theta = r\sin\Theta\ , \qquad \sigma\cos\theta = r\cos\Theta\ , \qquad \phi = \Phi\ . \tag{23}$$

The radial elliptical coordinate, σ and the radial spherical coordinate, r, are interrelated

$$r^2 = \sigma^2 + \sin^2\theta\ . \tag{24}$$

Solving (23) and (24) we get a direct transformation between the elliptical and spherical coordinates in explicit form [6, equation (20.24)]

$$\sigma = \sqrt{\frac{r^2-1}{2}\left(1+\sqrt{1+\frac{4r^2\cos^2\Theta}{(r^2-1)^2}}\right)}\,,$$

$$\cos\theta = \frac{r\cos\Theta}{\sqrt{\frac{r^2-1}{2}\left(1+\sqrt{1+\frac{4r^2\cos^2\Theta}{(r^2-1)^2}}\right)}}\,. \tag{25}$$

The approximate form of the relations (25) for relatively large values of the radial coordinate, $r \gg 1$, reads

$$\sigma \simeq r - \frac{\sin^2\Theta}{2r} + \ldots\,, \qquad \cos\theta = \cos\Theta\left(1+\frac{\sin^2\Theta}{2r}+\ldots\right)\,. \tag{26}$$

2.3 *Green's Function of the Poisson Equation in the Ellipsoidal Coordinates*

The Einstein equations (7), (8) represent the Poisson equations with the known right-hand side. The most straightforward solution of these equations can be achieved with the technique of the Green function $\mathcal{G}(\boldsymbol{x},\boldsymbol{x}')$ that satisfies the Poisson equation,

$$\Delta\mathcal{G}(\boldsymbol{x},\boldsymbol{x}') = -4\pi\delta^{(3)}(\boldsymbol{x}-\boldsymbol{x}')\,, \tag{27}$$

where, $\Delta = \partial_x^2+\partial_y^2+\partial_z^2$, is the Laplace operator, and $\delta^{(3)}(\boldsymbol{x}-\boldsymbol{x}')$ is the Dirac delta-function in the harmonic coordinates $\{x, y, z\}$. We need the Green function in the oblate ellipsoidal coordinates, $\{\sigma,\theta,\phi\}$. In these coordinates the Laplace operator reads

$$\Delta \equiv \frac{1}{\alpha^2\left(\sigma^2+\cos^2\theta\right)}\left[(1+\sigma^2)\frac{\partial^2}{\partial\sigma^2}+2\sigma\frac{\partial}{\partial\sigma}+\frac{\partial^2}{\partial\theta^2} + \cot\theta\frac{\partial}{\partial\theta}+\frac{\sigma^2+\cos^2\theta}{(1+\sigma^2)\sin^2\theta}\frac{\partial^2}{\partial\phi^2}\right]. \tag{28}$$

After substituting this form of the operator to the left side of (27), and applying a standard procedure of finding a Green function [69], we get the Green function, $\mathcal{G}(\boldsymbol{x},\boldsymbol{x}')$, in the ellipsoidal coordinates. It is represented in the form of expansion with respect to the ellipsoidal harmonics [70, 71]

Table 1 The modified associated Legendre functions.

$p_0(\sigma) = 1$	$q_0(\sigma) = \operatorname{arccot}\sigma$
$p_1(\sigma) = \sigma$	$q_1(\sigma) = -p_1(\sigma)q_0(\sigma) + 1$
$p_2(\sigma) = \frac{3}{2}\sigma^2 + \frac{1}{2}$	$q_2(\sigma) = p_2(\sigma)q_0(\sigma) - \frac{3}{2}\sigma$
$p_3(\sigma) = \frac{\sigma}{2}\left(5\sigma^2 + 3\right)$	$q_3(\sigma) = -p_3(\sigma)q_0(\sigma) + \frac{5\sigma^2}{2} + \frac{2}{3}$
$p_4(\sigma) = \frac{35}{8}\sigma^4 + \frac{15}{4}\sigma^2 + \frac{3}{8}$	$q_4(\sigma) = p_4(\sigma)q_0(\sigma) - \frac{35}{8}\sigma^3 - \frac{55}{24}\sigma$
$p_{11}(\sigma) = \sqrt{1+\sigma^2}$	$q_{11}(\sigma) = p_{11}(\sigma)q_0(\sigma) - \frac{\sigma}{\sqrt{1+\sigma^2}}$
$p_{31}(\sigma) = \frac{3}{2}\sqrt{1+\sigma^2}\left(1 + 5\sigma^2\right)$	$q_{31}(\sigma) = p_{31}(\sigma)q_0(\sigma) - \frac{\sigma}{2}\frac{13 + 15\sigma^2}{\sqrt{1+\sigma^2}}$

$$\mathcal{G}(\boldsymbol{x}, \boldsymbol{x}') = \frac{1}{|\boldsymbol{x} - \boldsymbol{x}'|}$$
$$= \begin{cases} \frac{1}{\alpha}\sum_{\ell=0}^{\infty}\sum_{m=-\ell}^{m=\ell}\frac{(\ell-|m|)!}{(\ell+|m|)!}q_{\ell|m|}\left(\sigma'\right)p_{\ell|m|}\left(\sigma\right)Y^*_{\ell m}(\hat{\boldsymbol{x}}')Y_{\ell m}(\hat{\boldsymbol{x}}) : (\sigma \le \sigma') \\ \frac{1}{\alpha}\sum_{\ell=0}^{\infty}\sum_{m=-\ell}^{m=\ell}\frac{(\ell-|m|)!}{(\ell+|m|)!}p_{\ell|m|}\left(\sigma'\right)q_{\ell|m|}\left(\sigma\right)Y^*_{\ell m}(\hat{\boldsymbol{x}}')Y_{\ell m}(\hat{\boldsymbol{x}}) : (\sigma' \le \sigma). \end{cases} \tag{29}$$

Here, $p_{\ell m}(u)$ and $q_{\ell m}(u)$ are the modified (real-valued) associated Legendre functions of a *real* argument u, that are related to the associated Legendre functions of an *imaginary* argument, $P_{\ell m}(iu)$ and $Q_{\ell m}(iu)$, by the following definition[2]

$$P_{\ell m}(iu) = i^n p_{\ell m}(u)\ , \qquad Q_{\ell m}(iu) = \frac{(-1)^m}{i^{\ell+1}}q_{\ell m}(u)\ , \tag{30}$$

where i is the imaginary unit, $i^2 = -1$. In case, when the index $m = 0$ we shall use notations, $p_\ell(u) \equiv p_{\ell 0}(u)$, and, $q_\ell(u) \equiv q_{\ell 0}(u)$. We shall also use special notation for the associated Legendre functions taken on the surface of ellipsoid of rotation having a fixed radial coordinate $\sigma = 1/\varkappa$. More specifically, we shall simply omit the argument of the surface functions, for example, we shall denote $p_\ell \equiv p_\ell(1/\varkappa)$ and $q_\ell \equiv q_\ell(1/\varkappa)$. Several modified associated Legendre functions which are ubiquitously used in the present chapter, are shown in Table 1.

Functions $Y_{\ell m}(\hat{\boldsymbol{x}})$ in (29) are the standard spherical harmonics[3]

$$Y_{\ell m}(\hat{\boldsymbol{x}}) \equiv C_{\ell m}P_{\ell|m|}(\cos\theta)e^{im\phi}\ , \tag{31}$$

---

[2]We remind that the associated Legendre functions of the imaginary argument, $z = x + iy$, are defined for all z except at a cut line along the real axis, $-1 \le x \le 1$. The associated Legendre functions of a real argument are defined only on the cut line, $-1 \le x \le 1$ [69, Section 12.10].

[3]Definition of the associated Legendre polynomials adopted in the present chapter follows [72, Sec. 8.81]. It differs by a factor $(-1)^m$ from the definition of the associated Legendre polynomials adopted in the book [10].

where $P_{\ell m}(\cos\theta)$ are the associated Legendre polynomials, and the normalization coefficient

$$C_{\ell m} \equiv \sqrt{(2\ell+1)\frac{(\ell-|m|)!}{(\ell+|m|)!}} \,. \tag{32}$$

The spherical harmonics are complex, $Y^*_{\ell m}(\hat{\boldsymbol{x}}) = Y_{\ell,-m}(\hat{\boldsymbol{x}})$, and form an orthonormal basis in the Hilbert space, that is for $|m| \le \ell$, $|m'| \le \ell'$ the integral over a unit sphere $\mathcal{S}^2$,

$$\oint\limits_{\mathcal{S}^2} Y^*_{\ell m}(\hat{\boldsymbol{x}}) Y_{\ell' m'}(\hat{\boldsymbol{x}}) d\Omega = 4\pi \delta_{\ell\ell'}\delta_{mm'} \,, \tag{33}$$

where $\delta_{\ell m} = \mathrm{diag}(1, 1, \ldots, 1)$ is a unit matrix (the Kronecker symbol).

In case when there is no dependence on the angle ϕ, the index $m = 0$ in (31) and $C_{\ell 0} = \sqrt{2\ell+1}$. Then, Green's function (29) takes on a more simple form,

$$\mathcal{G}(\boldsymbol{x}, \boldsymbol{x}') = \frac{1}{|\boldsymbol{x}-\boldsymbol{x}'|}$$
$$= \begin{cases} \dfrac{1}{\alpha}\sum\limits_{\ell=0}^{\infty}(2\ell+1) q_\ell\left(\sigma'\right) p_\ell\left(\sigma\right) P_\ell(\cos\theta') P_\ell(\cos\theta) & : (\sigma \le \sigma') \\[2ex] \dfrac{1}{\alpha}\sum\limits_{\ell=0}^{\infty}(2\ell+1) p_\ell\left(\sigma'\right) q_\ell\left(\sigma\right) P_\ell(\cos\theta') P_\ell(\cos\theta) & : (\sigma' \le \sigma) \,, \end{cases} \tag{34}$$

where the Legendre polynomials $P_\ell(\cos\theta)$ are normalized such that

$$\int\limits_0^\pi P_\ell(\cos\theta) P_m(\cos\theta) \sin\theta d\theta = \frac{2}{2\ell+1}\delta_{\ell m} \,. \tag{35}$$

In what follows the following expressions are used for connecting different Legendre polynomials between themselves and with the trigonometric functions,

$$\cos^2\theta = \frac{1}{3}\left[1 + 2P_2\left(\cos\theta\right)\right] ,$$
$$\sin^2\theta = \frac{2}{3}\left[1 - P_2\left(\cos\theta\right)\right] , \tag{36}$$

$$\sin\theta P_{11}(\cos\theta) = -\frac{2}{3}\left[1 - P_2(\cos\theta)\right] ,$$
$$\sin\theta P_{31}(\cos\theta) = -\frac{12}{7}\left[P_2(\cos\theta) - P_4(\cos\theta)\right] , \tag{37}$$
$$\sqrt{1+\sigma^2} q_{11}(\sigma) = \frac{2}{3}\left[q_0(\sigma) + q_2(\sigma)\right] ,$$

$$\sqrt{1+\sigma^2}q_{31}(\sigma) = \frac{12}{7}\left[q_2(\sigma) + q_4(\sigma)\right], \tag{38}$$

in order to make transformations of integrands in the process of calculation of gravitational potentials.

3 Mathematical Model of Matter Distribution and Geometry of the Post-Newtonian Reference Spheroid

3.1 Modeling Matter Distribution

Gravitational field of a stationary-rotating matter is fully described by the particular solutions of the Einstein equations (7), (8) for the metric tensor (6) which include solution of the Poisson-type equation for scalar potential

$$V(\boldsymbol{x}) = V_N(\boldsymbol{x}) + \frac{1}{c^2}V_{pN}(\boldsymbol{x}), \tag{39}$$

$$V_N(\boldsymbol{x}) = G\int\limits_{\mathcal{V}} \frac{\rho(\boldsymbol{x}')}{|\boldsymbol{x}-\boldsymbol{x}'|}d^3x', \tag{40}$$

$$V_{pN}(\boldsymbol{x}) = G\int\limits_{\mathcal{V}} \frac{\rho(\boldsymbol{x}')}{|\boldsymbol{x}-\boldsymbol{x}'|}\left[2v^2(\boldsymbol{x}') + 2V(\boldsymbol{x}') + \Pi(\boldsymbol{x}') + \frac{3p(\boldsymbol{x}')}{\rho(\boldsymbol{x}')}\right]d^3x', \tag{41}$$

and that for a vector (often called gravitomagnetic [73–75]) potential

$$V^i(\boldsymbol{x}) = G\int\limits_{\mathcal{V}} \frac{\rho(\boldsymbol{x}')v^i(\boldsymbol{x}')}{|\boldsymbol{x}-\boldsymbol{x}'|}d^3x', \tag{42}$$

where the field point has harmonic coordinates $\boldsymbol{x}$. In order to calculate the above integrals we have to know the distribution of mass density ρ, pressure p, velocity v^i, and the internal energy density of the fluid Π, as well as the boundary of the volume $\mathcal{V}$ occupied by the fluid.

Real Earth is near equilibrium shape. Small measurable changes in shape of the equipotential surface are from post-glacial viscous rebound, elastic adjustments to the shifting mass from melting glaciers, plate tectonics, and other long-wavelength geoid variations [76]. These factors are important in studying the problem of the dynamic Earth. However, our goal in the present chapter is more pragmatic and relates to the study of relativistic corrections in Earth's gravity field. Therefore, we shall neglect the dynamic changes in the distribution of masses and Earth's shape. According to previous studies [1, 54] the surfaces of the equal gravity potential, density, pressure, and the internal energy coincide both in the Newtonian and the post-Newtonian approximations so that in order to find out their shape it is enough to find out the surfaces of the equal potential. In what follows, we shall follow the model of the normal gravity field in classic geodesy and assume that the density of the fluid is almost uniform with a small post-Newtonian deviation from the homogeneity.

$$\rho(\boldsymbol{x}) = \rho_c\left[1 + \kappa F(\boldsymbol{x})\right] , \tag{43}$$

where ρ_c is a constant density at the center of the spheroid, and $F(\boldsymbol{x})$ is a homothetic function of ellipsoidal distribution with respect to its center,

$$F(\boldsymbol{x}) = \mathcal{A}\frac{q_1}{\varkappa^2}\left(\frac{x^2 + y^2}{a^2} + \frac{z^2}{b^2}\right) , \tag{44}$$

$q_1 \equiv q_1(1/\varkappa)$, and the constant parameter $\mathcal{A} = \mathcal{A}(\epsilon)$ is kept arbitrary in the course of the calculations that follow. We shall find the equations constraining the value of the parameter $\mathcal{A}$ later. Such type of the density distribution has been chosen because it is consistent with the distribution of pressure, at least in the post-Newtonian approximation (see below). The ratio $q_1/\varkappa^2$ was introduced to (44) explicitly to make the subsequent formulas look less cumbersome. We notice that the choice of the distribution (44) allows us to handle calculations analytically in a closed form without series expansion while other assumptions on the mass distribution would lead to analytical results that are more complicated than the results given in this chapter.

Distribution (44) in the ellipsoidal coordinates takes on the following form

$$F(\boldsymbol{x}) = \mathcal{A}q_1(1 - \epsilon^2)R(\sigma, \theta) , \tag{45}$$

where the function

$$R(\sigma, \theta) \equiv \left(1 + \sigma^2\right)\sin^2\theta + \left(1 + \varkappa^2\right)\sigma^2\cos^2\theta . \tag{46}$$

It is worth noticing that the ellipsoidal distribution of density (45) means that the surfaces of constant density are not the same as the surfaces of constant value of the radial coordinate σ. The density ρ remains dependent on the angular coordinate θ everywhere inside the ellipsoid except at its surface, where $\sigma = 1/\varkappa$ with the post-Newtonian accuracy, and $R(\varkappa^{-1}, \theta) = \epsilon^{-2}$. We also draw attention of the reader that in the limiting case of vanishing oblateness, $\varkappa \to 0$, the post-Newtonian correction to the density is not singular because $\lim_{\varkappa\to 0} \varkappa^{-2}q_1 = 1/3$.

Distribution of pressure inside the rotating homogeneous fluid is obtained by integrating the law of the hydrostatic equilibrium. Pressure enters calculations of the integrals characterizing the gravitational field, only at the post-Newtonian terms. Hence, it is sufficient to know its distribution to the Newtonian approximation which is easily obtained by solving the equation of hydrostatic equilibrium [41, 77]

$$p(\boldsymbol{x}) = 2\pi G\rho_c^2 a^2 \frac{q_1}{\varkappa^2}\left[1 - \epsilon^2 R(\sigma, \theta)\right] , \tag{47}$$

where we have denoted $q_1 \equiv q_1(1/\varkappa)$ once again. The internal energy is also required for calculation of the integrals only in terms of the post-Newtonian order of magnitude. In this approximation the internal energy can be considered as constant,

$$\Pi(\boldsymbol{x}) = \Pi_0 , \tag{48}$$

in correspondence with the thermodynamic equation (10) solved for the constant density, $\rho = \rho_c$. From now on we incorporate the constant thermodynamic energy to the central density and will not show Π_0 explicitly in our calculations. Because the fluid rotates uniformly in accordance with the law (12), we have for the distribution of the velocity squared,

$$v^2(\boldsymbol{x}) = \omega^2\alpha^2\left(1 + \sigma^2\right)\sin^2\theta . \tag{49}$$

3.2 *Post-Newtonian Reference Spheroid*

All integrals are calculated over a (yet unknown) volume occupied by the rotating fluid. The surface of the rotating, self-gravitating fluid is a surface of vanishing pressure that coincides with the surface of an equal gravitational potential [1, 53, 54]. In classical geodesy the reference figure for calculation of geoid's undulation is the Maclaurin ellipsoid which is a surface of the second order formed by a rigidly rotating fluid of constant density ρ. Maclaurin's ellipsoid is described by a polynomial [41]

$$\frac{x^2 + y^2}{a^2} + \frac{z^2}{b^2} = 1 , \tag{50}$$

where a and b are semi-major and semi-minor axes of the ellipsoid. We also assume $a > b$, and define the eccentricity of the Maclaurin ellipsoid as

$$e \equiv \frac{\sqrt{a^2 - b^2}}{a} . \tag{51}$$

Physically, the ellipsoidal shape of rotating, self-gravitating fluid is formed because the Newtonian gravity potential is a scalar function represented by a polynomial of the second order with respect to the Cartesian spatial coordinates, and the differential Euler equation defining the equilibrium of the gravity and pressure is of the first order partial different equation which leads to the quadratic (w.r.t. the coordinates) equation of the level surface [41].

We shall demonstrate in the following sections that in the post-Newtonian approximation the gravity potential, W, of the rotating fluid is a polynomial of the fourth order as was first noticed by Chandrasekhar [53]. Hence, the post-Newtonian equation of the level surface of a rigidly-rotating fluid is expected to be a surface of the fourth order. We shall assume that the surface remains axisymmetric in the post-Newtonian approximation and dubbed the body with such a surface as PN-spheroid.

Let coordinates $x^i = \{x, y, z\}$ denote a point on the surface of the PN-spheroid with the axis of symmetry directed along the rotational axis and with the origin located at its post-Newtonian center of mass.[4] Let the rotational axis coincide with the direction of z axis. Then, the most general equation of the PN-spheroid is

$$\frac{x^2+y^2}{a^2}+\frac{z^2}{b^2}=1+\kappa Q(\boldsymbol{x}) , \tag{52}$$

where $\kappa \equiv \pi G\rho a^2/c^2$, and function

$$Q(\boldsymbol{x}) \equiv \mathcal{K}_1\frac{x^2+y^2}{a^2}+\mathcal{K}_2\frac{z^2}{b^2}+\mathcal{B}_1\left(\frac{x^2+y^2}{a^2}\right)^2+\mathcal{B}_2\frac{z^4}{b^4}+\mathcal{B}_3\frac{\left(x^2+y^2\right)z^2}{a^2b^2} , \tag{53}$$

and $\mathcal{K}_1$, $\mathcal{K}_2$, $\mathcal{B}_1$, $\mathcal{B}_2$, $\mathcal{B}_3$ are arbitrary numerical coefficients.

Each cross-section of the PN-spheroid being orthogonal to the rotational axis, represents a circle. The equatorial cross-section has an equatorial radius, $\sigma = r_e$, being determined from (52) by the condition $z = 0$. It yields

$$r_e = a\left[1+\frac{1}{2}\kappa\left(\mathcal{K}_1+\mathcal{B}_1\right)\right] . \tag{54}$$

The meridional cross-section of the PN-spheroid is no longer an ellipse (as it was in case of the Maclaurin ellipsoid) but a curve of the fourth order. Nonetheless, we can define the polar radius, $z = r_p$, of the PN-spheroid by the condition, $x = y = 0$. Equation (52) yields

$$r_p = b\left[1+\frac{1}{2}\kappa\left(\mathcal{K}_2+\mathcal{B}_2\right)\right] . \tag{55}$$

The equatorial and polar radii of the PN-spheroid should be used in the post-Newtonian approximation instead of the equatorial and polar radii of the Maclaurin reference-ellipsoid for calculation of observable physical effects like the normal gravity force. We characterize the 'oblateness' of the PN-spheroid by the post-Newtonian eccentricity

$$\epsilon \equiv \frac{\sqrt{r_e^2-r_p^2}}{r_e} . \tag{56}$$

It differs from the eccentricity (51) of the Maclaurin ellipsoid by relativistic correction

[4] Post-newtonian definitions of mass, center of mass, and other multipole moments can be found, for example, in [5].

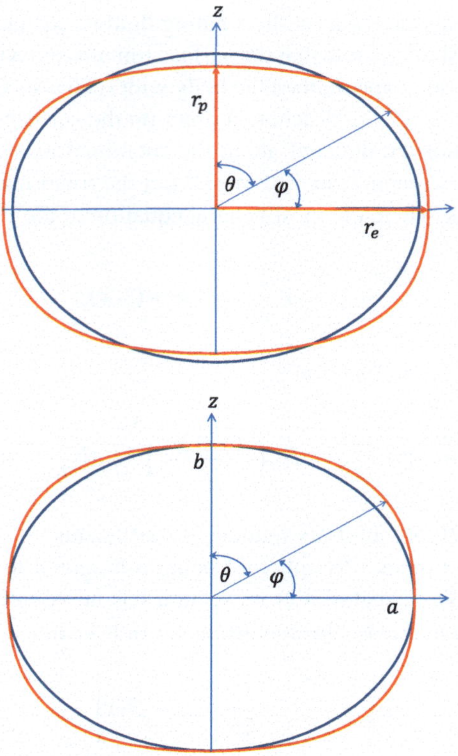

Fig. 1 Meridional cross-section of the PN-spheroid (a red curve in on-line version) versus the Maclaurin ellipsoid (a blue curve in on-line version). The top panel represents the most general case with arbitrary values of the PN-spheroid shape parameters $\mathcal{K}_1$, $\mathcal{K}_2$, $\mathcal{B}_1$, $\mathcal{B}_2$ when the equatorial, r_e, and polar, r_p, radii of the PN-spheroid differ from the semi-major, a, and semi-minor, b, axes of the Maclaurin ellipsoid, $r_e \neq a$, $r_p \neq b$. The bottom panel shows the most important physical case of $\mathcal{B}_1 = \mathcal{K}_1$, $\mathcal{B}_2 = \mathcal{K}_2$ when the equatorial and polar radii of the PN-spheroid and the Maclaurin ellipsoid are equal. The angle φ is the geographic latitude ($-90° \leq \varphi \leq 90°$), and the angle θ is a complementary angle (co-latitude) used for calculation of integrals in appendix of the present chapter ($0 \leq \theta \leq \pi$). In general, when $\mathcal{B}_1 \neq \mathcal{K}_1$, and/or $\mathcal{B}_2 \neq \mathcal{K}_2$, the maximal radial difference (the 'height' difference) between the surface of the PN-spheroid and that of the Maclaurin ellipsoid can amount to a relatively large value of several centimeters, and even more. In case of $\mathcal{B}_1 = \mathcal{K}_1$, $\mathcal{B}_2 = \mathcal{K}_2$ the radial undulation between the two surfaces is defined by the parameter $\mathcal{B}_3 \equiv \mathcal{B}$, and it does not exceed one centimeter

$$\epsilon = e - \kappa \frac{1 - e^2}{2e} \Big[(\mathcal{K}_2 - \mathcal{K}_1) + (\mathcal{B}_2 - \mathcal{B}_1) \Big] . \tag{57}$$

PN-spheroid versus the Maclaurin ellipsoid is visualized in Fig. 1.

Theoretical formalism for calculating the post-Newtonian level surface can be worked out in arbitrary coordinates. For mathematical and historic reasons the most convenient are harmonic coordinates which are also used by the IAU [27] and IERS [11]. The class of the harmonic coordinates is selected by the gauge condition (4).

Different harmonic coordinates are interrelated by coordinate transformations which are not violating the gauge condition (4). This freedom is known as a residual gauge (or coordinate) freedom. The field equations (7), (8) and their solutions are form-invariant with respect to the residual gauge transformations.

The residual gauge freedom is described by a post-Newtonian coordinate transformation,

$$x'^{\alpha} = x^{\alpha} + \kappa\xi^{\alpha}(x) , \tag{58}$$

where functions, ξ^{α}, obey the Laplace equation,

$$\Delta\xi^{\alpha} = 0 . \tag{59}$$

Solution of the Laplace equation which is convergent at the origin of the coordinate system, is given in terms of the harmonic polynomials which are selected by the conditions imposed by the statement of the problem. In our case, the problem is to determine the shape of the PN-spheroid which has the surface described by the polynomial of the fourth order (62) with yet unknown coefficients $\mathcal{B}_1, \mathcal{B}_2, \mathcal{B}_3$. The form of the Eq. (62) does not change (in the post-Newtonian approximation) if the functions ξ^{α} in (58) are polynomials of the third order. It is straightforward to show that the admissible harmonic polynomials of the third order have the following form

$$\xi^1 = hx + p\frac{x}{a^2}\left(\sigma^2 - 4z^2\right) , \tag{60a}$$

$$\xi^2 = hy + p\frac{y}{a^2}\left(\sigma^2 - 4z^2\right) , \tag{60b}$$

$$\xi^3 = kz + q\frac{z}{b^2}\left(3\sigma^2 - 2z^2\right) , \tag{60c}$$

where h, k, p and q are arbitrary constant parameters. Polynomials (60a)–(60c) represent solutions of the Laplace equation (59). We choose $\xi^0 = 0$ because we consider stationary spacetime so all functions are time-independent.

Coordinate transformation (58) with ξ^i taken from (60a)–(60c) does not violate the harmonic gauge condition (4) but it changes the numerical post-Newtonian coefficients in the mathematical form of Eqs. (52) and (53)

$$\mathcal{K}_1 \to \mathcal{K}_1 + 2h , \tag{61a}$$

$$\mathcal{K}_2 \to \mathcal{K}_2 + 2k , \tag{61b}$$

$$\mathcal{B}_1 \to \mathcal{B}_1 + 2p , \tag{61c}$$

$$\mathcal{B}_2 \to \mathcal{B}_2 - 4q , \tag{61d}$$

$$\mathcal{B}_3 \to \mathcal{B}_3 - 8p\frac{b^2}{a^2} + 6q\frac{a^2}{b^2} , \tag{61e}$$

Thus, it makes evident that only one out of the five coefficients $\mathcal{K}_1, \mathcal{K}_2, \mathcal{B}_1, \mathcal{B}_2, \mathcal{B}_3$ is algebraically independent while the four others can be chosen arbitrary. To simplify our calculations and eliminate the gauge-dependent terms from mathematical equations we decide to fix the numerical values of four parameters $\mathcal{K}_1, \mathcal{K}_2, \mathcal{B}_1, \mathcal{B}_2$. The constant $\mathcal{B}_3$ is left free. It is fixed by the condition of a hydrostatic equilibrium of the rotating fluid body (see Sect. 7.3).

One of the most simple and attractive choice of fixing the residual gauge freedom is $\mathcal{K}_1 = \mathcal{K}_2 = \mathcal{B}_1 = \mathcal{B}_2 = 0$. This choice of the residual gauge has been employed in our papers [3, 4]. It is particularly useful for conducting calculations in the Cartesian coordinates. With such a choice of the coordinates the polar radius $r_p = b$, the equatorial radius $r_e = a$, the eccentricity of the PN-spheroid $\epsilon = e$, and function

$$Q(\boldsymbol{x}) \equiv \mathcal{B}_3 \frac{\sigma^2 z^2}{a^2 b^2} \,. \tag{62}$$

In the ellipsoidal coordinates it is more convenient to define the surface of the rotating fluid by the following equation,

$$\sigma_s = \frac{1}{\varkappa}\left[1 + \mathcal{B}\frac{\omega^2 a^2}{c^2 \epsilon^2} P_2(\cos\theta)\right] , \tag{63}$$

where $\mathcal{B} = \mathcal{B}(\epsilon)$ is a constant arbitrary parameter which possible numerical value will be discussed below at the end of Sect. 6.1 and in Sect. 7.3. The reason for picking up equation of the surface of PN-spheroid in the form of (63) is a matter of mathematical convenience. It is worth making two remarks. First, the appearance of ϵ^2, in the denominator in the right side of (63) does not lead to divergence as the angular velocity of rotation $\omega^2 \sim \epsilon^2$. Second, Eq. (63) corresponds to the following choice of parameters in (53):

$$\mathcal{B}_1 = \mathcal{B}_2 = 0, \qquad \mathcal{K}_1 = -\frac{4q_2}{\varkappa^3}\mathcal{B}, \qquad \mathcal{K}_2 = \frac{16q_2}{\varkappa^3}\mathcal{B}, \qquad \mathcal{B}_3 = \frac{4q_2}{\varkappa^3}\mathcal{B} \,, \tag{64}$$

where parameter $\mathcal{B}$ is the same as in (63), $q_2 \equiv q_2(1/\varkappa)$, and the angular velocity ω is related to q_2 by means of the Maclaurin relationship (86). Parameterization (64) does not allow us to keep the equatorial, r_e, and polar, r_p, axes of the PN-spheroid as well as its eccentricity equal to the parameters of the reference Maclaurin ellipsoid (see Eqs. (54)–(57)). However, parameterization (64) simplifies calculation of integrals in the equatorial coordinates and will be used throughout this chapter.

With the definition (63) of the upper limit of the integration with respect to the radial coordinate σ, the volume integral from an arbitrary function

$$\mathcal{F}(\boldsymbol{x}, \boldsymbol{x}') = \begin{cases} \mathcal{F}_<(\sigma, \theta, \sigma', \theta') & \text{if } \sigma' \le \sigma \,, \\ \mathcal{F}_>(\sigma, \theta, \sigma', \theta') & \text{if } \sigma' \ge \sigma \,. \end{cases} \tag{65}$$

can be calculated with sufficient accuracy as a sum of two terms:

(1) in case when the radial coordinate σ of the field point $\boldsymbol{x}$ is taken inside the body

$$\int_{\mathcal{V}} \mathcal{F}(\boldsymbol{x},\boldsymbol{x}')d^3x' = 2\pi\alpha^3 \int_0^{\sigma}\int_0^{\pi} \mathcal{F}_<(\sigma,\theta,\sigma',\theta')(\sigma'^2+\cos^2\theta')d\sigma' d\theta'$$
$$+2\pi\alpha^3 \int_{\sigma}^{1/\varkappa}\int_0^{\pi} \mathcal{F}_>(\sigma,\theta,\sigma',\theta')(\sigma'^2+\cos^2\theta')d\sigma' d\theta'$$
$$+2\pi\mathcal{B}\frac{\omega^2a^4b}{c^2}\int_0^{\pi} \mathcal{F}_>\left(\sigma,\theta,\varkappa^{-1},\theta'\right)\left(\varkappa^{-2}+\cos^2\theta'\right)P_2(\cos\theta')d\theta' , \quad (66)$$

(2) in case when the radial coordinate σ of the field point $\boldsymbol{x}$ is taken outside the body

$$\int_{\mathcal{V}} \mathcal{F}(\boldsymbol{x},\boldsymbol{x}')d^3x' = 2\pi\alpha^3 \int_0^{1/\varkappa}\int_0^{\pi} \mathcal{F}_<(\sigma,\theta,\sigma',\theta')(\sigma'^2+\cos^2\theta')d\sigma' d\theta'$$
$$+2\pi\mathcal{B}\frac{\omega^2a^4b}{c^2}\int_0^{\pi} \mathcal{F}_<\left(\sigma,\theta,\varkappa^{-1},\theta'\right)\left(\varkappa^{-2}+\cos^2\theta'\right)P_2(\cos\theta')d\theta' . \quad (67)$$

The very last integral in the right hand side of (66) and (67) is of the post-Newtonian order of magnitude and will be treated as a post-Newtonian correction to the Newtonian gravitational potential.

4 Newtonian Gravitational Potential

The Newtonian gravitational potential V_N is given by (40) where the density distribution, $\rho(\boldsymbol{x})$, is defined in (43) and the integration is performed over the volume bounded by the radial coordinate σ_s in (63). The integral can be split in three parts:

$$V_N = V_N[\rho_c, \mathcal{S}] + V_N[\delta\rho, \mathcal{S}] + V_N[\rho_c, \delta\mathcal{S}] , \quad (68)$$

where $V_N[\rho_c, \mathcal{S}]$ denotes contribution from the constant density ρ_c, $V_N[\delta\rho, \mathcal{S}]$ is contribution from the variation $\delta\rho \equiv c^{-2}F(\boldsymbol{x})$ of the density given by (45), and $V_N[\rho_c, \delta\mathcal{S}]$ represents contribution from the fraction of the constant density ρ_c enclosed in the part of the volume lying between the real boundary and that of the Maclaurin ellipsoid. The integrals $V_N[\rho_c, \mathcal{S}]$ and $V_N[\delta\rho, \mathcal{S}]$ are volume integrals taken over the Maclaurin ellipsoid with the fixed value of the radial coordinate on its boundary, $\sigma = 1/\varkappa$. The term $V_N[\rho_c, \delta\mathcal{S}]$ comes from the very last integrals in

(66) and (67). Below we provide specific details of calculations of the three terms entering the right hand side of (68).

4.1 Integral Contribution to the Newtonian Potential from Constant Density

Contribution from the constant density ρ_c to the Newtonian potential is given by integral

$$V_N[\rho_c, \mathcal{S}] = G\rho_c \int\limits_{\mathcal{V}} \frac{d^3x'}{|\boldsymbol{x} - \boldsymbol{x}'|} , \tag{69}$$

which can be calculated by making use of the Green function (29). Depending on position of the field point $\boldsymbol{x}$ in space we distinguish the internal and external solutions.

4.1.1 The Internal Solution

The internal solution is valid for the field point $\boldsymbol{x}$ with the radial coordinate, $0 \le \sigma \le 1/\varkappa$. With the help of the Green function (34) it reads,

$$\begin{aligned} V_N[\rho_c, \mathcal{S}] = 2\pi G\rho_c\alpha^2 \sum_{\ell=0}^{\infty}(2\ell+1)q_\ell(\sigma)\, P_\ell(\cos\theta) \times \\ \int\limits_0^{\sigma} d\sigma' \int\limits_0^{\pi} \left(\sigma'^2 + \cos^2\theta'\right) p_\ell\left(\sigma'\right) P_\ell(\cos\theta') \sin\theta' d\theta' \\ +2\pi G\rho_c\alpha^2 \sum_{\ell=0}^{\infty}(2\ell+1)p_\ell(\sigma)\, P_\ell(\cos\theta) \times \\ \int\limits_\sigma^{1/\varkappa} d\sigma' \int\limits_0^{\pi} \left(\sigma'^2 + \cos^2\theta'\right) q_\ell\left(\sigma'\right) P_\ell(\cos\theta') \sin\theta' d\theta' . \end{aligned} \tag{70}$$

We, first, integrate with respect to the angular variable θ and, then, with respect to the radial coordinate σ. We also use the relation

$$\sigma^2 + \cos^2\theta = \frac{2}{3}\left[p_2(\sigma) + P_2(\cos\theta)\right] , \tag{71}$$

that allows us to operate with the Legendre polynomials instead of the trigonometric functions, and use the condition of orthogonality (35). Then, after substituting (71) to (70) and making use of the normalization condition (35), we obtain

$$V_N[\rho_c, \mathcal{S}] = V_0(\sigma) + V_2(\sigma) P_2(\cos\theta) \,, \tag{72}$$

where

$$V_0(\sigma) = \frac{8\pi}{3} G\rho_c \alpha^2 \left[q_0(\sigma) \int_0^{\sigma} d\sigma' p_2(\sigma') + \int_{\sigma}^{1/\varkappa} d\sigma' p_2(\sigma') q_0(\sigma') \right] \,, \tag{73}$$

$$V_2(\sigma) = \frac{8\pi}{3} G\rho_c \alpha^2 \left[q_2(\sigma) \int_0^{\sigma} d\sigma' p_2(\sigma') + p_2(\sigma) \int_{\sigma}^{1/\varkappa} d\sigma' q_2(\sigma') \right] \,. \tag{74}$$

Calculation of the integrals yields

$$V_0(\sigma) = \frac{2\pi G\rho_c \alpha^2}{3\varkappa^3} \left[\varkappa \left(1 - \varkappa^2 \sigma^2\right) + 2\left(1 + \varkappa^2\right) \arctan \varkappa \right] , \tag{75}$$

$$V_2(\sigma) = -\frac{2\pi G\rho_c \alpha^2}{3\varkappa^3} \left[\varkappa + \varkappa \left(3 + 2\varkappa^2\right) \sigma^2 - \left(1 + 3\sigma^2\right) \left(1 + \varkappa^2\right) \arctan \varkappa \right] . \tag{76}$$

Adding up the two expressions according to (72) and making the inverse transformation from the ellipsoidal to Cartesian coordinates, we get

$$V_N[\rho_c, \mathcal{S}] = \pi G\rho_c \alpha^2 \left[2\frac{1+\varkappa^2}{\varkappa^3} \arctan \varkappa - 2\frac{z^2}{\alpha^2} + \frac{x^2 + y^2 - 2z^2}{\alpha^2 \varkappa^2} \left(1 - \frac{1+\varkappa^2}{\varkappa} \arctan \varkappa \right) \right] . \tag{77}$$

This expression for the Newtonian potential $V_N[\rho_c]$ inside the Maclaurin ellipsoid is well-known from the classic theory of figures of rotating fluid bodies [41, 78] (see also [4, eq. 50]). It is straightforward to check by direct differentiation that (77) satisfies the Poisson equation $\Delta V_N[\rho_c] = -4\pi G\rho_c$, in accordance with (69).

4.1.2 The External Solution

Making use of the Green function (34) we get for the field point $\boldsymbol{x}$ with the radial ellipsoidal coordinate, $\sigma > \varkappa^{-1}$, the following external solution for the Newtonian potential of the homogeneous Maclaurin ellipsoid,

$$V_N[\rho_c, \mathcal{S}] = 2\pi G\rho_c \alpha^2 \sum_{\ell=0}^{\infty} (2\ell+1) q_\ell(\sigma) P_\ell(\cos\theta) \times \int_0^{1/\varkappa} d\sigma' \int_0^{\pi} \left(\sigma'^2 + \cos^2\theta'\right) p_\ell(\sigma') P_\ell(\cos\theta') \sin\theta' d\theta' \,. \tag{78}$$

We again integrate over the angular variable θ and, then, with respect to the radial variable σ. It yields,

$$V_N[\rho_c, \mathcal{S}] = \frac{4\pi G\rho_c\alpha^2}{3\varkappa^3}\left(1+\varkappa^2\right)\left[q_0(\sigma)+q_2(\sigma)P_2(\cos\theta)\right] . \tag{79}$$

In the asymptotic regime at spatial infinity, when the radial coordinate σ is very large, the Legendre functions have the following asymptotic behavior

$$q_0(\sigma) = \frac{1}{r} + \mathcal{O}\left(\frac{1}{r^3}\right) , \qquad q_2(\sigma) = \frac{2}{15r^3} + \mathcal{O}\left(\frac{1}{r^5}\right) , \tag{80}$$

so that the asymptotic expression for the Newtonian external gravitational potential at large distances from the body, is

$$V_N[\rho_c, \mathcal{S}] = \frac{Gm_N}{r} + \mathcal{O}\left(\frac{1}{r^3}\right) , \tag{81}$$

where the notation $m_N \equiv M_N/\alpha$, and M_N is the Newtonian mass of the Maclaurin ellipsoid

$$M_N = \frac{4\pi\rho_c\alpha^3}{3}\frac{1+\varkappa^2}{\varkappa^3} = \frac{4\pi\rho_c a^2}{3}\frac{\alpha}{\varkappa} = \frac{4\pi\rho_c a^2 b}{3} . \tag{82}$$

Therefore, expression (79) can be simplified to

$$V_N[\rho_c, \mathcal{S}] = Gm_N\Big[q_0(\sigma) + q_2(\sigma)P_2(\cos\theta)\Big] . \tag{83}$$

It is worth noticing that on the surface of the rotating body the two expressions for the internal and external gravitational potential, (72) and (83) match smoothly for the gravitational potential is a continuous function. It is also useful to remark that for a fixed value of the fluid's density, ρ_c, the normalized mass, m_N, decreases inversely proportional to the eccentricity: $m_N \sim \varkappa^{-1} \sim \epsilon^{-1}$.

The surface of the Maclaurin ellipsoid is equipotential, and it is defined by equation

$$\frac{1}{2}v^2 + V_N[\rho_c, \mathcal{S}] = W_0 = \text{const.} , \tag{84}$$

where v^2 is defined in (49). After taking into account (36) and (83), equation of the ellipsoid reads,

$$W_0 = \frac{\omega^2\alpha^2}{3\varkappa^2}\left(1+\varkappa^2\right)\left[1-P_2(\cos\theta)\right] + Gm_N\big[q_0 + q_2P_2(\cos\theta)\big] , \tag{85}$$

where we have introduced shorthand notations $q_0 \equiv q_0(1/\varkappa)$ and $q_2 \equiv q_2(1/\varkappa)$. Since the left hand side of (85) is constant, the right hand side of it must be constant as well. It yields two relationships

$$\omega^2 = \frac{4\pi G\rho_c}{\varkappa} q_2 , \tag{86}$$

$$W_0 = \frac{1}{3}\omega^2 a^2 + G m_N q_0 . \tag{87}$$

Equations (86), (87) can be recast to yet another form,

$$\omega^2 = \frac{3 G m_N}{a^2} q_2 , \tag{88}$$

$$W_0 = G m_N \left[q_0 + q_2 \right] . \tag{89}$$

Equation (86) yields relation between the angular velocity of rotation of the homogeneous fluid and oblateness of the Maclaurin ellipsoid while (87) or, equivalently, (89) defines the gravity potential on its surface. For small values of the second eccentricity $\varkappa \ll 1$, when the deviation of the ellipsoid from a sphere is very small, we can expand the Legendre function q_2 in the Taylor series, which yields the asymptotic expression for the angular velocity

$$\omega^2 = \frac{8}{15} G\pi\rho_c \varkappa^2 \left(1 - \frac{6}{7}\varkappa^2\right) + \mathcal{O}\left(\varkappa^6\right) , \qquad (\varkappa \ll 1) . \tag{90}$$

On the other hand, when the ellipsoid has a disk-like shape, we have $\varkappa \gg 1$, and the asymptotic expression of the angular velocity takes on another form,

$$\omega^2 = \frac{G\pi\rho_c}{\varkappa} \left(\pi - \frac{2}{\varkappa}\right) + \mathcal{O}\left(\frac{1}{\varkappa^3}\right) , \qquad (\varkappa \gg 1) . \tag{91}$$

Equation (86) tells us that the angular velocity of rotation of the Maclaurin ellipsoid, $\omega = \omega(\varkappa)$, considered as a function of the eccentricity, $\varkappa$, has a maximum which is reached for $\varkappa \simeq 2.52931$ [78]. The maximal value of the angular velocity of the Maclaurin ellipsoid at this point is $\omega^2 \simeq 0.45\pi G\rho_c$ [77] .

4.2 *Integral Contribution to the Newtonian Potential from the Density Inhomogeneity*

Contribution from the non-homogeneous part of the mass density to the Newtonian potential is given by the integral

$$V_N[\delta\rho_c, \mathcal{S}] = \frac{G\rho_c}{c^2} \int_{\mathcal{V}} \frac{F(\boldsymbol{x}') d^3x'}{|\boldsymbol{x} - \boldsymbol{x}'|} , \tag{92}$$

where function $F(\boldsymbol{x})$ is given in (43)–(45). Making use of (45) in the integral (92), brings it to the form,

$$V_N[\delta\rho_c, \mathcal{S}] = \mathcal{A}\frac{\pi G^2\rho_c^2\alpha^2 b^2 q_1}{c^2} I_1(\boldsymbol{x}) \;, \tag{93}$$

where we have introduced a notation

$$I_1(\boldsymbol{x}) \equiv \frac{1}{\alpha^2}\int\limits_{\mathcal{V}} \frac{R(\sigma', \theta')d^3x'}{|\boldsymbol{x}-\boldsymbol{x}'|} \;. \tag{94}$$

The integral (94) is calculated with making use of the Green function (34). We consider the internal, ($\sigma \le 1/\varkappa$), and external, ($\sigma \ge 1/\varkappa$), solutions separately.

4.2.1 The Internal Solution

Making use of the Green's functions (34) we get,

$$\begin{aligned}
I_1(\sigma,\theta) &= 2\pi\sum_{\ell=0}^{\infty}(2\ell+1)q_\ell\,(\sigma)\,P_\ell(\cos\theta) \\
&\quad\times\int\limits_0^\sigma d\sigma'\int\limits_0^\pi\left(\sigma'^2+\cos^2\theta'\right)R(\sigma',\theta')p_\ell\,(\sigma')\,P_\ell(\cos\theta')d\theta' \\
&\quad+2\pi\sum_{\ell=0}^{\infty}(2\ell+1)p_\ell\,(\sigma)\,P_\ell(\cos\theta) \\
&\quad\times\int\limits_\sigma^{1/\varkappa} d\sigma'\int\limits_0^\pi\left(\sigma'^2+\cos^2\theta'\right)R(\sigma',\theta')q_\ell\,(\sigma')\,P_\ell(\cos\theta')d\theta' \;.
\end{aligned} \tag{95}$$

We, first, integrate with respect to the angular variable θ and, then, with respect to the radial variable σ. The integrand of the above integral is

$$\begin{aligned}
\left(\sigma^2+\cos^2\theta\right)R(\sigma,\theta) &= \frac{2}{105}\left[\left(5-3\varkappa^2\right)p_2(\sigma)+4\left(3+\varkappa^2\right)p_4(\sigma)\right] \\
&\quad+\frac{2}{105}\left[5-3\varkappa^2+8\varkappa^2p_4(\sigma)\right]P_2(\cos\theta) \\
&\quad-\frac{8}{105}\left[3+\varkappa^2-2\varkappa^2p_2(\sigma)\right]P_4(\cos\theta) \;.
\end{aligned} \tag{96}$$

After integration over the angular variable θ, the integral can be represented as a linear combination of several terms,

$$I_1(\sigma,\theta) = I_{10}(\sigma) + I_{12}(\sigma)P_2(\cos\theta) + I_{14}(\sigma)P_4(\cos\theta)\,, \tag{97}$$

where each part corresponds to its own Legendre polynomial,

$$I_{10}(\sigma) = \frac{8\pi}{105}\left\{q_0(\sigma)\int_0^\sigma d\sigma'\left[\left(5-3\varkappa^2\right)p_2(\sigma') + 4\left(3+\varkappa^2\right)p_4(\sigma')\right]\right.$$
$$\left. + \int_\sigma^{1/\varkappa} d\sigma'\left[\left(5-3\varkappa^2\right)p_2(\sigma') + 4\left(3+\varkappa^2\right)p_4(\sigma')\right]q_0(\sigma')\right\}, \tag{98}$$

$$I_{12}(\sigma) = \frac{8\pi}{105}\left\{q_2(\sigma)\int_0^\sigma d\sigma'\left[5-3\varkappa^2+8\varkappa^2 p_4(\sigma')\right]p_2(\sigma')\right.$$
$$\left. + p_2(\sigma)\int_\sigma^{1/\varkappa} d\sigma'\left[5-3\varkappa^2+8\varkappa^2 p_4(\sigma')\right]q_2(\sigma')\right\}, \tag{99}$$

$$I_{14}(\sigma) = -\frac{32\pi}{105}\left\{q_4(\sigma)\int_0^\sigma d\sigma'\left[3+\varkappa^2-2\varkappa^2 p_2(\sigma')\right]p_4(\sigma')\right.$$
$$\left. + p_4(\sigma)\int_\sigma^{1/\varkappa} d\sigma'\left[3+\varkappa^2-2\varkappa^2 p_2(\sigma')\right]q_4(\sigma')\right\}. \tag{100}$$

Calculation of the integrals reveals

$$I_{10}(\sigma) = \frac{\pi}{15\varkappa^5}\left[\varkappa\left(1-\varkappa^2\sigma^2\right)\left(3+5\varkappa^2+3\varkappa^2\sigma^2+\varkappa^4\sigma^2\right)\right.$$
$$\left. + 12\left(1+\varkappa^2\right)^2\arctan\varkappa\right], \tag{101}$$

$$I_{12}(\sigma) = -\frac{2\pi}{21\varkappa^4}\left[3+5\varkappa^2+\sigma^2\left(9+15\varkappa^2+2\varkappa^4\right)\right.$$
$$\left. + 2\varkappa^6\sigma^4 - 3\left(1+\varkappa^2\right)^2\left(1+3\sigma^2\right)\frac{\arctan\varkappa}{\varkappa}\right], \tag{102}$$

$$I_{14}(\sigma) = \frac{\pi}{105\varkappa^4}\left[9+15\varkappa^2+6\sigma^2\left(15+25\varkappa^2+8\varkappa^4\right)\right.$$
$$+\sigma^4\left(105+175\varkappa^2+56\varkappa^4-8\varkappa^6\right)$$
$$\left. - 3\left(1+\varkappa^2\right)^2\left(3+30\sigma^2+35\sigma^4\right)\frac{\arctan\varkappa}{\varkappa}\right]. \tag{103}$$

4.2.2 The External Solution

The external solution is obtained by making use of the Green function (34)

$$I_1(\sigma,\theta) = 2\pi \sum_{\ell=0}^{\infty} (2\ell+1) q_\ell(\sigma) P_\ell(\cos\theta) \int_0^{1/\varkappa} d\sigma' \int_0^{\pi} \left(\sigma'^2 + \cos^2\theta'\right) \times R(\sigma',\theta') p_\ell\left(\sigma'\right) P_\ell(\cos\theta') d\theta' \,. \tag{104}$$

The result is

$$\begin{aligned} I_1(\sigma,\theta) = & \frac{8\pi}{105} q_0(\sigma) \int_0^{1/\varkappa} d\sigma' \left[\left(5 - 3\varkappa^2\right) p_2(\sigma') + 4\left(3 + \varkappa^2\right) p_4(\sigma')\right] \\ & + \frac{8\pi}{105} q_2(\sigma) P_2(\cos\theta) \int_0^{1/\varkappa} d\sigma' \left[5 - 3\varkappa^2 + 8\varkappa^2 p_4(\sigma')\right] p_2(\sigma') \\ & - \frac{32\pi}{105} q_4(\sigma) P_4(\cos\theta) \int_0^{1/\varkappa} d\sigma' \left[3 + \varkappa^2 - 2\varkappa^2 p_2(\sigma')\right] p_4(\sigma') \,. \end{aligned} \tag{105}$$

After performing the integrals it results in

$$I_1(\sigma,\theta) = \frac{4\pi}{35} \frac{\left(1+\varkappa^2\right)^2}{\varkappa^5} \left[7q_0(\sigma) + 5q_2(\sigma) P_2(\cos\theta) - 2q_4(\sigma) P_4(\cos\theta)\right] \,. \tag{106}$$

4.3 Integral Contribution to the Newtonian Potential from the Difference Between the Volumes of PN Spheroid and Maclaurin Ellipsoid

Contribution $V_N[\rho_c, \delta\mathcal{S}]$ from the spheroidal deviation of the shape of the rotating fluid from the Maclaurin ellipsoid is given by Eqs. (66), (67), where function $\mathcal{F}$ is proportional to the Green function (34). More specifically,

$$V_N[\rho_c, \delta\mathcal{S}] = \frac{4\pi}{c^2} \mathcal{B} G \rho_c \omega^2 \alpha^4 I_2(\sigma,\theta) \,, \tag{107}$$

where

$$I_2(\sigma,\theta)=\begin{cases}\frac{1}{2}\sum_{\ell=0}^{\infty}(2\ell+1)q_\ell p_\ell(\sigma)P_\ell(\cos\theta) \\ \times\int\limits_0^\pi\left(\varkappa^{-2}+\cos^2\theta'\right)P_2(\cos\theta')P_\ell(\cos\theta')d\theta' & :(\sigma\le 1/\varkappa)\\ \\ \frac{1}{2}\sum_{\ell=0}^{\infty}(2\ell+1)p_\ell q_\ell(\sigma)P_\ell(\cos\theta) \\ \times\int\limits_0^\pi\left(\varkappa^{-2}+\cos^2\theta'\right)P_2(\cos\theta')P_\ell(\cos\theta')d\theta' & :(\sigma\ge 1/\varkappa)\end{cases} \tag{108}$$

Calculation of the integral in (108) is performed with the help of (35) yielding

$$\int\limits_0^\pi\left(\varkappa^{-2}+\cos^2\theta'\right)P_2(\cos\theta')P_\ell(\cos\theta')d\theta'$$
$$=\frac{2}{2\ell+1}\left[\frac{2}{15}\delta_{0\ell}+\left(\frac{1}{\varkappa^2}+\frac{11}{21}\right)\delta_{2\ell}+\frac{12}{35}\delta_{4\ell}\right]. \tag{109}$$

It yields,

$$I_2(\sigma,\theta)=\begin{cases}\frac{2q_0}{15}+\left(\frac{1}{\varkappa^2}+\frac{11}{21}\right)q_2p_2(\sigma)P_2(\cos\theta)\\ +\frac{12q_4}{35}p_4(\sigma)P_4(\cos\theta) & :(\sigma\le 1/\varkappa)\\ \\ \frac{2}{15}q_0(\sigma)+\left(\frac{1}{\varkappa^2}+\frac{11}{21}\right)p_2q_2(\sigma)P_2(\cos\theta)\\ +\frac{12p_4}{35}q_4(\sigma)P_4(\cos\theta) & :(\sigma\ge 1/\varkappa)\end{cases} \tag{110}$$

where $q_0\equiv q_0(1/\varkappa)$, $q_2\equiv q_2(1/\varkappa)$, $q_4\equiv q_4(1/\varkappa)$, $p_2\equiv p_2(1/\varkappa)$, $p_4\equiv p_4(1/\varkappa)$.

5 Post-Newtonian Potentials

5.1 *Scalar Potential V_{pN}*

The post-Newtonian correction (41) to the Newtonian gravity potential obeys the Poisson equation

$$\Delta V_{pN}(\boldsymbol{x})=-4\pi G\rho_{pN}(\boldsymbol{x}), \tag{111}$$

where

$$\rho_{pN}(\boldsymbol{x})\equiv\rho_c\left[2v^2(\boldsymbol{x})+2V_N(\boldsymbol{x})+\Pi(\boldsymbol{x})\right]+3p(\boldsymbol{x}), \tag{112}$$

and the functions entering the right hand side of (112) are defined by Eqs. (47)–(49) and (72). Fock had proved (see [66, Eq. 73.26]) that for any (including a homogeneous) distribution of mass density the following equality holds

$$\rho(\boldsymbol{x})\left[\frac{1}{2}v^2(\boldsymbol{x}) + V_N(\boldsymbol{x})\right] = p(\boldsymbol{x}) + \rho(\boldsymbol{x})\Pi(\boldsymbol{x}) \ . \tag{113}$$

It can be used in order to re-write (112) as follows

$$\rho_{pN}(\boldsymbol{x}) \equiv \rho_c\left[v^2(\boldsymbol{x}) + 3\Pi(\boldsymbol{x})\right] + 5p(\boldsymbol{x}) \ . \tag{114}$$

This allows to eliminate the Newtonian gravitational potential V_N from the calculation of the post-Newtonian gravitational potential V_{pN} by solving (111).

After making use of expressions (47)–(49) and including the constant term $3\Pi = 3\Pi_0$, to the constant density (re-normalizing the central density ρ_c) we get in the elliptical coordinates,

$$\rho_{pN}(\boldsymbol{x}) = \rho_c\omega^2\alpha^2\left(1+\sigma^2\right)\sin^2\theta - 10\pi G\rho_c^2 b^2 q_1 R(\sigma,\theta) \ . \tag{115}$$

Integrating (111) directly with the help of the Green function (34), yields

$$V_{pN}(\boldsymbol{x}) = -10\pi G^2\rho_0^2\alpha^4\frac{q_1}{\varkappa^2}I_1(\sigma,\theta) + G\rho_c\omega^2\alpha^4 I_3(\sigma,\theta) \ , \tag{116}$$

where the integral $I_1(\sigma,\theta)$ has been calculated in Sect. 4.2, and $I_3(\sigma,\theta)$ is the integral from function $(1+\sigma^2)\sin^2\theta$ to the post-Newtonian gravitational potential V_{pN}. The integral $I_3(\sigma,\theta)$ is performed as follows.

5.2 *Integral from the Source* $(1+\sigma^2)\sin^2\theta$

The explicit form of the integral $I_3(\sigma,\theta)$ is as follows,

$$I_3(\sigma,\theta) = \frac{1}{\alpha^2}\int\limits_{\mathcal{V}}\frac{(1+\sigma'^2)\sin^2\theta' d^3x'}{|\boldsymbol{x}-\boldsymbol{x}'|} \ , \tag{117}$$

where

$$\left(\sigma^2+\cos^2\theta\right)(1+\sigma^2)\sin^2\theta = \frac{16}{105}\Big\{p_2(\sigma)+p_4(\sigma)+[1-p_4(\sigma)]\,P_2(\cos\theta) - [1+p_2(\sigma)]\,P_4(\cos\theta)\Big\}. \tag{118}$$

Substituting (118) to (117) and integrating with respect to the angular variables we get the internal and external solutions.

5.2.1 The Internal Solution

Making use of the Green's functions (34) the internal solution of (117) takes on the following form

$$
\begin{aligned}
I_3(\sigma,\theta) = 2\pi \sum_{\ell=0}^{\infty} (2\ell+1) q_\ell(\sigma) P_\ell(\cos\theta) \int_0^{\sigma} d\sigma' \int_0^{\pi} \left(\sigma'^2 + \cos^2\theta'\right) \\
\times (1+\sigma'^2) \sin^2\theta' p_\ell\left(\sigma'\right) P_\ell(\cos\theta') d\theta' \\
+ 2\pi \sum_{\ell=0}^{\infty} (2\ell+1) p_\ell(\sigma) P_\ell(\cos\theta) \int_{\sigma}^{1/\varkappa} d\sigma' \int_0^{\pi} \left(\sigma'^2 + \cos^2\theta'\right) \\
\times (1+\sigma'^2) \sin^2\theta' q_\ell\left(\sigma'\right) P_\ell(\cos\theta') d\theta' \ ,
\end{aligned}
\tag{119}
$$

which can be represented as a linear combination of the Legendre polynomials,

$$
I_3(\sigma,\theta) = I_{20}(\sigma) + I_{22}(\sigma) P_2(\cos\theta) + I_{24}(\sigma) P_4(\cos\theta) \ , \tag{120}
$$

where the coefficients are functions of the radial coordinate σ,

$$
\begin{aligned}
I_{20}(\sigma) = & \frac{64\pi}{105} \left\{ q_0(\sigma) \int_0^{\sigma} d\sigma' \left[p_2(\sigma') + p_4(\sigma')\right] \right. \\
& \left. + \int_{\sigma}^{1/\varkappa} d\sigma' \left[p_2(\sigma') + p_4(\sigma')\right] q_0(\sigma') \right\} ,
\end{aligned}
\tag{121a}
$$

$$
\begin{aligned}
I_{22}(\sigma) = & \frac{64\pi}{105} \left\{ q_2(\sigma) \int_0^{\sigma} d\sigma' \left[1 - p_4(\sigma')\right] p_2(\sigma') \right. \\
& \left. + p_2(\sigma) \int_{\sigma}^{1/\varkappa} d\sigma' \left[1 - p_4(\sigma')\right] q_2(\sigma') \right\} ,
\end{aligned}
\tag{121b}
$$

$$
\begin{aligned}
I_{24}(\sigma) = & -\frac{64\pi}{105} \left\{ q_4(\sigma) \int_0^{\sigma} d\sigma' \left[1 + p_2(\sigma')\right] p_4(\sigma') \right. \\
& \left. + p_4(\sigma) \int_{\sigma}^{1/\varkappa} d\sigma' \left[1 + p_2(\sigma')\right] q_4(\sigma') \right\} .
\end{aligned}
\tag{121c}
$$

Calculation of the integrals in (121) yields

$$I_{20}(\sigma) = \frac{2\pi}{15\varkappa^4}\left[1 + 2\varkappa^2 - 2\varkappa^4\sigma^2 - \varkappa^4\sigma^4 + 4\left(1+\varkappa^2\right)^2 \frac{\arctan\varkappa}{\varkappa}\right], \tag{122}$$

$$I_{22}(\sigma) = \frac{2\pi}{21\varkappa^6}\Big[3 + 4\varkappa^2 - \varkappa^4 + \left(9 + 12\varkappa^2 - 3\varkappa^4 - 2\varkappa^6\right)\sigma^2 + 2\varkappa^6\sigma^4$$
$$- \left(1+\varkappa^2\right)^2\left(3-\varkappa^2\right)\left(1+3\sigma^2\right)\frac{\arctan\varkappa}{\varkappa}\Big], \tag{123}$$

$$I_{24}(\sigma) = \frac{\pi}{70\varkappa^6}\Big[15 + 34\varkappa^2 + 23\varkappa^4 + \left(150 + 340\varkappa^2 + 230\varkappa^4 + 32\varkappa^6\right)\sigma^2$$
$$+ \left(175 + \frac{1190}{3}\varkappa^2 + \frac{805}{3}\varkappa^4 + \frac{128}{3}\varkappa^6\right)\sigma^4$$
$$- \left(1+\varkappa^2\right)^2\left(5+3\varkappa^2\right)\left(3+30\sigma^2+35\sigma^4\right)\frac{\arctan\varkappa}{\varkappa}\Big]. \tag{124}$$

5.2.2 The External Solution

The external solution of (117) is obtained by making use of the Green function (34)

$$I_3(\sigma,\theta) = 2\pi\sum_{\ell=0}^{\infty}(2\ell+1)q_\ell(\sigma)P_\ell(\cos\theta)$$
$$\times \int_0^{1/\varkappa} d\sigma' \int_0^{\pi}\left(\sigma'^2 + \cos^2\theta'\right)\left(1+\sigma'^2\right)\sin^2\theta' p_\ell\left(\sigma'\right) P_\ell(\cos\theta')d\theta' , \tag{125}$$

which is reduced after implementing (118) and integrating over the angular variable θ to

$$I_3(\sigma,\theta) = \frac{64\pi}{105}q_0(\sigma)\int_0^{1/\varkappa} d\sigma'\left[p_2(\sigma') + p_4(\sigma')\right]$$
$$+\frac{64\pi}{105}q_2(\sigma)P_2(\cos\theta)\int_0^{1/\varkappa} d\sigma'\left[1 - p_4(\sigma')\right]p_2(\sigma')$$
$$-\frac{64\pi}{105}q_4(\sigma)P_4(\cos\theta)\int_0^{1/\varkappa} d\sigma'\left[1 + p_2(\sigma')\right]p_4(\sigma') . \tag{126}$$

Performing the integrals over the radial variable yields the external solution

$$I_3(\sigma,\theta) = \frac{4\pi\left(1+\varkappa^2\right)^2}{\varkappa^5}\left[\frac{2}{15}q_0(\sigma) - \frac{3-\varkappa^2}{21\varkappa^2}q_2(\sigma)P_2(\cos\theta) - \frac{5+3\varkappa^2}{35\varkappa^2}q_4(\sigma)P_4(\cos\theta)\right]. \tag{127}$$

5.3 Vector Potential V^i

Vector potential V^i is defined above by Eq. (42). As we need it only in the post-Newtonian approximation, the density of the fluid entering (42) can be treated as constant ρ_c. Each element of a rigidly rotating fluid has velocity, $v^i(x) = \varepsilon^{ijk}\omega^j x^k$, where ε^{ijk} is the Levi-Civita symbol, so that (42) can be written as follows

$$V^i(\boldsymbol{x}) = \varepsilon^{ijk}\hat{k}^j\mathcal{D}^k(\boldsymbol{x}), \tag{128}$$

where $\hat{k}^i = \omega^i/\omega$ is the unit vector along z-axis which coincides with the direction of the angular velocity vector, ω^i, and the Cartesian vector $\mathcal{D}^k = \{\mathcal{D}^x, \mathcal{D}^y, \mathcal{D}^z\}$ is given by

$$\mathcal{D}^k(\boldsymbol{x}) = G\omega\rho_c\int_{\mathcal{V}}\frac{x'^k d^3x'}{|\boldsymbol{x}-\boldsymbol{x}'|}. \tag{129}$$

We denote, $\mathcal{D}^+ \equiv \mathcal{D}^x + i\mathcal{D}^y$. In the ellipsoidal coordinates one has[5]

$$\mathcal{D}^+(\boldsymbol{x}) = G\omega\rho_c\alpha\int_{\mathcal{V}}\frac{\sqrt{1+\sigma'^2}\sin\theta' e^{i\phi'}d^3x'}{|\boldsymbol{x}-\boldsymbol{x}'|}. \tag{130}$$

Because the angular velocity, $\omega^i = (0, 0, \omega)$, the vector potential $V^i = (V^x, V^y, V^z)$ has $V^z = 0$. The remaining two components of the vector potential can be combined together

$$V^+ = V^x + iV^y = i\mathcal{D}^+. \tag{131}$$

Equation (131) reveals that calculation of the vector potential is reduced to calculation of the integral in the right hand side of (130) which depends on the point of integration and is separated into the internal and external solutions. We discuss these solutions below.

[5]Notice that $\mathcal{D}^z \neq 0$ but we don't need this component for calculating V^+.

5.3.1 The Internal Solution

The internal solution is obtained for the field points located inside the volume occupied by the rotating fluid. Making use of the Green function (29) we have

$$
\begin{aligned}
\frac{\mathcal{D}^+}{G\omega\rho_c\alpha^3} = &\sum_{\ell=0}^{\infty}\sum_{m=-\ell}^{m=+\ell}\frac{(\ell-|m|)!}{(\ell+|m|)!}q_{\ell|m|}(\sigma)\,Y_{\ell m}(\hat{\boldsymbol{x}}) \\
&\int_0^{\sigma} d\sigma' \oint_{S^2} d\Omega' \left(\sigma'^2+\cos^2\theta'\right)\sqrt{1+\sigma'^2}\sin\theta' e^{i\phi'}\,p_{\ell|m|}\left(\sigma'\right)Y^*_{\ell m}(\hat{\boldsymbol{x}}') \\
+&\sum_{\ell=0}^{\infty}\sum_{m=-\ell}^{m=+\ell}\frac{(\ell-|m|)!}{(\ell+|m|)!}p_{\ell|m|}(\sigma)\,Y_{\ell m}(\hat{\boldsymbol{x}}) \\
&\int_{\sigma}^{1/\varkappa} d\sigma' \oint_{S^2} d\Omega' \left(\sigma'^2+\cos^2\theta'\right)\sqrt{1+\sigma'^2}\sin\theta' e^{i\phi'}\,q_{\ell|m|}\left(\sigma'\right)Y^*_{\ell m}(\hat{\boldsymbol{x}}')\,,
\end{aligned}
\tag{132}
$$

where, $d\Omega' = \sin\theta' d\theta' d\phi'$, is the element of the solid angle, and we integrate in (132) over the unit sphere. We expand functions under the sign of integrals in terms of the spherical harmonics

$$
\begin{aligned}
&\sqrt{1+\sigma^2}\left(\sigma^2+\cos^2\theta\right)\sin\theta e^{i\phi} = \\
&-\sqrt{1+\sigma^2}\left[\frac{2}{15}\frac{Y_{31}(\hat{\boldsymbol{x}})}{C_{31}}+\left(\sigma^2+\frac{1}{5}\right)\frac{Y_{11}(\hat{\boldsymbol{x}})}{C_{11}}\right],
\end{aligned}
\tag{133}
$$

and perform calculations of (132) with the help of the orthogonality relation (33). Finally, we obtain the internal solution of the potential (131) in the following form

$$
\begin{aligned}
\mathcal{D}^+ = &\frac{\pi G\omega\rho_c\alpha^3\sqrt{1+\sigma^2}}{30\varkappa^4}\Bigg\{12(1+\sigma^2\varkappa^4)P_{11}(\cos\theta) \\
&+\left[8\sigma^2\varkappa^4+(1+5\sigma^2)(3+5\varkappa^2)\right]P_{31}(\cos\theta) \\
&-3(1+\varkappa^2)^2\left[4P_{11}(\cos\theta)+(1+5\sigma^2)P_{31}(\cos\theta)\right]\frac{\arctan\varkappa}{\varkappa}\Bigg\}e^{i\phi}.
\end{aligned}
\tag{134}
$$

5.3.2 The External Solution

The external solution of (131) is obtained for the field points lying outside the volume occupied by the rotating fluid. Making use of the Green functions (29), we obtain

$$\frac{\mathcal{D}^+}{G\omega\rho_c\alpha^3} = \sum_{\ell=0}^{\infty}\sum_{m=-\ell}^{m=+\ell} \frac{(\ell-|m|)!}{(\ell+|m|)!} q_{\ell|m|}(\sigma)\, Y_{\ell m}(\hat{\boldsymbol{x}}) \int_0^{1/\varkappa} d\sigma'$$
$$\times \oint_{S^2} d\Omega' \left(\sigma'^2 + \cos^2\theta'\right)\sqrt{1+\sigma'^2}\sin\theta' e^{i\phi'} p_{\ell|m|}(\sigma')\, Y^*_{\ell m}(\hat{\boldsymbol{x}}') \,. \quad (135)$$

After making use of (133) and integrating over the angular variables, we get

$$\mathcal{D}^+ = -2\pi G\omega\rho_c\alpha^3 \left[\frac{1}{45} q_{31}(\sigma)\, P_{31}(\cos\theta) \int_0^{1/\varkappa} \sqrt{1+\sigma'^2}\, p_{31}(\sigma') d\sigma' \right.$$
$$\left. + q_{11}(\sigma)\, P_{11}(\cos\theta) \int_0^{1/\varkappa} \sqrt{1+\sigma'^2}\left(\sigma'^2 + \frac{1}{5}\right) p_{11}(\sigma') d\sigma' \right] e^{i\phi} \,. \quad (136)$$

Integration with respect to the radial coordinate σ yields

$$\mathcal{D}^+ = \mathcal{D} e^{i\phi} \,, \quad (137)$$

where

$$\mathcal{D} = -2\pi G\omega\rho_c\alpha^3 \frac{(1+\varkappa^2)^2}{5\varkappa^5}\left[q_{11}(\sigma)\, P_{11}(\cos\theta) + \frac{1}{6} q_{31}(\sigma)\, P_{31}(\cos\theta)\right] \,. \quad (138)$$

Similar result has been obtained in [58, 59].

We notice that (138) can be transformed to yet another (differential) form

$$\mathcal{D} = -\frac{3S}{4\alpha^2}\sqrt{1+\sigma^2}\sin\theta \frac{d}{d\sigma}\left[q_1(\sigma) + \frac{1}{6} q_3(\sigma) \frac{dP_3(\cos\theta)}{d\cos\theta}\right] \,, \quad (139)$$

where $S = 2Ma^2\omega/5$ is the angular momentum of the rotating spheroid.

6 Relativistic Multipole Moments of a Uniformly Rotating Spheroid

Expansion of gravitational field of an extended massive body to multipoles is ubiquitously used in celestial mechanics and geodesy in order to study the distribution of the matter inside the Earth and other planets of the solar system [6, 8, 10]. General relativity brings about several complications making the multipolar decomposition of gravitational fields more difficult. First, all multipole moments of the gravitational field should include the relativistic corrections to their definition. Second, besides the multipole moments of a single gravitational potential V of the Newtonian theory, one has to include the multipole moments of all components of the metric tensor. There is a vast literature devoted to clarification of various aspects of the multipolar decomposition of relativistic gravitational fields but it goes beyond the scope of the present chapter (see , for example, [5, 79–84]). We need two types of the post-Newtonian multipole moments which appear in the decomposition of the scalar potential V, and the vector potential V^i. The scalar and vector multipoles are more commonly known as mass and spin multipole moments [81] following the names of the leading terms in the multipolar decompositions of the potentials V and V^i respectively.

6.1 Mass Multipole Moments

Mass multipole moments of the external gravitational field of the rotating spheroid are defined by expanding the scalar potential V entering g_{00} component of the metric tensor (6a), in the asymptotic series for a large radial distance r in the spherical coordinates. The scalar potential V takes into account the post-Newtonian contributions from the internal energy Π, pressure p, the kinetic energy of rotation and the internal gravitational energy as well as the spheroidal shape of matter distribution and its inhomogeneity,

$$V = V_N[\rho_c, \mathcal{S}] + V_N[\delta\rho_c, \mathcal{S}] + V_N[\rho_c, \delta\mathcal{S}] + \frac{1}{c^2} V_{pN} \,, \tag{140}$$

where the terms standing in the right hand side of this formula have been provided above in Sects. 4 and 5. As we consider the multipolar expansion of V outside the body, we need only the external solutions which are

$$V_N[\rho_c, \mathcal{S}] = G m_N \left[q_0(\sigma) + q_2(\sigma) P_2(\cos\theta)\right] \,, \tag{141}$$

$$V_N[\delta\rho_c, \mathcal{S}] = \frac{9\mathcal{A}}{20}\frac{G^2 m_N^2}{c^2}\frac{q_1}{\varkappa}\left[q_0(\sigma) + \frac{5}{7} q_2(\sigma) P_2(\cos\theta) - \frac{2}{7} q_4(\sigma) P_4(\cos\theta)\right] \,, \tag{142}$$

$$V_N[\rho_c, \delta\mathcal{S}] = \frac{2\mathcal{B}}{5}\frac{G m_N}{c^2}\omega^2 a^2 \left[q_0(\sigma) + \frac{15}{2}\left(\frac{11}{42} + \frac{9}{7\varkappa^2} + \frac{3}{2\varkappa^4}\right) q_2(\sigma) P_2(\cos\theta) \right.$$
$$\left. + \frac{9}{28}\left(3 + \frac{30}{\varkappa^2} + \frac{35}{\varkappa^4}\right) q_4(\sigma) P_4(\cos\theta)\right] \,, \tag{143}$$

$$V_{pN} = \frac{2}{5} G m_N \omega^2 a^2 \left[q_0(\sigma) - \frac{5}{2}\frac{3 - \varkappa^2}{7\varkappa^2} q_2(\sigma) P_2(\cos\theta) \right.$$
$$\left. - \frac{3}{2}\frac{5 + 3\varkappa^2}{7\varkappa^2} q_4(\sigma) P_4(\cos\theta)\right]$$
$$- \frac{9}{2}\frac{q_1}{\varkappa} G m_N \left[q_0(\sigma) + \frac{5}{7} q_2(\sigma) P_2(\cos\theta) - \frac{2}{7} q_4(\sigma) P_4(\cos\theta)\right] \,, \tag{144}$$

where $m_N = M_N/\alpha$, and the constant Newtonian mass, M_N, is given by (82).

After replacing (141)–(144) in (140) and reducing similar terms, the scalar potential takes on the following form,

$$V = \mathcal{E}_0 q_0(\sigma) + \mathcal{E}_2 q_2(\sigma) P_2(\cos\theta) + \frac{1}{c^2}\mathcal{E}_4 q_4(\sigma) P_4(\cos\theta)\,, \tag{145}$$

where the constant numerical coefficients are

$$\mathcal{E}_0 \equiv Gm_N\left\{1 + \frac{1}{10c^2}\left[(\mathcal{A}-10)\frac{9q_1}{2\varkappa}Gm_N + 4(\mathcal{B}+1)\omega^2 a^2\right]\right\}\,, \tag{146}$$

$$\mathcal{E}_2 \equiv Gm_N\left\{1 + \frac{1}{c^2}\left[(\mathcal{A}-10)\frac{9q_1}{28\varkappa}Gm_N + 3\mathcal{B}\left(\frac{11}{42} + \frac{1}{7\varkappa^2} + \frac{3}{2\varkappa^4}\right)\omega^2 a^2\right.\right.$$
$$\left.\left. + \frac{1}{7}\left(1 - \frac{3}{\varkappa^2}\right)\omega^2 a^2\right]\right\}\,, \tag{147}$$

$$\mathcal{E}_4 \equiv \frac{9Gm_N}{70}\left\{(10-\mathcal{A})\frac{q_1}{\varkappa}Gm_N + \mathcal{B}\left(3 + \frac{30}{\varkappa^2} + \frac{35}{\varkappa^4}\right)\omega^2 a^2\right.$$
$$\left. - \frac{2}{3}\left(3 + \frac{5}{\varkappa^2}\right)\omega^2 a^2\right\}\,. \tag{148}$$

It is remarkable that the three constants, $\mathcal{E}_0, \mathcal{E}_2, \mathcal{E}_4$ are interrelated. Indeed, by direct inspection of (146)–(148), we obtain

$$\mathcal{E}_2 = \mathcal{E}_0 + \frac{\mathcal{E}_4}{c^2}\,. \tag{149}$$

Therefore, Eq. (145) takes on a more simple form,

$$V = \mathcal{E}_0\left[q_0(\sigma) + q_2(\sigma)P_2(\cos\theta)\right] + \frac{1}{c^2}\mathcal{E}_4\left[q_2(\sigma)P_2(\cos\theta) + q_4(\sigma)P_4(\cos\theta)\right]\,. \tag{150}$$

The scalar potential (150) is given in the ellipsoidal coordinates in terms of the ellipsoidal harmonics which are the modified Legendre functions $q_0(\sigma)$, $q_2(\sigma)$ and $q_4(\sigma)$. The advantage in using the ellipsoidal harmonics is that it allows us to represent the post-Newtonian scalar potential V with a finite number of a few terms only. The residual terms in (150) are of the post-post-Newtonian order of magnitude ($\sim 1/c^4$) which are systematically neglected.

In spite of the finite form of the expansion (150) in terms of the ellipsoidal functions it is a more common practice to discuss the multipolar structure of external gravitational field of an isolated body in terms of spherical coordinates (20). Mass multipole moments of the gravitational field are defined in general relativity similarly to the Newtonian gravity as coefficients in the expansion of scalar potential V with respect to the spherical harmonics [80]. For axially-symmetric body the spherical multipolar expansion of the scalar potential reads as follows [10, 80],

$$V = \frac{GM}{R}\left[1 - \sum_{\ell=1}^{\infty} J_{2\ell}\left(\frac{a}{R}\right)^{2\ell} P_{2\ell}(\cos\Theta)\right], \tag{151}$$

where M is the relativistic mass, and $J_{2\ell}$ are the relativistic multipole moments of the gravitational field that are defined (in terms of the spherical coordinates) by integrals over the body's volume

$$M = \int_{\mathcal{V}} \left[\rho(\boldsymbol{x}) + \frac{1}{c^2}\rho_{pN}(\boldsymbol{x})\right] R^2 dR dO , \tag{152}$$

$$J_{2\ell} = -\frac{2}{Ma^{2\ell}} \int_{\mathcal{V}} \left[\rho(\boldsymbol{x}) + \frac{1}{c^2}\rho_{pN}(\boldsymbol{x})\right] R^{2\ell+2} P_{2\ell}(\cos\Theta) dR dO , \quad (n \geq 1) , \tag{153}$$

with $dO \equiv \sin\Theta d\Theta d\Phi$ is the infinitesimal element of the solid angle in the spherical coordinates.

In order to read the multipole moments of the potential V out of (145) we have to transform (145) to spherical coordinates. This can be achieved with the help of the auxiliary formulas representing expansions of the ellipsoidal harmonics in series with respect to the spherical harmonics. Exact transformations between ellipsoidal and spherical harmonic expansions have been derived by Jekeli [85] for numerical computations. However, Jekeli's transformation lacks a convenient analytic form and are not suitable for our purposes. Therefore, below we present a general idea of calculation of the series expansion of the ellipsoidal harmonics in terms of the spherical harmonics.[6]

The ellipsoidal harmonics are solutions of the Laplace equation and are represented by the products of the modified Legendre functions $q_{\ell m}(\sigma)$ or $p_{\ell m}(\sigma)$ with the associated Legendre polynomials $P_{\ell m}(\cos\theta)$. We are interested in the expansion of the ℓth ellipsoidal harmonic $q_\ell(\sigma)P_\ell(\cos\theta)$ in series of the spherical harmonics which are also solutions of the Laplace equation. The most general expansion of this type reads

$$q_\ell(\sigma)P_\ell(\cos\theta) = \sum_{n=\ell}^{\infty} \frac{A_n P_n(\cos\Theta)}{r^{n+1}} , \tag{154}$$

where A_n are the numerical coefficients depending on n. As both sides of (154) are analytic harmonic functions, they are identical at any value of the coordinates. In order to calculate the numerical coefficients A_n, it is instructive to take the point with $\theta = 0$. At this point, we also have $\Theta = 0$, while $\sigma = r$, so that the expansion (154) is reduced to

$$q_\ell(\sigma)\Big|_{\sigma=r} = \sum_{n=\ell}^{\infty} \frac{A_n}{r^{n+1}} , \tag{155}$$

[6] Our method is partially overlapping with a similar development given in [10, Section 2.9].

which means that the coefficients A_n are simply the coefficients of the asymptotic expansion of the modified Legendre function $q_\ell(r)$ for large values of its argument. These coefficients are found by writing down the modified Legendre function $q_\ell(\sigma)$ in the left side of (155) in terms of the hypergeometric function ${}_2F_1$ (see [86, Eq. VI-56b])

$$q_\ell(\sigma)\Big|_{\sigma=r} = \frac{\sqrt{\pi}}{2^{\ell+1}}\frac{\Gamma(\ell+1)}{\Gamma\left(\ell+\frac{3}{2}\right)}\frac{{}_2F_1\left(1+\frac{\ell}{2};\frac{\ell+1}{2};\ell+\frac{3}{2};-\frac{1}{r^2}\right)}{r^{\ell+1}}\,, \tag{156}$$

and equating the coefficients of the expansion to A_n in the right side of (155).

After applying the above procedure, we get for the first several elliptic harmonics the following series,

$$q_0(\sigma) = +\sum_{\ell=0}^{\infty}\frac{(-1)^\ell}{2\ell+1}\frac{P_{2\ell}(\cos\Theta)}{r^{2\ell+1}}\,, \tag{157}$$

$$q_2(\sigma)P_2(\cos\theta) = -\sum_{\ell=1}^{\infty}\frac{2\ell(-1)^\ell}{(2\ell+1)(2\ell+3)}\frac{P_{2\ell}(\cos\Theta)}{r^{2\ell+1}}\,, \tag{158}$$

$$q_4(\sigma)P_4(\cos\theta) = +\sum_{\ell=2}^{\infty}\frac{4\ell(\ell-1)(-1)^\ell}{(2\ell+1)(2\ell+3)(2\ell+5)}\frac{P_{2\ell}(\cos\Theta)}{r^{2\ell+1}}\,. \tag{159}$$

Replacing expansions (157)–(159) in (150), reducing terms of the same power in $1/r^{2\ell+1}$, and comparing the terms of the expansion obtained with similar terms in (151), we conclude that

$$GM = \mathcal{E}_0\,, \tag{160}$$

$$J_{2\ell} = \frac{3(-1)^{\ell+1}\epsilon^{2\ell}}{(2\ell+1)(2\ell+3)}\left[1-\frac{14}{3c^2}\frac{\ell}{2\ell+5}\frac{\mathcal{E}_4}{\mathcal{E}_0}\right]\,, \qquad (n\geq 1) \tag{161}$$

where the second term in the square brackets yields the post-Newtonian correction to the Newtonian multipole moments of the Maclaurin ellipsoid which are defined as the coefficient standing in front of the square brackets in (161).

It is convenient from practical point of view to express the relativistic multipole moments (161) in terms of the dynamical form factor $J_2^{\rm dyn}$ of an extended body with an arbitrary internal distribution of mass density. The dynamical form factor is expressed in terms of the difference between the polar, C and equatorial, A, moments of inertia,

$$J_2^{\rm dyn} = \frac{C-A}{Ma^2}\,. \tag{162}$$

We follow the technique developed by Heiskanen and Moritz [10, Section 2.9] according to which the quadrupole moment of the homogeneous ellipsoid must be exactly equal to the dynamical form factor

$$J_2 = J_2^{\rm dyn} \,. \tag{163}$$

It is rather straightforward to prove that in terms of the dynamical form factor equation (161) reads

$$J_{2\ell} = \frac{3(-1)^{\ell+1}\epsilon^{2\ell}}{(2\ell+1)(2\ell+3)}\left[1-\ell+5\ell\frac{J_2^{\rm dyn}}{\epsilon^2}+\frac{4\ell}{3c^2}\frac{\ell-1}{2\ell+5}\frac{\mathcal{E}_4}{\mathcal{E}_0}\right] , \tag{164}$$

where C and A are the principal moments of inertia. Equation (164) looks quite different from (161) but the difference is illusory since the coefficient in the square brackets of (164) is identically equal to the corresponding term in (161) because for the model of the (almost) homogeneous spheroid accepted in the present chapter, the ratio

$$\frac{C-A}{Ma^2} = \frac{\epsilon^2}{5}\left(1-\frac{2}{3c^2}\frac{\mathcal{E}_4}{\mathcal{E}_0}\right) , \tag{165}$$

that can be easily checked by direct calculation of the integrals defining the moments of inertia. Equation (164) is a relativistic generalization of the result obtained previously by Heiskanen and Moritz [10, Equation 2-92].

Post-Newtonian equation (164) allows to calculate the multipole moments of the normal gravity field at any order as soon as the other parameters of the spheroid are defined. As a particular example we adopt the model of GRS80 international ellipsoid that is characterized by the following parameters (see [8, Section 4.3] and [11, Table 1.2]):

$$\begin{aligned}
GM &= 398600.5 \times 10^9\ {\rm m^3 s^{-2}} \,, \\
a &= 6378137\ {\rm m} \,, \\
J_2^{\rm dyn} &= 1082.63 \times 10^{-6} \,, \\
\omega &= 7.292115 \times 10^{-5}\ {\rm rad\, s^{-1}} \,, \\
1/f &= 298.257222101 \,.
\end{aligned}$$

Corresponding (derived) values for the first and second eccentricities of GRS80 are

$$\begin{aligned}
\epsilon &= 0.08181919104282 \,, \\
\varkappa &= 0.08209443815192 \,.
\end{aligned}$$

The value of the ratio $\mathcal{E}_4/\mathcal{E}_0$ can be calculated on the basis of Eq. (209) which is derived below in Sect. 7.3 from the condition of the hydrostatic equilibrium. Making use of the numerical values of the parameters of GRS80 model, we get

$$\frac{\mathcal{E}_4}{c^2\mathcal{E}_0} = -5.58671 \times 10^{-7}\,(\mathcal{B} - 0.0023445)\ , \tag{166}$$

where $\mathcal{B}$ is the parameter defining in our spheroidal model a small deviation from the shape of the Maclaurin ellipsoid. The value of this parameter can be chosen arbitrary in the range being compatible with the limitations imposed by the post-Newtonian approximation, say, $-10 \lesssim \mathcal{B} \lesssim 10$. Calculating the scalar multipole moments in our spheroidal model by means of (164) yields for the case of pure ellipsoid,

$$\begin{aligned} J_4 &= -2.37091222014 \times 10^{-6}\ , \quad J_6 = +6.08347064201 \times 10^{-9}\ , \\ J_8 &= -1.42681406953 \times 10^{-11}\ , \qquad (\mathcal{B} = 0) \end{aligned} \tag{167}$$

and for spheroid

$$\begin{aligned} J_4 &= -2.37091158429 \times 10^{-6}\ , \quad J_6 = +6.08346483750 \times 10^{-9}\ , \\ J_8 &= -1.42680988487 \times 10^{-11}\ , \qquad (\mathcal{B} = 1)\ . \end{aligned} \tag{168}$$

These values can be compared with the corresponding values of the GSR80 geodetic model [8, Equation 4.77c] which does not take into account relativistic corrections,

$$\begin{aligned} J_4^{\rm GRS80} &= -2.37091221865 \times 10^{-6},\ J_6^{\rm GRS80} = +6.08347062840 \times 10^{-9}, \\ J_8^{\rm GRS80} &= -1.42681405972 \times 10^{-11} \end{aligned} \tag{169}$$

One can see that accounting for relativistic corrections gives slightly different numerical values of the multipole moments for different models of the normal gravity field generated by rotating spheroid. Significance of these deviations for practical applications in geodynamics is a matter of future theoretical and experimental studies. Nonetheless, already now we can state that general relativity changes the classic model of the normal gravity field. Hence, the separation of the observed value of the field into the normal gravity and its perturbation differs from the Newtonian theory and has certain consequences for interpretation of the gravity field anomalies.

6.2 *Spin Multipole Moments*

The spin multipole moments are defined as coefficients in the expansion of vector potential V^i with respect to vector spherical harmonics [80]

$$\begin{aligned} V^i(r,\Theta,\Phi) = \sum_{\ell=0}^{\infty}\sum_{m=-\ell}^{m=+\ell}\Big[& E^{\ell m}(r)Y^i_{E,\ell m}(\Theta,\Phi) + B^{\ell m}(r)Y^i_{B,\ell m}(\Theta,\Phi) \\ & + R^{\ell m}(r)Y^i_{R,\ell m}(\Theta,\Phi)\Big]\ , \end{aligned} \tag{170}$$

where $E^{\ell m}$, $B^{\ell m}$, $R^{\ell m}$ are the spin multipole moments depending on the radial coordinate r, and $Y^i_{E,\ell m}$, $Y^i_{B,\ell m}$, $Y^i_{R,\ell m}$ are the Cartesian components of the three vector spherical harmonics, $\boldsymbol{Y}_{E,\ell m}$, $\boldsymbol{Y}_{B,\ell m}$, $\boldsymbol{Y}_{R,\ell m}$. The harmonics $\boldsymbol{Y}_{E,\ell m}$ and $\boldsymbol{Y}_{R,\ell m}$ are of "electric-type" parity $(-1)^\ell$, while $\boldsymbol{Y}_{B,\ell m}$ have "magnetic-type" parity $(-1)^{\ell+1}$ [80]. Only the "magnetic-type" harmonics present in the expansion of the vector potential in case of an axially-symmetric gravitational field [58], hence, we don't consider the "electric-type" harmonics below.

The "magnetic-type" harmonics are defined as follows [69]

$$\boldsymbol{Y}_{B,\ell m}(\Theta,\Phi) = i\frac{\boldsymbol{L}Y_{\ell m}(\Theta,\Phi)}{\sqrt{\ell(\ell+1)}}\,, \tag{171}$$

where $\boldsymbol{L} = -i\boldsymbol{x}\times\boldsymbol{\nabla}$ is the operator of the angular momentum, the cross '$\mathbf{x}$' denotes the Euclidean product of vectors, and $\boldsymbol{\nabla}$ is the gradient operator. The Cartesian components (L_x, L_y, L_z) of the vectorial operator of the angular momentum $\boldsymbol{L}$ expressed in terms of the spherical coordinates, are [69, Exercise 2.5.14]

$$L_x = i\left(\cos\Phi\cot\Theta\frac{\partial}{\partial\Phi} + \sin\Phi\frac{\partial}{\partial\Theta}\right)\,, \tag{172a}$$

$$L_y = i\left(\sin\Phi\cot\Theta\frac{\partial}{\partial\Phi} - \cos\Phi\frac{\partial}{\partial\Theta}\right)\,, \tag{172b}$$

$$L_z = -i\frac{\partial}{\partial\Phi}\,. \tag{172c}$$

We have found in Sect. 5.3 that all of the non-vanishing components of the vector potential V^i are included to the potential V^+ defined in (131). This potential is proportional to the components of the vector spherical harmonics, $\boldsymbol{Y}_{+,\ell m} \sim L_+ Y_{\ell m}$ where the action of the operator L_+ on the standard spherical harmonics is as follows [69, Exercise 12.6.7]

$$L_+ Y_{\ell m}(\Theta,\Phi) = \sqrt{(\ell-m)(\ell+m+1)}\,Y_{\ell,m+1}(\Theta,\Phi)\,, \tag{173}$$

which tells us that $V^+ \sim Y_{\ell,m+1}$. On the other hand, due to the fact that the angular coordinates of the ellipsoidal and spherical coordinates coincide, $\Phi = \phi$, and $V^+ = i\mathcal{D}e^{i\phi}$ as follows from (131) and (137), we conclude that the multipolar expansion (170) of V^+ with respect to the spherical harmonics contains only the spherical harmonics with $m = 1$, that is $V^+ = i\mathcal{D}^+ \sim Y_{\ell 1} \sim P_{\ell 1}e^{i\phi}$. This can be seen directly after applying the Green function in spherical coordinates and taking into account the rotational symmetry with respect to the angle Φ which yields Eq. (137) with function $\mathcal{D}$ having the following form

$$\mathcal{D} = \frac{GS}{2R^2}\left[\sin\Theta + \sum_{\ell=1}^{\infty}\frac{S_{2\ell+1}}{2\ell+1}\left(\frac{a}{R}\right)^{2\ell}P_{2\ell+1,1}(\cos\Theta)\right]\,, \tag{174}$$

where

$$S \equiv S_1 = \quad \omega \int_{\mathcal{V}} \rho(\boldsymbol{x}) R^4 \sin^2 \Theta d R d O \,, \tag{175}$$

$$S_{2\ell+1} = \frac{1}{\ell+1} \frac{\omega}{S a^{2\ell}} \int_{\mathcal{V}} \rho(\boldsymbol{x}) R^{2\ell+4} \sin \Theta P_{2\ell+1,1}(\Theta) d R d O, \quad (\ell > 1) \tag{176}$$

are the absolute value of the angular momentum (spin) of the rotating spheroid and the spin multipole moments of the higher order, and $\mathcal{V}$ is the volume bounded by the surface of the Maclaurin ellipsoid. It is sufficient to perform calculation of the integral in (175) with the constant value of density, $\rho(\boldsymbol{x}) = \rho_c$. It yields

$$S = \frac{2}{5} M a^2 \omega \,, \tag{177}$$

which coincides with the result obtained in textbooks on classic mechanics.

Calculation of the spin multipoles $S_{2\ell+1}$ can be also performed directly but it will be more instructive to find them out from the Taylor expansion of function $\mathcal{D}$ given in the ellipsoidal coordinates by Eq. (138). It is convenient to write down this equation by replacing the central density ρ_c with the total mass, M, as follows

$$\mathcal{D} = -\frac{3}{10} G M \omega \frac{(1+\varkappa^2)}{\varkappa^2} \left[q_{11}(\sigma) P_{11}(\cos\theta) + \frac{1}{6} q_{31}(\sigma) P_{31}(\cos\theta) \right] . \tag{178}$$

In order to calculate the spin multipole moments, we have to transform (178) from the ellipsoidal to spherical harmonics. For we have in (178) the ellipsoidal harmonics $q_{\ell m}(\sigma) P_{\ell m}(\cos\theta)$ with the index $m = 1$, and the odd index $\ell = 2k+1$, we have to apply a slightly different approach to get the transformation formula as compared with that employed in the previous Sect. 6.1. More specifically, because both the ellipsoidal and spherical harmonics are solutions of the Laplace equation, we have

$$q_{2\ell+1,1}(\sigma) P_{2\ell+1,1}(\cos\theta) = \sum_{n=\ell}^{\infty} \frac{B_n P_{2n+1,1}(\cos\Theta)}{r^{2n+2}} \,, \tag{179}$$

where B_n are the numerical coefficients depending on n. As both sides of (179) are analytic harmonic functions, they are identical at any value of the coordinates. In order to calculate the numerical coefficients B_n, we take the point with $\theta = \pi/2$. At this point we also have $\Theta = \pi/2$ and $\cos\Theta = 0$, while the radial coordinate, $\sigma = \sqrt{r^2-1}$. The Legendre polynomials

$$P_{2n+1,1}(0) = (-1)^{n+1} \frac{2}{\sqrt{\pi}} \frac{\Gamma(n+\frac{3}{2})}{\Gamma(n+1)} = (-1)^{n+1} \frac{(2n+1)!!}{2^n n!} \tag{180}$$

so that the expansion (179) is reduced to

$$(-1)^{\ell+1}\frac{(2\ell+1)!!}{2^\ell\ell!}q_{2\ell+1,1}(\sigma)\bigg|_{\sigma=\sqrt{r^2-1}}=\sum_{n=\ell}^{\infty}(-1)^{n+1}\frac{(2n+1)!!}{2^n n!}\frac{B_n}{r^{2n+2}}\,, \tag{181}$$

which means that the coefficients B_n are simply the coefficients of the asymptotic expansion of the modified Legendre function $q_{\ell 1}(\sqrt{r^2-1})$ for large values of its argument, $r \gg 1$. These coefficients are found by writing down the modified Legendre function in the left side of (181) in terms of the hypergeometric function (see [86, Eq. VI-57b])

$$q_{2\ell+1,1}(\sigma)\bigg|_{\sigma=\sqrt{r^2-1}}=\frac{\sqrt{\pi}}{2^{2\ell+2}}\frac{\Gamma(2\ell+3)}{\Gamma\left(2\ell+\frac{5}{2}\right)}\frac{{}_2F_1\left(\ell+\frac{3}{2};\ell+\frac{1}{2};2\ell+\frac{5}{2};\frac{1}{r^2}\right)}{r^{2\ell+2}}\,. \tag{182}$$

Comparing the coefficients of the expansion of the right side of (182) with the numerical coefficients in the right side of (181), we can find coefficients B_n.

After applying this procedure, we get for the first two ellipsoidal harmonics the following series expansions,

$$q_{11}(\sigma)P_{11}(\cos\theta)=\;2\sum_{\ell=0}^{\infty}\frac{(-1)^\ell}{(2\ell+1)(2\ell+3)}\frac{P_{2\ell+1,1}(\cos\Theta)}{r^{2\ell+2}}\,, \tag{183}$$

$$q_{31}(\sigma)P_{31}(\cos\theta)=-24\sum_{\ell=1}^{\infty}\frac{\ell(-1)^\ell}{(2\ell+1)(2\ell+3)(2\ell+5)}\frac{P_{2\ell+1,1}(\cos\Theta)}{r^{2\ell+2}}\,. \tag{184}$$

Replacing these expansions to (178) and reducing similar terms, we obtain

$$\mathcal{D}=\frac{GMa^2\omega}{5R^2}\left[\sin\Theta-15\sum_{\ell=1}^{\infty}\frac{(-1)^\ell\epsilon^{2\ell}}{(2\ell+1)(2\ell+3)(2\ell+5)}\left(\frac{a}{R}\right)^{2\ell}P_{2\ell+1,1}(\cos\Theta)\right]. \tag{185}$$

Comparing expansion (185) with (174), we conclude that the coefficients of the multipolar expansion of the vector potential in (174) are

$$S_{2\ell+1}=\frac{15(-1)^{\ell+1}\epsilon^{2\ell}}{(2\ell+3)(2\ell+5)}\,,\qquad(\ell\geq 1). \tag{186}$$

The spin multipole moments, $S_{2\ell+1}$, are uniquely related to the mass multipole moments, $J_{2\ell}$, of a homogeneous and uniformly rotating Maclaurin ellipsoid as follows

$$S_{2\ell+1}=5\,\frac{2\ell+1}{2\ell+5}J_{2\ell}\,. \tag{187}$$

Relations (186) and (187) have been also derived by Teyssandier [58] by making use of a different mathematical technique.

It is convenient to give, yet another form of the expansion (185) in terms of the derivatives of the Legendre polynomials. To this end, we employ the relation [72, Eq. 8.752-1]

$$P_{\ell m}(u) = (-1)^m \left(1 - u^2\right)^{m/2} \frac{d^m P_\ell(u)}{du^m} , \tag{188}$$

which allows us to recast (185) to the following form

$$\mathcal{D} = \frac{GMa^2\omega \sin\Theta}{5R^2}\left[1 + 15\sum_{\ell=1}^{\infty} \frac{(-1)^\ell \epsilon^{2\ell}}{(2\ell+1)(2\ell+3)(2\ell+5)}\left(\frac{a}{R}\right)^{2\ell} \frac{dP_{2\ell+1}(\cos\Theta)}{d\cos\Theta}\right] . \tag{189}$$

After accounting for relations (131), (137), the vector potential V^i can be written down explicitly in a vector form

$$V^i = \frac{G}{2}\frac{(\boldsymbol{S}\times\boldsymbol{x})^i}{R^3}\left[1 + 15\sum_{\ell=1}^{\infty} \frac{(-1)^\ell \epsilon^{2\ell}}{(2\ell+1)(2\ell+3)(2\ell+5)}\left(\frac{a}{R}\right)^{2\ell} \frac{dP_{2\ell+1}(\cos\Theta)}{d\cos\Theta}\right] , \tag{190}$$

where the angular momentum vector $\boldsymbol{S} = \{0, 0, S\}$, and S is defined in (177). We notice that a similar expansion formula given by Soffel and Frutos [44, Eq. 23] for the vector potential has a typo and should be corrected in accordance with (190).

7 Relativistic Normal Gravity Field

7.1 Equipotential Surface

Harmonic coordinates introduced in Sect. 2.1 represent an inertial reference frame in space which is used to describe the motion of probe masses (satellites) and light (radio) signals in metric (6a)–(6c). Let us consider a continuous ensemble of observers rotating rigidly in space with respect to z axis of the inertial reference frame with the angular velocity of the rotating spheroid, ω^i. Each observer moves with respect to the inertial reference frame along a world line $x^i \equiv \{x(t), y(t), z(t)\}$ such that $z(t)$ and $x^2(t) + y^2(t)$ remain constant. We assume that each observer carries out a clock measuring its own proper time $\tau = \tau(t)$ where $t = x^0/c$ is the coordinate time of the harmonic coordinates x^α introduced in Sect. 2.1.

The proper time of the clock is defined by equation $-c^2 d\tau^2 = ds^2$ where the interval ds is calculated along the world line of the clock [63, 87]. In terms of the metric tensor (5) the interval $d\tau$ of the proper time reads,

$$d\tau = \left(-g_{00}(t, \boldsymbol{x}) - \frac{2}{c} g_{0i}(t, \boldsymbol{x}) v^i - \frac{1}{c^2} g_{ij}(t, \boldsymbol{x}) v^i v^j\right)^{1/2} dt \;, \tag{191}$$

where, $\boldsymbol{x} = \{x^i(t)\}$ is taken on the world line of the clock, and $v^i = dx^i/dt = (\boldsymbol{\omega} \times \boldsymbol{x})^i$ is a constant linear velocity of the clock with respect to the inertial reference frame. The ensemble of the observers is static with respect to the rotating spheroid and represents a realization of a rigidly rotating reference frame extending to the outer space outside the spheroid. It should be understood that the rigidly rotating observers are generally not in a free fall except of those which are at the radial distance corresponding to the orbit of geostationary satellites. The rotating reference frame is local - it does not go to a spatial infinity and is limited by the radial distance at which the linear velocity equates to the speed of light, $v \le c$, that is $|\boldsymbol{x}| \le c/\omega$. For the Earth this distance does not exceed 27.5 AU - a bit less than the radius of Neptune's orbit.

After replacing the metric (6a)–(6c) in (191) and extracting the root square, we get the fundamental time delay equation in the post-Newtonian approximation [20]

$$\frac{d\tau}{dt} = 1 - \frac{W}{c^2} + \mathcal{O}\left(c^{-6}\right) \;, \tag{192}$$

where the time-independent function, W is given by

$$W = \frac{1}{2}v^2 + V + \frac{1}{c^2}\left(\frac{1}{8}v^4 + \frac{3}{2}v^2 V - 4v^i V^i - \frac{1}{2}V^2\right) . \tag{193}$$

Function W is the post-Newtonian potential of the *normal gravity field* taken at the point of localization of the clock [1, 5].

The equipotential surface is defined by the condition of the constant rate of clock's proper time with respect to the coordinate time, that is [2, 20, 45]

$$\frac{d\tau}{dt} = W = \text{const} \;. \tag{194}$$

In case of a stationary spacetime generated by a rigidly rotating body through Einstein's equations, the equipotential surface is orthogonal at each point to the direction of the gravity force (the plumb line) [1, 2, 45, 46]. Inside the rotating fluid the equipotential surface also coincides with the levels of equal density - ρ, pressure - p, and thermodynamic energy - Π [1, 5].

7.2 Normal Gravity Field Potential

The post-Newtonian potential, W, of the normal gravity field inside the rigidly rotating fluid body has been derived in detail in our previous publications [3, 4], and its derivation corresponds to the internal solution for the potentials worked out in the previous sections. The present chapter focuses on the structure of the normal gravity field outside the rotating spheroid which is described by the external solutions of the metric tensor coefficients discussed above.

The external solution of the scalar potential V entering (193) has been given in (145). Vector (gravitomagnetic) potential outside the body is given by (128) or, more explicitly,

$$V^x = -\sin\phi\mathcal{D}\ , \qquad V^y = \cos\phi\mathcal{D}\ , \qquad V^z = 0\ , \tag{195}$$

where $\mathcal{D}$ is given in (178). It is straightforward to show that the Euclidean dot product, $v^i V^i = v^x V^x + v^y V^y$, entering the post-Newtonian part of W, is

$$v^i V^i = \omega\alpha\sqrt{1+\sigma^2}\mathcal{D}\sin\theta\ , \tag{196}$$

or, by making use of (38), (37)

$$\begin{aligned} v^i V^i = \frac{2}{15}Gm_N\omega^2 a^2 \Bigg\{ & q_0(\sigma) + q_2(\sigma) - \left[q_0(\sigma) - \frac{5}{49}q_2(\sigma) - \frac{54}{49}q_4(\sigma)\right] \\ & \times P_2(\cos\theta) - \frac{54}{49}\left[q_2(\sigma) + q_4(\sigma)\right]P_4(\cos\theta)\Bigg\}\ . \end{aligned} \tag{197}$$

The rest of the terms entering expression (193) for the normal gravity potential W are

$$v^2 = \omega^2\alpha^2\left(1+\sigma^2\right)\sin^2\theta = \frac{2}{3}\omega^2\alpha^2\left(1+\sigma^2\right)\left[1 - P_2(\cos\theta)\right]\ , \tag{198}$$

$$\begin{aligned} v^4 &= \omega^4\alpha^4\left(1+\sigma^2\right)^2\sin^4\theta = 8\omega^4\alpha^4\left(1+\sigma^2\right)^2 \\ &\times \left[\frac{1}{15} - \frac{2}{21}P_2(\cos\theta) + \frac{1}{35}P_4(\cos\theta)\right]\ , \end{aligned} \tag{199}$$

$$\begin{aligned} v^2 V_N = \frac{2}{3}Gm_N\omega^2\alpha^2\left(1+\sigma^2\right)\Bigg\{ & q_0(\sigma) - \frac{1}{5}q_2(\sigma) \\ & - \left[q_0(\sigma) - \frac{5}{7}q_2(\sigma)\right]P_2(\cos\theta) - \frac{18}{35}q_2(\sigma)P_4(\cos\theta)\Bigg\}\ , \end{aligned} \tag{200}$$

$$\begin{aligned} V_N^2 = G^2 m_N^2\Bigg\{ & q_0^2(\sigma) + \frac{1}{5}q_2^2(\sigma) + 2q_2(\sigma)\left[q_0(\sigma) + \frac{1}{7}q_2(\sigma)\right]P_2(\cos\theta) \\ & + \frac{18}{35}q_2^2(\sigma)P_4(\cos\theta)\Bigg\}. \end{aligned} \tag{201}$$

Summing up all terms in (145) we can reduce it to a polynomial

$$W(\sigma,\theta) = \mathcal{W}_0(\sigma) + \mathcal{W}_2(\sigma)P_2(\cos\theta) + \frac{1}{c^2}\mathcal{W}_4(\sigma)P_4(\cos\theta)\ , \tag{202}$$

which coefficients are functions of the radial coordinate σ,

$$\mathcal{W}_0(\sigma) = \frac{1}{3}\omega^2\alpha^2(1+\sigma^2) + Gmq_0(\sigma) \\ +\frac{1}{c^2}\omega^2\alpha^2(1+\sigma^2)\left\{\frac{1}{15}\omega^2\alpha^2(1+\sigma^2) + Gm\left[q_0(\sigma) - \frac{1}{5}q_2(\sigma)\right]\right\} \\ -\frac{Gm}{c^2}\left\{\frac{8}{15}\omega^2 a^2\left[q_0(\sigma)+q_2(\sigma)\right] + \frac{1}{2}Gm\left[q_0^2(\sigma) + \frac{1}{5}q_2^2(\sigma)\right]\right\}, \quad (203)$$

$$\mathcal{W}_2(\sigma) = -\frac{1}{3}\omega^2\alpha^2(1+\sigma^2) + Gmq_2(\sigma) \\ +\frac{1}{c^2}\left\{\mathcal{E}_4 q_2(\sigma) - \omega^2\alpha^2(1+\sigma^2)\left[\frac{2}{21}\omega^2\alpha^2(1+\sigma^2)\right.\right. \\ \left.\left.+Gm\left(q_0(\sigma) - \frac{5}{7}q_2(\sigma)\right)\right]\right\} \\ +\frac{Gm}{c^2}\left\{\frac{8}{15}\omega^2 a^2\left[q_0(\sigma) - \frac{5}{49}q_2(\sigma) - \frac{54}{49}q_4(\sigma)\right]\right. \\ \left.-Gmq_2(\sigma)\left[q_0(\sigma) + \frac{1}{7}q_2(\sigma)\right]\right\}, \quad (204)$$

$$\mathcal{W}_4(\sigma) = \mathcal{E}_4 q_4(\sigma) + \frac{1}{35}\omega^2\alpha^2(1+\sigma^2)\left[\omega^2\alpha^2(1+\sigma^2) - 18Gmq_2(\sigma)\right] \\ -\frac{9}{35}Gm\left\{Gmq_2^2(\sigma) - \frac{16}{7}\omega^2 a^2\left[q_2(\sigma)+q_4(\sigma)\right]\right\}, \quad (205)$$

and we have denoted, $m \equiv M/\alpha$, where M is the relativistic mass (152) that is related to the Newtonian mass m_N through Eqs. (160) and (146).

7.3 The Figure of Equilibrium

The surface of a rotating fluid body is defined by the boundary condition of vanishing pressure, $p = 0$. This surface coincides with the level of the constant gravitational potential [2, 5] that is defined by the condition,

$$\mathcal{W}(\sigma_s, \theta) = W_0 = \text{const.}, \quad (206)$$

for the value of the radial coordinate $\sigma_s = \sigma_s(\theta)$ defined above in (63). After expanding the left hand side of (206) around the constant value of the radial coordinate $1/\varkappa$, the equation of the level surface takes on the following form

$$\bar{\mathcal{W}}_0 - \frac{\gamma_2\omega^2 a^3}{5\varkappa\epsilon c^2}\mathcal{B} + \left[\bar{\mathcal{W}}_2 - \left(\gamma_0 + \frac{2}{7}\gamma_2\right)\frac{\omega^2 a^3}{\varkappa\epsilon c^2}\mathcal{B}\right] P_2(\cos\theta)$$
$$+\frac{1}{c^2}\left[\bar{\mathcal{W}}_4 - \frac{18}{35}\frac{\gamma_2\omega^2 a^3}{\varkappa\epsilon}\mathcal{B}\right] P_4(\cos\theta) = W_0, \tag{207}$$

where γ_0 and γ_2 are the components of the gravity force defined below in (219), and $\bar{\mathcal{W}}_0 \equiv \mathcal{W}_0(\varkappa^{-1})$, $\bar{\mathcal{W}}_2 \equiv \mathcal{W}_2(\varkappa^{-1})$, $\bar{\mathcal{W}}_4 \equiv \mathcal{W}_4(\varkappa^{-1})$.

Because the potential is constant on the level surface the left hand side of (207) cannot depend on the angle θ which means that the coefficients in front of the Legendre polynomials, $P_2(\cos\theta)$ and $P_4(\cos\theta)$, must vanish. Equating,

$$\bar{\mathcal{W}}_4 - \frac{18}{35}\frac{\gamma_2\omega^2 a^3}{\varkappa\epsilon}\mathcal{B} = 0\,, \tag{208}$$

yields

$$q_4\mathcal{E}_4 + \frac{108}{35}\frac{G^2m^2}{\varkappa}q_1q_2\mathcal{B} = -\frac{54}{245}G^2m^2q_2\left(q_2 + 8q_4\right)\,, \tag{209}$$

where $\mathcal{E}_4$ has been defined in (148), $q_1 = q_1(1/\varkappa)$, $q_2 \equiv q_2(1/\varkappa)$, $q_4 \equiv q_4(1/\varkappa)$, and we have made use of (88). Equation (209) determines the coefficient $\mathcal{B}$ in the equation of the spheroidal surface (63) of the rotating fluid body as a function of the coefficient $\mathcal{A}$ defining the deviation of the internal density of the fluid, ρ, from the uniform distribution by Eqs. (43), (44).

Equating,

$$\bar{\mathcal{W}}_2 - \left(\gamma_0 + \frac{2}{7}\gamma_2\right)\frac{\omega^2 a^3}{\varkappa\epsilon c^2}\mathcal{B} = 0\,, \tag{210}$$

yields a relationship generalizing the Maclaurin equation (86) to the post-Newtonian approximation,

$$\omega^2 a^2 = 3Gmq_2\left[1 - \frac{4}{245}\frac{Gm}{c^2}\left(20q_2 + 49q_0 + 36q_4\right)\right]$$
$$+\frac{q_2}{c^2}\left[\mathcal{E}_4 - \frac{3G^2m^2}{7\varkappa^3}\left(21 + 11\varkappa^2\right)q_1\mathcal{B}\right]\,, \tag{211}$$

where q_0, q_1, q_2 and q_4 have the same meaning as in (209) above, and the constant coefficient $\mathcal{B}$ relates to $\mathcal{E}_4$ by means of (209). In its own turn the coefficient $\mathcal{E}_4$ is given by Eq. (148).

Finally, the constant value of the gravity potential on the surface of the rotating ellipsoid is,

$$W_0 \equiv \bar{\mathcal{W}}_0 - \frac{\gamma_2\omega^2 a^3}{5\varkappa\epsilon c^2}\mathcal{B}\,, \tag{212}$$

or, more explicitly,

$$W_0 = Gm(q_0 + q_2) - \frac{G^2m^2}{2c^2}\left(q_0^2 + \frac{7}{5}q_0q_2 + \frac{17}{5}q_2^2 + \frac{12}{\varkappa}q_1q_2\mathcal{B}\right). \tag{213}$$

In the small eccentricity approximation the value of the gravity potential on the level surface is

$$W_0 = Gm(q_0 + q_2) - \frac{G^2M^2}{2a^2c^2}\left[1 + \left(\frac{13}{25} + \frac{8\mathcal{B}}{15}\right)\varkappa^2 + \mathcal{O}\left(\varkappa^4\right)\right]. \tag{214}$$

The condition of the hydrostatic equilibrium (206) imposes a constraint on the linear combination of the constant parameters $\mathcal{E}_4$ and $\mathcal{B}$ through (209) which establishes the correspondence between the constants $\mathcal{A}$ and $\mathcal{B}$ defining the distribution of mass density (44) inside the rotating body and the shape of its surface (63). There are not any other limitations on these parameters. Hence, one of them can be chosen arbitrary. One choice is to accept $\mathcal{A} = 0$ that is to admit that the density of the fluid is homogeneous at any order of the post-Newtonian approximations. This makes the figure of the equilibrium of the rotating fluid deviate from the ellipsoid of revolution. This choice was made, for example, in papers [4, 54, 56, 88–90] that consider the corresponding figures of the post-Newtonian rotating homogeneous spheroids with emphasis on astrophysical applications. On the other hand, one can postulate the equipotential surface to be exactly the Maclaurin ellipsoid at any post-Newtonian approximation which is achieved by choosing the parameter $\mathcal{B} = 0$. Such an ellipsoidal figure of equilibrium of a rotating fluid body has a non-homogeneous distribution of mass density so that the parameter $\mathcal{A} \neq 0$. This case has been considered in our paper [3]. There is also a possibility to choose the constant parameter $\mathcal{B}$ in such a way that the post-Newtonian formula (211) connecting the angular velocity of rotation, ω, with the geometric parameters of the figure of equilibrium, will formally coincide with the classic Maclaurin relationship (88). This case deserves a special attention but we shall not dwell upon it over here as it relates to the problem of consistency of a set of astronomical constants which is a prerogative of the International Astronomical Union [http://asa.usno.navy.mil/SecK/Constants.html].

The gravity potential W_0 is defined by formula (213). The value of W_0 is the gravity potential on the surface of geoid, that is currently chosen as a defining constant[7] *without taking into account* the post-Newtonian contribution as follows, $W_0 = 62636856.0 \pm 0.5$ m^2 s^{-2} [11, Table 1.1]. The fractional uncertainty in the IERS Convention 2010 value of W_0 is $\delta W_0/W_0 \simeq 8 \times 10^{-9}$. This uncertainty is significantly less than that of the global geometric reference system and is no longer accurate enough to match the operational precision of VLBI and SLR measurements as well as satellite laser altimetry of ocean's surface which have reached the level of one millimeter. This accuracy is at the level of the post-Newtonian correction to the Newtonian value of W_0 (the first term in a right hand side of (213)) that can be easily evaluated on the basis of the approximate formula (214), and is about

[7] It is equivalent to a constant $L_G = W_0/c^2 = 6.969290134 \times 10^{-10}$ that determines the difference between TT and TCG time scales (see [20, 27] or [5, Appendix C.2, Resolution B1.9.]).

$GM_{\oplus}/2a_{\oplus}c^2 \simeq 3.5 \times 10^{-10}$ or, in terms of height, 2.2 mm on the Earth surface. This is within the operational precision of current geodetic techniques. We suggest that the significance of the post-Newtonian correction to the defining constant W_0 should be thoroughly discussed by corresponding IAU/IAGG working groups developing a new generation of the system of the geopotential-based geodetic constants (see discussion in [8, Section 4.3]).

7.4 The Somigliana Formula

The Somigliana formula in classic geodesy gives the value of the normal gravity γ^i on the reference ellipsoid [6, 8, 10]. Vector of the normal gravity is perpendicular to the equipotential surface, and is calculated in the post-Newtonian approximation in accordance with equation [2, 5]

$$\gamma^i = -c^2 \Lambda^{ij} \frac{\partial}{\partial x^j} \log\left(1 - \frac{1}{c^2}\mathcal{W}\right) , \tag{215}$$

where the matrix operator

$$\Lambda^{ij} = \delta^{ij} - \frac{1}{c^2}\left(\frac{1}{2}v^i v^j + \delta^{ij} V_N\right) + \mathcal{O}\left(\frac{1}{c^4}\right) , \tag{216}$$

defines transformation to the inertial frame of a local observer being at rest with respect to the rotating frame of reference.

We are looking for the normal component, $\gamma_n = \hat{n}^i \gamma^i$, of the vector γ^i in the direction of the plumb line that is given by the unit vector $\hat{\boldsymbol{n}}$ defined in (19). A particular interest represents the value of γ_n taken on the surface of the ellipsoid which corresponds to the classical derivation of the formula of *Somigliana* [8, 10]. After taking the partial derivative in (215), and making use of the ellipsoidal coordinates, it reads

$$\gamma_n = -\frac{1}{\alpha}\left[\left(1 + \frac{1}{2c^2}\omega^2\alpha^2(1+\sigma^2)\sin^2\theta\right)\left(\frac{1+\sigma^2}{\sigma^2+\cos^2\theta}\right)^{1/2} \frac{\partial \mathcal{W}}{\partial \sigma}\right]_{\sigma=\sigma_s} , \tag{217}$$

or, more explicitly

$$\gamma_n = \left(1 + \frac{1}{2c^2}\omega^2 a^2 \sin^2\theta\right)\left(\frac{1+{\sigma_s}^2}{{\sigma_s}^2+\cos^2\theta}\right)^{1/2}$$
$$\left[\gamma_0(\sigma) + \gamma_2(\sigma)P_2(\cos\theta) + \frac{1}{c^2}\gamma_4(\sigma)P_4(\cos\theta)\right]_{\sigma=\sigma_s} , \tag{218}$$

where we have introduced the following notations for the partial derivatives of the components of the normal gravity potential

$$\gamma_0(\sigma) \equiv -\frac{1}{\alpha}\frac{\partial \mathcal{W}_0}{\partial \sigma}\;, \qquad \gamma_2(\sigma) \equiv -\frac{1}{\alpha}\frac{\partial \mathcal{W}_2}{\partial \sigma}\;, \qquad \gamma_4(\sigma) \equiv -\frac{1}{\alpha}\frac{\partial \mathcal{W}_4}{\partial \sigma}\;, \tag{219}$$

and the radial coordinate, $\sigma_s = \sigma_s(\theta)$, in accordance with (63). It means that $\gamma_0(\sigma_s)$, $\gamma_2(\sigma_s)$, and $\gamma_4(\sigma_s)$ depend on the angular coordinate θ.

In what follows, it is more convenient to expand all functions entering the right hand side of (218) into the Taylor series expanded around the value of the radial coordinate $\sigma = \varkappa^{-1}$. This brings (218) to the following form,

$$\gamma_n = \left(\frac{1+\varkappa^2}{1+\varkappa^2\cos^2\theta}\right)^{1/2}\left[1+\frac{1}{2c^2}\omega^2 a^2\sin^2\theta\left(1+\mathcal{B}\frac{1-3\cos^2\theta}{1+\varkappa^2\cos^2\theta}\right)\right]$$
$$\left[\bar{\gamma}_0+\bar{\gamma}_2 P_2(\cos\theta)+\frac{1}{c^2}\bar{\gamma}_4 P_4(\cos\theta)\right]\;, \tag{220}$$

where now we have

$$\bar{\gamma}_0 = \left[\gamma_0 + \frac{1}{5}\frac{\omega^2 a^2}{5\epsilon^2 c^2}\frac{\partial\gamma_2}{\partial\sigma}\mathcal{B}\right]_{\sigma=\varkappa^{-1}}\;, \tag{221}$$

$$\bar{\gamma}_2 = \left[\gamma_2 + \frac{\omega^2 a^2}{\epsilon^2 c^2}\left(\frac{\partial\gamma_0}{\partial\sigma}+\frac{2}{7}\frac{\partial\gamma_2}{\partial\sigma}\right)\mathcal{B}\right]_{\sigma=\varkappa^{-1}}\;, \tag{222}$$

$$\bar{\gamma}_4 = \left[\gamma_4 + \frac{18}{35}\frac{\omega^2 a^2}{\epsilon^2 c^2}\frac{\partial\gamma_2}{\partial\sigma}\mathcal{B}\right]_{\sigma=\varkappa^{-1}}\;. \tag{223}$$

According to *Somigliana* [10] it is more convenient to write down the normal gravity force γ_n in terms of two constants which are the values of the normal gravity force taken at two particular positions on the surface: (1) a point a on the equator with $\theta = \pi/2$, and (2) a point b at the pole with $\theta = 0$. At these points (218) takes on the following forms,

$$\gamma_a = \sqrt{1+\varkappa^2}\left[1+(\mathcal{B}+1)\frac{\omega^2 a^2}{2c^2}\right]\left(\bar{\gamma}_0-\frac{\bar{\gamma}_2}{2}+\frac{3\bar{\gamma}_4}{8c^2}\right)\;, \tag{224}$$

$$\gamma_b = \bar{\gamma}_0+\bar{\gamma}_2+\frac{\bar{\gamma}_4}{c^2}\;. \tag{225}$$

Solving these equations with respect to $\bar{\gamma}_0$ and $\bar{\gamma}_2$ we get the post-Newtonian generalization of the theorem of *Pizzetti* [8, 78],

$$\bar{\gamma}_0 = \frac{2b}{3a}\left(1-\frac{\omega^2 a^2}{2c^2}\right)\gamma_a+\frac{1}{3}\gamma_b-\frac{7}{12c^2}\bar{\gamma}_4\;, \tag{226}$$

and the theorem of *Clairaut* [10],

$$\bar{\gamma}_2 = -\frac{2b}{3a}\left(1 - \frac{\omega^2 a^2}{2c^2}\right)\gamma_a + \frac{2}{3}\gamma_b - \frac{5}{12c^2}\bar{\gamma}_4 \,. \tag{227}$$

Equations (226) and (227) coincide with the corresponding post-Newtonian formulations of *Pizzetti* and *Clairaut* theorems given in our paper [3, Sections 11 and 12] after making transformation of the parameters to the new gauge defined by the parametrization (63) of the shape of the rotating spheroid.[8]

Theorems *Pizzetti* and *Clairaut* are used to derive the formula of *Somigliana* describing the magnitude of the normal gravity field vector in terms of the forces of gravity measured at equator and at pole [10]. This is achieved by replacing (226) and (227) back into (220) and expanding it with respect to $1/c^2$. It results in the post-Newtonian generalization of the formula of *Somigliana*,

$$\gamma_n = \frac{a\gamma_b \cos^2\theta + b\gamma_a \sin^2\theta}{\sqrt{a^2\cos^2\theta + b^2\sin^2\theta}} + \frac{1}{32c^2}\frac{4\omega^2 a^2\left(a\gamma_b - b\gamma_a\right) - 35a\gamma_4}{\sqrt{a^2\cos^2\theta + b^2\sin^2\theta}}\sin^2 2\theta$$
$$+ \mathcal{B}\frac{\omega^2 b}{8c^2}\frac{3b(a\gamma_b - b\gamma_a)\sin^2\theta - a(a\gamma_a + 2b\gamma_b)}{\left(a^2\cos^2\theta + b^2\sin^2\theta\right)^{3/2}}\sin^2 2\theta \,. \tag{228}$$

The first term in the right side of (228) is the canonical formula of *Somigliana* used ubiquitously in classic geodesy,[9] and the second and third terms being proportional to $1/c^2$, are the explicit post-Newtonian corrections. It is worth noticing that the post-Newtonian corrections to the *Somigliana* formula (228) are also included implicitly to the canonical (first) term through the values of the normal gravity force at equator, γ_a, and at the pole, γ_b, as follows from (221)–(225).

It is instructive to compare the results of the post-Newtonian formalism of the previous sections with the post-Newtonian approximations of axially-symmetric *exact* solutions of the Einstein equations. There are plenty of the known solutions (see, e.g., [91]) and some of their aspects have been analyzed in the application to relativistic geodesy in [44, 45]. Below, we focus on the post-Newtonian approximation of the Kerr metric and comment on its practical usefulness in geodesy.

8 Normal Gravity Field of the Kerr Metric

The Kerr metric is an exact, axisymmetric, stationary solution of the Einstein equations found by Roy Kerr [91]. The Kerr metric is a vacuum solution representing rotating black hole. It is often assumed in relativistic mechanics that the Kerr metric can be used to describe the external gravitational field of rotating extended body as

---

[8]For more details about the gauge transformations of the post-Newtonian spheroid the reader is referred to [3, Section 4].

[9]One should notice that in classic geodesy the *Somigliana* formula is usually expressed in terms of the geographic latitude Φ on ellipsoid that is related to the ellipsoidal angle θ by $\theta = \beta - \pi/2$, and, $a \tan\beta = b\tan\Phi$, [10, Eq. 2-77].

well. However, the exact internal solution generating the external Kerr metric has not yet been found, and may not exist [92]. The internal Kerr solution having been recently found by Hernandez-Pastora and Herrera [93] requires further study for verification. Nonetheless, it is instructive to investigate the geometric properties of the Kerr metric from the point of view of its application in relativistic geodesy. Preliminary study of this problem has been performed in [44, 45]. The present chapter focuses on the comparison of the multipolar structure of gravitational field of the Kerr metric with that of gravitational field of rotating spheroid. Because we employed the harmonic coordinates for description of the normal gravitational field, we will need the Kerr metric expressed in the harmonic coordinates defined by the condition (4).

8.1 Harmonic and Ellipsoidal Coordinates

The Kerr metric in harmonic coordinates, $x^\alpha = \{x, y, z\}$, has been derived in [94, 95]. We introduce the Kerr ellipsoidal coordinates, $\{\varsigma, \vartheta, \varphi\}$, connected to the harmonic coordinates x^α by equations

$$x = \alpha_K \sqrt{1+\varsigma^2} \sin\vartheta \cos\varphi \,, \tag{229a}$$

$$y = \alpha_K \sqrt{1+\varsigma^2} \sin\vartheta \sin\varphi \,, \tag{229b}$$

$$z = \alpha_K \varsigma \cos\vartheta \,, \tag{229c}$$

which look similar but not equal to the ellipsoidal coordinates (15) for uniformly rotating fluid body. The difference between the two types of the ellipsoidal coordinates is due to the fact that the geometric meaning of the parameter α in (15) is different from that of the Kerr parameter α_K which is equal (by definition of the Kerr geometry) to the ratio of the angular momentum, S, to mass, M, of the rotating body: $\alpha_K = S/Mc$. Dimension of the Kerr parameter α_K is the same (length) as one of the parameter $\alpha = a\epsilon$ which is used in definition (15) of the ellipsoidal coordinates in classic geodesy. Nonetheless, the two parameters, α and α_K, have different numerical values in the most general case making the two types of the ellipsoidal coordinates related to each other by transformation

$$\alpha_K \sqrt{1+\varsigma^2} \sin\vartheta = \alpha\sqrt{1+\sigma^2} \sin\theta \,, \qquad \alpha_K \varsigma \cos\vartheta = \alpha\sigma\cos\theta \,, \qquad \varphi = \phi \,. \tag{230}$$

We can establish a connection between two parameters, α_K and α, by making use of relationship, $S = I\omega$, that expresses the angular momentum S of rotating extended body with its rotational moment of inertia, I, and the angular velocity of rotation, ω. The moment of inertia can be expressed in terms of mass of the body and its equatorial radius, $I = \lambda M a^2$, where λ is a dimensionless integral parameter that depends on the distribution of matter inside the rotating body and is determined by an equation of state. For example, in case of a homogeneous ellipsoid, $\lambda = 2/5$ [77, 96].

Let us now assume that the Kerr metric is generated by an extended, rigidly rotating ellipsoid with some non-homogeneous distribution of mass density. Homogeneous distribution of density is excluded as the multipolar expansion of gravitational field of the Kerr metric does not coincide with that of the homogeneous ellipsoid of rotation as we show below. Then, in accordance with definition of the Kerr parameter, we should accept $\alpha_K = S/Mc = a\epsilon_K$, where $\epsilon_K = \lambda\omega a/c$ is the effective oblateness of the extended body that supposedly generates the Kerr metric (231). Thus, parameter α_K can be equal to $\alpha = a\epsilon$, if and only if, $\epsilon = \epsilon_K$. This condition imposes a certain limitation on the oblateness ϵ of rotating extended body which, on the other hand, is also a function of the rotational angular velocity and the average density, ρ, of the body, $\epsilon = \epsilon(\rho, \omega)$, as follows from the condition of hydrostatic equilibrium of the body's matter in the rotating frame.[10] It means that for a given total mass M and oblateness ϵ of extended rotating body, its angular velocity cannot be adjusted independently of the other parameters to make $\alpha_K = \alpha$. This is because the other parameters of the body like density, ρ, or a semi-major axis, a, must be appropriately chosen to maintain the condition of hydrostatic equilibrium in the rotating frame. We proceed by assuming that $\alpha_K \neq \alpha$. At the same time, we postulate that the total mass M and angular momentum S of the Kerr metric exactly coincide with the total mass M and angular momentum S of rotating body. This can be always done as these parameters are defined in terms of conserved integrals given at the spatial infinity of asymptotically-flat spacetime [42].

8.2 *Post-Newtonian Approximation of the Kerr Metric*

In the ellipsoidal coordinates (229) the exterior solution of Einstein's equations for the Kerr metric reads [94, 95]

$$
\begin{aligned}
ds^2 = &- c^2dt^2 + \alpha_K^2\left[(\varsigma+\mu_K)^2 + \cos^2\vartheta\right]\left(\frac{d\varsigma^2}{1+\varsigma^2-\mu_K^2} + d\vartheta^2\right) \\
&+ \frac{2\mu_K(\varsigma+\mu_K)}{(\varsigma+\mu_K)^2+\cos^2\vartheta}\left[\frac{\alpha_K\mu_K^2\sin^2\vartheta d\varsigma}{(1+\varsigma^2-\mu_K^2)(1+\varsigma^2)} - \alpha_K\sin^2\vartheta d\phi + cdt\right]^2 \\
&+ \alpha_K^2\sin^2\vartheta\left[(\varsigma+\mu_K)^2+1\right]\left[\frac{\mu_K^2 d\varsigma}{(1+\varsigma^2-\mu_K^2)(1+\varsigma^2)} - d\phi\right]^2 ,
\end{aligned}
\tag{231}
$$

where the Kerr mass parameter $\mu_K \equiv Gm_K/c^2$, and $m_K \equiv M/\alpha_K$. The interior solution for the Kerr metric is still unknown despite of numerous attempts to find it out [50, 82, 91, 97, 98]. Recently, a certain progress has been made by Hernandez-Pastora and Herrera [93] who used a model of a viscous, anisotropic distribution of mass density inside rotating body.

[10] For example, in case of a rigidly rotating homogeneous perfect fluid the relation, $\epsilon = \epsilon(\rho, \omega)$, is simply given by the Maclaurin formula (86).

The post-Newtonian approximation of the Kerr metric (231) in the ellipsoidal coordinates is

$$ds^2 = \left[-1 + \frac{2\mu_K\varsigma}{\varsigma^2+\cos^2\vartheta} - 2\mu_K^2\frac{\varsigma^2-\cos^2\vartheta}{(\varsigma^2+\cos^2\vartheta)^2}\right]c^2dt^2 - \frac{4\alpha_K\mu_K\varsigma\sin^2\vartheta}{\varsigma^2+\cos^2\vartheta}cdtd\phi$$
$$+\alpha_K^2\left[\frac{\varsigma^2+\cos^2\vartheta+2\mu_K\varsigma}{1+\varsigma^2}d\varsigma^2 + (1+\varsigma^2)\sin^2\vartheta\left(1+\frac{2\mu_K\varsigma}{\varsigma^2+\cos^2\vartheta}\right)d\phi^2\right.$$
$$\left.+\left(\varsigma^2+\cos^2\vartheta+2\mu_K\varsigma\right)d\vartheta^2\right]. \tag{232}$$

Harmonic coordinates are asymptotically Cartesian with the Euclidean metric δ_{ij} at spatial infinity. Components of the Euclidean metric in the ellipsoidal coordinates can be obtained directly from the coordinate transformation (229), and the result reads

$$\delta_{ij}dx^idx^j = \alpha_K^2\left[\frac{\varsigma^2+\cos^2\vartheta}{1+\varsigma^2}d\varsigma^2 + (1+\varsigma^2)\sin^2\vartheta d\phi^2 + \left(\varsigma^2+\cos^2\vartheta\right)d\vartheta^2\right]. \tag{233}$$

Comparing (233) with (232) allows us to recast the spacetime metric (232) to the following form

$$ds^2 = \left[-1+\frac{2V_K}{c^2} - \frac{2V_K^2}{c^4} + \frac{1}{c^4}\frac{2G^2m_K^2\cos^2\vartheta}{(\varsigma^2+\cos^2\vartheta)^2}\right]c^2dt^2$$
$$-\frac{8V_K^i}{c^2}dtdx^i + \left(1+\frac{2V_K}{c^2}\right)\delta_{ij}dx^idx^j, \tag{234}$$

where

$$V_K = \frac{Gm_K\varsigma}{\varsigma^2+\cos^2\vartheta}, \tag{235}$$
$$V_K^i = \frac{1}{2}\frac{(\boldsymbol{S}\times\boldsymbol{x})^i}{1+\varsigma^2}\frac{V_K}{\alpha_K^3 m_K}, \tag{236}$$

and $\boldsymbol{S} = \{0, 0, S\}$ is a vector of the total angular momentum (spin) of the body directed along the z axis of the harmonic coordinates which coincides with the direction of the rotational axis, $S = \alpha_K Mc$. Potentials V_K and V_K^i are harmonic functions and satisfy the Laplace equation

$$\Delta V_K = 0, \qquad \Delta V_K^i = 0, \tag{237}$$

where the Laplace operator in the ellipsoidal coordinates is defined in (28) after a corresponding replacement $\sigma \to \varsigma$.

The very last term in time-time component of the metric (234) is, actually, of the 2-nd post-Newtonian order of magnitude, and can be dropped out in the post-Newtonian approximation when comparing with the post-Newtonian metric (6) of extended body. The thing is that the metric (234) is written in dimensionless coordinates, while the parameter $\alpha_K = S/Mc$, that has been used to make them dimensionless, includes a relativistic factor of $1/c$ which is the main parameter of the post-Newtonian expansions. When we go back to the dimensional coordinates and also use the angular momentum, S and mass M, instead of $m = M/\alpha_K$, we have

$$\frac{1}{c^4}\frac{2G^2 m_K^2 \cos^2\vartheta}{(\varsigma^2+\cos^2\vartheta)^2} = \frac{1}{c^6}\frac{2G^2 S^2 \cos^2\vartheta}{\left(R^2+\alpha_K^2\cos 2\vartheta\right)^2}, \tag{238}$$

where $R^2 = x^2 + y^2 + z^2 = \alpha_K^2\left(\varsigma^2 + \sin^2\vartheta\right)$, and $S = \alpha_K M c$ is the angular momentum of the rotating body. The right hand side of (238) is apparently of the order of $1/c^6$ which is the post-post-Newtonian term not entering the first post-Newtonian approximation [5].

8.3 Normal Gravity Field Potential of the Kerr Metric

The normal gravity field represented by the Kerr metric can be expressed in a closed form similarly to the normal gravity field of uniformly rotating perfect fluid. Potential W of the normal field in case of the Kerr metric is defined by formula (193) where we have to use V_K and V_K^i for scalar and vector gravitational potentials. We have

$$\begin{aligned} W = &\frac{1}{2}\omega^2\alpha_K^2\left(1+\varsigma^2\right)\sin^2\vartheta + \frac{Gm_K\varsigma}{\varsigma^2+\cos^2\vartheta} \\ &+\frac{1}{c^2}\left\{\frac{1}{8}\omega^4\alpha_K^4\left(1+\varsigma^2\right)^2\sin^4\vartheta + \left[\frac{3}{2}(1+\varsigma^2) - 2\lambda\left(1+\frac{1}{\varkappa^2}\right)\right]\right. \\ &\left.\times\frac{Gm_K\varsigma}{\varsigma^2+\cos^2\vartheta}\omega^2\alpha_K^2\sin^2\vartheta - \frac{1}{2}\frac{G^2m_K^2\varsigma^2}{\left(\varsigma^2+\cos^2\vartheta\right)^2}\right\}, \end{aligned} \tag{239}$$

where λ is the parameter introduced above in Sect. 8.1, to connect the moment of inertia of a rotating body with its mass and the equatorial radius.

8.4 Multipolar Expansion of Scalar Potential

Analytic comparison of the Kerr metric with an external solution of uniformly rotating ellipsoid is achieved by comparing the multipolar expansions of the corresponding gravitational potentials. The multipolar expansion of the Kerr metric is obtained by applying the partial fraction decomposition,

$$\frac{\varsigma}{\varsigma^2+\cos^2\vartheta} = -\frac{1}{2i}\left(\frac{1}{i\varsigma+\cos\vartheta}+\frac{1}{i\varsigma-\cos\vartheta}\right), \tag{240}$$

and a generating function for the Legendre functions of the 2-nd type [72, formula 8.791]

$$\frac{1}{u-v} = \sum_{\ell=0}^{\infty}(2\ell+1)P_\ell(v)Q_\ell(u)\,. \tag{241}$$

It allows us to represent the Newtonian potential V_K of the Kerr metric in the following form

$$\begin{aligned} V_K &= Gm_K\sum_{\ell=0}^{\infty}(-1)^\ell(4\ell+1)q_{2\ell}(\varsigma)P_{2\ell}(\cos\vartheta) \\ &= m_K\Big[q_0(\varsigma)-5q_2(\varsigma)P_2(\cos\vartheta)+9q_4(\varsigma)P_4(\cos\vartheta)+\ldots\Big]. \end{aligned} \tag{242}$$

The post-Newtonian terms in time-time component of the metric (234) can be written as a partial derivative of the Newtonian potential

$$-V_K^2+\frac{G^2m_K^2\cos^2\vartheta}{(\varsigma^2+\cos^2\vartheta)^2} = m_K\frac{\partial V_K}{\partial\varsigma}\,. \tag{243}$$

Derivative of the Legendre function $q_\ell(\varsigma)$ is [72, formula 8.832], [71, formula (78)]

$$\frac{dq_\ell(\varsigma)}{d\varsigma} = -\frac{1+\ell}{1+\varsigma^2}\left[q_{\ell+1}(\varsigma)+\varsigma q_\ell(\varsigma)\right]\,. \tag{244}$$

It yields for the partial derivative of the Newtonian potential

$$\frac{\partial V_K}{\partial\varsigma} = -\frac{Gm}{1+\varsigma^2}\sum_{\ell=0}^{\infty}(-1)^\ell(4\ell+1)(2\ell+1)\left[q_{2\ell+1}(\varsigma)+\varsigma q_{2\ell}(\varsigma)\right]P_{2\ell}(\cos\vartheta)\,. \tag{245}$$

For analytical comparison of the Kerr metric with the post-Newtonian metric generated by extended rotating body it is sufficient to compare the multipolar expansion of potential V of the metric of the rotating body and that of V_K of the Kerr metric, and to deduce the correspondence between the parameters of the expansions [82]. Any kind of a multipolar expansion performed in either ellipsoidal or spherical coordinates can be used. Nonetheless, in theoretical practice, the comparison of multipolar expansions of gravitational fields is usually done in spherical coordinates. We follow this practice and introduce the spherical coordinates $\{\varrho, \Theta, \Phi\}$

$$x = \alpha_K \varrho \sin \Theta \cos \Phi \,, \tag{246a}$$

$$y = \alpha_K \varrho \sin \Theta \sin \Phi \,, \tag{246b}$$

$$z = \alpha_K \varrho \cos \Theta \,, \tag{246c}$$

where the angular coordinates Θ, Φ are the same as in (15) and the radial coordinate ϱ is related to R by a scale transformation, $R = \alpha_K \varrho$ so that, $\varrho = (\alpha/\alpha_K) r$.

To get the multipolar expansion of potential V_K we use the following expansion

$$\frac{\varsigma}{\varsigma^2 + \cos^2 \vartheta} = \sum_{\ell=0}^{\infty} (-1)^\ell \frac{P_{2\ell}(\cos \Theta)}{\varrho^{2\ell+1}} \,, \tag{247}$$

that can be easily confirmed by direct inspection for the point with angular coordinate, $\theta = \Theta = 0$, when $\varsigma = \varrho$, and applying expansion of the elementary function

$$\frac{\varrho}{\varrho^2 + 1} = \sum_{\ell=0}^{\infty} \frac{(-1)^\ell}{\varrho^{2\ell+1}} \,. \tag{248}$$

The expansion (247) can be re-written in the form of a multipolar expansion of a scalar potential,

$$V_K = \frac{GM}{R} \left[1 - \sum_{\ell=1}^{\infty} J_{2\ell}^K \left(\frac{a}{R} \right)^{2\ell} P_{2\ell}(\cos \Theta) \right] \,, \tag{249}$$

where the mass multipole moments of the Kerr metric

$$J_{2\ell}^K = (-1)^{\ell+1} \epsilon_K^{2\ell} = (-1)^{\ell+1} \left(\frac{\lambda \omega a}{c} \right)^{2\ell} \,, \qquad (\ell \geq 1) \tag{250}$$

It should be compared with the multipolar expansion (151) of the potential V of rotating homogeneous spheroid where the multipole moments J_{2n} are defined in (161). Comparing V_K and V and assuming that each of the potentials is generated by a corresponding axially-symmetric body, one can see that monopole terms ($\sim 1/R$) in (249) and (151) match perfectly so that the total mass of the Kerr metric can, indeed, be equated to the post-Newtonian mass of rotating spheroid, as it has been postulated above. The second order (quadrupole) moments can be matched as well. Indeed, we are allowed to equate $J_2^K = J_2$, under condition

$$\epsilon_K = \frac{\epsilon}{\sqrt{5}} \,, \qquad (\ell = 1) \,. \tag{251}$$

This result means that the Kerr metric can imitate gravitational field of a uniformly rotating Maclaurin ellipsoid in the quadrupole approximation. However, as soon as the condition (251) is satisfied, the multipole moments of the higher order (octupole, decapole, etc.) of multipole expansions of two potentials, V and $V_{\rm K}$, cannot be matched as follows from comparison of two expressions – (161) and (250). It means that the mass multipole moments of the Kerr metric, $J^{\rm K}_{2\ell}$, are, in general, different from the multipole moments $J_{2\ell}$ of rotating ellipsoid for $\ell \geq 2$.

8.5 *Multipolar Expansion of Vector Potential*

Gravitomagnetic vector-potential, $V^i_{\rm K}$ of the Kerr metric is given in (236). It has only two non-vanishing components, $V^i_{\rm K} = \left(V^x_{\rm K}, V^y_{\rm K}, 0\right)$, which can be combined together in the complex potential, $V^+_{\rm K} \equiv V^x_{\rm K} + iV^y_{\rm K}$, where i is the imaginary unit, c.f. (131). The explicit form of the potential is

$$V^+_{\rm K} = i\mathcal{D}_{\rm K} e^{i\phi} , \tag{252}$$

where

$$\mathcal{D}_{\rm K} = \frac{c}{2}\frac{\sin\vartheta}{\sqrt{1+\varsigma^2}} V_{\rm K} = \frac{c}{2}\frac{\sin\vartheta}{\sqrt{1+\varsigma^2}}\frac{Gm_{\rm K}\varsigma}{\varsigma^2+\cos^2\vartheta} . \tag{253}$$

We make use of the expansion (242) for $V_{\rm K}$, and equation

$$\sin\vartheta P_\ell(\cos\vartheta) = \frac{1}{4\ell+1}\left[P_{2\ell-1,1}(\cos\vartheta) - P_{2\ell+1,1}(\cos\vartheta)\right] , \tag{254}$$

that is given in [72, formula 8.733-4], to bring (253) to the following form

$$\mathcal{D}_{\rm K} = -\frac{Gm_{\rm K}c}{2}\sum_{\ell=0}^{\infty}(-1)^\ell\frac{q_{2\ell+2}(\varsigma)+q_{2\ell}(\varsigma)}{\sqrt{1+\varsigma^2}}P_{2\ell+1,1}(\cos\vartheta) . \tag{255}$$

We are now use [72, formula 8.734-5]

$$\sqrt{1+\varsigma^2}q_{2\ell+1,1}(\varsigma) = \frac{(2\ell+1)(2\ell+2)}{4\ell+3}\left[q_{2\ell}(\varsigma)+q_{2\ell+2}(\varsigma)\right] , \tag{256}$$

to obtain the expansion of $\mathcal{D}_{\rm K}$ in the final form

$$\begin{aligned}\mathcal{D}_{\rm K} &= -\frac{Gm_{\rm K}c}{4}\sum_{\ell=0}^{\infty}\frac{(-1)^\ell(4\ell+3)}{(\ell+1)(2\ell+1)}q_{2\ell+1,1}(\varsigma)P_{2\ell+1,1}(\cos\vartheta)\\ &= -\frac{3m_{\rm K}c}{4}\left[q_{11}(\varsigma)P_{11}(\cos\vartheta) - \frac{7}{18}q_{31}(\varsigma)P_{31}(\cos\vartheta)\right.\\ &\left.+\frac{11}{45}q_{51}(\varsigma)P_{51}(\cos\vartheta)+\dots\right] .\end{aligned} \tag{257}$$

Expansion of function $\mathcal{D}_K$ in spherical harmonics reads

$$\mathcal{D}_K = -\frac{Gm_K c}{2}\sum_{\ell=0}^{\infty}\frac{(-1)^\ell}{2\ell+1}\frac{P_{2\ell+1,1}(\cos\Theta)}{\varrho^{2\ell+2}} . \tag{258}$$

This formula can be checked by taking a point with coordinate, $\vartheta = \Theta = \pi/2$, where $\varsigma = \sqrt{\varrho^2 - 1}$ in accordance with (229) and (246). At this value of the angular coordinate expression (253) yields,

$$\mathcal{D}_K = \frac{Gm_K c}{2}\frac{1}{\varrho\sqrt{\varrho^2-1}} = \frac{Gm_K c}{2}\sum_{\ell=0}^{\infty}\frac{(2\ell-1)!!}{2^\ell \ell!}\frac{1}{\varrho^{2\ell+2}} , \tag{259}$$

but this is exactly the same formula as (258) for $\Theta = \pi/2$ with a special value of the polynomial $P_{2\ell+1,1}(0)$ taken from (180).

Coming back to a physical domain of the dimensional coordinates, and expressing the Legendre polynomial, $P_{2\ell+1,1}$ in terms of the first derivative with respect to its argument, we obtain

$$\mathcal{D}_K = \frac{GS}{2}\frac{\sin\Theta}{R^2}\left[1+\sum_{\ell=1}^{\infty}\frac{(-1)^\ell \epsilon_K^{2\ell}}{2\ell+1}\left(\frac{a}{R}\right)^{2\ell}\frac{dP_{2\ell+1}(\cos\Theta)}{d\cos\Theta}\right] , \tag{260}$$

where $S = \alpha_K Mc$ is a spin of the body generating the Kerr metric. Comparison of the multipole expansion (260) with (174) allows us to read out spin multipole moments, $S^K_{2\ell+1}$, of the Kerr metric,

$$S^K_{2\ell+1} = (-1)^{\ell+1}\epsilon_K^{2\ell} , \qquad (\ell \geq 1) . \tag{261}$$

As one can see, the spin multipole moments, $S^K_{2\ell+1}$, of the Kerr metric given in (261) are significantly different from those, S_{2n+1}, of the homogeneous rotating ellipsoid given in (186). Relation between spin and mass multipole moments of the Kerr metric,

$$S^K_{2\ell+1} = J^K_{2\ell} , \tag{262}$$

does not coincide with similar relation (187) between spin and mass multipole moments of a rotating spheroid made of homogeneous fluid. We also notice that relation (262) corresponds to a well-known relation between Geroch-Hansen mass and spin moments of the Kerr black hole [99, Chapter 14].

Replacing (260) to vector potential of the Kerr metric yields

$$V^i_K = \frac{G}{2}\frac{(\mathbf{S}\times\mathbf{x})^i}{R^3}\left[1+\sum_{\ell=1}^{\infty}\frac{(-1)^\ell \epsilon_K^{2\ell}}{2\ell+1}\left(\frac{a}{R}\right)^{2\ell}\frac{dP_{2\ell+1}(\cos\Theta)}{d\cos\Theta}\right] . \tag{263}$$

We can compare the multipole expansion (263) of vector potential of the Kerr metric with similar expansion (190) given for rotating spheroid. We notice that the first terms of these expansions match exactly each other so that the angular momentum, $\boldsymbol{S}$, of the body generating the Kerr geometry can be equated to the angular momentum of the rotating spheroid. As we have already equated the mass of the body generating gravitational field of the Kerr metric to that of the spheroid, we have to conclude that equating the angular momenta of the two bodies also requires imposing a limitation, $\alpha_K = \alpha$, or, in other words, $\epsilon_K = \epsilon$. However, this equation is not compatible with the matching condition (251) for the quadrupoles. It means that the Kerr metric is compatible with the normal gravity field of rotating ellipsoid merely in the mass-monopole spin-dipole approximation. This approximation is not suitable for geodetic applications. We conclude that the Kerr metric should not be used for the purposes of relativistic geodesy.

Acknowledgements I thank Physikzentrum Bad Honnef for hospitality and Wilhelm and Else Heraeus Stiftung for providing generous travel support to deliver a talk at 609 WE-Heraeus-Seminar "Relativistic Geodesy: Foundations and Applications" (13.03. - 19.03.2016). This work contributes to the research project "Spacetime Metrology, Clocks and Relativistic Geodesy" [http://www.issibern.ch/teams/spacetimemetrology/] sponsored by the International Space Science Institute (ISSI) in Bern, Switzerland.

References

1. S.M. Kopejkin, Relativistic manifestations of gravitational fields in gravimetry and geodesy. Manuscr. Geod. **16**, 301–312 (1991)
2. S.M. Kopeikin, E.M. Mazurova, A.P. Karpik, Towards an exact relativistic theory of Earth's geoid undulation. Phys. Lett. A **379**, 1555–1562 (2015)
3. S.M. Kopeikin, Reference ellipsoid and geoid in chronometric geodesy. Front. Astron. Space Sci. **3**(5), 5 (2016)
4. S. Kopeikin, W. Han, E. Mazurova, Post-Newtonian reference ellipsoid for relativistic geodesy. Phys. Rev. D **93**(4), 044069 (2016)
5. S. Kopeikin, M. Efroimsky, G. Kaplan, *Relativistic Celestial Mechanics of the Solar System* (Wiley, Berlin, 2011), xxxii+860 pp
6. P. Vaníček, E.J. Krakiwsky, *Geodesy, the Concepts*, 2nd edn. (Amsterdam, North Holland, 1986), xv+697 pp
7. B. Hofmann-Wellenhof, H. Moritz, *Physical Geodesy* (Springer, Berlin, 2006)
8. W. Torge, J. Müller, *Geodesy*, 4th edn. (De Gruyter, Berlin, 2012), 433 pp
9. L.D. Landau, E.M. Lifshitz, *The Classical Theory of Fields* (Pergamon Press, Oxford, 1975), xiii+402 pp
10. W.A. Heiskanen, H. Moritz, *Physical Geodesy* (W. H. Freeman, San Francisco, 1967), 364 pp
11. G. Petit, B. Luzum, IERS conventions. IERS Tech. Note **36**, 179 pp. (2010)
12. K. Sośnica, D. Thaller, A. Jäggi, R. Dach, G. Beutler, Sensitivity of Lageos orbits to global gravity field models. Artif. Satell. **47**, 47–65 (2012)
13. P.L. Bender, R.S. Nerem, J.M. Wahr, Possible future use of laser gravity gradiometers. Space Sci. Rev. **108**, 385–392 (2003)
14. N.K. Pavlis, S.A. Holmes, S.C. Kenyon, J.K. Factor, The development and evaluation of the Earth gravitational model 2008 (EGM2008). J. Geophys. Res. Solid Earth **117**, B04406 (2012)

15. N.K. Pavlis, S.A. Holmes, S.C. Kenyon, J.K. Factor, Correction to "the development and evaluation of the Earth gravitational model 2008 (EGM2008)". J. Geophys. Res. Solid Earth **118**, 2633–2633 (2013)
16. Fu Lee-Lueng, On the decadal trend of global mean sea level and its implication on ocean heat content change. Front. Mar. Sci. **3**, 37 (2016)
17. H.-P. Plag, M. Pearlman (eds.), *Global Geodetic Observing System* (Springer, Dordrecht, 2009), 322 pp
18. L.-L. Fu, B.J. Haines, The challenges in long-term altimetry calibration for addressing the problem of global sea level change. Adv. Space Res. **51**, 1284–1300 (2013)
19. Z. Altamimi, P. Rebischung, L. Métivier, X. Collilieux, ITRF2014: a new release of the international Terrestrial Reference Frame modeling nonlinear station motions. J. Geophys. Res. Solid Earth **121**, 6109–6131 (2016)
20. J. Müller, D. Dirkx, S.M. Kopeikin, G. Lion, I. Panet, G. Petit, P.N.A.M. Visser, High performance clocks and gravity field determination. Space Sci. Rev. **214**(5), 1–31 (2018)
21. R. Bondarescu, M. Bondarescu, G. Hetényi, L. Boschi, P. Jetzer, J. Balakrishna, Geophysical applicability of atomic clocks: direct continental geoid mapping. Geophys. J. Int. **191**, 78–82 (2012)
22. E. Mai, J. Müller, General remarks on the potential use of atomic clocks in relativistic geodesy. ZFV - Zeitschrift fur Geodasie, Geoinformation und Landmanagement **138**(4), 257–266 (2013)
23. E. Mai, Time, atomic clocks, and relativistic geodesy. Report No 124, Deutsche Geodätische Kommission der Bayerischen Akademie der Wissenschaften (DGK) (2014), 128 pp., http://dgk.badw.de/fileadmin/docs/a-124.pdf
24. E. Hackmann, C. Lämmerzahl, Generalized gravitomagnetic clock effect. Phys. Rev. D **90**(4), 044059 (2014)
25. J.M. Cohen, B. Mashhoon, Standard clocks, interferometry, and gravitomagnetism. Phys. Lett. A **181**, 353–358 (1993)
26. V.F. Fateev, S.M. Kopeikin, S.L. Pasynok, Effect of irregularities in the earth's rotation on relativistic shifts in frequency and time of earthbound atomic clocks. Meas. Tech. **58**, 647–654 (2015)
27. M. Soffel, S.A. Klioner, G. Petit, P. Wolf, S.M. Kopeikin, P. Bretagnon, V.A. Brumberg, N. Capitaine, T. Damour, T. Fukushima, B. Guinot, T.-Y. Huang, L. Lindegren, C. Ma, K. Nordtvedt, J.C. Ries, P.K. Seidelmann, D. Vokrouhlický, C.M. Will, C. Xu, The IAU 2000 resolutions for astrometry, celestial mechanics, and metrology in the relativistic framework: explanatory supplement. Astron. J. (USA) **126**, 2687–2706 (2003)
28. M. Soffel, S. Kopeikin, W.-B. Han, Advanced relativistic VLBI model for geodesy. J. Geod. **91**(7), 783–801 (2017)
29. V.A. Brumberg, S.M. Kopeikin, Relativistic equations of motion of the earth's satellite in the geocentric frame of reference. Kinematika i Fizika Nebesnykh Tel **5**, 3–8 (1989)
30. V.A. Brumberg, S.M. Kopejkin, Relativistic reference systems and motion of test bodies in the vicinity of the Earth. Nuovo Cim. B Ser. **103**, 63–98 (1989)
31. T. Damour, M. Soffel, C. Xu, General-relativistic celestial mechanics. IV. Theory of satellite motion. Phys. Rev. D **49**, 618–635 (1994)
32. A. San Miguel, Numerical integration of relativistic equations of motion for Earth satellites. Celest. Mech. Dyn. Astron. **103**, 17–30 (2009)
33. U. Kostić, M. Horvat, A. Gomboc, Relativistic positioning system in perturbed spacetime. Class. Quantum Gravity **32**(21), 215004 (2015)
34. K.-M. Roh, B.-K. Choi, The effects of the IERS conventions (2010) on high precision orbit propagation. J. Astron. Space Sci. **31**, 41–50 (2014)
35. K.-M. Roh, S.M. Kopeikin, J.-H. Cho, Numerical simulation of the post-Newtonian equations of motion for the near Earth satellite with an application to the LARES satellite. Adv. Space Res. **58**, 2255–2268 (2016)
36. I. Ciufolini, E.C. Pavlis, A confirmation of the general relativistic prediction of the Lense-Thirring effect. Nature **431**, 958–960 (2004)
37. I. Ciufolini, Dragging of inertial frames. Nature **449**, 41–47 (2007)

38. I. Ciufolini, E.C. Pavlis, A. Paolozzi, J. Ries, R. Koenig, R. Matzner, G. Sindoni, K.H. Neumayer, Phenomenology of the lense-thirring effect in the solar system: measurement of frame-dragging with laser ranged satellites. New Astron. **17**, 341–346 (2012)
39. V.G. Gurzadyan, I. Ciufolini, A. Paolozzi, A.L. Kashin, H.G. Khachatryan, S. Mirzoyan, G. Sindoni, Satellites testing general relativity: residuals versus perturbations. Int. J. Mod. Phys. D **26**, 1741020 (2017)
40. K.S. Thorne, R.D. Blandford, Black holes and the origin of radio sources, in *Extragalactic Radio Sources. IAU Symposium*, vol. 97, ed. by D.S. Heeschen, C.M. Wade (1982), pp. 255–262
41. S. Chandrasekhar, *Ellipsoidal Figures of Equilibrium* (Yale University Press, New Haven, 1969), ix+252 pp
42. A.N. Petrov, S.M. Kopeikin, R.R. Lompay, B. Tekin, *Metric Theories of Gravity: Perturbations and Conservation Laws* (De Gruyter, Berlin, 2017), xxiv+597 pp
43. L. Blanchet, Gravitational radiation from post-Newtonian sources and inspiralling compact binaries. Living Rev. Relativ. **17**, 2 (2014)
44. M. Soffel, F. Frutos, On the usefulness of relativistic space-times for the description of the Earth's gravitational field. J. Geod. **90**(12), 1345–1357 (2016)
45. D. Philipp, V. Perlick, D. Puetzfeld, E. Hackmann, C. Lämmerzahl, Definition of the relativistic geoid in terms of isochronometric surfaces. Phys. Rev. D **95**(10), 104037 (2017)
46. M. Oltean, R.J. Epp, P.L. McGrath, R.B. Mann, Geoids in general relativity: geoid quasilocal frames. Class. Quantum Gravity **33**(10), 105001 (2016)
47. K.D. Krori, P. Borgohain, Uniform-density cold neutron stars in general relativity. J. Phys. Math. Gen. **8**, 512–520 (1975)
48. J. Ponce de León, Fluid spheres of uniform density in general relativity. J. Math. Phys. **27**, 271–276 (1986)
49. L. Lindblom, Static uniform-density stars must be spherical in general relativity. J. Math. Phys. **29**, 436–439 (1988)
50. J.N. Islam, *Rotating Fields in General Relativity* (Cambridge University Press, Cambridge, 1985), 127 pp
51. E. Gourgoulhon, An introduction to the theory of rotating relativistic stars. *Lectures Given at the Compstar 2010 School (Caen, 8–16 Feb 2010)* (2010)
52. J.L. Friedman, N. Stergioulas, *Rotating Relativistic Stars* (Cambridge University Press, Cambridge, 2013), 438 pp
53. S. Chandrasekhar, The post-Newtonian effects of general relativity on the equilibrium of uniformly rotating bodies. I. The Maclaurin spheroids and the virial theorem. Astrophys. J. **142**, 1513–1518 (1965)
54. S. Chandrasekhar, The post-Newtonian effects of general relativity on the equilibrium of uniformly rotating bodies. II. The deformed figures of the Maclaurin spheroids. Astrophys. J. **147**, 334–352 (1967)
55. N.P. Bondarenko, K.A. Pyragas, On the equilibrium figures of an ideal rotating liquid in the post-Newtonian approximation of general relativity. II: Maclaurin's P-ellipsoid. Astrophys. Space Sci. **27**, 453–466 (1974)
56. D. Petroff, Post-Newtonian Maclaurin spheroids to arbitrary order. Phys. Rev. D **68**(10), 104029 (2003)
57. G.L. Clark, The gravitational field of a rotating nearly spherical body. Philos. Mag. **39**(297), 747–778 (1948)
58. P. Teyssandier, Rotating stratified ellipsoids of revolution and their effects on the dragging of inertial frames. Phys. Rev. D **18**, 1037–1046 (1978)
59. H. Cheng, G.-X. Song, C. Huang, The internal and external metrics of a rotating ellipsoid under post-Newtonianian approximation. Chin. Astron. Astrophys. **31**, 192–204 (2007)
60. C.M. Will, *Theory and Experiment in Gravitational Physics* (Cambridge University Press, Cambridge, UK, 1993), xi+396 pp
61. S. Kopeikin, I. Vlasov, Parametrized post-Newtonian theory of reference frames, multipolar expansions and equations of motion in the N-body problem. Phys. Rep. **400**, 209–318 (2004)

62. M.H. Soffel, *Relativity in Astrometry, Celestial Mechanics and Geodesy* (Springer, Berlin, 1989), xiv+208 pp
63. V.A. Brumberg, *Essential Relativistic Celestial Mechanics* (Adam Hilger, Bristol, 1991), x+263 pp
64. A.D. Rendall, Convergent and divergent perturbation series and the post-Minkowskian approximation scheme. Class. Quantum Gravity **7**(5), 803–812 (1990)
65. J. Müller, M. Soffel, S.A. Klioner, Geodesy and relativity. J. Geod. **82**, 133–145 (2008)
66. V.A. Fock, *The Theory of Space, Time and Gravitation*, 2nd edn. (Macmillan, New York, 1964); Translated from the Russian by N. Kemmer, xii+448 pp
67. S. Weinberg, *Gravitation and Cosmology: Principles and Applications of the General Theory of Relativity* (Wiley, New York, 1972)
68. A.P. Lightman, W.H. Press, R.H. Price, S.A. Teukolsky, *Problem Book in Relativity and Gravitation* (Princeton University Press, Princeton, 1975), xiv+603 pp
69. G.B. Arfken, H.J. Weber, *Mathematical Methods for Physicists*, 4th edn. (Academic Press, San Diego, 1995), xviii+1029 pp
70. E.W. Hobson, *The Theory of Spherical and Elliptical Harmonics* (Cambridge University Press, Cambridge, 1931), vi+500 pp
71. V. Pohánka, Gravitational field of the homogeneous rotational ellipsoidal body: a simple derivation and applications. Contrib. Geophys. Geod. **41**, 117–157 (2011)
72. I.S. Gradshteyn, I.M. Ryzhik, in *Table of Integrals, Series and Products*, 4th edn., ed. by Y.V. Geronimus, M.Y. Tseytlin (Academic Press, New York, 1965); First appeared in 1942 as MT15 in the Mathematical tables series of the National Bureau of Standards
73. I. Ciufolini, J.A. Wheeler, *Gravitation and Inertia* (Princeton University Press, Princeton, 1995), 512 pp
74. S.M. Kopeikin, Gravitomagnetism and the speed of gravity. Int. J. Mod. Phys. D **15**, 305–320 (2006)
75. S.M. Kopeikin, The gravitomagnetic influence on Earth-orbiting spacecrafts and on the lunar orbit, in *General Relativity and John Archibald Wheeler*, vol. 367, Astrophysics and Space Science Library, ed. by I. Ciufolini, R.A.A. Matzner (Springer, Berlin, 2010)
76. B.H. Hager, M.A. Richards, Long-wavelength variations in Earth's geoid - physical models and dynamical implications. Philos. Trans. R. Soc. Lond. Ser. A **328**, 309–327 (1989)
77. J.-L. Tassoul, *Theory of Rotating Stars* (Princeton University Press, Princeton, 1979), xvi+508 pp
78. P. Pizzetti, *Principii della teoria meccanica della figura dei pianeti* (E. Spoerri, Pisa, 1913), xiii+251 pp
79. R.O. Hansen, Multipole moments of stationary space-times. J. Math. Phys. **15**, 46–52 (1974)
80. K.S. Thorne, Multipole expansions of gravitational radiation. Rev. Mod. Phys. **52**, 299–339 (1980)
81. L. Blanchet, T. Damour, Radiative gravitational fields in general relativity. I - general structure of the field outside the source. Philos. Trans. R. Soc. Lond. Ser. A **320**, 379–430 (1986)
82. H. Quevedo, Multipole moments in general relativity - static and stationary vacuum solutions. Fortschritte der Physik **38**, 733–840 (1990)
83. T. Damour, B.R. Iyer, Multipole analysis for electromagnetism and linearized gravity with irreducible Cartesian tensors. Phys. Rev. D **43**, 3259–3272 (1991)
84. L. Blanchet, On the multipole expansion of the gravitational field. Class. Quantum Gravity **15**, 1971–1999 (1998)
85. C. Jekeli, The exact transformation between ellipsoidal and spherical harmonic expansions. Manuscr. Geod. **13**, 106–113 (1988)
86. C. Snow, *Hypergeometric and Legendre Functions with Applications to Integral Equations of Potential Theory*, 2nd edn. (US Government Printing Office, Washington, 1952), xi+427 pp
87. M. Soffel, R. Langhans, *Space-Time Reference Systems* (Springer, Berlin, 2013), xiv+314 pp
88. S. Chandrasekhar, J.C. Miller, On slowly rotating homogeneous masses in general relativity. Mon. Not. Roy. Astron. Soc. **167**, 63–80 (1974)

89. J.M. Bardeen, A reexamination of the post-Newtonian Maclaurin spheroids. Astrophys. J. **167**, 425 (1971)
90. R. Meinel, M. Ansorg, A. Kleinwächter, G. Neugebauer, D. Petroff, *Relativistic Figures of Equilibrium* (Cambridge University Press, Cambridge, 2008), p. ix+218 pp
91. H. Stephani, D. Kramer, M. MacCallum, C. Hoenselaers, E. Herlt, *Exact Solutions of Einstein's Field Equations*, 2nd edn. (Cambridge University Press, Cambridge, 2003), xxix+701 pp
92. W.C. Hernandez, Material sources for the Kerr metric. Phys. Rev. **159**, 1070–1072 (1967)
93. J.L. Hernandez-Pastora, L. Herrera, Interior solution for the Kerr metric. Phys. Rev. D **95**(2), 024003 (2017)
94. C. Jiang, W. Lin, Harmonic metric for Kerr black hole and its post-Newtonian approximation. Gen. Relativ. Gravit. **46**, 1671 (2014)
95. W. Lin, C. Jiang, Exact and unique metric for Kerr-Newman black hole in harmonic coordinates. Phys. Rev. D **89**(8), 087502 (2014)
96. H. Essén, The physics of rotational flattening and the point core model. Int. J. Geosci. **5**, 555–570 (2014)
97. A. Krasinski, Ellipsoidal space-times, sources for the Kerr metric. Ann. Phys. **112**, 22–40 (1978)
98. T. Wolf, G. Neugebauer, About the non-existence of perfect fluid bodies with the Kerr metric outside. Class. Quantum Gravity **9**, L37–L42 (1992)
99. C. Bambi, *Black Holes: A Laboratory for Testing Strong Gravity* (Springer, Singapore, 2017), xv+340 pp

Anholonomity in Pre-and Relativistic Geodesy

Erik W. Grafarend

Abstract I was invited to speak about anholonomity or the problem to find coordinate reference systems which are differentiable. In general non-differentiable functions like (pseudo) orthonormal reference systems are differentiable forms being not classical functions. These differentiable forms are the basis of Elie Cartan's "exterior calculus". Geodetic examples are extensively reviewed in the context of the pre-and relativistic Geodesy.

1 Motivation: Anholonomity and the Axisymmetric Gravity Field

Geometric Geodesy as well as Physical Geodesy in spacetime are the fundament of Geodetic Science. From the beginning "Relativistic Positioning" as well as the "Somigliana–Pizzetti Gravity Field"- since 1930 the legal official IAG reference gravity field (International Association of Geodesy) - now the "Kerr metrical relativistic Gravity Field" - my proposal to the IAG, Member of the International Union of the Geodesy and Geophysics of ICSU - in spacetime played a dominant role: it balances the gravitational field and the rotational field in a unique way. It takes into account that the planet Earth and all other terrestrial planets/moons rotate.

From the birth of Geodetic Science, anholonomity of Geodetic Reference Frames was a central topic. It is the problem of integration and differentiation of special geodetic differentiable forms which generalize the notion of functions. In school we learn differentiation and integration, but not differential forms, special functions which cannot be integrated in a closed form. We have learned that mixed second order differentials of functions commute, for instance, $f_{ij} = f_{ji} = 0$, namely $f_{12} : \partial^2 f/\partial x^1 \partial x^2$, $f_{21} =: \partial^2 f/\partial x^2 \partial x^1$. It might be a surprise for Geodesists and Physicists that this conditions is not fulfilled, in general. Indeed there are many

---

E. W. Grafarend (✉)
Department of Geodesy and Geoinformatics, Faculty of Aerospace Engineering and Geodesy, Faculty of Mathematics and Physics, Geschwister Scholl Strasse 24 D, 70174 Stuttgart, Germany
e-mail: grafarend@gis.uni-stuttgart.de

D. Puetzfeld and C. Lämmerzahl (eds.), *Relativistic Geodesy*,
Fundamental Theories of Physics 196, https://doi.org/10.1007/978-3-030-11500-5_7

geodetic functions where this condition does not hold. For those functions -a type of generalized functions- we note the characteristic property $f_{ij} - f_{ji} = \omega_{ij} \neq 0$.

A first geodetic example is the scalar-valued gravity potential W, the sum of the gravitational potential U and the centrifugal potential, also called rotational potential V. It relates to the differential dW of the gravity potential W called holonomic potential to the height displacement $đH^{\gamma}$ and the modulus of the gravity vector $\|\Gamma\|_2$ in terms of $dW = \|\Gamma\|_2\, đH^{\gamma}$. We have used here the notation of M. Planck documenting that "đ" is not a "perfect differential", also called anholonomic coordinate. Such a notion is plausible since $đH^{\gamma} \neq 0$ in a closed loop, also called ring integral. The argument is that in a local geodetic reference frame we have to refer to the terrestrial gravity field, a deeper argument we bring here to your attention.

"Imperfect differentials" gave raise to the notion of differential forms. Our example refers to the differential 1-form. The modulus of the Cartesian coordinates $\|\Gamma\|_2$ (ℓ_2 norm) is called "Frobenius" integrating factor due to $đH^{\gamma} \neq 0$. Integrability is a key property of classical functions. But p-forms are extensions towards anholonomity. Elie Cartan based his theory of "exterior calculus" on non-integrability.

Here we start with a review of anholonomity: before and after relativity. In Chap. 2 we begin with the geodetic inventor of generalized geodetic functions: F. R. Helmert. Being a Geodesist, he had studied two years in Mathematics and Physics where he took lectures at the University of Leipzig of F. G. Frobenius and H. Grassmann.

It was there that he learned about "integrability" and "differentiable forms": he showed that the gravity potential of rigid bodies is integrable, but geodetic heights are not, in general. "Geodätische Kote" was his term used for potential heights. In Chap. 2 we therefore review anholonomity, before and after Relativity. The two reports of the Ohio State University (OSU), Department of Geodetic Science (1972, 1973) of the author build up the basis and applied to three-dimensional Geodesy and gravity gradients. The highlights are (i) the indicator diagrams and (ii) Cartan's pseudo-torsion or anholonomity forms in the "natural reference system", $\tau_{ij}{}^{k}$.

The mathematician J. Zund in his latest book [1] realized the importance of our anholonomity formulae. The five characteristic curvature parameters $\{k_1, k_2, t_1, \kappa_1, \kappa_2\}$ of Hotine-Marussi's three-dimensional Geodesy were the highlights of contemporary geodetic science. A detailed computation of the Frobenius matrix of integrating factors in terms of the five arbitrary curvature parameters was demonstrated for "natural coordinates" holonomic $\{\Lambda, \Phi, W\}$ (astronomical longitude, astronomical latitude, potential of gravity) and anholonomic Cartesian coordinates $\{X, Y, Z\}$. A special section is concerned with the curvature (first curvature) and torsion (second curvature) of field lines of the gravity field and its orthonormal "plumbline". We document the special role of the Marussi gauge emphasizing the central role of gauge theory, in general. The duality between vertical and horizontal fields equipped with an orthonormal (in Relativity "pseudo-orthonormal") metric δ_{ij} leads to the second order differential equation of the plumbline manifold in Marussi gauge. Basic is the proof that spherical symmetric gravity fields lead to zero curvature/torsion or zero anholonomity or "holonomic coordinates", but not non-spherical, for instance axisymmetric gravity field. Alternatively, a departure from non-spherical symmetry, conventionally called "anormal potential", "gravity disturbance", "vertical deflections"- standard geodetic terms - lead to anholonomity as already stated by P. Teunissen [2]. Or standard spherical /ellipsoidal series expansions to the order/degree 180/180, 360/360, or 720/720 up to 7200/7200 illustrate perfectly "anholonomity". Chapter 3 introduces "Real null frames and coframes" within General Relativity" in order to prove anholonomity. We take reference to Blagojevic, M., Garecki, J., Hehl, F. W. and Obukhov, Y.N. [3]. More specifically, in pseudo-Riemannian spacetime we define a non-unique pseudo-orthonormal reference frame indexed $\alpha, \beta, \ldots, \in \{1, 2, 3, 4\}$. We choose a timelike 4-leg by $\mathbf{e}_4$ and $\{\mathbf{e}_1, \mathbf{e}_2, \mathbf{e}_3\}$ spacelike $1, 2, 3$-legs. Such a popular frame of reference is pseudo-orthonormal, but unfortunately anholonomic, in general. Its metric is given by $g(e_\alpha, e_\beta) = diag(1, 1, 1, -1)$, namely locally Minkowski. With orthonormal frames of reference we have previously experienced anholonomity: Please, study the subject of "Null frames" called $\{l, m, n, \overline{m}\}$.

> First, surface geometry as a two-dimensional Riemann manifold we intent to orthonormalize. We arrive at a non-unique orthogonal reference frame, a special Cartan frame of reference. We have shown this when we analyze an ellipsoidal frame of reference in E. Grafarend and F. W. Krumm [4].
> Second, in analyzing the Euler Kinematical equations as well as the dynamical Euler equation for rigid bodies and the dynamical Euler-Louisville equations in terms of Euler or Cartan angle we found the well known anholonomity in terms of a Frobenius matrix. We take reference to E. Grafarend and W. Kühnel [5].

Chapter 4 is specifying the notion Killing vectors of symmetry, namely for the sphere (3 Killing vectors) and for the ellipsoid of revolution (1 Killing vector). It was needed to understand spherical symmetry versus ellipsoidal symmetry, in particular for the geodetic Somigliana–Pizzetti reference field.

Of particular importance is Chap. 5: we study the influence of local vertical nets which cause indeed anholonomity due to the reference of the physical local vertical. At the end we analyze the famous object of anholonomity in terrestrial networks, namely an example of Cartan's exterior calculus.

We conclude in Chap. 6 with special comments in the special role of anholonomity for Geodesy, namely on the irregular boundary of the planet Earth. Our final highlight is the literature list of geodetic contributions of the topic anholonomity, by no means a forgotten subject. Finally we recommend to the International Association of Geodesy to adopt the axisymmetric Kerr metric or its linear approximation Lense-Thirring for a rotating - gravitating Earth, namely replacing the axisymmetric Somigliana–Pizzetti reference field to include Relativistic Geodesy.

2 Anholonomity: Before and After Relativity

First, we study geodetic anholonomity as being established by Friedrich Robert Helmert. He is the real founder of Physical Geodesy, for instance of gravimetry.

His concept of physical heights in terms of the gravity potential W, the sum of the gravitational potential U and the centrifugal potential V, as part of the scalar-valued gravity field established holonomic coordinates.

> What are holonomic heights, better: holonomic height differences, what are holonomic coordinates in contrast to anholonomic ones? Such questions we will answer!

It is worth to study the arguments of F.R. Helmert why he transformed geometric heights or height differences called "anholonomic" or not integrable in a closed loop to "holonomic heights" or their differences in terms of potential numbers. Metrical heights are constructed by dividing by a constant value, for instance the Global Mean Value of Gravity (a proposal by S. Heitz) of by the external Somigliana–Pizzetti gravity field, the standard gravity field known to sub-nano-Gal accuracy according to A. Ardalan and E. Grafarend [6], also called "orthometric height" (F. Sanso and P. Vanicek [7].

Born 31 July 1843 in Freiberg/Sachsen/F.R. Helmert attended in his hometown the "Bürgerschule", at the age of 14 he joined the "St. Annen Schule". in Dresden. The started his university studies at the age 16 years in joining the "Polytechnical School" nowadays called "Technical University Dresden". For his Ph.D. studies he continued working in Dresden with A. Nagel as his supervisor. As a benefit of his excellent studies a stipend was offered to him: he chose to study two years at the University of Leipzig, namely to study physics and mathematics. He attended the lectures of F.G. Frobenius and read H. Grassmann on the topic of integrable differential forms, nowadays called "Cartan calculus". He finished his Ph.D. studies in Dresden: "Studium über rationelle Vermessungen im Gebiet der Höheren Geodäsie". In the year 1869 he was appointed as the "observer" at the Hamburg Observatory. His first publication of the year 1874 dealt with "Vermessung und rechnerische Ausgleichung eines Sternhaufens". At the age of 29 years he was appointed "full professor" at the "Polytechnische Schule" nowadays called "Aachen University". It was in Aachen where he wrote the legendary monumental works:

(i) Ausgleichungsrechnung nach der Methode der kleinsten Quadrate mit Anwendungen auf die Geodäsie und die Theorie, 1872. (This basic work saw many reprints within the 20th century.)
(ii) Die Mathematischen und Physikalischen Theorien der Höheren Geodäsie, 1880 and 1884, two volumes. (This basic works was often reprinted and translated in the 20th century.)

F. R. Helmert was from the year 1886 up to his death in the year 1917 Director of the "Königlich Preussischens Geodätischens Institut" which was founded in the year 1870 in Berlin from 1890 onward up to this day located at the Telegraphenberg in the city of Potsdam . At the same time he was appointed Full Professor to the Chair of Geodetic Sciences at the "Friedrich-Wilhelms-Universität" in Berlin. In the year 1900 he was elected to ordinary member at the "Prussian Academy of Sciences", a colleague of Albert Einstein in Berlin.

A lot of effects in Physical Geodesy bear his name:

- Helmert ellipsoid transformation
- Helmert projection
- Helmert level ellipsoid
- Helmert heights
- Helmert deflection of vertical
- Polar height variation
- Geodynamics

Here we specialize on "holonomic versus anholonomic heights" or the key question:

Why is Geodesy part of Physics?

During his studies in Leipzig he became acquainted with "integrating factors" and the celebrated Frobenius Lemma: Frobenius was teaching in Leipzig at that time.

Example:

$$\oint dW = -\oint \Gamma \, đH^{\gamma} = 0 \text{ versus } \oint đH^{\gamma} \neq 0$$

The notion "đH" was introduced by M. Planck: the ring integral of geometric heights đH is not zero, illustrated by "-" and by "dH" combined to đH. We also say "loop integral". In contrast, the ring integral of the differential of "potential heights" - the closed loop integral-is zero. W is called "the Gauss-Green potential", later on subject to "Cartan calculus" or "differential forms".

In a co-rotating reference system-a system rotating with the Earth- we enjoy four types of forces:

(i) Gravitational forces:
"conservative" derived from a scalar potential

(ii) Centrifugal force:
"conservative" derived from a scalar potential

(iii) Euler force:
"produced by angular momentum", non-conservative
- responsible for "Polar Motion"(POM) and "Length-of-Day" variation (LOD), or
- Precession-Nutation, vector valued force

(iv) Coriolis force:
"produced currents at Sea", non-conservative

Summary

"Conservative": field equations

$$\text{(i) grad } W = \text{grad} U + \text{grad} V$$
$$\text{(ii) div grad } W = -4\pi G\rho + 2\boldsymbol{\Omega}^2$$

"Non-conservative": field equations

$rot\ \Gamma = 2\boldsymbol{\Omega}$
"for rigid bodies, similar to
Maxwell equations of a deforming
body"

What are our subjects?

At first we review the contents of my OSU reports (The Ohio State University Columbus/Ohio/USA) from the year 1972 and 1973 of 81 pages and 126 pages on anholonomity. Second, we shortly review my contributions on the subject of field lines of gravity, their curvature and torsion, the Lagrange and the Hamilton equations of the plumbline. Finally, third, we present examples of anholonomity.

2.1 *Two Reports form the Ohio State University (OSU): 1972–1973*

There are not too many books on the subject of Differential Geomathematics, Differential Geometry, Differential Topology and Theoretical Physics where you study "holonomity" and "anholonomity". We advice the interested reader to consult J. A. Schouten (1954: "Ricci Calculus") [8] with an index on "anholonomity") as well as R. Cushman et al. [9] : "Geometry on non-holonomically constraint systems", World Scientific, New Jersey. My favorite is S. Sternberg [10]: read Chap. 6 on Cartan calculations in semi-Riemann geometry, in particular "frame fields and co-frame fields", as well as Chap. 15 on "The Frobenius theorem", in particular pages 312–313 on "A dual formulation of the Frobenius theorem". In this remarkable book you find also a review of

(i) Relativity
(ii) Higgs et al.

A photo of F.G. Frobenius (1849–1917) is available on page 303 of the book by S. Sternberg [10].

My first report of the Department of Geodetic Science, No. 174, The Ohio State University, Columbus/Ohio/USA, 81 pages, of the year 1972 treats "Three-Dimensional Geodesy and Gravity Gradients" based on the Frobenius integration theorem. The anholonomity or non-integrability is due to measurements in the "natural" or astronomic reference system. The characteristic Cartan object of anholonomity depends on either the gradients of the terrestrial gravity field or their gradients. The specific integration factors are determined, namely six integrating factors in the astronomic reference frame or nine integrating factors in the geodesic reference frame. They depend either on the gradients of the standard gravity field or on the Euclidean norm of the gravity field/vertical deflections of the disturbing gravity field.

> At this time, I expressed my opinion that with the availability of gravity gradients would open a new era of three-dimensional geodesy, namely developed by Antonio Marussi.

It makes sense to have a more detailed look into the subject of the first Ohio State University Report (OSU report).

After Sect. 1, introduction and summary, Sect. 2 dealt with "The Forbenius Integration Theorem": review of the literature, H. Grassmann and exterior algebra, forms, differential forms, exterior forms, E. Cartan or exterior analysis, H. Poincare Lemma, G. Stokes Lemma G. Frobenius integration theorem, Pfaffian forms, Pfaffian equation, integrating factors, examples, Sect. 3 with "Differential Geometry in the Cartan Symbolism": review of literature, vector, tensor, object, dual basis, holonomic and anholonomic coordinate system, E. Cartan's object of anholonomity, G. Ricci rotation coefficients, G. Ricci, T. Levi-Civita, J. A. Schouten: parallel displacement, structure equations, torsion and curvature forms, integrability conditions, E. Cartan's theorems de conservations de la courbure et de la torsion as well a Sect. 4: "The Astronomic Coordinate System and The A. Marussi Integration Theorem": Review of literature, astronomic reference system, geocentric and local astronomic coordinates, E. Cartan's object of anholonomity and the G. Frobenius theorem, A. Marussi integration theorem, gravity gradients. Sect. 5 concentrates on "The Geodetic Coordinate System": geodetic coordinate system, E. Cartan's object of anholonomity and the G. Frobenius Theorem applied to geodetic coordinate systems, integration theorem, gravity gradients. The important transformation of one reference system to another is dealt with in Sect. 6: "Transformation Between Astronomic and Geodetic Coordinate System": Transformation matrices, E. Cartan's object of anholonomity, vertical deflection vector, astronomical and geodetic azimuth, astronomical and geodetic vertical angles, generalized Laplace condition, integration theorem. Finally, Sect. 7 enjoys "The Geodetic Integration Theorem and the Concept of Boundary Value Problems": G. Frobenius integration theorem applied to geodesy, integrating factors, gravity gradients, geodetic boundary value problem, free boundary value problems, pseudo-differential operators. Finally, Sect. 8 lists 141 references, namely for relativists: F. Hehl [11] as well as (1970) [12]. Finally the joint work of F. Hehl and E. Kröner [13], had to acknowledge! E. Kröner had been quoted by 4 publications, partially in books. (Both of them were my supervisors in my studies on Theoretical Physics at Clausthal Techn. University.)

My second report of the Department of Geodetic Science, No. 202, The Ohio State University, Columbus/Ohio/USA, 126 pages, of the year 1973 treats "Gravity Gradients and Three-Dimensional Net Adjustments without Ellipsoidal Reference", the practical applications of the Marussi transformation form "natural" non-unique coordinates into the unique coordinates astronomical longitude and latitude, and the gravity potential by means of gravity gradients as well as the Euclidean norm of the gravity vector as integrating factors. If we tolerate point errors of about $\pm 10\,\text{m}$, the r.m.s. error of gravity gradients for terrestrial data is found to be about $\pm 1\, E$ ("Eötvös") for point distances of about 15 km and pole free latitudes. Analogous values are $\pm 1\,\text{m}$ and $\pm 0.1\, E$, $\pm 10\,\text{cm}$ and $\pm 0.01\, E$ approximately. A detailed error analysis for the Marussi transformation is given, completed by an example for a three-dimensional net adjustment based on the Frobenius theorem and available terrestrial gravity gradient information.

Here, we review only the first sections of the OSU report number 202, namely

§2: Geodesy, general consideration

and

§3: Geodesy being properly posed,

only indicator diagrams, integrability and its five parameters $\{k_1, k_2, t_1, \kappa_1, \kappa_2\}$ as integrating factors.

The First Example:

In the beginning or 2d-Riemann geometry Geodesy was two-dimensional surface geometry being embedded into three dimensional Euclidean space. Gravity was a new item to Geodesy due to anholonomity! The Science of Geodesy introduces a new concept into geometry, that of gravitation: it is absent from classical geometry. Our sense of gravitation furnishes a qualitative measure and a quantitative measure, albeit fortuitous to a certain extent, it is given by any balance. In I. Newton's theory, gravity is described by a potential W as the sum of (i) gravitational potential (U) and of (ii) centrifugal potential (V) as the scalar potential part of gravity, the gradient of which, taken negative, it is the acceleration imparted to a small test-body.

The gravitational potential is a property or parameter-of-state. it is associated with the behavior of the body at the instant under consideration, or else, it is measured with reference to the instantaneous indication of balance or gravimeter.

In order to give a rigorous mathematical definition of the gravitational "property" or "parameter-of-state", it is necessary to consider an example, with two independent variables x_1, y_1, which must be measurable properties or characteristics of the system, we can write

$$(i)\quad dW = Xdx + Ydy \quad \text{subject to} \tag{2.1}$$

$$(ii)\quad X = \frac{\partial W}{\partial x} \quad \text{and} \quad Y = \frac{\partial W}{\partial y} \tag{2.2}$$

Evidently, we then have the conditions

$$(iii)\quad \frac{\partial X}{\partial y} = \frac{\partial Y}{\partial x} \quad \text{or} \quad \frac{\partial^2 W}{\partial y \partial x} = \frac{\partial^2 W}{\partial x \partial y} \tag{2.3}$$

which is the necessary and sufficient condition for the expressions $Xdx + Ydy$ to be perfect differential. It is equivalent to the statement that W is a perfect property. The same condition can be written in integral form

$$(iv) \quad \oint dW = 0 \tag{2.4}$$

for any closed path in the x, y-plane. Denoting the two-dimensional vector which is defined by its components X and Y any symbol, we can apply G. Stokes theorem to the scalar product of a two dimensional field, obtaining

$$(v) \quad \oint dW = \oint Xdx + Ydy = \int \left(\frac{\partial Y}{\partial x} - \frac{\partial X}{\partial y} \right) dxdy = 0 \tag{2.5}$$

Since the rotor or curl $\frac{\partial Y}{\partial x} - \frac{\partial X}{\partial y}$ vanished, it is concluded that the statement $\oint dW$ is, in fact, equivalent to the assumption that W is a property.

The condition for a perfect differential with n-independent variable is the vanishing of the n-dimensional operator "rot" or "curl" and can be represented by $n(n-1) \div 2$ equation of the above form. A differential of the type $Xdx + Ydy$ is known as J. Pfaff's differential, in general. When there are two independent variable it is always possible to transform the expression $Xdx + Ydy$ into a perfect differential by dividing it by a denominator, even it it was not one originally. With three independent variables x, y, z and a vector differential form, the theory is more complex. Certain requirement of integrability impose certain conditions on the vector components. They will be found by the G. Frobenius theorem.

The integral $\oint dW = 0$ for any closed path can be represented graphically in a plane if the gradient grad$W$ is taken along the height line. $dW = -\Gamma dH$ including the pair acceleration $\Gamma = \|\text{grad}W\|$ and geometric height H - a mechanical and a physical - is used in the indicator diagram, introduced by J. Watt for thermodynamics.

The path of the "L.N. Carnot" cycle 1, 2, 3 and 4 consists of four individuals. In the left diagram of Fig. 1 $\Gamma_{12} = \Gamma_{34}$, $g_{23} = \Gamma_{41}$, $H_{12} = -H_{34}$ and $H_{23} = -H_{41}$ must hold, thus $\Gamma_{12}H_{12} + \Gamma_{23}H_{23} + \Gamma_{34}H_{34} + \Gamma_{41}H_{41}$ results zero. The paths 23 and 41 are isogravitational in the central diagram of Fig. 1. In addition $H_{12} = -H_{34}$ must be provided. In the right diagram of Fig. 1 the path 12 and 34 are isophysical. But $\Gamma_{23} = \Gamma_{41}$ has to be assumed, otherwise the cycle is nonintegrable. Now it is quite obvious that $\oint dH^{\gamma} \neq 0$.

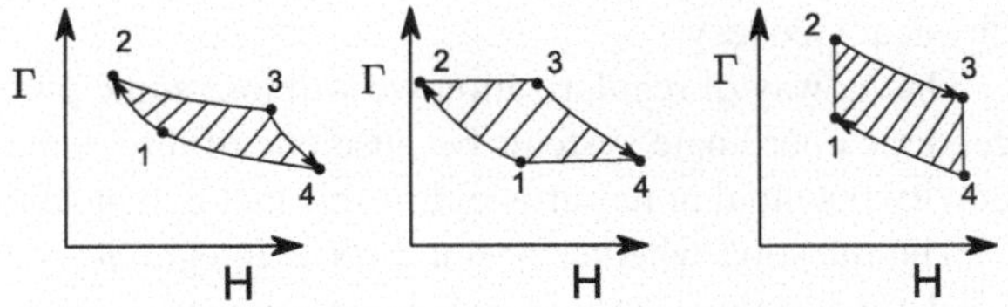

Fig. 1 Indicator diagram for $\oint dW = \oint \Gamma \, đH^{\gamma} = 0$

Fig. 2 Indicator diagram for $\oint đH^{\gamma} = \oint \Gamma^{-1} dW \neq 0$

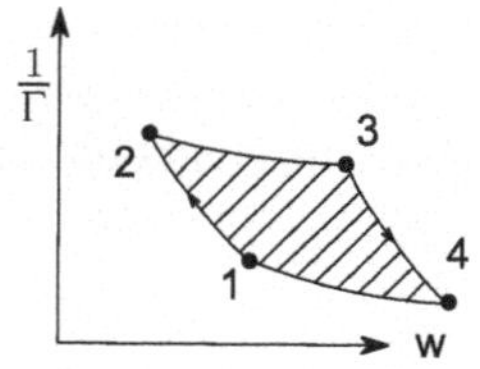

Γ is the well known integrating factor for height. In the next paragraph we prove the non-integrability for the position coordinates also. We keep in mind that the indicator diagram for an integrable pair of coordinates is perfectly symmetric (Fig. 2).

Geodesy: Properly posed.

Geodetic instruments (theodolites, leveling instruments, electronic distance measurement equipment, satellite cameras) operate in special coordinate systems which we will shortly review. We will find that it is impossible to construct a unique coordinate system if we measure only directions, angles, and distances. The nonuniqueness is caused by the Earth's gravity field, influencing the levels of the instruments, and the astronomic orientation of the coordinate axes; Geodesy is "an improperly posed problem".

In order to make geodesy properly posed we have to apply modern techniques of differential geometry. Why is differential geometry such an important tool? In general, the main idea of differential geometry is to apply the tools of analysis to the solution of geometrical problems. An important device in analysis is to study geometrical objects by studying their "infinitesimal parts". Thus a curve is studied by examining its tangent vectors, a function by studying its differentials, and so on. The crucial advantage of the passing to the "infinitesimal" is that it linearizes everything. Thus a curve is "infinitesimally" a straight line when we look at the tangent, every differentiable map is "infinitesimally" linear when we look at its Jacobian matrix,. etc. Since we will be interested in studying such objects as curves, surfaces,. and in general, higher dimensional "differentiable objects", we first study their "infinitesimally" analogues.

From an advanced geometric point of view the "geodetic" situation is typical for a nonsymmetric E. Cartan world with non vanishing pseudo-torsion. The originally improperly posed problem of Geodesy (non uniqueness of "natural" coordinates) is regulated by the G. Frobenius integration theorem via the A. Marussi tensor of gravity gradients. By the method of integrating factors we have a bijective transformation from the non-unique natural coordinate system into the unique coordinate system of Physical Geodesy.

On the way of regularization we will lose some pure geometric interpretations of geodetic coordinate systems because one of three unique coordinates is the Earth's gravity potential at its surface. But this fact is typical for regularization methods.

The question whether or not $\oint d\mathbf{x}$ is integrable is crucial. We refer to "natural" geodetic coordinates in a local astronomic triad, oriented by the orthonormal basis system $e^{\bullet}$ at any point of the Earth's surface. The 3-axis is the antidirection of the

Fig. 3 Local astronomic triad

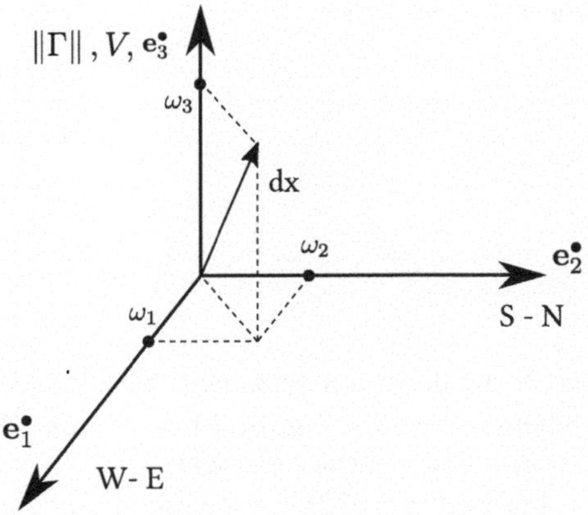

local gravity vector Γ, the 1-axis and the 2-axis are directed towards astronomic East and astronomic North. "•" reminds to the astronomic directions. Obviously the 3-axis coincides with the vertical axis of a perfectly leveled theodolite. Well known are the astronomic azimuth, the vertical angle and the "infinitesimally" Euclidean distance between two surface points as local polar coordinates. Any misclosure of a ring polygon $\oint d\mathbf{x}$ (vector differential one-form) will decide about uniqueness or nonuniqueness of these "natural" coordinates. The coordinates are taken along the local basis vectors $e^{\bullet}$.

$$d\mathbf{x} \overset{\bullet}{=} e_{W-E}\omega_{W-E} + e_{S-N}\omega_{S-N} + e_V\omega_V \tag{2.6}$$

$$d\mathbf{x} = e_1\omega^1 + e_2\omega^2 + e_3\omega^3 \tag{2.7}$$

e_{W-E} represents the East basis vector, e_{S-N} the North basis vector, and e_V the vertical basis vector (Fig. 3).

Orthonormality give $(e_i, e_j) \overset{\bullet}{=} \delta_{ij}, \quad \omega^i = (d\mathbf{x}, e^i)$. (,) indicates scalar product. Within the differential one-form $\oint d\mathbf{x}$ we apply "G. Stokes'identity", due to

$$\boxed{\oint d\mathbf{x} = \oint (e, \omega) = \int d(e, w) = \int (e, \tau) \neq 0} \tag{2.8}$$

The integral $\int (e, \tau)$ has to be taken over the surface included by the ringpolygon of the curve C. τ is the E. Cartan pseudo-torsion, often called the object of anholonomity, and responsible for the misclosure of the closed path relative to the natural triad (Fig. 4).

In the kernel-index G. Ricci calculus G. Stokes' Theorem reads

$$\boxed{\oint d\mathbf{x}^i = \oint e^i_j\omega^j = \int df^{mn} e^j_m e^k_n \partial_{[j} e^i_{k]}} \tag{2.9}$$

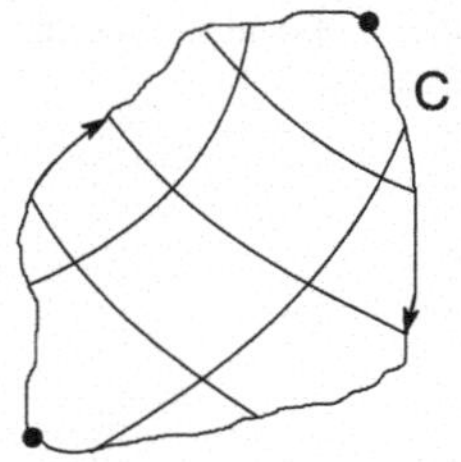

Fig. 4 Line integration

where we have not separated "dead" and "living" tensor indices. [] are the permutation brackets, e.g. $[a, b] = ab - ba$; ∂_i the analytical form of grad; df^{mn} the antisymmetric surface element.

I proved in (1972) that the pseudo-torsion form $\tau^i_{.mn} := e^j_m e^k_n \partial_{[j} e^i_{k]}$ (we apply the Einstein summation convention over repeated indices, running 1,2,3) does not vanish relative to the natural basis and is due to Box 1. In order to compute the object of anholonomity (nonintegrabilitly, nonuniqueness) $\tau := (e^\bullet, (e^\bullet, [\text{grad}, e^\bullet]))$ we have to differentiate the components of the triad and to "strangle" twice by the same basis vectors. The specialist in differential geometry can speculate that the pseudo-torsion form should depend on curvature and torsion of the surface. Thus we are led to the five characteristic quantities k_1, k_2, t_1, κ_1 and κ_2 which we are going to explain carefully. Λ, Ω resp. refer to the astronomic longitude, latitude resp. The sign $\overset{\bullet}{=}$ emphasizes that the right hand side equation holds only in the reference "•".

Box 1: E. Cartan's pseudo-torsion in the natural reference system. Grafarend representation

$$\tau_{ij\,.}^{\;\;1} \overset{*}{=} \frac{1}{2}\begin{bmatrix} 0 & -k_1 \tan\Phi & k_1 \\ k_1 \tan\Phi & 0 & -\kappa_1 \tan\Phi + t_1 \\ -k_1 & \kappa_1 \tan\Phi - t_1 & 0 \end{bmatrix}$$

$$\tau_{ij\,.}^{\;\;2} \overset{*}{=} \frac{1}{2}\begin{bmatrix} 0 & -t_1 \tan\Phi & \kappa_1 \tan\Phi \\ t_1 \tan\Phi & 0 & k_2 \\ -\kappa_1 \tan\Phi & -k_2 & 0 \end{bmatrix} \quad \tau_{ij\,.}^{\;\;3} \overset{*}{=} \frac{1}{2}\begin{bmatrix} 0 & 0 & -\kappa_1 \\ 0 & 0 & -\kappa_2 \\ \kappa_1 & \kappa_2 & 0 \end{bmatrix}$$

End Box 1

κ_1 and κ_2 are the projections of the gradient grad lng in the East and North direction. We denote $\|\text{grad} W\| = \|\Gamma\|$ where W is the real gravity potential at the Earth's surface (Fig. 5). The East and North components of $\|\Gamma\|^{-1}$ grad $\|\Gamma\|$ are responsible for the misclosure of the third component, the heights. If and only if $\|\Gamma\|$ const., the heights are integrable. $\|\Gamma\| = \text{const.}$ leads to $\kappa_1 = 0$ and $\kappa_2 = 0$. Or we say that the vertical component of the pseudo-torsion form.

The projections κ_1, κ_2 and t_1.

κ_1 and κ_2 are the normal curvatures in the East and North directions of any equipotential surface W passing through the observation point. t_1 is the geodetic torsion

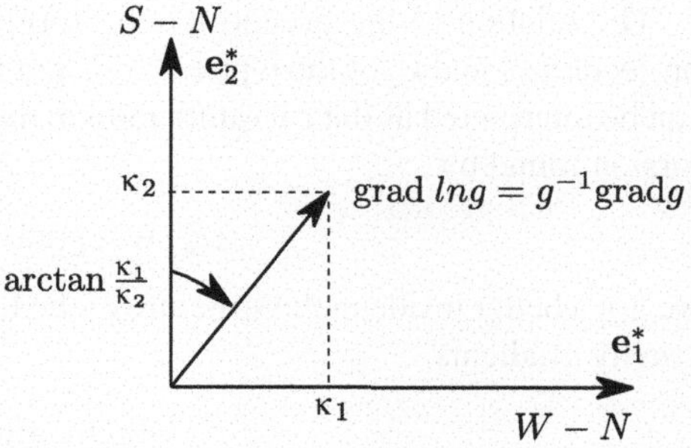

Fig. 5 The projections κ_1, κ_2

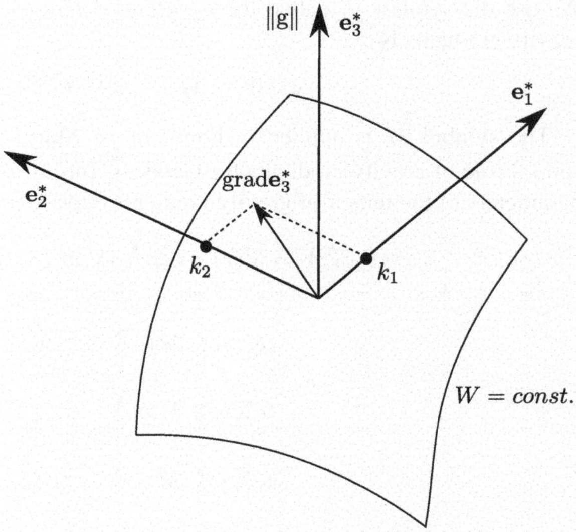

Fig. 6 The projections k_1, k_2

of the equipotential surface in the East direction. Let us transform the vertical basis vector $e_3^\bullet$ in the horizontal plane by grad $e_3^\bullet$. The projections of grad $e_3^\bullet$ onto the bases $e_1^\bullet$ and $e_2^\bullet$ are the components κ_1 and κ_2 (Fig. 6).

$$(e_1^\bullet, \text{grad } e_3^\bullet) = \kappa_1 e_1^\bullet \tag{2.10}$$

$$(e_2^\bullet, \text{grad } e_3^\bullet) = \kappa_2 e_2^\bullet \tag{2.11}$$

The procedure of normalization leads to

$$e_1^\bullet, (e_1^\bullet, \text{grad } e_3^\bullet) = k_1(e_1^\bullet, e_1^\bullet) = \kappa_1 \tag{2.12}$$

$$e_2^\bullet, (e_2^\bullet, \text{grad } e_3^\bullet) = k_2(e_2^\bullet, e_2^\bullet) = \kappa_2 \tag{2.13}$$

The relation to the pseudo-torsion form of the type $(e^\bullet, (e^\bullet, \text{grad } e^\bullet))$ and the projection κ_1 and κ_2 of the type $(e^\bullet, (e^\bullet, \text{grad } e^\bullet))$ is quite obvious $1 \div \kappa_1$ and $1 \div \kappa_2$ can be interpreted as the curvature radii in the East and North direction. The geodetic torsion found by

$$(e_1^\bullet, (e_2^\bullet, \text{grad } \ e_3^\bullet)) = t_1$$

we got a better inside in the structure of the E. Cartan pseudo-torsion if we look upon gravity gradients.

Gravity gradients

We use the symbol ∇_i called the covariant derivative. By definition the tensor of gravity gradients is

$$\nabla\nabla W = M, \qquad \nabla_i \nabla_j W = M_{ij}$$

The symbol M is applied in honor of A. Marussi [14] who first noticed the central role of gravity gradients for Geodesy. Introducing $-\nabla W = \Gamma \doteq \|\mathbf{\Gamma}\| \, e_3$ the parameters of the tensor of gravity gradients depend only on $\nabla\Gamma$ and $\nabla e_3^\bullet$.

$$-\nabla \ \nabla W \doteq \nabla(\|\Gamma\| \, e_3) \doteq (\nabla \, \|\Gamma\|) e_3 + \|\Gamma\| \, \nabla e_3$$

$$(e_1^\bullet, (e_1^\bullet, M)) \doteq - \, \|\Gamma\| \, k_1 \tag{2.14}$$

$$(e_2^\bullet, (e_2^\bullet, M)) \doteq - \, \|\Gamma\| \, k_2 \tag{2.15}$$

$$(e_1^\bullet, (e_2^\bullet, M)) \doteq - \, \|\Gamma\| \, t_1 \tag{2.16}$$

$$(e_1^\bullet, (e_3^\bullet, M)) \doteq - \, \|\Gamma\| \, \kappa_1 \tag{2.17}$$

$$(e_2^\bullet, (e_3^\bullet, M)) \doteq - \, \|\Gamma\| \, \kappa_2 \tag{2.18}$$

With the definitions of the five projections $k_1, k_2, \kappa_1, \kappa_2$ and t_1 the relation to the tensor M of gravity gradients are quite obvious.

If we recall the coordinates along the local natural triad x, y, z the five characteristic curvature parameters are given in the form

$$k_1 \doteq -\frac{1}{\|\Gamma\|} \frac{\partial^2 W}{\partial x^2} = -\frac{1}{\Gamma} W_{xx} \tag{2.19}$$

$$k_2 \doteq -\frac{1}{\|\Gamma\|} \frac{\partial^2 W}{\partial y^2} = -\frac{1}{\Gamma} W_{yy} \tag{2.20}$$

$$t_1 \doteq -\frac{1}{\|\Gamma\|} \frac{\partial^2 W}{\partial x \partial y} = -\frac{1}{\Gamma} W_{xy} \tag{2.21}$$

$$\kappa_1 \doteq -\frac{1}{\|\Gamma\|}\frac{\partial^2 W}{\partial x \partial z} = -\frac{1}{\Gamma} W_{xz} \tag{2.22}$$

$$\kappa_2 \doteq -\frac{1}{\|\Gamma\|}\frac{\partial^2 W}{\partial y \partial z} = -\frac{1}{\Gamma} W_{yz} \tag{2.23}$$

Thus the misclosure of a ringpolygon is due to

$$\begin{aligned} \oint dx \doteq &+\int dx\, dy \Gamma^{-1} W_{xx} \tan\Phi \\ &-\int dx\, dz \Gamma^{-1} W_{xx} \\ &-\int dy\, dz (+\Gamma^{-1} W_{xz} \tan\Phi \\ &+\Gamma^{-1} W_{xy}) \neq 0 \end{aligned} \tag{2.24}$$

$$\begin{aligned} \oint dy \doteq &+\int dx\, dy \Gamma^{-1} W_{xy} \\ &+\int dy\, dz \Gamma^{-1} W_{xz} \tan\Phi \\ &-\int dy\, dz \Gamma^{-1} W_{yy} \neq 0 \end{aligned} \tag{2.25}$$

$$\begin{aligned} \oint dz \doteq &-\int dy\, dz \Gamma^{-1} W_{xz} \\ &-\int dy\, dz \Gamma^{-1} W_{yz} \neq 0 \end{aligned} \tag{2.26}$$

We had discussed the role of the horizontal gradient of $W_z \doteq g$ related to the third component z, the height H. Here we have the influence of the other components of the gravity gradient tensor also. Their non-null causes the misclosure of the position coordinates x and y. Thus dx, dy, dz are imperfect differentials in the sense of J. Pfaff. The integration of coordinates between two surface points is path dependent. Therefore we cannot find unique coordinate differences between these two points.

The way of regularization of the a priori improperly posed problem (violation of uniqueness) is this. We have to find a coordinate system in which the pseudo-torsion form τ vanishes. This is a necessary and sufficient condition for applying the G. Frobenius Theorem in order to find integrating factors and a transformation from the a priori non-unique system to a unique one. A. Marussi [14] introduced for this reason a non-orthogonal triad by

$$e_1 = \frac{\partial \mathbf{x}}{\partial \Lambda}, \qquad e_2 = \frac{\partial \mathbf{x}}{\partial \Phi}, \qquad e_3 = \frac{\partial \mathbf{x}}{\partial W},$$

to which we refer with the symbol $e^{\bullet\bullet}$. Of course, this is by definition a gradient field, an integrable system. The second derivative permute, for instance

$$\partial_\Phi e_1 - \partial_\Lambda e_2 \doteq (\partial_\Phi \partial_\Lambda - \partial_\Lambda \partial_\Phi) x \doteq 0 \tag{2.27}$$

Thus a properly posed "geodetic" problem consists of coordinates Λ, Φ, W which have perfect differential $d\Lambda, d\Phi, dW$

$$\boxed{\oint d\Lambda = 0, \quad \oint d\Phi = 0, \quad \oint dW = 0} \tag{2.28}$$

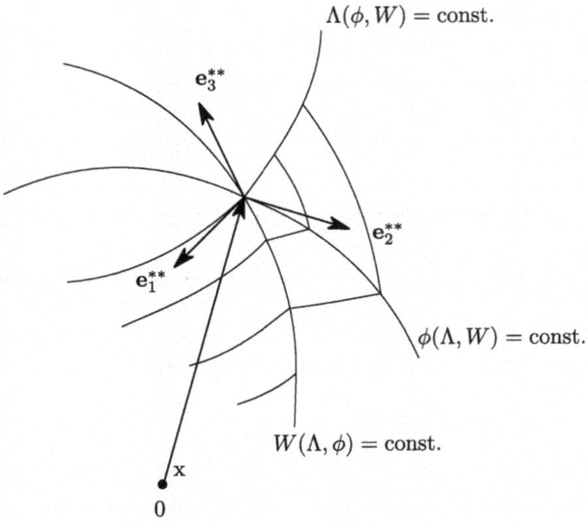

Fig. 7 Local regularized triad

Unique "state variables" are the astronomic longitude, the astronomic latitude, and the geopotential (Fig. 7).

What are the connections between the nonunique natural coordinates which we measure in and the unique ones? That is the question how to calculate $d\Lambda, d\Phi, , dW$ out of dx, dy, dz, or what is the G. Frobenius matrix, of integrating factors. From my report (1972d, p. 38) we take the transformation

$$e^{\bullet} = \begin{bmatrix} -k_1 \sec\Phi & -t_1 & 0 \\ -t_1 \sec\Phi & -k_2 & 0 \\ -\kappa_1 \sec\Phi & -\kappa_2 & \Gamma \end{bmatrix} \tag{2.29}$$

$$d\Lambda \stackrel{\bullet}{=} -k_1 \sec\Phi dx - t_1 \sec\Phi dy + \kappa_1 \sec\Phi dz \tag{2.30}$$

$$d\Phi \stackrel{\bullet}{=} -t_1 dx - k_2 dy + \kappa_2 dz \tag{2.31}$$

$$dW \stackrel{\bullet}{=} \Gamma dz \tag{2.32}$$

where we have used the inverse of the G. Frobenius matrix of rank 3. The derived transformation formulas are the most important ones of Physical Geodesy. They explain why natural coordinate differentials are non-integrable and non-unique. The gravity field of the Earth is involved in the transformation matrix. That explains why geodesy has to be Physical Geodesy. Furthermore we can analyze when the influence of gravity is excluded. This is the situation if and only if the whole tensor of gravity gradients vanishes, emphasizing the central role of gravity gradients. In other words, Gravity gradients are responsible for the non-uniqueness of natural coordinates which we determine by measuring distances, vertical and horizontal angles. Finally we rewrite the transformation in terms of gravity gradients explicitly,

$$d\Lambda \doteq +\frac{1}{\Gamma}W_{xx}\sec\Phi dx + \frac{1}{\Gamma}W_{xy}\sec\Phi dy + \frac{1}{\Gamma}W_{xz}\sec\Phi dz \tag{2.33}$$

$$d\Phi \doteq +\frac{1}{\Gamma}W_{xy}dx + \frac{1}{\Gamma}W_{yy}dy + \frac{1}{\Gamma}W_{yz}dz \tag{2.34}$$

$$dW \doteq -\Gamma dz \tag{2.35}$$

firstly derived by A. Marussi [14]. What is the useful result for practice?

Firstly, we can uniquely integrate the coordinate differentials between two points since $d\Lambda, d\Phi, dW$ are perfect differentials. For instance the coordinate difference between two points P and Q read

$$\oint_P^Q d\Lambda = \Lambda_Q - \Lambda_P = \oint_P^Q \Gamma^{-1}W_{xx}\sec\Phi d\mathbf{x} + \oint_P^Q \Gamma^{-1}W_{xy}\sec\Phi dy + \oint_P^Q \Gamma^{-1}W_{xz}\sec\Phi dz \tag{2.36}$$

$$\oint_P^Q d\Phi = \Phi_Q - \Phi_P = \oint_P^Q \Gamma^{-1}W_{xy}d\mathbf{x} + \Gamma^{-1}W_{yy}dy + \Gamma^{-1}W_{yz}dz \tag{2.37}$$

$$\oint_P^Q dW = W_Q - W_P = \oint_P^Q \Gamma dz \tag{2.38}$$

We repeat that

$$\oint_P^Q d\mathbf{x} \neq x_Q - x_P, \quad \oint_P^Q dy \neq y_Q - y_P, \quad \oint_P^Q dz \neq z_Q - z_P \tag{2.39}$$

Secondly, instead of measuring astronomic longitudes and latitudes, for a sufficient accuracy nearly impossible on oceans, we can calculate longitude and latitude differences if we measure distances, horizontal and vertical angles and gravity gradients. To get a better understanding for dx, dy, dz let us introduce local polar coordinates that is the azimuth A, the vertical angle B and the distance s (Fig. 8).

$$\Delta x \doteq s \ \sin A \sin B$$
$$\Delta y \doteq s \ \cos A \sin B$$
$$\Delta z \doteq s \ \cos B$$

It is well known that longitude, latitude, azimuth and vertical angles are not independent.

$$dA = d\Lambda \sin\Phi + \cot B(d\Phi - d\Lambda\cos\Phi\cos A) \tag{2.40}$$

Fig. 8 Azimuth A, vertical angle B, distance s

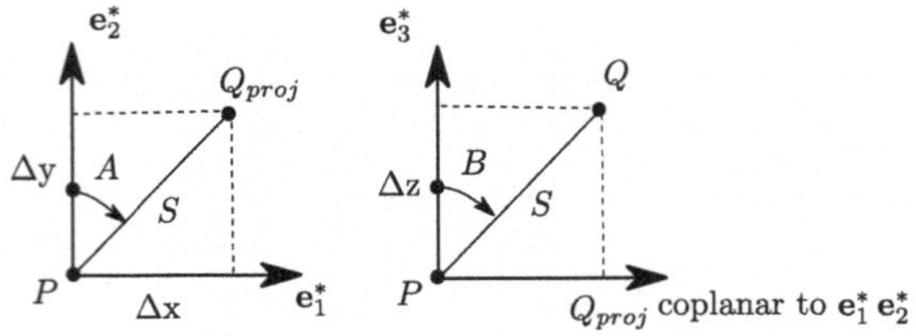

$$dB = -d\Lambda \cos\Phi \sin A - d\Phi \cos A \tag{2.41}$$

If ds is not the geometrical path, but the optical one, we have to replace ds by n d s where n is the actual refractive index of the observation line.

Indicator Diagrams:

The indicator diagram for $\oint dW = \oint \Gamma dz = 0$ was given in Fig. 9. Now we are going to represent graphically the same diagrams for $\oint d\Lambda = 0$ and $\oint d\Phi = 0$. These diagrams are multidimensional because $d\Lambda$ and $d\Phi$ are functions of $d\mathbf{x}$, dy, dz. The adjoint pairs of variables

$$\begin{array}{lll} \Gamma^{-1} W_{xx} \sec\Phi, x; & \Gamma^{-1} W_{xy} \sec\Phi, y; & \Gamma^{-1} W_{xz} \sec\Phi, z; \\ \Gamma^{-1} W_{yx}, x; & \Gamma^{-1} W_{yy}, y; & \Gamma^{-1} W_{yz}, z; \\ & \Gamma, z & \end{array}$$

connect one physical and one mechanical variable. Their indicator diagram, is six-dimensional, six-dimensional, and two-dimensional resp. We would like to mention the fact that the missing W_{zz} inside the integrating factors is not astonishing. At the Earth's surface the Poisson-Laplace equation div grad $W = tr\ M = 2\omega^2$ holds, explaining that W_{xx} and W_{yy} are necessary and sufficient to describe $W_{zz} = -(W_{xx} + W_{yy}) + 2\omega^2$. ω represents the mean rotation speed of the Earth. Therefore only the five independent components W_{xx}, W_{xy}, W_{xz}, W_{yy} and W_{yz} build the G. Frobenius matrix of integrating factors.

In order to get some insight into the multidimensional indicator diagrams let us assume $W_{xy} = W_{xz} = W_{yz} = 0$. This means vanishing off-diagonal elements in the A. Marussi tensor of gravity gradients or vanishing torsion $t_1 = 0$ and projections $\kappa_1 = \kappa_2 = 0$. In this case $\|\Gamma\|$= const. and the E. Cartan pseudo-torsion results zero. Thus the height component $dz = dh$ is integrable and unique. Again this fact emphasizes the value of our knowledge about the E. Cartan pseudo-torsion of surfaces.

What is the geodetic interpretation of this result? Only $k_1 \neq 0$ and $k_2 \neq 0$ is a polygon on an isogravitational surface (not an equipotential surface) with different vertical axes; the East and North axes are principal curvature directions. Within the indicator diagram $(\kappa_1 = \kappa_2 = t_1 = 0)(W_{xx} \sec\Phi)_{12}\kappa_{12} + (W_{xx} \sec\Phi)_{23}x_{23} + (W_{xx} \sec\Phi)_{34}$ $x_{34} + (W_{xx} \sec\Phi)_{41}x_{41} = 0$, $(W_{yy})_{12}y_{12} + (W_{yy})_{23}y_{23} + (W_{yy})_{34}y_{34} + (W_{yy})_{41}y_{41}$ $= 0$ and $H_{12} + H_{34} = 0$ have to be fulfilled, if we have a polygon net of four points. The Euclidean norm $\|\Gamma\|$ = const. The constrained equations are satisfied by the sufficient restrictions

$$(W_{xx}\sec\Phi)_{12} = (W_{xx}\sec\Phi)_{34}, \quad (W_{xx}\sec\Phi)_{23} = (W_{xx}\sec\Phi)_{41} \tag{2.42}$$

$$(W_{yy})_{12} = (W_{yy})_{34}, \quad (W_{yy})_{23} = (W_{yy})_{23} = (W_{yy})_{41} \tag{2.43}$$

$$x_{12} = -x_{34}, \quad x_{23} = -x_{41}, \quad x_{42} = -y_{34}, y_{23} = -y_{41}. \tag{2.44}$$

Of special interest is the same indicator diagram with some further restrictions: (a) the paths 23 and 41 are isogradient, or (b) the paths 12 and 34 are isocoordinate (Figs. 10 and 11).

We have to provide in the case (a) $x_{12} = -x_{34}$ and in the case (b) $(W_{yy})_{23} = (W_{yy})_{41}$. Another interpretation is this: In the case a) the point 2 and 3 or 1 and 4 lie on the same gravity gradient—Fig. 9—$W_{xx}sec\Phi$, W_{yy} resp. But in the case (b) the points 1 and 2 or 3 and 4 have identical position coordinates.

Finally we mention that there are a lot of other examples for "L. N. Carnot" cycles within the indicator diagram. Special cases are $k_1 = 0$, $t_1 = k_1 tan\Phi$ in which x is integrable, y and z not, and $\kappa_1 = t_1 = \kappa_2 = 0$ in which y is integrable, x and z not. A singular case is $\Phi = \pi/2$ where the E. Cartan pseudo-torsion and G. Frobenius matrix of integrating factors are unbounded. But this is well known because we

Fig. 9 Indicator diagram for $\oint d\Lambda = 0, \oint d\phi = 0, \oint dW = 0, \kappa_1, \kappa_2, t_1 = 0, k_1 \neq 0, k_2 \neq 0$

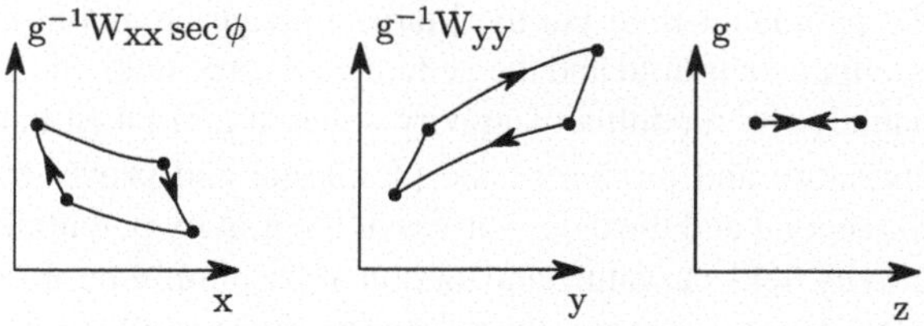

Fig. 10 Indicator diagram for $\oint d\Lambda = 0, \oint d\phi = 0, \oint dW = 0$, hypothesis $\kappa_1, \kappa_2, t_1 = 0, k_1 \neq 0, k_2 \neq 0$ isogradient

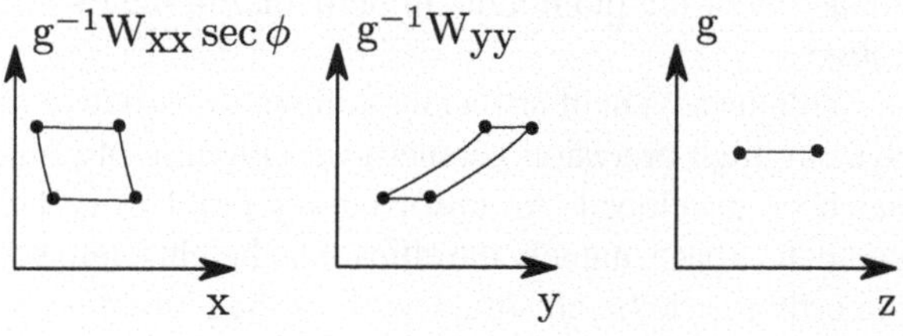

Fig. 11 Indicator diagram for $\oint d\Lambda = 0, \oint d\phi = 0, \oint dW = 0$, hypothesis $\kappa_1, \kappa_2, t_1 = 0, k_1 \neq 0, k_2 \neq 0$ isocoordinate

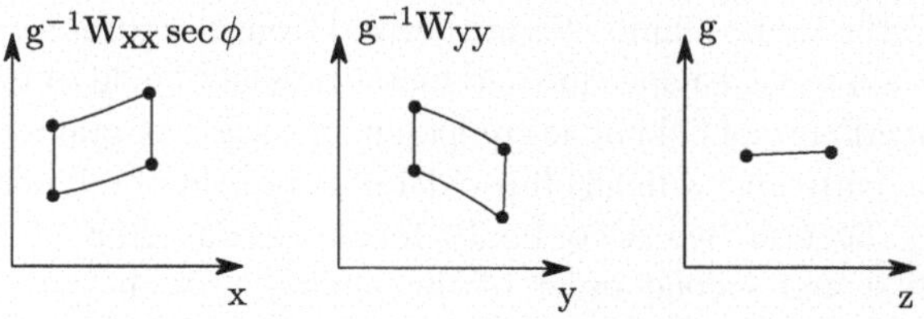

cannot find East and North at the Earth's poles, that is, there is no singularity-free coordinate system referring on any star-shaped surface.

A shortest version of these first and second OSU reports appeared also in E. Grafarend [15]. A more detailed version appeared in P. Defrise and E. Grafarend [16].

2.2 *Field Lines of Gravity, Their Curvature and Torsion, the Lagrange and Hamilton Equations of the Plumbline*

The length of the gravitational field lines of the orthogonal trajectories of a family of gravity equipotential surfaces of the plumbline between a terrestrial topographic point and a point on a reference equipotential surface like the Geoid - also known as the orthometric height - plays a central role in Satellite Geodesy as well as in Physical Geodesy. As soon as we determine the geometry of the Earth pointwise by means of a satellite GPS we are left with the problem of converting ellipsoidal heights (geometric heights) into orthometric heights (physical heights). For the computation of the plumbline we derive its three differential equations of first order as well as the three geodesic equations of second order. The three differential equations of second order take the form of a Newton differential equation when we introduce the parameter time via the Marussi gauge on a conformally flat three dimensional Riemann manifold and the generalized force field, the gradient of the super potential, namely the modulus of gravity squared and taken half. In particular, we compute curvature and torsion of the plumbline and prove their functional relationship to the second and third derivatives of the gravity potential. For a spherically symmetric gravity field, curvature and torsion of the plumbline are zero, the plumbline is straight. Finally we derive the three Lagrangian as well as the six Hamiltonian differential equations of the plumbline, in particular in their star form with respect to Marussi gauge.

With the advent of artificial satellites, in particular the satellite Global Positioning System, high precision geometric positioning of points of the surface of the Earth has been developed. An unsolved key problem is the transformation of heights in geometry space, namely the ellipsoidal heights, into heights in gravity space, namely the orthometric heights/the length of the plumbline with respect to the Geoid. The field lines of gravity / the orthogonal trajectories of a family of gravity equipotential surfaces/the plumblines are derived from a set of first order differential equations as soon as we balance the horizontal/tangential field of the plumbline with the vertical field/normal field of an equipotential surface of gravity as described by Caputo [17], in particular with an ellipsoidal gravity field of reference, for instance.

Section two accordingly focuses on a setup of the differential equations of first and second order of the plumbline with special reference to the transformation from the parameter arc length s to the ≪dynamic time parameter t ≫ according to the celebrated Marussi Gauge ([18–20]). The arc length squared

$ds^2 = \|\text{grad } w\|^2 (dx^2 + dy^2 + dz^2)$ has been represented in terms of conformal coordinates/isometric coordinates (e.g., Caputo, [17]) with the modulus of gravity squared $\|\gamma\|^2 = \|\text{grad } w\|^2 = \lambda^2$ as the factor of conformality squared λ^2. In particular we succeed in proving that the second order differential equations of the plumbline establish a geodesic in a three-dimensional Riemann manifold $\{\mathbb{M}^3, g_{kl}\}$, notably in the form of a Newton dynamical equation if the matrix g_{kl} of the metric is ≪conformally flat≫, $g_{kl} = \lambda^2(x)\delta_{kl}$ as restricted to the Marussi Gauge. In order to determine the departure of the plumbline from a straight line, we compute its curvature and torsion on the basis of the Frenet derivational equations. We aim at the proof that the curvature of the plumbline is a functional of the second derivatives of the gravity potential, its torsion of the third derivatives of the gravity potential, while straight if the gravity field was spherically symmetric. Section two is devoted to establishing the three Lagrangean differential equations of second order as well as the six Hamiltonian differential equations of first order, in particular in Marussi Gauge. In contrast to Moritz [21] we succeed in constructing non-degenerate ≪star Lagrangeans≫ and ≪ star Hamiltonians≫.

Curvature and torsion of the field lines of the gravity field, the plumbline

At the beginning let us set up the differential equations of the plumbline/the orthogonal trajectory with respect to a family of equipotential surfaces by means of Box 2. The quality between horizontal and vertical fields establishes the first order differential equations of the plumbline: The normalized tangent vector of the plumbline is identical to the normalized surface vector of an equipotential surface pointwise. The normalized surface vector of an equipotential surface agrees with the negative gravity vector, the gradient of the gravity vector. As soon as we differentiate the identity of the horizontal field of the plumbline and the vertical field of an equipotential surface once more, we arrive at the second order differential equation of a plumbline of inhomogeneous type. The inhomogeneity is generated by the quadrupole moment in gravity space. As soon as we introduce the parameter t in order to replace the curve arc length S via the Marussi gauge (Marussi 1979 [18], 1985 [19]) we are led to the first order differential eqs. and the second order differential eqs. of a plumbline/orthogonal trajectory of a family of equipotential surfaces. With respect to the Marussi gauge, the second order differential equations of a plumbline in $\{\mathbb{R}^3, g_{kl}\}$ coincide with the second order differential equations of a geodesic in Newton form in the Marussi manifold $\{\mathbb{M}^3, \gamma^2(x)\delta_{kl}\}$ with $\gamma^2(x)$ as the factor of conformality. Gravity squared taken half operates as a potential, according to a proposal by Chandrasekhar et al. called superpotential: The gradient of the superpotential $\gamma^2/2$ operates as the force field balanced by the acceleration vector $x^{\blacklozenge\blacklozenge}$.

Indeed we have to explain better the duality between a curve in $\{R^3, \delta_{kl}\}$, where the Kronecker δ_{kl}, relates to the canonical metric in a three-dimensional Euclidean space, and a curve in $\{\mathbb{M}^3, \gamma^2(x)\delta_{kl}\}$.$\{\mathbb{M}^3, \gamma^2(x)\delta_{kl}\}$ is an abbreviated notation for a three-dimensional Riemannian space parameterized by three conformal coordinates / isometric coordinates, whose canonical metric is given by the product of the factor of conformality γ^2, the modulus of gravity squared, and the Kronecker δ_{kl},. Such a Riemann manifold $\{\mathbb{M}^3, \gamma^2(x)\delta_{kl}\}$ will be called a Marussi manifold. Indeed there are many Marussi manifolds dependent on the various representations of the gravity field of the Earth. While the differential equations of second order generate a curve in the chart $\{\mathbb{R}^3, \delta_{kl}\}$, at the same time this curve can be considered as a geodesic in the Riemann manifold in terms of special coordinates, the ones of conformal/isometric type. This conception of the curve as a geodesic will become clearer in the next chapter. We should mention that the duality described earlier has already been applied by Goenner et al. [22] in order to interpret Newton mechanics as geodesic flow on a Maupertuis' manifold.

Secondly we are going to derive the Frenet equations of the plumbline, a curve in $\{\mathbb{R}^3, \delta_{kl}\}$, a three-dimensional Euclidean space completely covered by one chart of Cartesian coordinates $\{x^1, x^2, x^3\}$. As outlined by means of Box 3, we establish by the Frenet frame {normalized tangent vector, normalized normal vector, normalized binormal vector} called $\{\mathbf{f}_1, \mathbf{f}_2, \mathbf{f}_3\}$ (x), which is subject to the coupling to the gravity field, thanks to the first order differential equations of the plumbline in Marussi gauge. The derivational equations of the Frenet frame are built on the celebrated anti-symmetric Ω-matrix which contains as structure elements the curvature κ and the torsion τ, also called first and second curvature of the plumbline. A straightforward computation of curvature and torsion of the plumbline subject to the coupling of the gravity field. Here we took advantage of the cross product identity $a \times b = \|a\|^2 - \|b\|^2 - \langle a|b\rangle^2$ where $\|a\|$ indicates the Euclidean norm of the vector a as well as $\langle a|b\rangle$ the Euclidean scalar product/inner product of two vectors a and b. Obviously the curvature κ of the plumbline is proportional to gravity gradients or second derivatives of the gravity potential. This can be seen by means of as soon as we apply, namely grad $\gamma^2/2$, $(\partial_k\gamma^2)/2 = \gamma_1\partial_k\gamma_1 + \gamma_2\partial_k\gamma_2 + \gamma_3\partial_k\gamma_3 = \partial_1 w \partial_k \partial_1 w + \partial_2 w \partial_k \partial_2 w + \partial_3 w \partial_k \partial_3 w$.

Box 2

Duality between horizontal and vertical fields in $\{R^3, \delta_{ij}\}$ equipped with a Euclidean matrix δ_{ij}

Normalized tangent vector of the plumbline is identical to the surface normal vector of an equipotential surface 1st order differential equations

$$\frac{d\mathbf{x}}{dS} = -\frac{\text{grad } w}{\|\text{grad } w\|} \sim \frac{dx^k}{dS} = -\ \partial_k\ w/\sqrt{\delta^{lm}\partial_l w \partial_m w} \tag{2.45}$$

second order differential equations

$$\frac{d^2x^k}{dS^2} + \frac{\gamma_{,l}}{\gamma^3}(\gamma^2\delta^{kl} - \gamma^k\gamma^l) = 0 \tag{2.46}$$

Marussi gauge

$$\left\|\mathbf{x}^{\blacklozenge}\right\| = \|\text{grad}\, w\|\,, \frac{d\mathbf{x}}{dS} = \left\|\mathbf{x}^{\blacklozenge}\right\|^{-1}\mathbf{x}^{\blacklozenge} \tag{2.47}$$

fist order differential equation of the plumbline in Marussi gauge

$$\boxed{\mathbf{x}^{\blacklozenge} = -\text{grad}\, w} \tag{2.48}$$

second order differential equation of the plumbline in Marussi gauge

$$\mathbf{x}^{\blacklozenge\blacklozenge k} = (-\partial_l \gamma^k) x^{\blacklozenge l} = (\partial_l \gamma^k)\partial_l w = (\partial_l \gamma^k)\gamma^l \tag{2.49}$$

$$\boxed{\mathbf{x}^{\blacklozenge\blacklozenge k} - \tfrac{1}{2}\partial_k \gamma^2 = 0} \tag{2.50}$$

End Box 2

In contrast, the torsion τ of the plumbline is proportional to the second derivatives of the gravity vector or the third derivatives of the gravity potential. Such a result is motivated by the identity $\gamma^2_{,ij}/2 = \gamma_{k,ij}\gamma_k + \gamma_{(k,i)}\gamma_{(k,j)} = (\partial_i\partial_j\partial_k w)\partial_k w + (\partial_i\partial_k w)(\partial_i\partial_k w)$. Finally as a corollary we report the result that curvature and torsion of the plumbline amount to zero if the gravity field has spherical symmetry. Or we may say that $\kappa_0 = 0$, $\tau_0 = 0$, if the gravity field $w_0(x, y, z) = w_0(r)$ depends only on the radial coordinate r. This result gave the motivation for a decomposition of curvature and torsion $\kappa = \kappa_0 + \delta\kappa$, $\tau = \tau_0 + \delta\tau$ subject to $\kappa_0 = \tau_0 = 0$ in terms of the gravity field $\gamma = \gamma_0 + \delta\gamma$ the normal gravity field $\gamma_0 = \gamma_0(r)$ and the disturbing gravity field $\{\delta\gamma(\lambda, \varphi, r)\}$ which depends on the lateral variation of lengthy spherical coordinates. Since the representations $\delta\kappa$, $\delta\tau$ are lengthy, we drop them here.

Box 3

Curvature and torsion of the plumbline in $\{\mathbb{R}^3, \delta_{ij}\}$ equipped with a Euclidean metric δ. The Frenet frame as the natural triad of the plumbline

$$f_1 := \frac{\mathbf{x}^{\blacklozenge}}{\left\|\mathbf{x}^{\blacklozenge\blacklozenge}\right\|} \tag{2.51}$$

$$f_2 := \frac{\mathbf{x}^{\blacklozenge\blacklozenge} - \left\langle \mathbf{x}^{\blacklozenge\blacklozenge} | f_1 \right\rangle f_1}{\left\|\mathbf{x}^{\blacklozenge\blacklozenge} - \left\langle \mathbf{x}^{\blacklozenge\blacklozenge} | f_1 \right\rangle f_1\right\|} \tag{2.52}$$

$$f_3 := \frac{\mathbf{x}^{\blacklozenge\blacklozenge\blacklozenge} - \left\langle \mathbf{x}^{\blacklozenge\blacklozenge\blacklozenge} | f_1 \right\rangle f_1 - \left\langle \mathbf{x}^{\blacklozenge\blacklozenge\blacklozenge} | f_2 \right\rangle f_2}{\left\|\mathbf{x}^{\blacklozenge\blacklozenge\blacklozenge} - \left\langle \mathbf{x}^{\blacklozenge\blacklozenge\blacklozenge} | f_1 \right\rangle f_1 - \left\langle \mathbf{x}^{\blacklozenge\blacklozenge\blacklozenge} | f_2 \right\rangle f_2\right\|} \tag{2.53}$$

subject to

$$\boxed{(i)\mathbf{x}^{\blacklozenge i} = -\gamma^i\,, \quad (ii)\mathbf{x}^{\blacklozenge\blacklozenge i} = \tfrac{1}{2}\gamma^2_{,i}\,, \quad (iii)\mathbf{x}^{\blacklozenge\blacklozenge\blacklozenge i} = -\tfrac{1}{2}\gamma^2_{,ij}\gamma^j} \tag{2.54}$$

The derivational equations of the Frenet frame

$$f_1^{\blacklozenge} = \kappa S^{\blacklozenge} f_2 \tag{2.55}$$

$$f_2^{\blacklozenge} = -\kappa S^{\blacklozenge} f_1 + \tau S^{\blacklozenge} f_3 \tag{2.56}$$

$$f_3^{\blacklozenge} = -\tau S^{\blacklozenge} f_2 \tag{2.57}$$

$$\begin{bmatrix} f_1^{\blacklozenge} \\ f_2^{\blacklozenge} \\ f_3^{\blacklozenge} \end{bmatrix} = \begin{bmatrix} 0 & \omega_{12} & 0 \\ -\omega_{12} & 0 & \omega_{23} \\ 0 & -\omega_{23} & 0 \end{bmatrix} \begin{bmatrix} f_1 \\ f_2 \\ f_3 \end{bmatrix} \tag{2.58}$$

$$\begin{bmatrix} f_1^{\blacklozenge} \\ f_2^{\blacklozenge} \\ f_3^{\blacklozenge} \end{bmatrix} = \begin{bmatrix} 0 & \kappa S^{\blacklozenge} & 0 \\ -\kappa S^{\blacklozenge} & 0 & \tau S^{\blacklozenge} \\ 0 & -\tau S^{\blacklozenge} & 0 \end{bmatrix} \begin{bmatrix} f_1 \\ f_2 \\ f_3 \end{bmatrix} \tag{2.59}$$

Curvature κ and torsion τ

$$\kappa = \frac{\left\| \mathbf{x}^{\blacklozenge} \times \mathbf{x}^{\blacklozenge\blacklozenge} \right\|}{\left\| \mathbf{x}^{\blacklozenge} \right\|^3}, \quad \tau = \frac{\left\langle \mathbf{x}^{\blacklozenge} | \mathbf{x}^{\blacklozenge\blacklozenge} \times \mathbf{x}^{\blacklozenge\blacklozenge} \right\rangle}{\left\| \mathbf{x}^{\blacklozenge} \mathbf{x}^{\blacklozenge\blacklozenge} \right\|^2}, \tag{2.60}$$

$$\begin{aligned} \kappa &= \tfrac{1}{2\gamma^3} \sqrt{\gamma^2 \left\| \text{grad}\gamma^2 \right\|^2 - \left\langle \gamma | \text{grad}\gamma^2 \right\rangle^2} \\ &= \frac{\sqrt{(\gamma_1^2+\gamma_2^2+\gamma_3^2)(\gamma_{,1}^2+\gamma_{,1}^2+\gamma_{,1}^2)^2-(\gamma_1\gamma_{,1}^2+\gamma_2\gamma_{,2}^2+\gamma_3\gamma_{,3}^2)^2}}{2(\gamma_1^2+\gamma_2^2+\gamma_3^2)^{3/2}} \end{aligned} \tag{2.61}$$

$$\tau = \frac{\gamma_{,1j}^2 \gamma^j (\gamma_2 \gamma_{,3}^2 - \gamma_3 \gamma_{,2}^2) + \gamma_{,2j}^2 \gamma^j (\gamma_3 \gamma_{,1}^2 - \gamma_1 \gamma_{,3}^2) + \gamma_{,3j}^2 \gamma^j (\gamma_1 \gamma_{,2}^2 - \gamma_2 \gamma_{,1}^2)}{\gamma^2 \delta^{kl} \gamma_{,k}^2 \gamma_{,l}^2 - (\delta^{kl} \gamma_k \gamma_{,l}^2)^2} \tag{2.62}$$

Corollary

$$\boxed{\kappa_0 = 0, \tau_0 = 0, \quad \text{if } w_0(x, y, z) = w_0(r)} \tag{2.63}$$

End Box 3

Instead, thirdly, we compute by means of Box 4 the plumbline in a spherically symmetric gravity field subject to Marussi gauge. Let us depart from the first order differential equations of a plumbline subject to Marussi gauge. Indeed by means of we restrict the gravity field to be spherically symmetric: the gravity potential $w(x, y, z) = f(r)$ has been chosen to be a function of the radial coordinate only. The appropriate coordinate system in which to solve the first order differential equations is the spherical coordinate system $\{\lambda, \varphi, r\}$. We have used the forward transformations ≪Cartesian coordinates into spherical coordinates≫ in order to represent the first order differential equations of the plumbline subject to Marussi gauge in spherical coordinates, to prove $\lambda^{\blacklozenge} = 0$, $\varphi^{\blacklozenge} = 0$ and $r^{\blacklozenge} = -f(r)$, a result collected in the corollary. Finally as an example we have chosen the potential and the gravity field of a homogeneous, massive sphere in the inner zone A and the outer zone B in order to solve the ordinary differential equation of the radial component of the plumbline. The function $r = r_0 \exp\, gm/R^s\,(t - t_0)$ is a representation of the solution of $r^{\blacklozenge} = -f(r)$ in case 1, $R > r$, while $r = \sqrt[3]{r_0^3 + 3\,gm(t - t_0)}$ in case 2, $R < r$. For $R = r$, both solutions agree with each other. gm denotes the product of the gravitational constant g and the mass m of the homogeneous, massive sphere.

Box 4

Computation of a plumbline in a spherical symmetric gravity field. Marussi gauge.

$$\mathbf{x}^{\blacklozenge} = -\text{grad}\, w \Leftrightarrow \begin{bmatrix} \mathbf{x}^{\blacklozenge} = -\frac{\partial w}{\partial x} \\ \mathbf{y}^{\blacklozenge} = -\frac{\partial w}{\partial y} \\ \mathbf{z}^{\blacklozenge} = -\frac{\partial w}{\partial z} \end{bmatrix} \tag{2.64}$$

spherically symmetric gravity field

$$w(x, y, z) = f(r) \text{ subject to } r^2 = x^2 + y^2 + z^2 \tag{2.65}$$

$$\partial_k w = \frac{dw}{dr}\frac{\partial r}{\partial x^k} = f'(r)\frac{x^k}{r},\ f'(r) := \frac{df}{dr} \tag{2.66}$$

forward transformation: Cartesian coordinates into spherical coordinates

$$\begin{aligned} &\lambda = arctan(y/x) + \left(-\tfrac{1}{2}sgn(y) - \tfrac{1}{2}sgn(y)sgn(x) + 1\right)\pi,\ \lambda \in \{\mathbb{R}|0 \le \lambda < 2\pi\} \\ &\phi = arctan\left(\tfrac{z}{\sqrt{x^2+y^2}}\right),\ \ \phi \in \{\mathbb{R}| -\pi/2 < \phi < +\pi/2\} \end{aligned} \tag{2.67}$$

representation of the first order differential equation of the plumbline in spherical coordinates $d\tan\lambda = \frac{x^2+y^2}{x^2}d\lambda = \frac{-ydx+xdy}{x^2}$

$$d\tan\phi = \frac{1}{(x^2 + y^2)^{3/2}}((x^2 + y^2)dz - zxdx - zydy) = \frac{r^2}{x^2 + y^2}d\phi \tag{2.68}$$

$$\mathbf{x}^{\blacklozenge k} = -f'(r)\frac{x^k}{r} \Leftrightarrow \begin{bmatrix} \mathbf{x}^{\blacklozenge} = -f'(x)\frac{x}{r} \\ \mathbf{y}^{\blacklozenge} = -f'(r)\frac{y}{r} \\ \mathbf{z}^{\blacklozenge} = -f'(r)\frac{z}{r} \end{bmatrix} \tag{2.69}$$

$$\lambda^{\blacklozenge} = \frac{1}{x^2+y^2}(-yx^{\blacklozenge}+xy^{\blacklozenge}) \tag{2.70}$$

$$\phi^{\blacklozenge} = \frac{1}{(x^2+y^2)^{1/2}}\frac{1}{2}(-zxx^{\blacklozenge}-zyy^{\blacklozenge}+(x^2+y^2)z^{\blacklozenge}) \tag{2.71}$$

$$r^{\blacklozenge} = \frac{1}{r}(xx^{\blacklozenge}+yy^{\blacklozenge}+zz^{\blacklozenge}) \tag{2.72}$$

$$\lambda^{\blacklozenge} = \frac{1}{x^2+y^2}(yf^{'}(r)x/r - xf^{'}(r)y/r) = 0 \tag{2.73}$$

$$\phi^{\blacklozenge} = \frac{1}{(x^2+y^2)^{1/2}}\frac{1}{2}(zxf^{'}(r)x/r + zyf^{'}(r)y/r - (x^2+y^2)f^{'}(r)z/r) \tag{2.74}$$

$$r^2 r^{\blacklozenge} = -(x^2 f^{'}(r) + y^2 f^{'}(r) + z^2 f^{'}(r)) = -r^2 f^{'}(r) \tag{2.75}$$

Corollary

$$\lambda^{\blacklozenge} = 0, \quad \phi^{\blacklozenge} = 0, \quad r^{\blacklozenge} = -f^{'}(r), \quad if \;\; w(x,y,z) = f(r) \tag{2.76}$$

Example: massive sphere

$$w_0(r) = \left[\begin{array}{c} \frac{gm}{2R}\left(3-\frac{r^2}{R^2}\right)\forall 0 \le r < R : \text{zone A} \\ \frac{gm}{r}\forall R \le r < \infty : \text{zone B} \end{array}\right. \tag{2.77}$$

or in terms of the Heaviside function $H(R, r)$

$$w_0(r) = H(R-r)\frac{gm}{2R}\left(3-\frac{r^2}{R^2}\right) + H(r-R)\frac{gm}{r} \tag{2.78}$$

$$\text{grad } w_0(r) = -e_r \left[\begin{array}{c} \frac{gm}{R^3}r\forall 0 \le r < R : \text{zone A} \\ \frac{gm}{r^2}\forall R \le r < \infty : \text{zone B} \end{array}\right. \tag{2.79}$$

or

$$\text{grad } w_0(r) = -e_r H(R-r)\frac{gm}{R^3}r - e_r H(r-R)\frac{gm}{r^2} \tag{2.80}$$

$$-\partial_k \, w_0 = H(R-r)\frac{gm}{R^3}x^k + H(r-R)\frac{gm}{r^3}x^k = x^{\blacklozenge k} \tag{2.81}$$

Corollary

$$\lambda^{\blacklozenge} = 0, \quad \phi^{\blacklozenge} = 0, \quad r\, r^{\blacklozenge} = r^2(H(r-r))\frac{gm}{R^3} + H(r-R)\frac{gm}{r^3} \tag{2.82}$$

if $w(x, y, z) = w_0(r)$

$$\boxed{\text{Case1:}}\ R > r : r^\blacklozenge = \frac{gm}{R^3} r \Rightarrow \frac{dr}{r} = \frac{gm}{R^3} dt \Rightarrow$$
$$\Rightarrow \ln r - \ln r_0 = \frac{gm}{R^3}(t - t_0) \Rightarrow$$
$$\Rightarrow \frac{\ln r}{\ln r_0} = \frac{gm}{R^3}(t - t_0) \Rightarrow$$
$$\Rightarrow \frac{r}{r_0} = \exp \frac{gm}{R^3}(t - t_0) \Rightarrow$$

$$\boxed{r = r_0 \exp \frac{gm}{R^3}(t - t_0)} \tag{2.83}$$

$$\boxed{\text{Case2:}}\ r > R : r^\blacklozenge = \frac{gm}{r^2} \Rightarrow r^2 dr = gmdt \Rightarrow$$
$$\Rightarrow \frac{r^3}{3} - \frac{r_0^3}{3} = gm(t - t_0) \Rightarrow$$
$$\Rightarrow \frac{1}{3} r^3 = \frac{1}{3} r_0^3 + gm(t - t_0) \Rightarrow$$

$$\boxed{r = \sqrt[3]{r_0^3 + 3gm(t - t_0)}} \tag{2.84}$$

End Box 4

Finally we illustrate by the solution of the first order differential eqs. of the plumbline subject to Marussi gauge, namely the bundle of straight lines with the mass center as the focal point for a spherically symmetric gravity field. For a more realistic gravity field in the crust of the Earth, Svensson in Grafarend [23] has computed a sample plumbline in the Alpes by a Runge–Kutta numerical computation of the solution of the first order differential equation of the plumbline in Marussi gauge and a gravity field given by a set of homogeneous massive spheres around the plumbline representing the local gravity field. Reference is made to Fig. 12. In addition, by Fig. 13 we illustrate the computation of a realistic plumbline at the

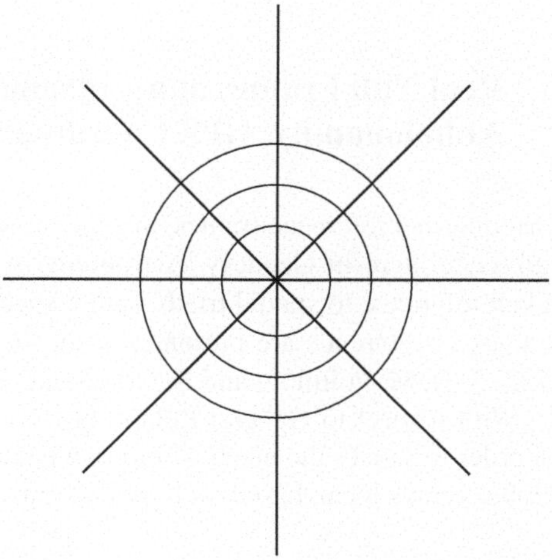

Fig. 12 A set of plumblines for a spherically symmetric gravity field γ_0; straight lines with the mass center of the Earth as a focal point

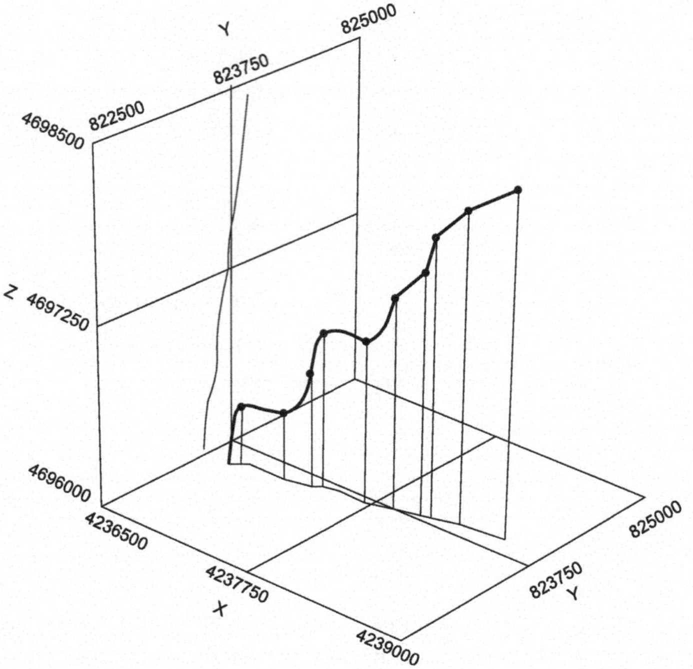

Fig. 13 Computation of a realistic plumbline at a mountain point in the Alpes according to Svensson in Grafarend [23]

Swiss high mountain point Jungfraujoch performed by Hunziker [24]. An alternative procedure for the gravitational field in the crust is outlined in Engels and Grafarend [25], Engels et al. [26] and in Grafarend et al. [27]. Finally we refer to Grossman [28] and [29] for the focal point of plumblines (Fig. 14).

3 Real Null Frames and Coframes in General Relativity, Anholonomity, GPS Coordinates

Orthonormal reference frames, in general, are a typical example of the anholonomic reference, here in Geodesy, Astronomy, Astrometry and Space Science: they are taken reference to spatial triads with respect to Newton time. For instance, various frames of reference are presented as in I.I. Mueller et al. ([30, 31]), E. Grafarend ([32, 33]), M. Fujimoto and E. Grafarend [34].

With respect to General Relativity we introduce real null frames and coframes in order to satisfy the needs of modern Positioning Systems, namely GPS and LPS: global versus local, based on light and radar signals.

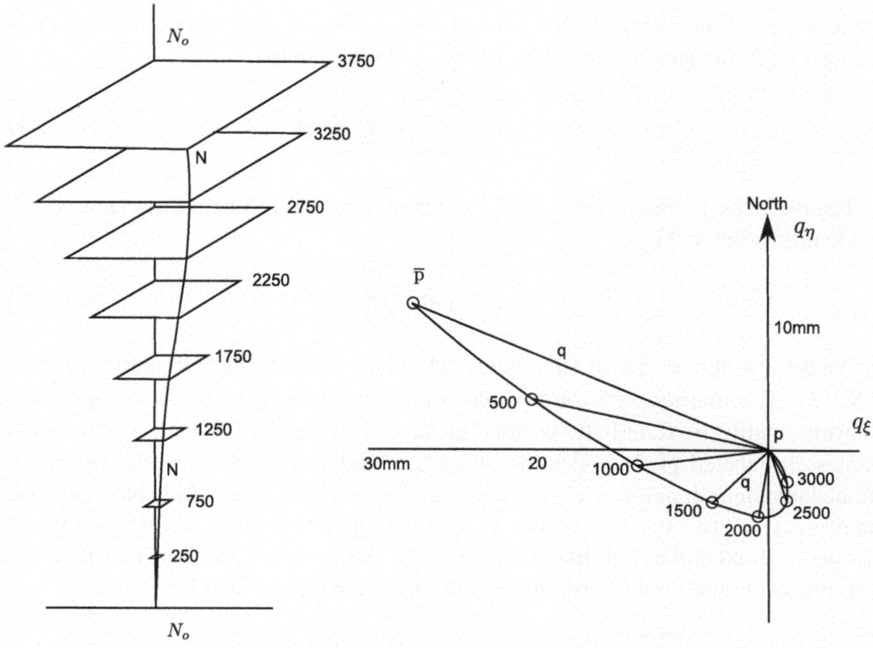

Fig. 14 Computation of a realistic plumbline at the mountain point ≪Jungfraujoch≫ at the Swiss Alpes by Hunziker (1960, p. 151), departure from a straight line projection onto the reference Fig. 13 mm/30.8 mm in the North/West direction

More specifically, in pseudo-Riemannian spacetime we introduce at each point in spacetime a local pseudo-orthonormal frame of reference $\mathbf{e}_\alpha(x)$ with respect to $\alpha, \beta, \cdots \in \{1, 2, 3, 4\}$. We chose a timelike 4-leg $\mathbf{e}_4$, where as $\{\mathbf{e}_1, \mathbf{e}_2, \mathbf{e}_3\}$ are spacelike. specifically, this frame of reference is pseudo-orthonormal, but unfortunately anholonomic. Its metric is given by $g(\mathbf{e}_\alpha, \mathbf{e}_\beta) = diag(1, 1, 1, -1)$, the local Minkowski metric.

Two complex half null and two real half null reference frames, in total null, tetrads play now a dominant role, in particular to study gravitational waves, advanced an retarded ones:

two real null vectors $\mathbf{e}_2, \mathbf{e}_3$
and
two complex conjugate null vectors $\mathbf{e}_1, \mathbf{e}_4$

Frames consisting of four null vectors

At each point of the four-dimensional spacetime characterized by coordinates x^i subject to $i, j, \cdots \in \{1, 2, 3, 4\}$ we associate a four-dimensional tangent space. For linear independent vectors $\mathbf{e}_\alpha$ constitute a basis, also called a frame. Dual to this frame is the coframe ω^β which consists of 4 covectors, also called one-forms. Let us decompose frame and coframe with respect to local frames

$$\boxed{\mathbf{e}_\alpha = f^i_\alpha \partial_i \sim \omega^\beta = f^\beta_j dx^j} \tag{3.1}$$

The matrices f^i_α and f^β_j are called Frobenius matrices of integrating factors. Duality is defined by

$$f^\alpha_i \ f^j_\alpha = \delta^j_i \quad \text{and} \quad f^\alpha_i f^i_\beta = \delta^\alpha_\beta \tag{3.2}$$

We refer to the review of Blagojevic, M., Garecki, J., Hehl, F.W. and Obukhov, Y.N. [3]: The matrices $f^i{}_{\cdot\alpha}$ are called frame components or tetrad components. Conventionally, in Relativity vectors in an orthonormal or pseudo-orthonormal frames are labeled $\{1, 2, 3, 4\}$ underlining the fundamental difference between $\mathbf{e}_4$, the negative length $g_{44} = g(\mathbf{e}_4, \mathbf{e}_4) = -1$ and $\mathbf{e}_a, a \in \{1, 2, 3\}$ which have positive lengths $g_{aa} = g(\mathbf{e}_a, \mathbf{e}_a) = +1$. We call a vector timelike if $g(u, u) < 0$, spacelike if $g(u, u) > 0$ and null of lightlike if $g(u, u) = 0$. The components of the metric tensor with respect to the pseudo-orthonormal frame $\mathbf{e}_\alpha$ are represented by

$$g_{\alpha\beta} \overset{*}{:=} \begin{bmatrix} 1 & 0 & 0 & 0 \\ 0 & 1 & 0 & 0 \\ 0 & 0 & 1 & 0 \\ 0 & 0 & 0 & -1 \end{bmatrix} \tag{3.3}$$

The star equal sign indicates that this equation holds only in this specific frame, the pseudo-orthonormal one.

Null frames

We begin with a pseudo-orthonormal reference frame $\mathbf{e}_\alpha$ whose metric will take the standard form

$$g_{\alpha\beta} := \begin{bmatrix} 1 & 0 & 0 & 0 \\ 0 & 1 & 0 & 0 \\ 0 & 0 & 1 & 0 \\ 0 & 0 & 0 & -1 \end{bmatrix} \tag{3.4}$$

We build an alternative frame of reference which is well suited to investigate gravitational waves, so-called null coordinates, more specifically advanced and retarded one. For instance, define

$$\boxed{\begin{aligned} l &:= \tfrac{1}{\sqrt{2}}(\mathbf{e}_1 + \mathbf{e}_2), & n &:= \tfrac{1}{\sqrt{2}}(\mathbf{e}_1 - \mathbf{e}_2) \\ m &:= \tfrac{1}{\sqrt{2}}(\mathbf{e}_3 + i\mathbf{e}_4), & \overline{m} &:= \tfrac{1}{\sqrt{2}}(\mathbf{e}_3 - i\mathbf{e}_4) \end{aligned}} \tag{3.5}$$

$m, \overline{m}$ have been introduced by E. T: Newman and R. Penrose, see R. Penrose and W. Rindler [35]. i is the imaginary unit, the over bar means complex conjugation subject to $i^2 = -1$.

Let us analyze the new frame $(l, n, m, \overline{m}) := \mathbf{e}_\alpha$ as a functional relation of the old frame $(\mathbf{e}_1, \mathbf{e}_2, \mathbf{e}_3, \mathbf{e}_4,) := \mathbf{e}_\alpha$ for spacetime. Both frames of references, $\mathbf{e}_\alpha$, respectively $\mathbf{e}'_\alpha$ are elements of the set $\{1, 2, 3, 4\}$. We note the condition of null frames in terms of metric coefficients

$$\boxed{g(l, l) = g(n, n) = g(m, m) = g(\overline{m}, \overline{m}) = 0} \tag{3.6}$$

We note also the transformation of the metric in terms of pseudo-orthonormal frames of reference of type

$$\begin{aligned} ds^2 &\overset{*}{=} [\omega_1, \omega_2, \omega_3, \omega_4] \begin{bmatrix} 1 & 0 & 0 & 0 \\ 0 & 1 & 0 & 0 \\ 0 & 0 & 1 & 0 \\ 0 & 0 & 0 & -1 \end{bmatrix} \begin{bmatrix} \omega_1 \\ \omega_2 \\ \omega_3 \\ \omega_4 \end{bmatrix} \\ &\overset{**}{=} [\omega_1, \omega_2, \omega_3, \omega_4] \begin{bmatrix} 0 & 0 & 0 & 0 \\ 1 & 0 & 0 & 0 \\ 0 & 0 & 1 & -1 \\ 0 & 0 & -1 & 0 \end{bmatrix} \begin{bmatrix} \omega_1 \\ \omega_2 \\ \omega_3 \\ \omega_4 \end{bmatrix} \end{aligned} \tag{3.7}$$

"*" as well as "**" indicate that these identities hold only in this special frame of reference.

Transformation of the Metric for Null Frames of Reference

Let us study the transformation of the metric for these special frames of reference of type null frame.

$$ds^2 \overset{*}{=} g_{\mu\nu}(u, u)\omega^\mu \omega^\nu \overset{**}{=} g_{\mu'\nu'}(u' u')\omega^{\mu'} \omega^{\nu'} \tag{3.8}$$

$$g_{\mu\nu}(u, u) j^\mu_{\mu'} j^\nu_{\nu'} \omega^{\mu'} \omega^{\nu'} = g_{\mu'\nu'} \left[(u'(u), u'(u)\right] \omega^{\mu'} \omega^{\nu'} \tag{3.9}$$

$j^{\mu'}_\mu$ and $j^{\nu'}_\nu$ are Jacobi matrices of first derivatives.

$$g_{\mu\nu} \left[u(u'), u(u')\right] j^\mu_{\mu'} j^\nu_{\nu'} = g_{\mu'\nu'}(u', u') \tag{3.10}$$

Proof half-null frame "l, π"

Let us begin with the first case: for all $\mu, \nu \in \{1, 2\}$ we assume $g_{\mu\nu} = diag(1, 1)$ and $g(l, l) = g(n, n) = 0$ or $g_{1'1'} = g_{2'2'} = 0$ as well as $g(l, n) = g(n, l)$ or $g_{1'2'} = g_{2'1'}$ a result which we call "symmetry of the metric".

Problem
Given $g_{\mu\nu} = diag(1, 1)$ and $g(l, l) = g(n, n) = 0$ or $g_{1'1'} = g_{2'2'} = 0$.
Find $g(l, n) = g(n, l)$ or $g_{1'2'} = g_{2'1'}$

Since $g(l, l) = g(n, n)$ or $g_{1'1'} = g_{2'2'} = 0$ which holds by definition of the half-null frame we are left with the problem to construct $g(l, n) = g(n, l)$ or $g_{1'2'} = g_{2'1'}$. Please, note

$$l \;:= \frac{1}{\sqrt{2}}(\mathbf{e}_1 + \mathbf{e}_2), \quad n \;:= \frac{1}{\sqrt{2}}(\mathbf{e}_1 - \mathbf{e}_2)$$
$$\omega^{1'} := \frac{1}{\sqrt{2}}(\omega e^1 + \omega^2), \;\; \omega^{2'} := \frac{1}{\sqrt{2}}(\omega e^1 - \omega^2)$$

versus

$$\mathbf{e}_1 = \frac{1}{\sqrt{2}}(l + n), \mathbf{e}_2 = \frac{1}{\sqrt{2}}(l - n)$$
$$\omega^1 = \frac{1}{\sqrt{2}}(\omega^{1'} + \omega^{2'}), \omega^2 = \frac{1}{\sqrt{2}}(\omega^{1'} - \omega^{2'})$$

or

$$(\omega^1)^2 = \frac{1}{2}\left[(\omega^{1'})^2 + 2\omega^{1'}\omega^{2'} + (\omega^{2'})^2\right]$$
$$(\omega^2)^2 = \frac{1}{2}\left[(\omega^{1'})^2 + 2\omega^{1'}\omega^{2'} + (\omega^{2'})^2\right]$$
$$g(l, l) = g(n, n) = 0 \Longleftrightarrow (\omega^{1'})^2 = (\omega^{2'})^2 = 0$$
$$(\omega^1)^2 + (\omega^2)^2 = 2\omega^{1'}\omega^{2'} (\omega^1)^2 = \omega^{1'}\omega^{2'}, (\omega^2)^2 = \omega^{1'}\omega^{2'}$$

$$g_{1'2'} = g_{ln} = +1 = g_{nl} = g_{2'1'} \tag{3.11}$$

At this end, we summarize: The first half-null frame accounts to

$$g_{\mu'\nu'} = \begin{bmatrix} 0 & +1 \\ +1 & 0 \end{bmatrix} \text{for all } \mu', \nu' \in \{1, 2\} \tag{3.12}$$

Proof
half-null frame "$m, \overline{m}$"

Let us continue with the second case: for all $\mu, \nu \in \{1, 2\}$ we define $g_{\mu\nu} = diag(+1, -1)$ and $g(m, m) = g(\overline{m}, \overline{m}) = 0$ or $g_{3''3''} = g_{4''4''} = 0$ as well as $g(m, \overline{m}) = g(m, m)$ or $g_{3''4''} = g_{4''3''}$ a result which we call "symmetry of the pseudo-metric".

Problem
Given $g_{\mu\nu} = diag(+1, -1)$ and $g(m, m) = g(\overline{m}, \overline{m}) = 0$ or
$g_{3''3''} = g_{4''4''} = 0$
Find $g(m, \overline{m}) = g(\overline{m}, m)$ or $g_{3''4''} = g_{4''3''}$

Since $g(m, m) = g(\overline{m}, \overline{m}) = 0$ or $g_{3''3''} = g_{4''4''} = 0$ by definition of the half-null frame we are left with the problem to construct $g(m, \overline{m}) = g(\overline{m}, m)$ or $g_{3''4''} = g_{4''3''}$. Please, note

$$m := \frac{1}{\sqrt{2}}(\mathbf{e}_3 + i\mathbf{e}_4), \quad \overline{m} := \frac{1}{\sqrt{2}}(\mathbf{e}_3 - i\mathbf{e}_4)$$
$$\omega^{3''} := \frac{1}{\sqrt{2}}(\omega e^3 + i\omega^4), \quad \omega^{4''} := \frac{1}{\sqrt{2}}(\omega e^3 - i\omega^4)$$

versus

$$\mathbf{e}_3 = \frac{1}{\sqrt{2}}(m + i\overline{m}), \mathbf{e}_4 = \frac{1}{\sqrt{2}}(m - i\overline{m})$$

and

$$\omega^3 = \frac{1}{\sqrt{2}}(\omega^{3''} + \omega^{2'}), \omega^4 = \frac{1}{\sqrt{2}}(\omega^{3''} - i\omega^{4''})$$

or

$$(\omega^3)^2 - (\omega^4)^2 = \frac{1}{2}(\omega^{3'}) + i\omega^{4'}(\omega^{3'} + i(\omega^{4'})+$$
$$+\frac{1}{2}(i\omega^{3'}) + \omega^{4'}(i\omega^{3'} + (\omega^{4'})$$
$$g(m, m) = g(\overline{m}, \overline{m}) = 0 \Longleftrightarrow (\omega^{3''})^2 = (\omega^{4''})^2 = 0$$
$$(\omega^3)^2 - (\omega^4)^2 = -2\omega^{3''}\omega^{4''}$$

$$g_{3''4''} = g_{m\overline{m}} = -1 = g_{\overline{m},m} = g_{4''3''} \tag{3.13}$$

Finally, we summarize: The second half-null frame accounts to

$$\boxed{g_{\mu''\nu''} = \begin{bmatrix} 0 & -1 \\ -1 & 0 \end{bmatrix} \text{for all } \mu''\nu'' \in \{3, 4\}} \tag{3.14}$$

Here, i is the imaginary part and over-bar means complex conjugation. The total transformation leads to the Lorentz metric in a Newnan–Penrose null frame

$$\mathbf{e}_\alpha := (l, n, m, \overline{m}) :$$

in detail

$$g_{\mu''\nu''} = \begin{bmatrix} 0 & 0 & 0 & 0 \\ 1 & 0 & 0 & 0 \\ 0 & 0 & 1 & -1 \\ 0 & 0 & -1 & 0 \end{bmatrix} \tag{3.15}$$

Such a frame is very well suited for investigating gravitating waves as well as electromagnetic waves. M. Blagojevic, J. Garecki, F.W. Hehl and Yu.N. Obukhov [3] developed alternatively Real Null Coframes to which we refer. Here we adopted the title Real null coframes in General Relativity and GPS type coordinates from M. Blagojevic et al. [3].

4 Examples

Our first example relates to Killing vectors of symmetry, in particular of the ellipsoid-of-revolution and the sphere. Second, we study anholonomity, namely the influence of the local vertical in computing geodetic networks in the form of three dimensional Geodesy. Our example discusses the misclosure of a triangular network observed by scaled distance measurements. The local vertical in gravity space depends on astronomic longitude and astronomic latitude. In detail, we present the analysis of a triangular network in the range of (i) 25 m and (ii) 500 m. The structure of this network causes miscloser in the range of (i) 1 mm and (ii) 300 mm. Anholonomity influences these misclosers. Third, we analyze anholonomity as the problem of integrability or differential forms. The tool in treating problematic orthonormal frames of reference is "exterior calculus", here introduced of analyzing the 2-sphere with respect to two frames of reference.

4.1 *Killing Vectors of Symmetry*

We begin with two questions:

First question

Let a transformation group act on the coordinate transformation of a surface-of-revolution. What are the transformation groups which leaves the first differential invariant ds^2 of a surface-of-revolution equivariant or form invariant? Answer 1: The transformation group which leaves the first differential invariant ds^2, also called "arc length", equivariant is the one-dimensional rotation group R_3 (longitude), a rotation "around the 3-axis" of the ambient space $\{\mathbb{R}^3, \delta_{ij}\}$. The 3-axis establishes the Killing vector of symmetry.

The proof

Our proof for the "Answer 1" is outlined in Box 5. First, we present a parameter representation of a surface-of-rotation defined by $\{u, v\}$ in a equatorial frame -of-reference and defined by $\{u^*, v^*\}$ in a rotated equatorial frame-of-reference. Second, we follow the action of the rotation group $\mathbb{R}_3(\Omega) \in SO(2)$. Third, we generate the forward and backward transformations

$$\{\mathbf{e}_1, \mathbf{e}_2, \mathbf{e}_3|0\} \to \{\mathbf{e}_{1*}, \mathbf{e}_{2*}, \mathbf{e}_{3*}|0\} \text{ as well as}$$

$$\{\mathbf{e}_{1*}, \mathbf{e}_{2*}, \mathbf{e}_{3*}|0\} \to \{\mathbf{e}_1, \mathbf{e}_2, \mathbf{e}_3|0\} \text{ of}$$

orthogonal base vectors which span the three-dimensional Euclidean ambient space. Fourth, we fill in the backward transformation of bases into the first parameter representation of the surface-of-revolution and compare with the second one. In this way, we find the "Kartenwechsel", [change form one chart to another chart] $\{u^* = u - \Omega, v^* = v\}$. Fifth, we compute the first differential invariant ds^2 of the surface-of-revolution, namely the matrix of the metric $\mathbf{G} = diag[f^2, f'^2 + g'^2]$. $\{f, g\}$ are representations of the parameters describing the surface-of-revolution given by the formula (4.1) and (4.2) in Box 5! "Cha-Cha-Cha" leads us via the Jacobi map $\mathbf{J}$ to the second representation d^{*2} of the first differential invariant which turns out to be equivariant or form invarinat. Indeed, we have shown that under the action of the rotation group $ds^2 = ds^{*2}$. Sixth, we identity $\mathbf{e}_3$ as the Killing vector of the symmetry of the surface of revolution.

Box 5: Surface of revolution, Killing vector of symmetry, equivariance of the arc length under the action of the special orthogonal group SO(2)

Surface-of-revolution parametrized in an equatorial frame of reference:

$$\mathbf{x}(u, v) = \mathbf{e}_1 F(v) \cos u + \mathbf{e}_2 f(v) \sin u + \mathbf{e}_3 g(v) \tag{4.1}$$

Surface-of-revolution parametrized in a rotated equatorial frame of reference:

$$\mathbf{x}(u^*, v^*) = \mathbf{e}_{1*} F(v^*) \cos u^* + \mathbf{e}_{2*} f(v^*) \sin u^* + \mathbf{e}_{3*} g(v^*) \tag{4.2}$$

Action of the special orthogonal group $SO(2)$:

$$R_3(\Omega) \in SO(2) := \{R_3 \in \mathbb{R}^{3\times3} | R_3^* R_3 = I_3, |R_3| = 1\},$$

$$\begin{bmatrix} \mathbf{e}_{1*} \\ \mathbf{e}_{2*} \\ \mathbf{e}_{3*} \end{bmatrix} = R_3(\Omega) \begin{bmatrix} \mathbf{e}_1 \\ \mathbf{e}_2 \\ \mathbf{e}_3 \end{bmatrix} = \begin{bmatrix} \cos\Omega & \sin\Omega & 0 \\ -\sin\Omega & \cos\Omega & 0 \\ 0 & 0 & 1 \end{bmatrix} \begin{bmatrix} \mathbf{e}_1 \\ \mathbf{e}_2 \\ \mathbf{e}_3 \end{bmatrix}$$

$$\begin{bmatrix} \mathbf{e}_1 \\ \mathbf{e}_2 \\ \mathbf{e}_3 \end{bmatrix} = R_3^*(\Omega) \begin{bmatrix} \mathbf{e}_{1*} \\ \mathbf{e}_{2*} \\ \mathbf{e}_{3*} \end{bmatrix} = \begin{bmatrix} \cos\Omega & -\sin\Omega & 0 \\ \sin\Omega & \cos\Omega & 0 \\ 0 & 0 & 1 \end{bmatrix} \begin{bmatrix} \mathbf{e}_{1*} \\ \mathbf{e}_{2*} \\ \mathbf{e}_{3*} \end{bmatrix} \tag{4.3}$$

$$\mathbf{e}_1 = \mathbf{e}_{1*} \cos\Omega - \mathbf{e}_{2*} \sin\Omega, \mathbf{e}_2 = \mathbf{e}_{2*} \sin\Omega + \mathbf{e}_{2*} \cos\Omega, \mathbf{e}_3 = \mathbf{e}_{3*}$$

Coordinate transformation

$$\mathbf{x}(u, v) = f(v)\mathbf{e}_{1*}(\cos\Omega \cos u + \sin\Omega \sin u) + f(v)\mathbf{e}_{2*}(-\sin\Omega \cos u + \cos\Omega \sin u) + \mathbf{e}_{3*} g(v)$$

$$\mathbf{x}(u, v) = f(v)\mathbf{e}_{1*} \cos u - \Omega + f(v)\mathbf{e}_{2*} \sin u - \Omega \mathbf{e}_{3*} g(v) \tag{4.4}$$

$$v = v^*, \mathbf{x}(u, v) = \mathbf{x}(u^*, v^*)$$

$$\cos u^* = \cos(u - \Omega), \sin u^* = \sin(u - \Omega), \tan u^* = \tan(u - \Omega) \tag{4.5}$$

$$u^* = u - \omega$$

Arc length (first differential invariant): $ds^2 = [du, dv] J_x^* J_x \begin{bmatrix} du \\ dv \end{bmatrix}$

$$J_x = \begin{bmatrix} D_u x & D_v x \\ D_u y & D_v y \\ D_u z & D_v z \end{bmatrix} = \begin{bmatrix} -f \sin u & f' \cos u \\ f \cos u & f' \sin u \\ 0 & g' \end{bmatrix}, G := J_x^* J_x = \begin{bmatrix} f^2 & 0 \\ 0 & f'^2 + g'^2 \end{bmatrix} \tag{4.6}$$

Killing vector of symmetry (rotational axis):

1st version: $ds^2 = f^2 du^2 + (f'^2 + g'^2) dv^2$ 2nd version: $ds^{*2} = f^{*2} du^{*2} + (f^{*'2} + g^{*'2}) dv^{*2}$

$$u^* = u - \Omega, \quad v^* = v \Leftrightarrow du^{*2} = du^2, \quad dv^{*2} = dv^2 \tag{4.7}$$

$$ds^2 = f^2 du^2 + (f'^2 + g'^2) dv^2 = f^2 du^{*2} + (f'^2 + g'^2) dv^{*2} = ds^{*2}$$

$$\mathbf{e}_3 = [\mathbf{e}_1, \mathbf{e}_2, \mathbf{e}_3] \begin{bmatrix} 0 \\ 0 \\ 1 \end{bmatrix} \sim \begin{bmatrix} 0 \\ 0 \\ 1 \end{bmatrix} \tag{4.8}$$

End Box 5

Second question

Let a transformation group act on the coordinate transformation of the sphere. Or we may say, we make a coordinate transformation. What are the transformation groups, the coordinate transformations which leave the first differential invariant ds^2 of a sphere equivariant or form invariant Answer 2: The transformation group which leaves the first differential invariant ds^2, also called "arc length", equivariant is the three-dimensional rotation group $R(\alpha, \beta, \gamma)$, a subsequent rotation "around the 1-axis, the 2-axis and the 3-axis" of the ambient space $\{\mathbb{R}^3, \delta_{ij}\}$. The three-axes establish three Killing vector of symmetry.

Box 6: Sphere, Killing vector of symmetry, equivariance of the arc length under the action of the special orthogonal group SO(3)

Sphere parametrized in an equatorial frame of reference:

$$\boldsymbol{x}(u, v) = \mathbf{e}_1 \cos v \cos u + \mathbf{e}_2 \cos v \sin u + \mathbf{e}_3 \sin v \tag{4.9}$$

Sphere parametrized in an oblique frame of reference:

$$\boldsymbol{x}(u^*, v^*) = \mathbf{e}_{1*} \cos v^* \cos u^* + \mathbf{e}_{2*} \cos v^* \sin u^* + \mathbf{e}_{3*} \sin v^* \tag{4.10}$$

Action of the special orthogonal group $SO(3)$:

$$R(\alpha, \beta, \gamma) \in SO(3) := \{R \in SO(3) | R^* R = I_3, |R| = 1\},$$

$$\begin{bmatrix} \mathbf{e}_{1^*} \\ \mathbf{e}_{2^*} \\ \mathbf{e}_{3^*} \end{bmatrix} = R_1(\alpha)R_2(\beta)R_3(\gamma) \begin{bmatrix} \mathbf{e}_1 \\ \mathbf{e}_2 \\ \mathbf{e}_3 \end{bmatrix} \tag{4.11}$$

$$\begin{bmatrix} \mathbf{e}_1 \\ \mathbf{e}_2 \\ \mathbf{e}_3 \end{bmatrix} = R_3^*(\alpha)R_2^*(\beta)R_1^*(\gamma) \begin{bmatrix} \mathbf{e}_{1^*} \\ \mathbf{e}_{2^*} \\ \mathbf{e}_{3^*} \end{bmatrix} \tag{4.12}$$

$$\mathbf{e}_1 = \mathbf{e}_{1^*}(\cos\gamma\cos\beta) - \mathbf{e}_{2^*}(\sin\gamma\cos\alpha + \cos\gamma\sin\beta\sin\alpha + \mathbf{e}_{3^*}(\sin\gamma\sin\alpha + \cos\gamma\sin\beta\cos\alpha),$$
$$\mathbf{e}_2 = \mathbf{e}_{1^*}(\sin\gamma\cos\beta) + \mathbf{e}_{2^*}(\cos\gamma\cos\alpha + \sin\gamma\sin\beta\sin\alpha + \mathbf{e}_{3^*}(-\cos\gamma\sin\alpha + \sin\gamma\sin\beta\cos\alpha),$$
$$\mathbf{e}_3 = \mathbf{e}_{1^*}(-\sin\beta) + \mathbf{e}_{2^*}(\cos\beta\sin\alpha) + \mathbf{e}_{3^*}(\cos\beta\cos\alpha).$$

Coordinate transformations:

$$\boldsymbol{x}(u,v) = \boldsymbol{x}(u^*,v^*)$$
$$\Leftrightarrow$$
$$\mathbf{e}_{1*}f_1(\alpha,\beta,\gamma|u,v) + \mathbf{e}_{2*}f_2(\alpha,\beta,\gamma|u,v) + \mathbf{e}_{3*}f_3(\alpha,\beta,\gamma|u,v) \tag{4.13}$$

$$\begin{aligned} &= \mathbf{e}_{1*}\cos v^*\cos u^* + \mathbf{e}_{2*}\cos v^*\sin u^* + \mathbf{e}_{3*}\sin v^*,\\ &\cos v^*\cos u^* = f_1(\alpha,\beta,\gamma|u,v)\\ &= \cos\gamma\cos\beta\cos v\cos u + \sin\gamma\cos\beta\cos v\sin u - \sin\beta\sin v,\\ &\cos v^*\sin u^* = f_2(\alpha,\beta,\gamma|u,v)\\ &-(\sin\gamma\cos\alpha + \cos\gamma\sin\beta\sin\alpha)\cos v\cos u +\end{aligned}$$

$$+ (\cos\gamma\cos\alpha + \sin\gamma\sin\beta\sin\alpha)\cos v\sin u + \cos\beta\sin\alpha\sin v, \tag{4.14}$$

$$\begin{aligned} &\sin v^* = f_3(\alpha,\beta,\gamma|u,v)\\ &(\sin\gamma\sin\alpha + \cos\gamma\sin\beta\cos\alpha)\cos v\cos u -\\ &-(\cos\gamma\sin\alpha + \sin\gamma\sin\beta\cos\alpha)\cos v\sin u + \cos\beta\cos\alpha\sin v,\end{aligned}$$

$$\tan u^* = \frac{f_2}{f_1}, \quad \sin v^* = f_3 \tag{4.15}$$

Arc length (first differential invariant):

$$ds^2 = [du, dv]\begin{bmatrix} r^2\cos v & 0 \\ 0 & r^2 \end{bmatrix}\begin{bmatrix} du \\ dv \end{bmatrix} (\text{“diffeomorphism”}), \begin{bmatrix} du^* \\ dv^* \end{bmatrix} = J\begin{bmatrix} du \\ dv \end{bmatrix}, J := \begin{bmatrix} D_u u^* & D_v u^* \\ D_u v^* & D_v v^* \end{bmatrix},$$

$$\begin{aligned} &d\tan u^* = (1+\tan^2 u^*)du^* \Rightarrow du^* = \cos^2 u^* d\tan u^*,\\ &d\sin v^* = \cos v^* dv^* \Rightarrow dv^* = \frac{1}{\sqrt{1-\sin^2 v^*}} a\sin v*,\\ &ds^2 = r^2\cos^2 v du^2 + r^2 dv^2 = r^2\cos^2 v^* du^{*2} + r^2 dv^{*2} = ds^{*2}\end{aligned} \tag{4.16}$$

Killing vector of symmetry (rotation axis):

$$\begin{aligned}
&\text{1- axis of symmetry: } \mathbf{e}_1 \sim \begin{bmatrix} 1 \\ 0 \\ 0 \end{bmatrix}, \\
&\text{2- axis of symmetry: } \mathbf{e}_1 \sim \begin{bmatrix} 0 \\ 1 \\ 0 \end{bmatrix}, \\
&\text{3- axis of symmetry: } \mathbf{e}_1 \sim \begin{bmatrix} 0 \\ 0 \\ 1 \end{bmatrix}.
\end{aligned} \tag{4.17}$$

End Box 6

The proof

First, we present a parameter representation of the sphere defined by $\{u, v\}$ in a equatorial frame-of-reference and defined by $\{u^*, v^*\}$ in a oblique frame-of-reference generated by the three-dimensional orthogonal group $SO(3)$. Second, the action of the transformation group $SO(3)$ is parametrized by Cardan angles $\{\alpha, \beta, \gamma\}$, namely a rotation $R_1(\alpha)$ by around the 1 axis, a rotation $R_2(\beta)$ around the 2-axis, and a rotation $R_3(\gamma)$ around the 3-axis. Third, we transform forward and backward the orthonormal system of base vectors $\{\mathbf{e}_1, \mathbf{e}_2, \mathbf{e}_3|0\}$ and $\{\mathbf{e}_{1*}, \mathbf{e}_{2*}, \mathbf{e}_{3*}|0\}$, which span the three-dimensional Euclidean space, the ambient space of the sphere $\mathbb{S}^2_r$. $\{\mathbf{e}_1, \mathbf{e}_2, \mathbf{e}_3|0\}$ establish the conventional equatorial frame-of-reference, but $\{\mathbf{e}_{1*}, \mathbf{e}_{2*}, \mathbf{e}_{3*}|0\}$ at the origin the meta-equatorial reference frame. Fourth, the backward transformation is substituted into the parameters representation of the placement vector $\mathbf{e}_1 r \cos v \cos u + \mathbf{e}_2 r \cos v \sin u + \mathbf{e}_3 r \sin v \in \mathbb{S}^2_r$, such that, $\mathbf{e}_{1*} f_1(\alpha, \beta, \gamma|u, v) + \mathbf{e}_{2*} f_2(\alpha, \beta, \gamma|u, v) + \mathbf{e}_{3*} f_3(\alpha, \beta, \gamma|u, v)$ is a materialization of the "Kartenwechsel". In this way, we are led to $\tan \alpha^* = f_2/f_1$ and $\sin v^* = f_3$, both functions of the parameters $\{\alpha, \beta, \gamma\} \in SO(3)$ of longitude u and the latitude v. Fifth, as soon as we substitute "Cha-Cha-Cha", namely the diffeomorphism $\{du, dv\} \to \{du^*, dv^*\}$ by means of the Jacobi matrix $\mathbf{J}$ in the first differential invariant ds^{*2}, namely the matrix of the metric $\mathbf{G} = diag[r^2 \cos^2 v, r]$, we are led to the first representation ds^2 of the first differential invariant of form-invariant: $ds^2 = r^2 \cos^2 v du^2 + r^2 dv^2 = r^2 \cos^2 v^* du^{*2} + r^2 dv^{*2}$. Indeed we have shown that under the action of the three-dimensional rotation group, namely $R(\alpha, \beta, \gamma) = R_1(\alpha) R_2(\beta) R_3(\gamma), ds^2 = ds^{*2}$. Sixth, we identify the three Killing vectors $\{\mathbf{e}_1, \mathbf{e}_2, \mathbf{e}_3\}$ or $[1, 0, 0]$, $[0, 1, 0]$ and $[0, 0, 1]$, respectively, the symmetry of the sphere $\mathbb{S}^2_r$. For more details refer to two volume book E. Grafarend, R.J. You R. Syffus (2012)

5 The Influence of the Local Vertical on the Analysis of Geodetic Networks Due to Anholonomity

The systematic error in geodetic coordinate computations of local networks - also called "engineering networks"- usually separated in horizontal and vertical parts and which is caused by neglected changes of the local vertical are estimated. We document that in local network of approximately 25 cm the systematic errors amount to about a millimeters but in local network of approximately 500 m extension about 10–30 m. Accordingly for the computation of precise geodetic networks, the influence of the local gravity field, caused by variations of the local physical vertical-measured by vertical deflections of gravity disturbances-cannot be neglected.

Why is the influence of the terrestrial gravity potential and its gravity field so small?

It is worth while to study the gravity field and the shape of the figure of the Earth more accurate.

Model Assumptions

First assumption

The figure of the Earth is assumed to be described to be the sphere of constant radius r_0, the radius of its boundary.

Second assumption

The mass distribution of the Earth is assumed to be homogenous.

Inversion $w_0 \longrightarrow r_0$

We depart form the assumption that geodetic measurements are able to yet information of the potential values on the spherical surface. The terrestrial potential w_0 with respect to the simple Earth model could be easily converted into the Earth r_0, namely

$$w_0 = gmr_0^{-1} \longrightarrow r_0 = gmw_0^{-1}$$

Inversion $\gamma_0 \longrightarrow r_0$

We depart form the assumption that geodetic measurements of type gravimetry γ_0 are able to get information of the gravity values on the spherical surface. Terrestrial gravity values γ_0, the length of the gravity vector, could be easily inverted to the Earth radius, especially with respect to the simple Earth model, namely for $\gamma_0 > 0$:

$$\gamma_0 = w_0' = gmr_0^{-2} \longrightarrow r_0 = \sqrt{gm}\gamma_0^{-2}$$

Inversion $\gamma_0' = w_0'' \longrightarrow r_0$

We depart from the assumption that geodetic measurements of type gravity gradient/of type first derivative of gravity or of type second derivative of the potential, are able to get information of relative gravity data on the spherical surface. The terrestrial values $\gamma_0' = w_0''$ with respect to the simple Earth model could be easily inverted in the Earth radius, namely

$$\gamma_0' = w_0'' = 2gmr_0^{-3} \longrightarrow r_0 = \sqrt[3]{2gm}\gamma_0'^{-1/3}$$

We have seen that for a simple Earth model (i) homogeneous mass distribution, (ii) spherical Earth - there is no integrability problem. But the real Earth has a heterogeneous, inhomogeneous, latitude and longitude dependent mass distribution and the figure and the gravity filed is neither spherical nor ellipsoidal, but irregular! There is the argument for small anholonomity.

Typical for surveying problems in local networks, for instance "deformation networks", is the adjustment in horizontal as well as vertical parts. Mainly planar geodetic nets are treated exclusively in the surveying literature. The influence of change in the local vertical is neglected. But the local vertical is the reference direction in positioning instruments especially in LPS ("Local Positing Systems"): Local networks are typically anholonomic since the local vertical due to irregular mass distribution varies form point to point, not being reflected by planar networks.

5.1 The Local Geodetic Reference Systems (LPS)

At first we take reference to the local moving reference frame systematically called

"Horizontal Reference Frame"
(HRF)
as well as the space-fixed

"Equatorial Reference Frame"
(ERF)

We want to document that the change of local vertical is of central importance. Let us assume that we reduce in the frame of a peripheral model or geodetic measurements, namely

- instrumental errors
- central error
- null point systematic errors
- refractional effects
- relativistic effects of first order.

By means of modern electronic measurement instruments in a local positioning system of type theodolite and distance measurement equipment we are able to determine the spherical coordinates of type

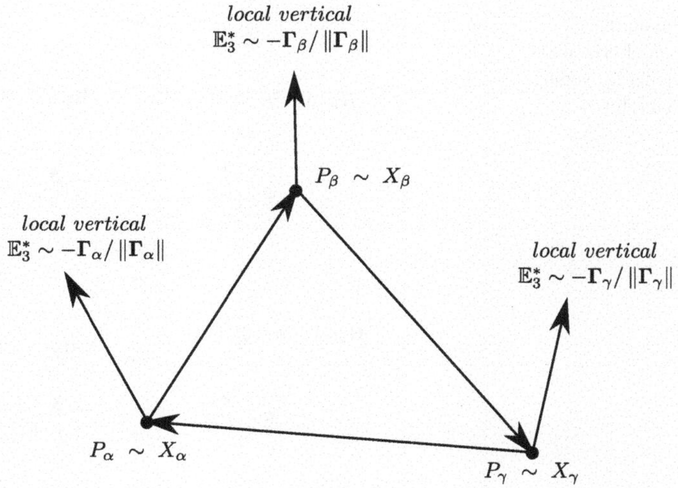

Fig. 15 Triangular network $\{P_\alpha, P_\beta, P_\gamma | 0\}$, placement vector at the origin O local vertical , $\mathbb{E}_3(P_\alpha), \mathbb{E}_3(P_\beta), \mathbb{E}_3(P_\gamma), \Gamma_\alpha, \Gamma_\beta, \Gamma_\gamma$ local gravity vectors

- horizontal direction $H_{\alpha\beta}, H_{\beta\gamma} H_{\gamma\alpha}$,
- vertical direction (the complement of the zenith distance) $V_{\alpha\beta}, V_{\beta\gamma} V_{\gamma\alpha}$,
- distance $S_{\alpha\beta}, S_{\beta\gamma} S_{\gamma\alpha}$,

between target points and station points $\{P_\alpha, P_\beta, P_\gamma\}$, in particular the relative position vectors $X_{\alpha\beta} := X_\beta - X_\alpha$, $X_{\beta\gamma} := X_\gamma - X_\beta$, and $X_{\gamma\alpha} = X_\alpha - X_\gamma$. Figure 15 illustrations the triangular network $\{P_\alpha, P_\beta, P_\gamma | 0\}$ with respect to vectors at the origin $\mathbf{0}$ and the local verticals $\mathbb{E}_3(P_\alpha)$, $\mathbb{E}_3(P_\beta)$ and $\mathbb{E}_3(P_\gamma)$ in a closed loop as well as local gravity vectors. It is a common practice to represent the relative position vector in an orthogonal local horizontal frame of reference.

Consult Fig. 16: Commutative diagram where we illustrate the moving horizontal reference frames, namely moving relative to the fixed equatorial reference frame "freely transported to the point $\{P_\alpha, P_\beta, P_\gamma\}$".

fixed orthonormal reference frame $\{\mathbb{F}_1^\bullet, \mathbb{F}_2^\bullet, \mathbb{F}_3^\bullet\}$	$\rightleftharpoons$	moving orthonormal reference frame $\{\mathbb{E}_1^*, \mathbb{E}_2^*, \mathbb{E}_3^*\}$

We have denoted the fixed equatorial frame of reference by $\{\mathbb{F}_1^\bullet, \mathbb{F}_2^\bullet, \mathbb{F}_3^\bullet\}$ as well as the moving horizontal frame of reference by $\{\mathbb{E}_1^*, \mathbb{E}_2^*, \mathbb{E}_3^*\}$.

Question

Fig. 16 Commutative diagram: moving horizontal reference systems $\mathbf{E}_\alpha$ versus fixed equatorial reference system

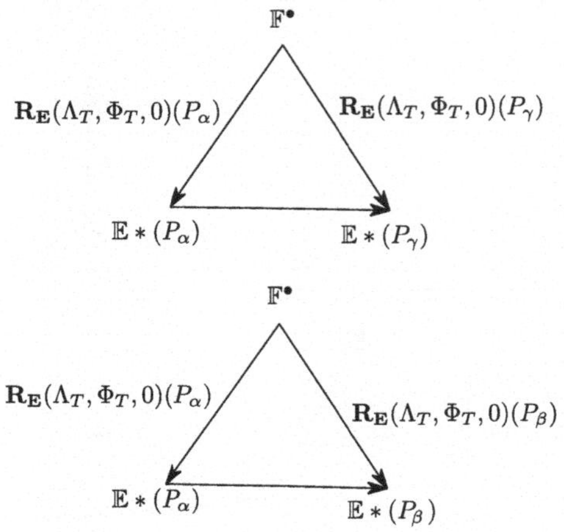

What is a "fixed equatorial frame of reference"?

What is a "moving horizontal frame of reference"?

Answer

Our point of view is "operational"! We start from observations: It is a custom to begin with the "Earth rotation" as all planets rotate: unfortunately the polar axis is moving, the effects are called polar motion (POM) and Length of day (LOD). Traditionally, we define the Mean Rotation Axis by averaging over time, for instance over a year, over ten years or a decades or centuries. There are daily data available from the International Earth Rotation Service, in particular for POM and LOD, also for time. Various excitation mechanisms are studied, worth mentioning "atmospheric tides", tides of the solid part as well as the fluid part (sea tides), also the core-mantle coupling and Earth Quakes.

The background of the other base vectors $\mathbb{F}_1^\bullet$ and $\mathbb{F}_2^\bullet$ in the equatorial plane, orthonormal to the base vector $\mathbb{F}_3^\bullet$ is also fascinating: it takes reference to the Greenwich zero meridian fixed to date it. In the center of time and space we define the origin of the fixed equatorial triad or 3-leg. A parallel transparent to the center of the Earth, we fix the Greenwich base vector $\mathbb{F}_1^\bullet$ as well as the orthogonal base vector $\mathbb{F}_2^\bullet$ orthonormal to $\mathbb{F}_1^\bullet$ and $\mathbb{F}_2^\bullet$

Answer

We have only discuses up to now the "fixed equatorial frame of reference" $\{\mathbb{F}_1^\bullet, \mathbb{F}_2^\bullet, \mathbb{F}_3^\bullet|0\}$. What is the definition of the "moving horizontal frame of reference" $\{\mathbb{F}_1^*, \mathbb{F}_2^*, \mathbb{F}_3^*|0\}$ or $\{\mathbb{E}_1^*, \mathbb{E}_2^*, \mathbb{E}_3^*|P\}$? A theodolite and a leveling instrument measure the local physical vertical as well as local physical horizontal plane. With respect to the so called "orientation unknown" the local vertical as well as the local horizontal plane at the station point P_α we measure the modulus of the gravity vector $\|\Gamma\| = \|\text{grad } W\|$ by gravimetry as well as astronomic latitude ϕ and astronomic longitude Λ, the spherical coordinate of the gravity vector. The moving horizontal frame of reference at placement P_α is defined by $\{\mathbb{F}_1^*, \mathbb{F}_2^*, \mathbb{F}_3^*|0\}$ or $\{\mathbb{E}_1^*, \mathbb{E}_2^*, \mathbb{E}_3|P\}^*$ in particular

- $\mathbb{F}_1^*$ orthogonal to the vector Γ in the horizontal plane "orientation unknown"
- $\mathbb{F}_2^*$ orthogonal to the vector Γ in the horizontal plane, orthogonal to $\mathbb{F}_1^*$ and $\mathbb{F}_3^*$
- $\mathbb{F}_3^* = \mathbb{E}_3^* = \Gamma(P_\alpha) \div \|\Gamma_\alpha(P_\alpha)\|$
- $\mathbb{E}_1^*$ orthogonal to the vector Γ in the horizontal plane, define azimuth and vertical angle
- $\mathbb{E}_2^*$ orthogonal to the vector Γ define azimuth and vertical angle
- $\mathbb{E}_3^* = \mathbb{F}_3^* = \Gamma(P_\alpha) \div \|\Gamma_\alpha(P_\alpha)\|$

The various definitions of reference frames $\{\mathbb{E}_1^*, \mathbb{E}_2^*, \mathbb{E}_3^*|P\}$ or $\{\mathbb{F}_1^*, \mathbb{F}_2^*, \mathbb{F}_3^*|P\}$ as well as $\{\mathbb{E}_1^\bullet, \mathbb{E}_2^\bullet, \mathbb{E}_3^\bullet|0\}$ or $\{\mathbb{F}_1^\bullet, \mathbb{F}_2^\bullet, \mathbb{F}_3^\bullet|0\}$ become more obvious when we study the observations from a station point P_α to a target point P_β, for instance. We introduce a new index, also called "a dummy index", for instance, P_α^α versus P_β^α. It is indicating that both reference frames P at the points P_α and P_β are the same! How to do this? We apply parallel transport from P_α and P_β of the reference frame attached to P! In the Euclidean space in terms of three-dimensional Geodesy we think in terms of the Euclidean Axiom 5 of the embedding space. Concretely we connect the orthonormal reference frame $\{\mathbb{F}_1^*, \mathbb{F}_2^*, \mathbb{F}_3^*|P\}$ freely transported from $O \longrightarrow P$ compared to the orthonormal reference frame $\{\mathbb{F}_1^*, \mathbb{F}_2^*, \mathbb{F}_3^*|P\}$

$$[\mathbb{F}_1^*, \mathbb{F}_2^*, \mathbb{F}_3^*|P] = [\mathbb{F}_1^\bullet, \mathbb{F}_2^\bullet, \mathbb{F}_3^\bullet|0]R_E^\top(\Lambda, \phi, \Sigma) \tag{5.1}$$

Σ denotes the "orientation unknown" typically for the chosen theodolite origin in the horizontal plane. The 3 parameters (Λ, Φ, Σ) with in $\mathbb{R}(\Lambda, \Phi, \Sigma) := R_3(\Sigma)R_2(\frac{\pi}{2} - \Phi)R_3(\Lambda)$ refer to Euler angles astronomical longitude Λ, the compliment of astronomical latitude Φ and the "orientation unknown" Σ. Let us use the transformation $\{\mathbb{F}_1^\bullet, \mathbb{F}_2^\bullet, \mathbb{F}_3^\bullet|0\}$ to $\{\mathbb{E}_1^\bullet, \mathbb{E}_2^\bullet, \mathbb{E}_3^\bullet|0\}$ reminding us the definition of $\frac{dw}{dt}$, namely the dot.

$$[\mathbb{F}_1^\bullet, \mathbb{F}_2^\bullet, \mathbb{F}_3^\bullet] = [\mathbb{E}_1^\bullet, \mathbb{E}_2^\bullet, \mathbb{E}_3^\bullet]R_E^\top(\Sigma) \tag{5.2}$$

a transformation caused by the orientation unknown. For local measurement tasks we apply the south-North orientation of the local horizontal reference system $[\mathbb{E}_1^\bullet, \mathbb{E}_2^\bullet, \mathbb{E}_3^\bullet]$. $\top$ is the sign of the transposition. In contrast, for global measurement problems the Greenwich orientation of the equatorial frame of reference is advantages, namely orthogonal to the rotation axis of the Earth.

The global fixed reference frame $\mathbb{F}^\bullet$ produces holonomic, but the local moving reference frame $\mathbb{E}^*$ anholonomic coordinates, also called differential forms or leg calculus (Fig. 17).

Fig. 17 Local reference system $\{\mathbb{E}_1^*, \mathbb{E}_2^*, \mathbb{E}_3^* | P\}$, $\{\mathbb{F}_1^*, \mathbb{F}_2^*, \mathbb{F}_3^* | P\}$ and global reference system $\{\mathbb{F}_1^\bullet, \mathbb{F}_2^\bullet, \mathbb{F}_3^\bullet | 0\}$, angles (Λ, Φ, Σ)

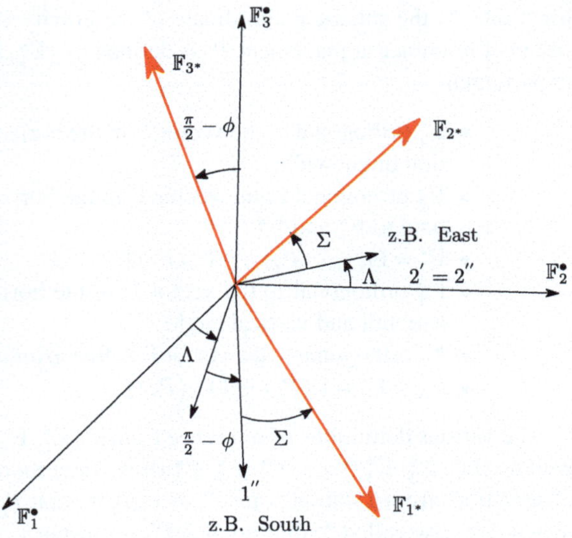

With respect to the moving horizontal reference system we are able now to represent the relative vector "station point to target point" on the surface of the Planet Earth

$$\mathbf{X}_{\alpha\beta} = \mathbf{E}_1.X_{\alpha\beta} + \mathbf{E}_2.Y_{\alpha\beta} + \mathbf{E}_3.Z_{\alpha\beta} \tag{5.3}$$

$$\begin{aligned} X_{\alpha\beta} &= S_{\alpha\beta} \cos A_{\alpha\beta} \cos B_{\alpha\beta} \\ Y_{\alpha\beta} &= S_{\alpha\beta} \sin A_{\alpha\beta} \cos B_{\alpha\beta} \\ Z_{\alpha\beta} &= S_{\alpha\beta} \sin B_{\alpha\beta} \end{aligned} \tag{5.4}$$

$$\boxed{\begin{aligned} A_{\alpha\beta} &= H_{\alpha\beta} + \Sigma_\alpha = \text{Arctan}\, Y_{\alpha\beta} \div Y_{\alpha\beta} \\ B_{\alpha\beta} &= V_{\alpha\beta} = \text{Arctan}\, Z_{\alpha\beta} \div \sqrt{X_{\alpha\beta}^2 + Y_{\alpha\beta}^2} \\ S_{\alpha\beta} &= \sqrt{X_{\alpha\beta}^2 + Y_{\alpha\beta}^2 + Z_{\alpha\beta}^2} \end{aligned}} \tag{5.5}$$

$A_{\alpha\beta}$ denotes the southern azimuth form the station point $\boldsymbol{X}_\alpha$ to the target point $\boldsymbol{X}_\beta$. The horizontal direction is abbreviated by $H_{\alpha\beta}$, namely form the point $\boldsymbol{X}_\alpha$ or P_α to the point $\boldsymbol{X}_\beta$. The orientation are unknown at the point $\boldsymbol{X}_\alpha$ or P_α is called Σ_α. The vertical direction directed upwards was denoted by the letter $B_{\alpha\beta}$. Finally, $S_{\alpha\beta}$ abbreviated the oblique distance. Figures 18 and 19 illustrate the relative position vector in the moving frame of reference in the horizon. Please, note $\boldsymbol{X}_{\beta\alpha} \neq \boldsymbol{X}_{\alpha\beta}$ because of different frames of reference.

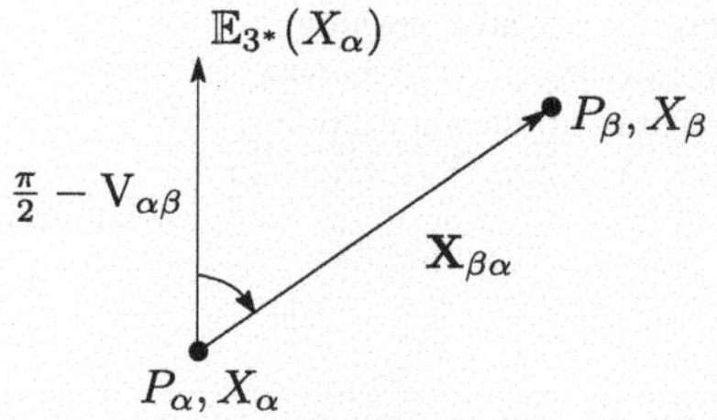

Fig. 18 Relative position vector $X_{\alpha\beta}$ in the horizontal frame of reference P_α

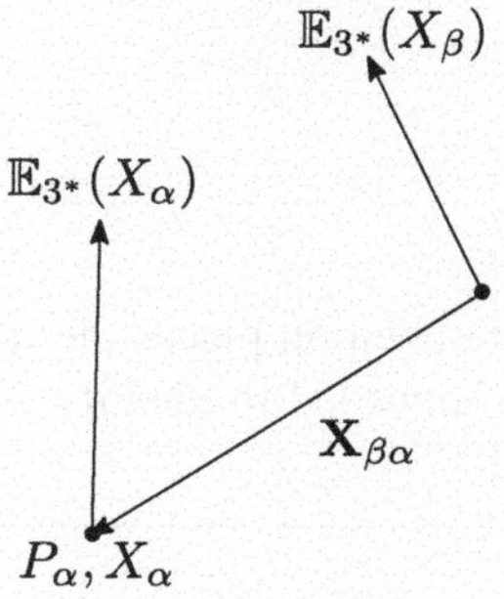

Fig. 19 Relative position vector $X_{\beta\alpha}$ in the horizontal frame of reference P_β : $X_{\beta\alpha} \neq X_{\alpha\beta}$

5.2 *Parallel Transport of Reference Frames*

Due to the parallel transport of frames of reference, namely $\mathbb{E}^*_\alpha(X_\beta) := \mathbb{E}^{\beta *}$ and $\mathbb{E}^*(X_\alpha) : \mathbb{E}^*$, the reference frames enjoy the inequality $\mathbb{E}^{\alpha *} := \mathbb{E}^{\beta *}$, in consequence

$$\boxed{A_{\alpha\beta} \neq A_{\beta\alpha} - \pi,\ B_{\alpha\beta} \neq -B_{\beta\alpha}} \tag{5.6}$$

as well as the anholonomic condition

$$-\begin{bmatrix} X_{\beta\alpha} \\ Y_{\beta\alpha} \\ Z_{\beta\alpha} \end{bmatrix}_{\mathbb{E}^*(X_\beta)} \neq \begin{bmatrix} X_{\alpha\beta} \\ Y_{\alpha\beta} \\ Z_{\alpha\beta} \end{bmatrix}_{\mathbb{E}^*(X_\alpha)}, \quad -\begin{bmatrix} \overset{\beta}{X}{}^*_{\beta\alpha} \\ \overset{\beta}{Y}{}^*_{\beta\alpha} \\ \overset{\beta}{Z}{}^*_{\beta\alpha} \end{bmatrix} \neq \begin{bmatrix} \overset{\alpha}{X}{}^*_{\alpha\beta} \\ \overset{\alpha}{Y}{}^*_{\alpha\beta} \\ \overset{\alpha}{Z}{}^*_{\alpha\beta} \end{bmatrix} \tag{5.7}$$

Now we are in a position to interpret the commutative diagram in Figs. 18 and 19 better. The fixed frame of reference $\mathbf{F}^\bullet$ is always on top, but the moving frames of reference not:

$$\overset{\alpha}{\mathbb{E}^*} \to \overset{\beta}{\mathbb{E}^*} = \overset{\alpha}{\mathbb{E}^*} \mathbf{R}_E(\Lambda_\alpha, \Phi_\alpha, 0)\mathbf{R}_E^T(\Lambda_\beta, \Phi_\beta, 0)$$

$$\overset{\alpha}{\mathbb{E}^*} \to \overset{\gamma}{\mathbb{E}^*} = \overset{\alpha}{\mathbb{E}^*} \mathbf{R}_E(\Lambda_\alpha, \Phi_\alpha, 0)\mathbf{R}_E^T(\Lambda_\gamma, \Phi_\gamma, 0)$$

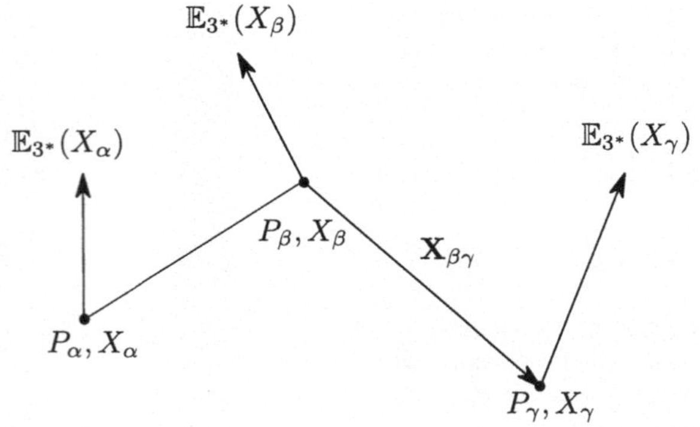

Fig. 20 Relative position vector $(P_\alpha, P_\beta, P_\gamma)$, moving reference frames at the points $\mathbb{E}_{3*}(X_\alpha)$, $\mathbb{E}_{3*}(X_\beta)$, $\mathbb{E}_{3*}(X_\gamma)$

The Heitz notation places the indicies $\{\alpha, \beta, \gamma\}$ on top of the kernel symbol. This notation is needed to identify the reference frame attached to the point where we measure.

Taylor expansions

We use the following Taylor expansions:

$$\Lambda_\beta = \Lambda_\alpha + \Delta\Lambda_{\beta\alpha}, \qquad \Phi_\beta = \Phi_\alpha + \Delta\Phi_{\beta\alpha}$$
$$\Lambda_\gamma = \Lambda_\alpha + \Delta\Lambda_{\gamma\alpha}, \qquad \Phi_\gamma = \Phi_\alpha + \Delta\Phi_{\gamma\alpha}$$

as well as

$$\left[\begin{array}{l} \cos\Lambda_\beta = \cos(\Lambda_\alpha + \Delta\Lambda_{\beta\alpha}) = \cos\Lambda_\alpha \cos\Delta\Lambda_{\beta\alpha} - \sin\Lambda_\alpha \sin\Delta\Lambda_{\beta\alpha} \doteq \cos\Lambda_\alpha - \Delta\Lambda_{\beta\alpha}\sin\Lambda_\alpha \\ \sin\Lambda_\beta = \sin(\Lambda_\alpha + \Delta\Lambda_{\beta\alpha}) = \sin\Lambda_\alpha \cos\Delta\Lambda_{\beta\alpha} - \cos\Lambda_\alpha \sin\Delta\Lambda_{\beta\alpha} \doteq \sin\Lambda_\alpha - \Delta\Lambda_{\beta\alpha}\cos\Lambda_\alpha, \end{array}\right.$$

$$\left[\begin{array}{l} \cos\Lambda_\gamma = \cos(\Lambda_\alpha + \Delta\Lambda_{\gamma\alpha}) = \cos\Lambda_\alpha \cos\Delta\Lambda_{\gamma\alpha} - \sin\Lambda_\alpha \sin\Delta\Lambda_{\gamma\alpha} \doteq \cos\Lambda_\alpha - \Delta\Lambda_{\gamma\alpha}\sin\Lambda_\alpha \\ \sin\Lambda_\gamma = \sin(\Lambda_\alpha + \Delta\Lambda_{\gamma\alpha}) = \sin\Lambda_\alpha \cos\Delta\Lambda_{\gamma\alpha} - \cos\Lambda_\alpha \sin\Delta\Lambda_{\gamma\alpha} \doteq \sin\Lambda_\alpha - \Delta\Lambda_{\gamma\alpha}\cos\Lambda_\alpha, \end{array}\right.$$

These are transformation elements of the astronomical longitude, a similar result we find for astronomical latitude. The left-over transformations apply now for the matrix products, for instance

$$\mathbf{R}_E(\Lambda_\alpha, \Gamma_\alpha, 0)\mathbf{R}_E^T(\Lambda_\gamma, \Gamma_\gamma, 0) \doteq \begin{bmatrix} 1 & -\Delta\Lambda \sin\Phi_\alpha & -\Delta\Phi \\ \Delta\Lambda \sin\Phi_\alpha & 1 & \Delta\Lambda \cos\Phi_\alpha \\ \Delta\Phi & -\Delta\Lambda\cos\Phi_\alpha & 1 \end{bmatrix},$$

namely the sum of (i) a unit matrix I and of (ii) antisymmetric matrix (Fig. 20)

$$\mathbf{A} := \begin{bmatrix} 0 & -\Delta\Lambda \sin\Phi_\alpha & -\Delta\Phi \\ \Delta\Lambda \sin\Phi_\alpha & 0 & \Delta\Lambda \cos\Phi_\alpha \\ \Delta\Phi & -\Delta\Lambda\cos\Phi_\alpha & 0 \end{bmatrix} = -\mathbf{A}^T.$$

Box 7: Transformation of rectangular and curvilinear coordinate differences in two reference systems of the gravity space, Case one rectangular coordinates in two moving reference systems

$$\alpha \longrightarrow \beta$$

$$\boldsymbol{X}_{\alpha\beta} = -\boldsymbol{X}_{\beta\alpha} = \left[\overset{\alpha}{E}_{1*}, \overset{\alpha}{E}_{2*}, \overset{\alpha}{E}_{3*}\right] \begin{bmatrix} \overset{\alpha}{X}{}^*_{\alpha\beta} \\ \overset{\alpha}{Y}{}^*_{\alpha\beta} \\ \overset{\alpha}{Z}{}^*_{\alpha\beta} \end{bmatrix} = -\left[\overset{\alpha}{E}_{1*}, \overset{\alpha}{E}_{2*}, \overset{\alpha}{E}_{3*}\right] \begin{bmatrix} \overset{\beta}{X}{}^*_{\beta\alpha} \\ \overset{\beta}{Y}{}^*_{\beta\alpha} \\ \overset{\beta}{Z}{}^*_{\beta\alpha} \end{bmatrix} =$$

$$= \left[\overset{\alpha}{E}_{1*}, \overset{\alpha}{E}_{2*}, \overset{\alpha}{E}_{3*}\right] (\mathbf{I} + \mathbf{A}) \begin{bmatrix} \overset{\beta}{X}{}^*_{\beta\alpha} \\ \overset{\beta}{Y}{}^*_{\beta\alpha} \\ \overset{\beta}{Z}{}^*_{\beta\alpha} \end{bmatrix} \tag{5.8}$$

$$\beta \longrightarrow \alpha$$

$$\begin{bmatrix} \overset{\beta}{X}{}^*_{\beta\alpha} \\ \overset{\beta}{Y}{}^*_{\beta\alpha} \\ \overset{\beta}{Z}{}^*_{\beta\alpha} \end{bmatrix} = \mathbf{R}_E(\Lambda_\beta, \Phi_\beta, 0)\mathbf{R}_E^T(\Lambda_\alpha, \Phi_\alpha, 0) \begin{bmatrix} \overset{\alpha}{X}{}^*_{\alpha\beta} \\ \overset{\alpha}{Y}{}^*_{\alpha\beta} \\ \overset{\alpha}{Z}{}^*_{\alpha\beta} \end{bmatrix} =$$

$$= (\mathbf{I} + \mathbf{A}^\top) \begin{bmatrix} \overset{\alpha}{X}{}^*_{\alpha\beta} \\ \overset{\alpha}{Y}{}^*_{\alpha\beta} \\ \overset{\alpha}{Z}{}^*_{\alpha\beta} \end{bmatrix} = (\mathbf{I} - \mathbf{A}) \begin{bmatrix} \overset{\alpha}{X}{}^*_{\alpha\beta} \\ \overset{\alpha}{Y}{}^*_{\alpha\beta} \\ \overset{\alpha}{Z}{}^*_{\alpha\beta} \end{bmatrix} \tag{5.9}$$

curvilinear coordinates in two moving reference systems
-horizontal direction, vertical direction-

$$\left[\begin{array}{l} A_{\beta\alpha} = H_{\beta\alpha} + \sum_\beta = \text{Arctan}\, \overset{\beta}{Y}{}^*_{\beta\alpha} \div \overset{\beta}{X}{}^*_{\beta\alpha} \\ B_{\beta\alpha} := V_{\beta\alpha} = \text{Arctan}\, {}^{\beta}Z^{*\,2}_{\beta\alpha} \div \sqrt{\overset{\beta}{X}{}^{*2}_{\beta\alpha} + \overset{\beta}{Y}{}^{*2}_{\beta\alpha}} \\ S_{\beta\alpha} = \sqrt{\overset{\beta}{X}{}^{*2}_{\beta\alpha} + \overset{\beta}{Y}{}^{*2}_{\beta\alpha} + \overset{\beta}{Z}{}^{*2}_{\beta\alpha}} \end{array}\right. \tag{5.10}$$

$$\begin{aligned} A_{\beta\alpha} &= H_{\beta\alpha} + \textstyle\sum_\beta \doteq \text{Arctan}\, \frac{-\Delta\Lambda \sin\Phi_\alpha \overset{\alpha}{X}{}^*_{\alpha\beta} + \overset{\alpha}{Y}{}^*_{\alpha\beta} - \Delta\Lambda\cos\Phi_\alpha \overset{\alpha}{Z}{}^*_{\alpha\beta}}{\overset{\alpha}{X}{}^*_{\alpha\beta} + \Delta\Lambda \sin\Phi_\alpha \overset{\alpha}{Y}{}^*_{\alpha\beta} + \Delta\Phi \overset{\alpha}{Z}{}^*_{\alpha\beta}} \\ B_{\beta\alpha} &:= V_{\beta\alpha} \doteq \text{Arctan}\, \frac{\Delta\Phi\ {}^{\alpha}X^*_{\alpha\beta} - \Delta\Lambda\cos\Phi_\alpha\ {}^{\alpha}Y^*_{\alpha\beta} -\ {}^{\alpha}Z^*_{\alpha\beta}}{\sqrt{\overset{\alpha}{X}{}^{*2}_{\alpha\beta} + \overset{\alpha}{Y}{}^{*2}_{\alpha\beta} - 2\Delta\Lambda\cos\Phi_\alpha \overset{\alpha}{Y}{}^*_{\alpha\beta} \overset{\alpha}{Z}{}^*_{\alpha\beta} + 2\Delta\Phi \overset{\alpha}{X}{}^*_{\alpha\beta} \overset{\alpha}{Z}{}^*_{\alpha\beta}}} \\ S_{\beta\alpha} &= S_{\alpha\beta} = \sqrt{\overset{\alpha}{X}{}^{*2}_{\alpha\beta} + \overset{\alpha}{Y}{}^{*2}_{\alpha\beta} + \overset{\alpha}{Z}{}^{*2}_{\alpha\beta}} \end{aligned} \tag{5.11}$$

End Box 7

Box 8: Transformation of rectangular and curvilinear coordinate differences in two reference systems of the gravity space, Case two

rectangular coordinates in two moving reference systems

$$\beta \longrightarrow \alpha$$

$$\left[\begin{array}{l} \overset{\beta}{X}_{\beta\gamma} = S_{\beta\gamma} \cos A_{\beta\gamma} \cos B_{\beta\gamma} \\ \overset{\beta}{Y}_{\beta\gamma} = S_{\beta\gamma} \sin A_{\beta\gamma} \cos B_{\beta\gamma} \\ \overset{\beta}{Z}_{\beta\gamma} = S_{\beta\gamma} \sin B_{\beta\gamma} \end{array}\right. \tag{5.12}$$

$$\alpha \longrightarrow \beta$$

$$\left[\begin{array}{l} \overset{\alpha}{X}_{\beta\gamma} = \overset{\beta}{X}_{\beta\gamma} - \Lambda_{\alpha\beta} \sin \Phi_\alpha \overset{\beta}{Y}_{\beta\gamma} - \Phi_{\alpha\beta} \overset{\beta}{Z}_{\beta\gamma} \\ \overset{\alpha}{Y}_{\beta\gamma} = \Lambda_{\alpha\beta} \sin \Phi_\alpha \overset{\beta}{X}_{\beta\gamma} - \Lambda_{\alpha\beta} \cos \Phi_\alpha \overset{\beta}{Z}_{\beta\gamma} \\ \overset{\alpha}{Z}_{\beta\gamma} = \Phi_{\alpha\beta} \overset{\beta}{X}_{\beta\gamma} - \Lambda_{\alpha\beta} \cos \Phi_\alpha \overset{\beta}{Y}_{\beta\gamma} + \overset{\beta}{Z}_{\beta\gamma} \end{array}\right. \tag{5.13}$$

curvilinear coordinates in two moving reference systems
-horizontal direction, vertical direction-

$$\Lambda_{\alpha\beta} := \Lambda_\beta - \Lambda_\alpha \qquad \Phi_{\alpha\beta} := \Phi_\beta - \Phi_\alpha$$

(Taylor expansion)

$$\left[\begin{array}{l} A_{\beta\alpha} = H_{\beta\alpha} + \sum_\beta = \text{Arctan } \overset{\beta}{Y}_{\beta\alpha} \div \overset{\beta}{X}_{\beta\alpha} \\ B_{\beta\alpha} = V_{\beta\alpha} = \text{Arctan } \overset{\beta}{Z}{}^2_{\beta\gamma} \div \sqrt{\overset{\beta}{X}{}^2_{\beta\gamma} + \overset{\beta}{Y}{}^2_{\beta\gamma}} \\ S_{\beta\alpha} = \sqrt{\overset{\beta}{X}{}^2_{\beta\gamma} + \overset{\beta}{Y}{}^2_{\beta\gamma} + \overset{\beta}{Z}{}^2_{\beta\gamma}} \end{array}\right. \tag{5.14}$$

$$\begin{array}{l} A_{\beta\gamma} = H_{\beta\gamma} + \sum_\beta \doteq \text{Arctan } \dfrac{-\Lambda_{\alpha\beta} \sin \Phi_\alpha \overset{\alpha}{X}_{\beta\gamma} + \overset{\alpha}{Y}_{\beta\gamma} - \Lambda_{\alpha\beta} \cos \Phi_\alpha \overset{\alpha}{Z}_{\beta\gamma}}{\overset{\alpha}{X}_{\beta\gamma} + \Lambda_{\alpha\beta} \sin \Phi_\alpha \overset{\alpha}{Y}_{\beta\gamma} + \Phi_{\alpha\beta} \overset{\alpha}{Z}_{\beta\gamma}} \\ B_{\beta\gamma} = V_{\beta\gamma} \doteq \text{Arctan } \dfrac{-\Phi_{\alpha\beta} \overset{\alpha}{X}_{\beta\gamma} + \Lambda_{\alpha\beta} \cos \Phi_\alpha \overset{\alpha}{Y}_{\beta\gamma} + \overset{\alpha}{Z}_{\beta\gamma}}{\sqrt{\overset{\alpha}{X}{}^2_{\beta\gamma} + \overset{\alpha}{Y}{}^2_{\beta\gamma} - 2\Lambda_{\alpha\beta} \cos \Phi_\alpha \overset{\alpha}{Y}_{\beta\gamma} \overset{\alpha}{Z}_{\beta\gamma} + 2\Phi_{\alpha\beta} \overset{\alpha}{X}_{\beta\gamma} \overset{\alpha}{Z}_{\beta\gamma}}} \\ S_{\beta\alpha} = \sqrt{\overset{\alpha}{X}{}^2_{\beta\gamma} + \overset{\alpha}{Y}{}^2_{\beta\gamma} + \overset{\alpha}{Z}{}^2_{\beta\gamma}} \end{array} \tag{5.15}$$

$$A_{\beta\gamma} \doteq H_{\beta\gamma} + \sum\nolimits_{\beta} = \operatorname{Arctan} \frac{\overset{\alpha}{Y}_{\beta\gamma}}{\overset{\alpha}{X}_{\beta\gamma}} - \left(\sin \Phi_\alpha + \frac{\overset{\alpha}{X}_{\beta\gamma} \overset{\alpha}{Z}_{\beta\gamma}}{\overset{\alpha}{X}{}^2_{\beta\gamma} + \overset{\alpha}{Y}{}^2_{\beta\gamma}} \cos_\alpha \right) \Lambda_{\alpha\beta} - \frac{\overset{\alpha}{Y}_{\beta\gamma} \overset{\alpha}{Z}_{\beta\gamma}}{\overset{\alpha}{X}{}^2_{\beta\gamma} + \overset{\alpha}{Y}{}^2_{\beta\gamma}} \Phi_{\alpha\beta} \tag{5.16}$$

$$B_{\beta\gamma} \doteq V_{\beta\gamma} \operatorname{Arctan} \frac{\overset{\alpha}{Z}_{\beta\gamma}}{\sqrt{\overset{\alpha}{X}{}^2_{\beta\gamma} + \overset{\alpha}{Y}{}^2_{\beta\gamma}}} + \frac{\overset{\alpha}{Y}_{\beta\gamma}}{\sqrt{\overset{\alpha}{X}{}^2_{\beta\gamma} + \overset{\alpha}{Y}{}^2_{\beta\gamma}}} \cos \Phi_\alpha \Lambda_{\alpha\beta} - \frac{\overset{\alpha}{X}_{\beta\gamma}}{\sqrt{\overset{\alpha}{X}{}^2_{\beta\gamma} + \overset{\alpha}{Y}{}^2_{\beta\gamma}}} \Phi_{\alpha\beta} \tag{5.17}$$

End Box 8

Obviously, the Euclidean distance is independent of the choice of the reference frame. The results is in contrast to the analysis of the observed directions of type horizontal and vertical. The observational equations for horizontal and vertical directions contain three orientation angles, indeed the classical orientation unknown Σ_β at the station point $\mathbb{P}_\beta$, or $\mathbb{P}_\beta$ as well as the longitudinal and latitudinal difference $\Lambda_{\alpha\beta}$, $\Phi_{\alpha\beta}$ or $\Lambda_{\alpha\gamma}$, $\Phi_{\alpha\gamma}$ between the horizontal reference system of the station point and those points which we chose as reference points in a triangular network. In addition, we have modeled the coordinate differences between station point and target point.

Two numerical examples

At the beginning of the section, we discuss two Earth models:

(i) If the Earth would be spherical or a sphere of constant radius, the result at the first pages $w_0 = gm/r_0$ are applicable: $r = gm/w_0$. indeed we would have no anholonomity.

(ii) If the Earth would be ellipsoidal or a rotational symmetric 2-axis ellipsoidal of type Somigliana–Pizzetti of radius (a_0, b_0) (semi-major radius, semi-minor radius), the result of the Chaps. 2–5 are applicable: we would have no anholonomity.

Obviously the real irregular Earth figure is more complex due to the heterogeneous, inhomogeneous mass distributions and the complex Earth rotation, namely observed Length-of-Day variations (LOD) and Polar Motion (POM). For our first example, we start with a triangular geodetic network between the point $\{P_\alpha, P_\beta, P_\gamma\}$ in the horizontal reference frame of the point P_α. The three holonomity conditions of Box 9 express the loop condition for our 25 m local network to zero. In addition, we assume vertical deflection variation of the order or $1''$, $05''$, $2.5''$ for longitude and latitude of an approximate latitude $\Phi_\alpha = 48.783°$.

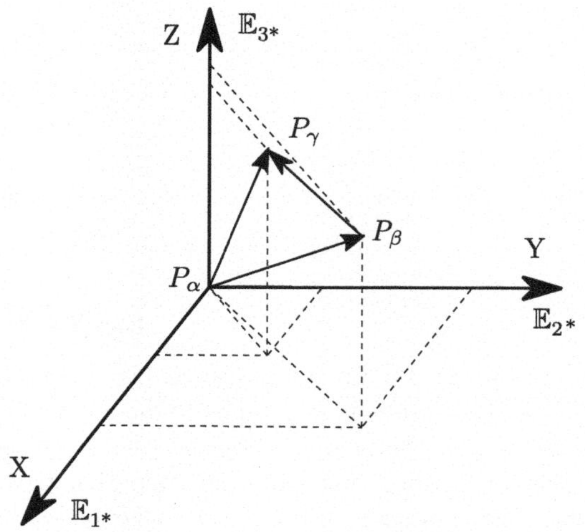

Fig. 21 Triangular network calculated in the horizontal frame of reference at the point P_α

Box 9: 25 m local network

$$\overset{\alpha}{X}_{\alpha\beta} = +30\,\text{m},\ \overset{\alpha}{X}_{\beta\gamma} = +50\,\text{m},\ \overset{\alpha}{X}_{\gamma\alpha} = -80\,\text{m},$$
$$\overset{\alpha}{Y}_{\alpha\beta} = +30\,\text{m},\ \overset{\alpha}{Y}_{\beta\gamma} = -50\,\text{m},\ \overset{\alpha}{Y}_{\gamma\alpha} = +20\,\text{m},$$
$$\overset{\alpha}{Z}_{\alpha\beta} = +05\,\text{m},\ \overset{\alpha}{Z}_{\beta\gamma} = +15\,\text{m},\ \overset{\alpha}{Z}_{\gamma\alpha} = -20\,\text{m},$$
$$\Lambda_{\alpha\beta} = 1'' \sim \quad 4.85 \cdot 10^{-6} RAD$$
$$\Phi_{\alpha\beta} = -0.5'' \sim -\ 2.42 \cdot 10^{-6} RAD$$
$$\Lambda_{\alpha\gamma} = -1'' \sim -\ 4.85 \cdot 10^{-6} RAD$$
$$\Phi_{\alpha\gamma} = -2.5'' \sim -12.12 \cdot 10^{-6} RAD$$
$$\Phi_\alpha = -48.788^\circ$$

End Box 9

Box 10: 25 m local network, detailed computation

$$\begin{bmatrix} X_{\beta\gamma} \\ Y_{\beta\gamma} \\ Z_{\beta\gamma} \end{bmatrix} = \begin{bmatrix} 1 & +\Lambda_{\alpha\beta} \sin\Phi_\alpha & +\Phi_{\alpha\beta} \\ -\Lambda_{\alpha\beta} & 1 & -\Lambda_{\alpha\beta}\cos\Phi_\alpha \\ -\Phi_{\alpha\beta} & +\Lambda_{\alpha\beta}\cos\Phi_\alpha & 1 \end{bmatrix} = \begin{bmatrix} \overset{\alpha}{X}_{\beta\gamma} \\ \overset{\alpha}{Y}_{\beta\gamma} \\ \overset{\alpha}{Z}_{\beta\gamma} \end{bmatrix} =$$
$$\begin{bmatrix} 1 & +3.65 \cdot 10^{-6} & -2.42 \cdot 10^{-6} \\ -3.65 \cdot 10^{-6} & 1 & -3.19 \cdot 10^{-6} \\ +2.42 \cdot 10^{-6} & +3.19 \cdot 10^{-6} & 1 \end{bmatrix} \begin{bmatrix} +50\,\text{m} \\ -50\,\text{m} \\ +15\,\text{m} \end{bmatrix}$$

$$\begin{bmatrix} \overset{\gamma}{X}_{\gamma\alpha} \\ \overset{\gamma}{Y}_{\gamma\alpha} \\ \overset{\gamma}{Z}_{\gamma\alpha} \end{bmatrix} = \begin{bmatrix} 1 & +\Lambda_{\alpha\gamma}\sin\Phi_\alpha & +\Phi_{\alpha\gamma} \\ -\Lambda_{\alpha\gamma} & 1 & -\Lambda_{\alpha\gamma}\cos\Phi_\alpha \\ -\Phi_{\alpha\gamma} & +\Lambda_{\alpha\gamma}\cos\Phi_\alpha & 1 \end{bmatrix} = \begin{bmatrix} \overset{\alpha}{X}_{\gamma\alpha} \\ \overset{\alpha}{Y}_{\gamma\alpha} \\ \overset{\alpha}{Z}_{\gamma\alpha} \end{bmatrix} =$$

$$\begin{bmatrix} 1 & -3.65\cdot 10^{-6} & -12.12\cdot 10^{-6} \\ +3.65\cdot 10^{-6} & 1 & +3.19\cdot 10^{-6} \\ +12.12\cdot 10^{-6} & -3.19\cdot 10^{-6} & 1 \end{bmatrix} \begin{bmatrix} -80\,\text{m} \\ +20\,\text{m} \\ -20\,\text{m} \end{bmatrix}$$

End Box 10

Box 11: 25 m local network, misclosures

$$\overset{\beta}{X}_{\beta\gamma} = +50\,\text{m} - 0.22\,\text{mm},\ \overset{\gamma}{X}_{\gamma\alpha} = -80\,\text{m} + 0.17\,\text{mm}$$
$$\overset{\beta}{Y}_{\beta\gamma} = -50\text{m} - 0.23\,\text{mm},\ \overset{\gamma}{Y}_{\gamma\alpha} = +20\,\text{m} - 0.36\,\text{mm}$$
$$\overset{\beta}{Z}_{\beta\gamma} = +15\,\text{m} - 0.04\,\text{mm},\ \overset{\gamma}{Z}_{\gamma\alpha} = -20\,\text{m} - 1.03\,\text{mm}$$

$$\overset{\alpha}{X}_{\alpha\beta} + \overset{\beta}{X}_{\beta\gamma} + \gamma X_{\gamma\alpha} \neq 0 : -0.22\,\text{mm} + 0.17\,\text{mm} = -0.05\,\text{mm}$$
$$\overset{\alpha}{Y}_{\alpha\beta} + \overset{\beta}{Y}_{\beta\gamma} + \gamma Y_{\gamma\alpha} \neq 0 : -0.23\,\text{mm} - 0.35\,\text{mm} = -0.59\,\text{mm}$$
$$\overset{\alpha}{Z}_{\alpha\beta} + \overset{\beta}{Z}_{\beta\gamma} + \gamma Z_{\gamma\alpha} \neq 0 : -0.04\,\text{mm} - 1.03\,\text{mm} = -1.07\,\text{mm}$$

End Box 11

Have a look at the triangular network of the Fig. 21. Box 10 reviews the transformation of the frame of reference at the point P^β and P^γ to the frame of reference at out datum point P^α. In addition, Box 11 summarizes the misclosures of the order of under 1mm. For our second example, we start again with a triangular geodetic network between the points $\{P_\alpha, P_\beta, P_\gamma\}$ in the horizontal frame at the point P_α. The three holonomity conditions for our 500 m local network to zero, see Box 12. In addition, we assumed vertical deflection variation of the order of $2'' - 5''$ for longitude and latitude of approximate latitude $\Phi_\alpha = 48.782°$

Box 12: 500 m local network

$$\overset{\alpha}{X}_{\alpha\beta} = +500m,\ \overset{\alpha}{X}_{\beta\gamma} = +800m,\ \overset{\alpha}{X}_{\gamma\alpha} = -1300m,$$
$$\overset{\alpha}{Y}_{\alpha\beta} = +500m,\ \overset{\alpha}{Y}_{\beta\gamma} = -800m,\ \overset{\alpha}{Y}_{\gamma\alpha} = +300m,$$
$$\overset{\alpha}{Z}_{\alpha\beta} = +50m,\ \overset{\alpha}{Z}_{\beta\gamma} = +150m,\ \overset{\alpha}{Z}_{\gamma\alpha} = -200m,$$
$$\Lambda_{\alpha\beta} = 25'' \sim \ 12.12\times 10^{-5} RAD$$
$$\Phi_{\alpha\beta} = -15'' \sim -\ 7.27\times 10^{-5} RAD$$
$$\Lambda_{\alpha\gamma} = -15'' \sim -\ 7.27\times 10^{-5} RAD$$
$$\Phi_{\alpha\gamma} = -45'' \sim -\ 2.18\times 10^{-4} RAD$$
$$\Phi_\alpha = 48.788°$$

End Box 12

Box 13: 500 m local network, detailed computation

$$\begin{bmatrix} \overset{\beta}{X}_{\beta\gamma} \\ \overset{\beta}{Y}_{\beta\gamma} \\ \overset{\beta}{Z}_{\beta\gamma} \end{bmatrix} = \begin{bmatrix} 1 & +\Lambda_{\alpha\beta}\sin\Phi_\alpha & +\Phi_{\alpha\beta} \\ -\Lambda_{\alpha\beta} & 1 & -\Lambda_{\alpha\beta}\cos\Phi_\alpha \\ -\Phi_{\alpha\beta} & +\Lambda_{\alpha\beta}\cos\Phi_\alpha & 1 \end{bmatrix} = \begin{bmatrix} \overset{\alpha}{X}_{\beta\gamma} \\ \overset{\alpha}{Y}_{\beta\gamma} \\ \overset{\alpha}{Z}_{\beta\gamma} \end{bmatrix} =$$

$$\begin{bmatrix} 1 & +9.12\cdot 10^{-5} & -7.27\cdot 10^{-5} \\ -9.12\cdot 10^{-5} & 1 & -7.99\cdot 10^{-5} \\ +7.27\cdot 10^{-5} & +7.99\cdot 10^{-5} & 1 \end{bmatrix} \begin{bmatrix} +800\,\mathrm{m} \\ -800\,\mathrm{m} \\ +150\,\mathrm{m} \end{bmatrix}$$

$$\begin{bmatrix} \overset{\gamma}{X}_{\gamma\alpha} \\ \overset{\gamma}{Y}_{\gamma\alpha} \\ \overset{\gamma}{Z}_{\gamma\alpha} \end{bmatrix} = \begin{bmatrix} 1 & +\Lambda_{\alpha\gamma}\sin\Phi_\alpha & +\Phi_{\alpha\gamma} \\ -\Lambda_{\alpha\gamma} & 1 & -\Lambda_{\alpha\gamma}\cos\Phi_\alpha \\ -\Phi_{\alpha\gamma} & +\Lambda_{\alpha\gamma}\cos\Phi_\alpha & 1 \end{bmatrix} = \begin{bmatrix} \overset{\alpha}{X}_{\gamma\alpha} \\ \overset{\alpha}{Y}_{\gamma\alpha} \\ \overset{\alpha}{Z}_{\gamma\alpha} \end{bmatrix} =$$

$$\begin{bmatrix} 1 & -5.47\cdot 10^{-5} & -2.18\cdot 10^{-4} \\ +5.47\cdot 10^{-5} & 1 & +4.79\cdot 10^{-5} \\ +2.18\cdot 10^{-4} & -4.79\cdot 10^{-5} & 1 \end{bmatrix} \begin{bmatrix} -1300\,\mathrm{m} \\ +300\,\mathrm{m} \\ -200\,\mathrm{m} \end{bmatrix}$$

End Box 13

Box 14: 500 m local network, misclosures

$$\overset{\beta}{X}_{\beta\gamma} = +800\,\mathrm{m} - 83.9\,\mathrm{mm},\ \overset{\gamma}{X}_{\gamma\alpha} = -1300\,\mathrm{m} + 27.2\,\mathrm{mm}$$

$$\overset{\beta}{Y}_{\beta\gamma} = -800\,\mathrm{m} - 84.9\,\mathrm{mm},\ \overset{\gamma}{Y}_{\gamma\alpha} = +300\,\mathrm{m} - 80.7\,\mathrm{mm}$$

$$\overset{\beta}{Z}_{\beta\gamma} = +150\,\mathrm{m} - 5.8\,\mathrm{mm},\ \overset{\gamma}{Z}_{\gamma\alpha} = -200\,\mathrm{m} - 297.8\,\mathrm{mm}$$

$$\overset{\alpha}{X}_{\alpha\beta} + \overset{\beta}{X}_{\beta\gamma} + \overset{\gamma}{X}_{\gamma\alpha} \neq 0 : -83.9\,\mathrm{mm} + 27.2\,\mathrm{mm} = -56.7\,\mathrm{mm}$$

$$\overset{\alpha}{Y}_{\alpha\beta} + \overset{\beta}{Y}_{\beta\gamma} + \overset{\gamma}{Y}_{\gamma\alpha} \neq 0 : -84.9\,\mathrm{mm} - 80.7\,\mathrm{mm} = -165.6\,\mathrm{mm}$$

$$\overset{\alpha}{Z}_{\alpha\beta} + \overset{\beta}{Z}_{\beta\gamma} + \overset{\gamma}{Z}_{\gamma\alpha} \neq 0 : -5.8\,\mathrm{mm} - 297.8\,\mathrm{mm} = -303.6\,\mathrm{mm}$$

End Box 14

Box 13 reviews the transformation of the frame of reference at the points P^β and P^γ to the frame of reference at the datum point P^α. In addition, Box 14 summarizes the misclosure of the order 250 mm. The results speak for themselves. For observation in the range of kilometers (3rd order networks) these effects cannot be neglected.

5.3 *The Object of Anholonomity in Terrestrial Networks, Cartan Exterior Calculus, an Example*

Our example documents the anholonomity of a spherical coordinate system once we use an orthonormal frame of reference, a Cartan 2-leg, namely an three-dimensional Euclidean space and its embedding into a two-dimensional Riemann space. It is the embedding of 2-space within Euclidean geometry. Our point of view in extrinsic.

Box 15: Gauss surface geometry, Cartan surface geometry orthonormal frame of reference, example of the sphere

$$\mathbf{x}(u,v) = r\cos u\cos v\mathbf{e}_1 + r\sin u\cos v\mathbf{e}_2 + r\sin v\mathbf{e}_3$$
$$\mathbf{g}_1 = \frac{\partial \mathbf{x}}{\partial u} = r\cos v(-\mathbf{e}_1\sin u + \mathbf{e}_2\cos u)$$
$$\mathbf{g}_2 = \frac{\partial \mathbf{x}}{\partial v} = -r\sin v\cos u\mathbf{e}_1 - r\sin v\sin u\mathbf{e}_2 + r\cos v\mathbf{e}_3$$
$$\mathbf{g}_3 = \cos v(\mathbf{e}_1\cos u + \mathbf{e}_2\sin u) + \mathbf{e}_3\sin v$$
$$\|\mathbf{g}_1\| = r\cos v,\ \|\mathbf{g}_2\| = r,\ \|\mathbf{g}_3\| = 1,\ \langle g_1|g_2\rangle = 0$$
$$\mathbf{c}_1 := \frac{\mathbf{g}_1}{\|\mathbf{g}_1\|},\ \mathbf{c}_2 := \frac{\mathbf{g}_2}{\|\mathbf{g}_2\|},\ \mathbf{c}_3 := \mathbf{g}_3$$

End Box 15:

Box 16: Cartan's first structure equations, displacement vector of the surface of the sphere Gaussian frame of reference versus Cartan frame of reference, integrability

"directional equations of first kind"

$$\mathbf{dx} = \frac{\partial \mathbf{x}}{\partial u}du + \frac{\partial \mathbf{x}}{\partial v}dv = \mathbf{c}_1 r\cos v du + \mathbf{c}_2 r dv = \sigma^1\mathbf{c}_1 + \sigma^2\mathbf{c}_2$$
$$\begin{bmatrix}\sigma^1\\ \sigma^2\end{bmatrix} = \begin{bmatrix} r\cos v & 0\\ 0 & r\end{bmatrix}\begin{bmatrix} du\\ dv\end{bmatrix} = \begin{bmatrix} a & b\\ c & d\end{bmatrix}\begin{bmatrix} du\\ dv\end{bmatrix}$$
$$\begin{bmatrix} du\\ dv\end{bmatrix} = \begin{bmatrix} a & b\\ c & d\end{bmatrix}^{-1}\begin{bmatrix}\sigma^1\\ \sigma^2\end{bmatrix} = \begin{bmatrix} (r\cos v)^{-1} & 0\\ 0 & r^{-1}\end{bmatrix}\begin{bmatrix}\sigma^1\\ \sigma^2\end{bmatrix}$$
$$du = (r\cos v)^{-1}\sigma^1,\ dv = r^{-1}\sigma^2 \leftrightarrow \sigma^1 = r\cos v du,\ \sigma^2 = r dv$$

integrability?

$$\frac{\partial\sigma^1}{\partial v} - \frac{\partial\sigma^2}{\partial u} = -r\sin v \neq 0$$
$$\begin{bmatrix}\sigma^1\\ \sigma^2\end{bmatrix} =: \boldsymbol{\sigma},\quad \boldsymbol{du} := \begin{bmatrix} du\\ dv\end{bmatrix}$$
$$\text{if } \boldsymbol{\sigma} = A\,\boldsymbol{du}\ ,\quad \text{then } d_C\boldsymbol{\sigma} = d\boldsymbol{A}\wedge\boldsymbol{du}$$

End Box 16

First, we span the tangent space-here denoted $\{\mathbf{g}_1, \mathbf{g}_2\}$- as well as the normal space-here denoted $\{\mathbf{g}_3\}$ in honor of C.F. Gauss (Mathematician, Geodesist, Physicist). The $L_2 - norms$ $\|\mathbf{g}_1\|$ and $\|\mathbf{g}_1\|$ as well as the unweighted scalar product $< \mathbf{g}_1, \mathbf{g}_2 >$, in addition $\|\mathbf{g}_3\| = 1$ prove orthogonality of the vectors of the tangent space elements, but miss normality. That is done by defining the Cartan 2-frame denoted by $\{\mathbf{e}_1, \mathbf{e}_2\}$ which is orthonormal. We have collected the details in Box 15.

Second, in Box 16 we present Cartan's "derivational equations of the first kind". Once we denote longitude by the letter u and latitude by the letter v we are in a position to design an orthonormal 2-leg by introducing the base $\{w^1, w^2\}$ by the transformation $(du, dv) \longrightarrow (\omega^1, \omega^2)$ by the matrix. $\begin{bmatrix} a & b \\ c & d \end{bmatrix}$.$\{\omega^1, \omega^2\}$ are the elements of the Cartan 2-leg. Here the matrix elements are differential factors $(a\, b) = (a, 0)$, $(c, d) = (0, d)$ and $a = r \cos v, \quad d = r$ or $\begin{bmatrix} a & b \\ c & d \end{bmatrix}^{-1} = \begin{bmatrix} (r \cos v)^{-1} & 0 \\ 0 & r^{-1} \end{bmatrix}$

$$\boxed{\omega^1 = r \cos v du, \quad \omega^2 = r dv}$$

The differential forms $\{\omega^1, \omega^2\}$ the question:

are the differential forms integrable?
Can we interchange the differentiation order?

$$\frac{\partial \omega^1}{\partial v} - \frac{\partial \omega^2}{\partial u} = -r \sin v \neq 0 \ \text{ for } v \neq 0 \tag{5.18}$$

The result is no. Such a result is typical for anholonomity. Let us collect the differential form $\begin{bmatrix} \omega^1 \\ \omega^2 \end{bmatrix}$ is the column vector, in contrast $\begin{bmatrix} du \\ dv \end{bmatrix}$ the column vector du as well as the matrix

$$\begin{bmatrix} a & b \\ c & d \end{bmatrix}. \tag{5.19}$$

We can summarize the result in the Cartan derivative or differential:

$$\text{if } \boldsymbol{\omega} = \mathbf{A}\, d\boldsymbol{u}, \quad \text{then} \quad d\boldsymbol{\omega} = d\mathbf{A} \wedge d\boldsymbol{u} \tag{5.20}$$

Box 17: Cartan's first structure equation 1-differential forms, exterior calculus Cartan-derivative

"1-differential form"

$$\sigma^1 = a\, du + b\, dv = a_\alpha du^\alpha$$
$$\sigma^2 = c\, du + d\, dv = b_\alpha du^\alpha$$

"exterior" or E. Cartan-derivative

$$d_C \sigma^\alpha = \sum_{\beta,\gamma=1}^{2} \tfrac{1}{2!} \Omega^\alpha_{\beta\gamma} \sigma^\beta \wedge \sigma^\gamma \sum_{\beta<\gamma} \Omega^\alpha_{\beta\gamma} \sigma^\beta \wedge \sigma^\gamma$$

!anti-symmetry!

$$du^\alpha \wedge du^\beta = -du^\beta \wedge du^\alpha$$
$$\sigma^\alpha = a^\alpha_\beta du^\beta \sim \underline{\sigma} = A\, d\underline{u}$$

three-index-symbol

$$d_C \sigma^\alpha = \sum_{\beta,\gamma}^{2} \sigma^\beta \wedge \sigma^\gamma$$

$$d\sigma^1_C = \tfrac{1}{2} \Omega^1_{12} \sigma^1 \wedge \sigma^2 + \tfrac{1}{2} \Omega^1_{21} \sigma^2 \wedge \sigma^1 = \tfrac{1}{2} (\Omega^1_{12} - \Omega^1_{21}) \sigma^1 \wedge \sigma^2 = \Omega^1_{12} \sigma^1 \wedge \sigma^2$$

$$d\sigma_C^2 = \tfrac{1}{2}\Omega_{12}^2\sigma^1 \wedge \sigma^2 + \tfrac{1}{2}\Omega_{21}^2\sigma^2 \wedge \sigma^1 = \tfrac{1}{2}(\Omega_{12}^2 - \Omega_{21}^2)\sigma^1 \wedge \sigma^2 = \Omega_{12}^2\sigma^1 \wedge \sigma^2$$

$$\boxed{\sigma^1 \wedge \sigma^2 = -\sigma^2 \wedge \sigma^1}$$

comparison of coefficients

$$\Omega_{12}^1 = -\Omega_{21}^1 = \tfrac{1}{r}\tan v$$
$$\Omega_{12}^2 = -\Omega_{21}^2 = 0$$

assumption : $d_C d\underline{u} = 0$

proof : $d_C d\underline{u} = (\frac{\partial du}{\partial v} - \frac{\partial dv}{\partial v})dv \wedge du = 0$

$$d_C\underline{\sigma} = dA \wedge du = \begin{bmatrix} -r\sin v dv \wedge du \\ 0 \end{bmatrix}$$

$$d_C\sigma = \begin{bmatrix} -\frac{1}{r}\tan v\sigma^2 \wedge \sigma^1 \\ 0 \end{bmatrix} \neq 0$$

"summary"

$$\boxed{\begin{aligned} d\sigma_C^\alpha &= \sum_{\beta,\gamma=1}^{2} \Omega_{\beta\gamma}^\alpha \sigma^\beta \wedge \sigma^\gamma \\ &= \sum_{\beta<\gamma} \Omega_{\beta\gamma}^\alpha \sigma^\beta \wedge \sigma^\gamma \end{aligned}}$$

End Box 17

Finally, we dig into the definition of the Cartan derivative. We ask what is the measure of anholonomity or what is the three-index-symbol of anholonomity. Please, for a fast information consult Box 17 at this end. In the case of spherical coordinates of type longitude and latitude, we can represent the Cartan derivative by $d_C\omega = [-\frac{1}{r}\tan v \;\; w^2 \wedge w^1]$

6 Final Comments

At this end, we want to illustrate the intimate relation between Geodesy and Physics:

> Our target is the anholonomity problem in the classical and relativistic gravity field including numerical examples. It applies for relativistic Real Null Frames and coframes, applying for instance for GPS. It is a measure how well the assumed physical and geometric field fits a sphere or an ellipsoid or an irregular boundary of the planet Earth.

We come back to the recommendation of the International Union of Geodesy and Geophysics (IUGG) to adopt the International Reference Ellipsoid as early as 1924. The International Gravity Formula in terms of the axisymmetric, non-harmonic Somigliana–Pizzetti gravity field was accepted as early as 1930 and recommended by the International Association of Geodesy (IAG).

Ever since geodesists did hesitate to use it, but instead the spherical symmetric for 99% the spherical symmetric Newton's gravitational field which is harmonic outside the spherical Earth of constant radius. Series expansions to degree/order 7200 are nowadays a standard. Even at our conference on Relativistic Geodesy two speakers declared in public that Geodesy does not need Relativity.

As everyone can see it: the Planet Earth as all terrestrial planets rotates!

We recommend therefore to the International Association of Geodesy (IAG) to adopt the axisymmetric Kerr metric or its linear approximation officially. Perhaps our detailed report helps.

Recommendation to the IAG

Adopt the axisymmetric Kerr metric or its linear approximation Lense-Thirring for a rotating-gravitating Earth

We are relatively certain that only few Geodesists as well as Physicists understand the topic anholonomity. Simply speaking, it is the problem to decide what is the correct reference figure in Geometry Space or in Gravity Space. For instance, if the Earth were not rotating and has a radial symmetric gravity field, namely of potential type gm/r_0, the Earth could not have an anholonomic coordinate system. But for a real Earth it would be a central question if the mass distribution of the Earth or any other planet would depend on longitude and latitude, not only radial. The figure of the Earth would be best described by a dependence of radial, longitudinal as well as latitudinal coordinates. Such an Earth model would be free of anholonomity. At least, the celebrated axisymmetric Somigliana–Pizzetti and a series development in terms of degree/order 3600 or 7200 of its perturbed gravity field, of course, in terms of ellipsoidal harmonics, could solve many topographic problems. In particular, the disturbances are the proper key for solving global, regional, nearly local geodetic problems. It has to be realized that the disturbance potential, vertical deflections and the modules of the second derivatives of the anomalous gravity field and its higher derivatives as well as higher derivatives cause the effect of anholonomity: surveying engineers as well as the common geodesists are not familiar with the concept of holonomic and anholonomic coordinates, of the leg-calculus or "differential forms" applied to geodetic sciences. It is for this community that we list important geodetic contributions on this subject:

A. Marrussi [14, 18, 19, 36–42], B. Chovitz [43–45], E. Doukakis [46], E. Grafarend [15, 47–55], E. Grafarend and W. Kuehnel [5], E. Livieratos [56], F. Bocchio [57–63], F. Sanso [64], F. Sanso, and P. Yanicek [7], G. Ricci [65, 66], G. Ricci and T. Levi-Civita [67], H. Flanders [68], H. Weyl, [69, 70], J. A. Schouten [8, 71], J. D. Zund [1, 72–78], J. D. Zund and W. Moor, [79], J. D. Zund and J. M. Wilkes [80]. J. M. Wilkes, and J. D. Zund [81], M. Blagojevic ,J. Garecki,F.W. Hehl, and Y.N. Obukhov [3], M. Fujimoto and E. Grafarend [34], M. Hotine [20, 82–86], N. Grossman [28, 29, 87, 88], P. Defrise [89, 90], P. Defrise and E. Grafarend [16, 91], P. Holota and Z. Nadenik [92], P. Pizzetti [93–95], P. Stäckel [96, 97], P. Vanicek [98], S. Roberts [99, 100], Teunissen [2], V. Schwarze [101], W. Neutsch [102], Y. Georgiadou and E. Livieratos [103], Z. Nadenik [104],

Dedication

Our contribution is dedicated to late Josef D. Zund, mathematician, in particular for the book "Foundations of Differential Geodesy", Springer Verlag, Berlin-Heidelberg-New York 1994, namely for introducing to geodesists the notion of the general Leg Calculus in terms of anholonomic coordinates, better called differential forms in 2-,3-, and 4-dimensions, the Hotine-Marussi approach, so-called "Three Dimensional Geodesy", 5 basic forms. To read his chapter VII, The "Fundamental Theorem of Differential Geodesy", in particular the celebrated Pizzetti sphere, the Pizzetti inequality, the bifurcation process for coupling gravitation and rotation as well as the algebraic theory of the Marussi tensor of gravity gradients in chapter VIII, is a special highlight for geodesists and physicists. His chapter X on the hierarchy of geodetic reference systems is magic.

Acknowledgements C. Lämmerzahl and D. Pützfeld, /Bremen/ invited me to speak about anholonomity in the context of Relativistic Geodesy within the WE-Heraeus Seminar. Special thanks go to D. Pützfeld, F.W. Hehl/Cologne/ and H. Quevedo/Mexico City/ for their helpful comments. In addition, I am grateful for the support of J. Müller (Hanover) on the International Reference Ellipsoid and to S. M. Kopeikin (Columbia/Missouri) on studying Relativistic Equilibrium Figures and the relativistic theory of the Geoid. Last, but not least, I am grateful to M.A. Javaid(Stuttgart) for his expert typing.

References

1. J.D. Zund, *Foundations of Differential Geodesy* (Springer, Berlin, 1994)
2. P.J.G. Teunissen, Anholonomity when using the development method for the reduction of observations to the reference ellipsoid. Bull. Geod. **56**, 356–363 (1982)
3. M. Blagojevic, F.W. Hehl, J. Garecki, Y.N. Obukhov, Real null coframes in general relativity and GPS coordinates. Phys. Rev. **65**(4), 044018 (2002)
4. E.W. Grafarend, F.W. Krumm, *Map Projections: Cartographic Infromation Systems* (Springer, Berlin, 2006)
5. E.W. Grafarend, W. Kuehnel, A minimum atlas for the rotation group SO(3). J. Geomath. **2**(2011), 113–122 (2011)
6. E.W. Grafarend, A.A. Ardalan, Ellipsoidal geoidal undulations (ellipsoidal Bruns formula): case study. J. Geod. **75**, 544–552 (2001)
7. F. Sanso, P. Vanicek, The orthometric height and the holonomity problem. J. Geod. **80**, 225–232 (2006)
8. J.A. Schouten, *Ricci-Calculus*, 2nd edn. (Springer, Heidelberg, 1954), p. 127
9. R. Cushmann, H. Duistermaat, J. Sniatycki, *Geometry of Nonholonomically Constrained Systems* (World Scientific, Singapore, 2010), p. 404
10. S. Sternberg, *Curvature in Mathematical and Physics* (Dover Publication, Mineola, 2012)
11. F. Hehl, Der Spin und Torsion in der Allgemeinen Relativitätstheorie. Abh. Braunschweigische Wiss. Ges. **18**, 98 (1966)
12. F. Hehl, *Der Spin und Torsion in der Allgemeinen Relativitätstheorie*. Universität Clausthal, Habilitationsschrift Techn (1970)
13. F. Heh, E. Kröner, Über den Spin in der Allgemeinen Relativitätstheorie. Z. Phys. **187**, 478 (1965)

14. A. Marussi, Fondements de geometrie differentielle absolue du champ potential terrestre. Bull. Geod. **14**, 411–439 (1949)
15. E.W. Grafarend, Three dimensional geodesy I: the holonomity problem. Z. Vermesssungswesen **100**, 269–281 (1975)
16. P. Defrise, E.W. Grafarend, Torsion and anholonomity of geodetic frames. Bollettino di Geodesia e Scienze Affini **35**, 153–160 (1976)
17. M. Caputo, *The Gravity Field of the Earth* (Academic, New York, 1967)
18. A. Marussi, Natural reference systems and their reciprocals in geodesy, Technical report, Publ. T.J, Kukkamaki 70th Birthday, Publ. Finnish Geodetic Institute, Nr. 89, Helsinki (1979)
19. A. Marussi, *Intrinsic Geodesy (trans: Reilly WI)* (Springer, Berlin, 1985)
20. M. Hotine, in *Differential Geodesy (Edited with commentary by J. D. Zund)* (Springer, Berlin, 1991)
21. H. Moritz, The Hamiltonian structure of refraction and the gravity field. Manuscripta Geod. **20**, 52–60 (1994)
22. H. Goenner, E.W. Grafarend, R.J. You, Newton mechanics as geodesic flow on maupertuis' manifold; the local isometric embedding into flat spaces. Manuscripta Geod. **19**, 339–345 (1994)
23. E.W. Grafarend, Differential geometry of the gravity field. Manuscripta Geod. **11**, 29–37 (1986)
24. E. Hunziker, Lotlinienkrünunung und Projektion eines Punktes oder einer Strecke auf das Geoid, Schweiz. Z Vermessung. Kulturtechnik und Photogrammetrie **58**, 144–152 (1960)
25. J. Engels, E. Grafarend, The gravitational field of topographic/isostatic masses and the hypothesis of mass condensation. Surv. Geophys. **140**, 495–524 (1993)
26. J. Engels, E. Grafarend, P. Sorcik, The gravitational field of topographic-isostatic masses and the hypothesis of mass condensation II - the topographic-isostatic Geoid. Sun Geophys. **17**, 41–66 (1996)
27. E.W. Grafarend, J. Engels, P. Sorcik, The gravitational field of topographic/isostatic masses and the hypothesis of mass condensation. Part I and II, Technical report, Department of Geodesy, Stuttgart (1995)
28. N. Grossman, Holonomic measurables in geodesy. J. Geophys. Res. **79**, 689–694 (1974)
29. N. Grossman, The pull-back operation in physical geodesy and the behaviour of plumblines. Manuscripta Geod. **3**, 55–105 (1978)
30. I.I. Mueller, E. Grafarend, H.B. Papo, B. Richter, Investigations on the hierarchy of reference frames in geodesy and geodynamic, Technical Report 289, Department of Geodetic Science, The Ohio State University (1979)
31. I.I. Mueller, E. Grafarend, H.B. Papo, B. Richter, Concepts for reference frames in geodesy and geodynamics: the reference directions. Bull. Geod. **53**(289), 195–213 (1979)
32. E.W. Grafarend, Spacetime geodesy. Boll, di Geodesia e Scienze Affini **38**, 551–589 (1975)
33. E.W. Grafarend, Die Beobachtungsgleichungen der dreidimen-sionalen Geodasie im Geometrie- and Schwereraum, ein Beitrag zur operationellen Geodasie. Zeitschrift für Vermessungswesen **106**, 411–429 (1981)
34. M. Fujimoto, E. Grafarend, Spacetime coordinates in the geodetic reference frame. in *Relativity in Celestial Mechanics and Astronomy*, ed. by J. Kovalevsky, V.A. Brumberg (IAU, 1986), pp. 269–276
35. R. Penrose, W. Rindler, *Spinors and Space-Time*, vol. 1 (Cambridge University Press, Cambridge, 1987)
36. A. Marussi, Les principes de la geodesie intrinseque. Bull. Geod. **19**, 68–76 (1951)
37. A. Marussi, Fondamenti di geodesia intrinseca. Publicazioni della Commissione Geodetica Italiana, Ser. HI **7**, 1–47 (1951)
38. A. Marussi, Su alcune propriety integrali delle rappresen-tazioni conformi di superfici su superfici, rendiconti della classe di scienze fisiche, matematiche e naturali. Accademia Nazionale dei Lincei (Roma), Ser. VIII **10**, 307–310 (1951)
39. A. Marussi, La coordination des system geodesiques. Bull. Geod. **43**, 16–19 (1957)

40. A. Marussi, Dalla geodesia classica alla geodesia in tre dimensioni. Bollettini di Geodesia e Scienze Affini, anno **XVIII**, 485–495 (1959)
41. A. Marussi, The tidal field of a planet and the related intrinsic reference systems. Geophys. J. R. Astron. Soc. **56**, 409–417 (1979)
42. A. Marussi, Intrinsic geodesy (a revised and edited version of his 1952 lectures by J.D. Zund), Technical report, Report No. 390, Department of Geodetic Science and Surveying, The Ohio State University, Columbus (1988)
43. B.H. Chovitz. Hotine's mathematical geodesy, in *IV Symposium on Mathematical Geodesy* (1969), pp. 159–172
44. B.H. Chovitz, Generalized three-dimensional conformal transformations. Bull. Geod. **104**, 159–163 (1972)
45. B.H. Chovitz, The influence of Hotine's mathematical geodesy. Bollettino di Geodesia e Scienze Affini, anno **XLI**, 57–64 (1982)
46. E. Doukakis, Remark on time and reference frames. Bull. Geod. **81** (1978)
47. E.W. Grafarend, The object of anholonomity and a generalized Riemannian geometry for geodesy. Bollettino di Geofisica Teorica ed Applicata **13**, 241–253 (1971)
48. E.W. Grafarend, Three dimensional geodesy and gravity gradients, Technical report, Ohio State University, report no. 174, Columbus, Ohio, USA (1972)
49. E.W. Grafarend, Le theoreme de conservation de la courbure et la torsion or attempts at a unified theory of geodesy. Bull. Geod. **109**, 237–260 (1973)
50. E.W. Grafarend, Gravity gradients and three dimensional net adjustment without ellipsoidal reference, Technical report, The Ohio State University, Report No. 202, Columbus (1973)
51. E.W. Grafarend, in *Cartan frames and a foundation of Physical Geodesy, Methoden and Verfahren der Mathematischen Physik, Bd 12*, ed. by B. Brosowski, E. Martensen. BI-Verlag, Mathematical Geodesy, Mannheim (1975), pp. 179–208
52. E.W. Grafarend, Threedimensional geodesy iii: refraction. Bollettino di Geodesia e Scienze Affini **35**, 153–160 (1976)
53. E.W. Grafarend, Geodasie - Gausssche oder Cartansche Flaähengeometrie? Allgemeine Vermessungs-Nachrichten **4**, 139–150 (1977)
54. E.W. Grafarend, Der Einfluss der Lotrichtung auf lokale geodätische Netze. Z. Vermessungswesen **112**, 413–424 (1987)
55. E.W. Grafarend, Tensor algebra, linear algebra, multi-linear algebra, Technical report, 344 references, Department of Geodesy and Geoinformatics, Stuttgart University Stuttgart (2004)
56. E. Livieratos, On the geodetic singularity problem. Manuscripta Geod. **1**, 269–292 (1976)
57. F. Bocchio, Su alcune applicazioni di interesse geode-tico delle connessioni non simmetriche, Rendiconti della classe di scienze Fisiche, matematiche e naturali. Accademia Nazionale dei lincei (Roma) Ser. VIII **48**, 343–351 (1970)
58. F. Bocchio, From differential geodesy to differential geophysics. Geophys. J. R. Astron. Soc. **39**, 1–10 (1974)
59. F. Bocchio, The holonomity problem in geophysics. Bollettino di Geodesia e Scienze Affini, anno XXXIV, 453–459 (1975)
60. F. Bocchio, Some of Marussi's contributions in the field of two-dimensional and three dimensional representation. Bollettino di Geodesia e Scienze Affini, anno XXXVII, 441–450 (1978)
61. F. Bocchio, An inverse geodetic singularity problem. Geophys. J. R. Astron. Soc. **67**, 181–187 (1981)
62. F. Bocchio, Geodetic singularities in the gravity field of a non-homogenious planet. Geophys. J. R. Astron. Soc. **68**, 643–652 (1982)
63. F. Bocchio, Geodetic singularities, reviews of geophysics and space physics, in *Advances in Geodesy*, vol. 20, ed. by R.H. Rapp, E.W. Grafarend (American Geophysical Union, Washington, 1982), pp. 399–409
64. F. Sanso, The geodetic boundary value problem in gravity space, *Memorie Scienze Fisiche* (Academia Nationale dei Lincei, Roma, 1997)
65. G. Ricci, *Lezioni sulla teoria delle superficie* (F. Drucker, Verona-Padova, 1989)

66. G. Ricci, Opere, in *Edizioni Cremonese*, 2 (1956/1957)
67. G. Ricci, T. Levi-Civita, Methodes de calcul differentiel absolu et leurs applications. Math. Ann. **54**, 125–201 (1901)
68. H. Flanders, in *Differential Forms with Applications to the Physical Sciences*. (Academic, New York, 1963)
69. H. Weyl, *Raum-Zeit-Materie: Vorlesungen überiber allgemeine Relativitätstheorie*, 5th edn. (Springer, Berlin, 1918)
70. H. Weyl, *Gruppentheorie and Quantenmechanik* (Verlag S. Hirzel, Leipzig, 1928)
71. J.A. Schouten, *Tensor Analysis for Physicists* (Clarendon, Oxford, 1951)
72. J.D. Zund, Tensorial methods in classical differential geometry - i: basic principles. Tensor NS **47**, 74–82 (1988)
73. J.D. Zund, Tensorial methods in classical differential geometry - i: basic surface tensors. Tensor NS **47**, 83–92 (1988)
74. J.D. Zund, Differential geodesy of the Eötvös torsion balance. Manuscripta Geod. **14**, 13–18 (1989)
75. J.D. Zund, The assertion of hotine on the integrabil-ity conditions in his general coordinate system. Manuscripta Geod. **15**, 373–382 (1990)
76. J.D. Zund, An essay on the mathematical foundations of the Marussi-Hotine approach to geodesy. Bollettino di Geodesia e Scienze Affini anno XLIX, 113–179 (1990)
77. J.D. Zund, The Hotine problem in differential geodesy. Manuscripta Geod. **15**, 373–382 (1990)
78. J.D. Zund, The mathematical foundations of the hotine-marussi approach to geodesy. Bollettino di Geodesia e Scienze Affini anno LI, 125–138 (1992)
79. J.D. Zund, W. Moore, Hotine's conjecture in differential geodesy. Bull. Goodesique **61**, 209–222 (1987)
80. J.D. Zund, J.M. Wilkes, The significance and generalization of two transformation formulas in Hotine's mathematical geodesy. *Bollettino di Geodesia e Scienze Affini anno XLVII*, 77–85 (1998)
81. J.M. Wilkes, J.D. Zund, Group-theoretical approach to the Schwarzschild solution. Am. J. Phys. **50**, 25–27 (1982)
82. M. Hotine, Metrical properties of the Earth's gravitational field, report to i.a.g, Technical report, Toronto Assembly 33-64 of Hotine (1957)
83. M. Hotine, Geodesic coordinate systems, Technical report, Venice Symposium, 65-89 of Hotine (1957)
84. M. Hotine, A primer on non-classical geodesy, Technical report, Venice Symposium, 91–130 (1959)
85. M. Hotine, The orthomorphic projection of the spheroid. Emp. Surv. Rev. **8**, 300–311 (1967)
86. M. Hotine, Mathematical geodesy, U.S. Department of Commerce Washington, D.C. (1969)
87. N. Grossman, Is the geoid a trapped surface? Bollettino di Geodesiae Scienze Affini anno XXXIV, 173–183 (1978)
88. N. Grossman, The nature of space near the Earth. Bollettino di Geodesia e Scienze Affini anno XXXV, 413–424 (1979)
89. P. Defrise, Meteorologie et geometrie differentiae. Institute Royal Moteorologique de Belgique, Bruxelles, A 91 (1975)
90. P. Defrise, Sur des applications de la geometrie differentielle en Meteorologie et en Goodesie. Bollettino di Geodesia e Scienze Affini **37**, 185–196 (1978)
91. P. Defrise, E.W. Grafarend, Torsion and Anholonomity of geodetic frames. Bollettino di Geodesia e Scienze Affini **35**, 81–92 (1976)
92. P. Holota, Z. Nadenik, Les formes differentielles exterieures dans la Geodesie ii: Courbure moyenne. Studia geoph. at geod. **15**, 106–112 (1971)
93. P. Pizzetti, Un principio fondamentale nello studio delle studio delle superfici di livello terrestri. Rendiconti della Reale Accademia dei Lincei (Roma) **10**, 35–39 (1901)
94. P. Pizzetti. *Höher Geodäsie, Enzyklopädie der Mathematischen Wissenschaften Band VI*. Geodasie und Geophysik, 125–239, B.G. Teubner, Leipzig, Erster Teil, 1906
95. P. Pizzetti, *Principii della teoria mecannica della figura dei pianeti* (E. Spoerri, Pisa, 1913)

96. P. Stäckel, *Über die integration der Hamilton-Jacobischen Differentialgleichung mittelst Separation der Variablen* (Habilitationsschrift, Halle, 1891)
97. P. Stäckel, Über die Bewegung eines Punktes in einer n-fachen Mannigfaltigkeit. Math. Ann. **42**, 537–563 (1893)
98. P. Vanicek, To the problem of holonomity of height systems in: *Letter to the Editor*, vol 36. The Canadian Surveos (1982), pp. 122–123
99. S. Roberts, On the parallel surfaces of conicoids and conics. Proc. Lond. Math. Soc. **1**(4), 57–91 (1872)
100. S. Roberts, On parallel surfaces. Proc. Lond. Math. Soc. **1**(4), 218–235 (1973)
101. V. Schwarze, Satellitengeodätische Positionierung in der Relativistischen Raum-Zeit. Ph.D. Thesis, Deutsche Geodätische Kommission, Bayerische Akademie der Wissenschaften, München (1995)
102. W. Neutsch, *Koordinaten-Theorie und Anwendungen* (Spektrum Akademischer Varlag, Heidelberg, 1995)
103. Y. Georgiadou, E. Livieratos, Anholonomity of the reciprocal natural frame in the normal gravity field of the Earth. Geophys. J. R. Astron. Soc **67**, 177–179 (1981)
104. Z. Nadenik, Les forms differentielles exterieures dans la Geodesie i: coubure de Gauss. Studia Geoph. et Geod. **15**, 1–6 (1971)

Epistemic Relativity: An Experimental Approach to Physics

Bartolomé Coll

Abstract The recent concept of *relativistic positioning system* (RPS) has opened the possibility of making Relativity the *general standard frame* in which to state any physical problem, theoretical or experimental. Because the velocity of propagation of the information is finite, *epistemic relativity* proposes to integrate the physicist as a truly component of every physical problem, taking into account explicitly *what* information, *when* and *where*, the physicist is able to know. This leads naturally to the concept of *relativistic stereometric system* (RSS), allowing to measure the intrinsic properties of physical systems. Together, RPSs and RSSs complete the notion of *laboratory* in general relativity, allowing to perform experiments in finite regions of any space-time. Epistemic relativity incites the development of relativity in new open directions: advanced studies in RPSs and RSSs, intrinsic characterization of gravitational fields, composition laws for them, construction of a finite-differential geometry adapted to RPSs and RSSs, covariant approximation methods, etc. Some of these directions are sketched here, and some open problems are posed.

1 Introduction

As everyone knows, General Relativity is a theory of the gravitational field but, also and primarily, a theory of the space-time. During the last century, it has been largely proved experimentally that the gravitational field, as well as the space-time, is better described by it than by Newtonian theory.

Nevertheless, by natural reasons, the environments in which most of the experiments have taken place have been either static (e.g. Earth surface), or stationary (e.g.

I want the organizers of this seminar, Dirk Puetzfeld and Claus Lämmerzahl, to know how much I appreciated their inviting me to talk about my ideas on this subject. It is also a pleasure to thanks the Wilhelm und Else Heraeus-Stiftung for their kind hospitality.

B. Coll (✉)
Relativistic Positioning Systems, Department of Astronomy & Astrophysics, University of Valencia, c/ Dr. Moliner 50, 46100 Burjassot, Valencia, Spain
e-mail: bartolome.coll@uv.es

D. Puetzfeld and C. Lämmerzahl (eds.), *Relativistic Geodesy*, Fundamental Theories of Physics 196, https://doi.org/10.1007/978-3-030-11500-5_8

aircrafts or satellites in circular orbits), or periodic (e.g. satellites in elliptic orbits), or quasi-periodic with almost-periodic or constant non-periodic parts (e.g. Solar system ephemeris or radiation by black hole accretion disks). These environments allow to trivialize some important specificities of the relativistic space-time, especially some related to the finite velocity of propagation of the information.

Thus, for example, in these particular environments, the deterministic character of Einstein equations seems to be *predictive*, like the (Laplace) determinism of Newtonian theory. But out of these environments, because Einstein equations can be associated to a hyperbolic system[1] with influence domain[2] attached to the velocity of light, the initial data on an initial instant cannot be *physically* known but at or after the cusp of the influence domain, so that Einstein determinism is generically *retrodictive*,[3] not predictive. Nevertheless, most of the problems related to initial conditions in relativity, even for generic environments, have been thought as predictive ones, as if someone at the initial instant were able to know the initial data. This manner of thinking, strongly attached to the evolutive Newtonian point of view[4] but physically inadmissible in relativistic generic environments, masks the concepts that we should develop in order to be able to *describe physically generic environments in relativity*.

The aim of *epistemic relativity* is to give us such ability.[5] Section 2 is devoted to specify its basic ingredients. The first of these ingredients is constituted by *relativistic positioning systems*,[6] of which *auto-locating systems* and *autonomous positioning systems* are the more interesting ones. They are the object of Sect. 3. The second basic ingredient is constituted by *relativistic stereometric systems*,[7] which allow to detect intrinsic properties of physical systems. Their notion and first physical results are presented in Sect. 4. By the way, epistemic relativity plays a heuristic role giving new meanings to already known subjects or opening new ones. Section 5 shows how the *intrinsic characterization* of metrics may help epistemic relativity to

[1]The differential operator associated to the $n(n+1)/2$ Einstein equations is not hyperbolic, but degenerate. Nevertheless, because n of these equations (constraint equations) are involutive with respect to the other $n(n-1)/2$ ones (evolution equations), one can supplement this last system with n suitable additional equations (coordinate conditions). It is the differential operator of this new system of $n(n+1)/2$ equations for the metric coefficients which may be made hyperbolic with a suitable choice of coordinate conditions (for example, with harmonic ones).

[2]The influence domain for an initial instant (local spatial hypersurface) is the domain where the solution exists and is unique for every initial (Cauchy) data on this initial instant.

[3]Retrodictive, as opposed to predictive, means here that one cannot but *verify afterwards* that the physical quantities measured *in* the influence domain agree with the initial data received *at or after* the cusp of this domain.

[4]In Newtonian physics, the contents of the three-dimensional space at every instant is supposed known or knowable at that instant, so that the evolutive equations describe the dependence in time of this three-dimensional contents.

[5]Epistemic relativity was first presented at the GraviMAS FEST workshop, in honor of Lluís Mas, Mallorca, Spain, 2008, http://www.uib.es/depart/dfs/GRG/GraviMAS_FEST/. See also [1].

[6]For the genesis of the concept of relativistic positioning systems, see for example [1].

[7]They were relativistic stereometric systems which, joined with relativistic positioning systems, suggested the idea of epistemic relativity. This is why relativistic stereometry and epistemic relativity were conceived conjointly.

identify gravitational fields and, finally, Sect. 6 develops a *finite-differential geometry* in which distance function and metric ought to play symmetric roles.

I believe that in epistemic theory, as an incipient theory, what is interesting are the basic concepts. This is why I have restricted the bibliography to them.[8]

2 Epistemic Relativity

Being the best theory of the space-time in which all the physical phenomena take place, relativity ought to be able to describe any physical experiment in terms of its proper, relativistic, concepts, regardless of the quantitative evaluation of the experiment for which, in many cases, Newtonian calculations could suffice.

Also, being the best theory of the gravitational field, relativity ought to propose experiences and methods of measurement of general gravitational fields.

But today, such descriptions or proposals are conspicuous by their absence. In fact, we do not know how to do them. Thus it seems evident that *relativity needs to develop a proper experimental approach to the physical world.*

2.1 The Notion of Epistemic Relativity

Fortunately, we already know the conceptual basic ingredients for such a development. Before to describe them, the idea of a 'relativistic experimental approach' needs to be more explicit.

In relativity, a large number of scientific works analyze physical and geometrical properties of the space-time, but

- do not integrate the physicist as a part of them, and
- forget implicitly that
  - information is energy,
  - neither the density of energy, nor its velocity of propagation can be infinite in relativity.

Many of these properties of the space-time may be analyzed by a geometer on his desk, but to be known by an experimental physicist,[9] they would require qualities

[8] Some subjects related to epistemic relativity have been the object of interesting developments (e.g. positioning systems, relativistic geometric optics, intrinsic characterization of non-vacuum metrics, Regge calculus) not directly related to the basis of this theory. For this reason the corresponding bibliography is absent here.

[9] In all this text, the worlds 'physicist', 'observer' or 'user' denote any person or device able to receive the pertinent information, to record and to analyze it and to perform the actions and computations needed for the problem in question. For short, we shall refer to any of them as 'it'.

of an omniscient[10] god ! For these reasons, we shall say that such works belong to *ontic relativity*.[11]

Ontic relativity is necessary to develop a relativistic experimental approach (and even sufficient in static, stationary, periodic or almost periodic gravitational fields!), but it is also generically *insufficient*.

Let us remember that the statement of any problem in general relativity implies:

- to describe it directly in the space-time, by means of intrinsic relativistic objects, i.e. by means of worldlines and worldtubes (the analogs to the histories or evolutions of the instantaneous objects of Newtonian theory),
- to banish generically the use of any physical extended present (due to high precision clocks or regions of non-constant inertial or gravitational fields).

The works in relativity that, in addition:

- integrate the physicist as an element of the problem considered,
- concern physical properties that the physicist can measure, and
- take into account explicitly what information, when and where, the physicist is able to know,

will be considered as characterizing *epistemic relativity*.[12]

Paradoxically, until now a very few number of relativistic problems have been solved explicitly under these assumptions.

2.2 The Ingredients of Epistemic Relativity

The main objective of epistemic relativity is to provide the physicist with the *knowledge* and *protocols* necessary to make relativistic gravimetry (chronometry in arbitrary directions) in any *unknown* space-time environment (Fig. 1).

This is the first and unavoidable step to develop experimental relativity as the *natural scientific approach to the physical world.*

The basis of relativistic gravimetry are Einstein equations, and these equations need a precise mathematical model of the space-time. The adequacy of this mathematical model and the physical space-time that it describes implies a one-to-one correspondence between the points of the model and the physical events they describe. And because in the mathematical model points are precisely identified (by their coordinates), we need to know how to locate the events in the physical space-time.

What else do we need in epistemic relativity? In fact, what we need is to transform the space-time region of physical interest in a (finite) laboratory.

But, what is a laboratory in relativity? In fact, a reflection on this matter shows that any laboratory, regardless of the specificity of its measurement devices, has to provide us with:

---

[10]Remember that relativity is retrodictive.

[11]From the Greek 'ontos', *being*, with the meaning of 'what is' as opposed to 'how one knows it'.

[12]From the Greek 'episteme', *knowledge*, with the meaning of 'how we obtain it'.

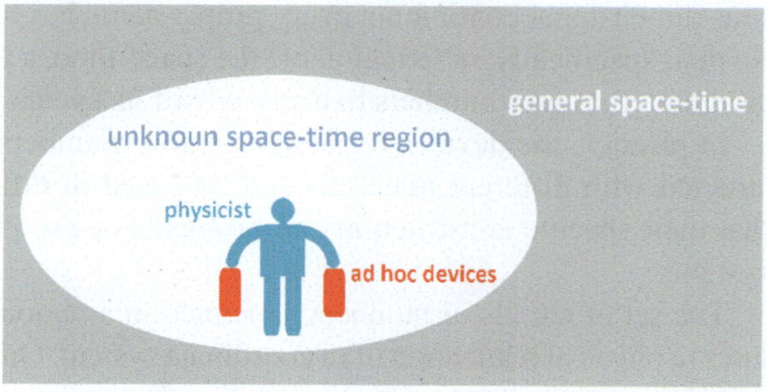

Fig. 1 Epistemic relativity wants to provide the physicist with the baggage and protocols necessary to make gravimetry in any unknown region of arbitrary space-times

- a precise physical location of the significant parts of the system, and
- a precise physical description of its intrinsic properties.

The devices able to carry out these two tasks are called *relativistic positioning systems* and *relativistic stereometric systems* respectively. Thus:

A finite *laboratory* in relativity is a *space-time region* endowed with:

- a relativistic positioning system and
- a relativistic stereometric system.

The first task of epistemic relativity is to transform the finite regions of interest of the space-time in laboratories or, equivalently, to construct these two relativistic systems.

These two relativistic systems are sufficient to describe any experiment in their region of space-time. Remember that, as already mentioned, we refer to physical systems as *relativistic objects*, i.e. as worldlines or worldtubes so that they naturally contain the changes of state that the Newtonian physical systems at every instant may suffer during an experiment (they already are the histories or evolutions of Newtonian systems). In the same point of view, the physical properties of relativistic objects cannot but be the *history* of the physical properties of the Newtonian objects at every instant of its constituent events. It is thus clear that the two above functions of locating and detecting are effectively able to describe any physical experiment in the space-time domain in question.

3 Relativistic Positioning Systems

3.1 What is to Locate

In general, to *locate* an object is to *identify the place* it takes up. To locate with accuracy objects of any size (avoiding vagueness such as near of, inside of, etc.), we need to assign a proper name to every event of the space in question. A *locating system* is a system of assignation of proper names to the events of a space.

In particular, because the space-time is a continuum, the assignation of a proper name to every one of its events has to be done with numbers. And because it is a four-dimensional continuum, every proper name has to be done with four numbers. So that, specifically, in (a region of) the space-time, a locating system is a system of assignation of four numbers to every one of its events.

In physics, the devices for assigning four numbers to every event may be constructed with different materials and be based in different physical properties, so that their specific construction and protocols of use give raise to different locating systems.

The set of tetrads of numbers generated in a region by a locating system constitutes (a physical realization of) a coordinate system. On the other hand, in topological manifolds, a coordinate system in a region $\mathcal{R}$ may be extracted from a local chart ($\mathcal{R}$, φ) by means of the coordinate functions associated to φ, where φ is a homeomorphism from $\mathcal{R}$ to the associated linear space. We see that, mathematically, locating systems are represented by homeomorphisms φ. Thus, a locating system in a finite region $\mathcal{R}$ of the space-time, may be seen as the physical construction of a local chart or, conversely, a local chart may be seen as a mathematical idealization of a locating system.[13]

3.2 *Properties of Locating Systems*

According to the characteristics of the assignation system of coordinates, a locating system may be *active* if its assignation system only operates for events that emit signals of presence, or *passive* if it operates for all events of the region irrespective of their emission state; and it may be *immediate* if the values of the coordinates of every event are obtained without delay, or *retarded* otherwise. Also, according to its use, a locating system is said *real* if its assignation system starts one for all in all the region, and it is said *virtual* if it is used case by case to obtain only the specific coordinates of particular events.

Another classification takes into account the *function* allocated to the locating system. Two locating systems are of particular importance for us: *reference systems*, which allow *one* observer to know the coordinates of the events of the region and *positioning systems*, which allow every event of the region to know its proper coordinates (Fig. 2).

It is then evident that real locating systems may be immediate and passive, meanwhile virtual locating systems are necessarily active and retarded. And, taking into account that in relativity information propagates at finite velocity, it result that *all* reference systems are retarded.

[13] It is to be noted, because frequently forgotten, that the local charts on a differentiable manifold may be *structural*, i.e. belonging to the atlas defining its differentiable structure, or *not*. Only in the first case, the region $\mathcal{R}$ has to be an open set. Nonstructural local charts may be of lesser differentiable class than structural ones. In particular, they may be simply continuous, although in this case, the natural frame being absent, one may be lead to complete the local chart with an independent field of vector tetrads in order to construct a basis for the tensor algebra.

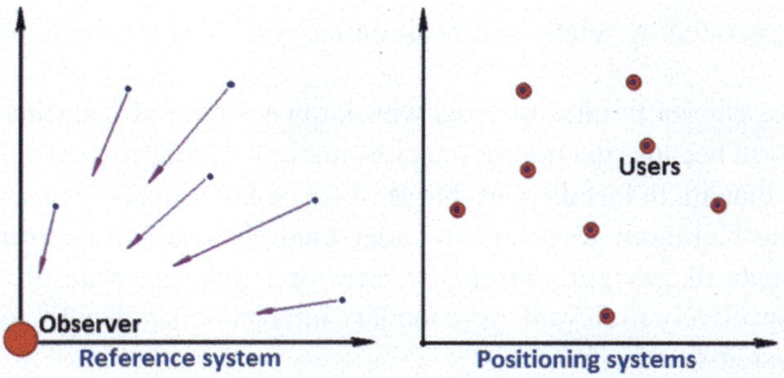

Fig. 2 A reference system allows one observer to know the coordinates of the events of the region. A positioning system allows every event of the region to know its proper coordinates

We can conclude that, when it is possible to construct them, real positioning systems are the locating systems that offer the best physical performances.

3.3 Relativistic Positioning Systems

We have just seen that, among all locating systems, the real positioning systems offer the performances of being passive and immediate. These performances are still insufficient for us, because the main objective of epistemic relativity is to make relativistic gravimetry in an *unknown* space-time environment.

Because the space-time environment is unknown, the locating system has also to be *generic*, that is to say, able to be constructed in regions of *any* space-time.[14] And because the main objective is to make relativistic gravimetry, it has also to be (gravitation-) *free*, i.e. able to be constructed without the previous knowledge of the space-time metric.

Relativistic positioning systems are the real, generic and free positioning systems of the space-time.

The construction of relativistic positioning systems is unexpectedly simple. A clock may be seen as a *continuous generator of numbers*, namely the time that it displays at every instant. Broadcasting this time by means of an electromagnetic signal, every event receiving the signal will know this number at the instant of reception. And because the proper name of every event of the space-time consists of four numbers (its coordinates), four clocks arbitrarily distributed will provide every event reached by the four signals with such numbers. It is thus easy to see that: Four clocks broadcasting their times constitute a relativistic positioning system[15] (see [3]). The

[14]It is understood that such regions have the appropriate accessibility characteristics to make gravimetry.

[15]Generically. Some distributions of clocks may associate same coordinates to different events, whatever the region considered (see [2]).

coordinates generated by relativistic positioning systems are called *emission coordinates* [4].

Note: I have chosen to introduce the well-known notion of coordinate system in this unusual form because the traditional presentations induce to give too importance to ingredients that are in fact dispensable or of secondary interest, but wrongly make many people to feel them necessary to understand physically a coordinate system. Thus, ingredients like *origin*, *coordinate lines* or *synchronization* of a coordinate system are respectively irrelevant, of secondary interest or inexistent[16] for relativistic positioning systems.

It is worthwhile to remark that relativistic positioning systems are a particular class of positioning systems but that, although similar in some aspects to positioning systems like GNSSs (global navigation satellite systems), differ clearly from them, whatever be the relativistic formalism with which these last ones are analyzed. These GNSS are *technical* objects, wonder technical objects, with a lot of technical, social and even scientific applications. But they are not *scientific objects*. The reason is that they have been thought, calculated and constructed using consciously a defective theory, the Newtonian one, to construct a *Newtonian space* in the region between the Earth surface and the constellation of satellites, in which the International Atomic Time on the Earth surface is imposed on all the region as a Newtonian absolute time. The "relativistic effects"[17] are not used to improve the approximate Newtonian calculations and to obtain more precise values of the *physical* times involved, but to deform the values of these physical times so as to mimic an unphysical Newtonian time everywhere (the calculated "relativistic effects" are *subtracted* to the physical values, not *added* to the Newtonian calculations of them!).

The above comment is *not at all* a criticism of GNSSs: as technical objects, they have to be useful to us and give us what we ask them to give. For relational (technical, social) convenience, it is the case, in particular, of a universal time for all of us, be we on the Earth surface or at 15.000 km height, at rest or at whatever velocity. But, I think that it would be better to have, first of all, a scientific object, a RGNSS (Relativistic GNSS) able to give us our proper physical time, value of the gravitational field or local distance, whatever our position and velocity with respect to the Earth surface. The reason is that, starting from such a scientific object, a simple software would be able also to mimic GNSSs with their absolute time everywhere. The (theoretical) question is not so much to improve GNSSs, but instead to construct cutting-edge scientific RGNSSs.

There are no specific coordinates associated to GNSSs.[18] Nevertheless, emission coordinates are an important tool in the study of relativistic positioning systems and in particular in RGNSS. Among others, they suggest new questions, provide new

---

[16]Or fourfold. Generically there does not exist a privileged synchronization of the form time = *constant* in relativistic positioning systems, but emission coordinates being constituted by four times τ^A, one could say that there exist four synchronizations $\tau^A = constant$.

[17]Euphemism for "Newtonian defaults".

[18]The pseudo-ranges are only considered as parameters able to calculate conventional coordinates on the Earth surface (WGS84, ITRS or others). The appellation 'GPS coordinates' refers to these conventional coordinates provided by the GPS.

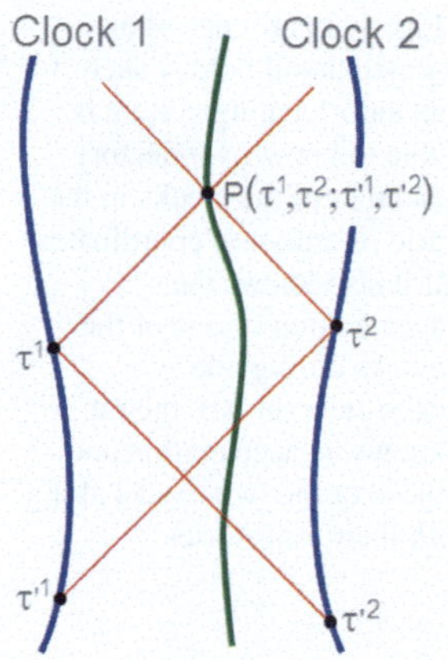

Fig. 3 An auto-locating system allows any user to know its proper trajectory in emission coordinates, but also the trajectories of the clocks of the system (here a two-dimensional scheme)

results and allow the constellation of satellites of a RGNSS to generate *autonomous* positioning systems for the Earth.

3.4 Auto-Locating Systems

As an intermediate step, *auto-locating systems* are those relativistic positioning systems whose clocks are endowed with a transponder broadcasting the times they receive from the other clocks of the system (see [3, 5] and references therein). So, every event P of the region receives the sixteen times τ^A and τ^{AB} $A, B = 1, \ldots, 4$, $A \neq B$. The four (τ^A) are the emission coordinates of P, meanwhile every one of the four sets A of four times (τ^A, τ^{AB}) are the emission coordinates of the clocks A of the system (see Fig. 3). Thus, any user that records these data is able to know its trajectory as well as the trajectories of the clocks in the *grid* of emission coordinates.[19]

An *echo-interval* for the clock A due to the clock B is the segment of world line of A from the instant of emission of a broadcast signal by A to the instant that it receives that signal echoed by B. In two dimensions, it is already proved a very interesting theorem for auto-locating systems (see [5]):

Theorem 3.4.1 *In a flat two-dimensional space-time, consider a user of an auto-locating system that knows its trajectory and those of the two clocks in the grid. If it also knows the acceleration of one of the clocks during one echo-interval, then it knows its acceleration and those of the two clocks along all their known trajectories (Fig. 4).*

[19]The grid $\mathcal{G}$ of emission coordinates is the Cartesian product of the segments $[\tau^A]$ of the times broadcast by the relativistic positioning system: $\mathcal{G} \equiv [\tau^1] \times [\tau^2] \times [\tau^3] \times [\tau^4]$.

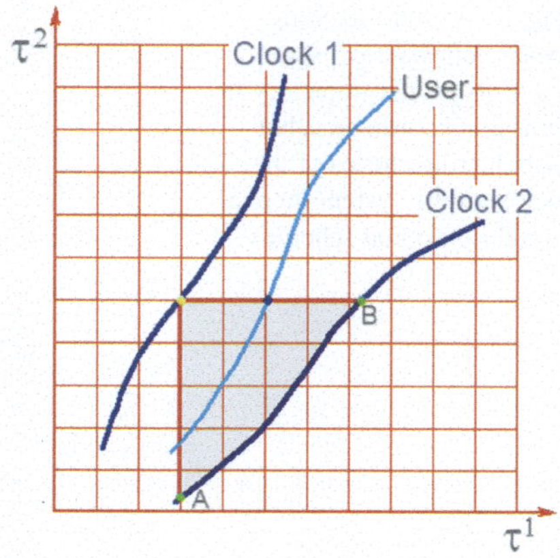

Fig. 4 In absence of a gravitational field, a user of an auto-locating system is able to know its trajectory and that of the clocks in the grid of emission coordinates. If it also knows the acceleration of one of the clocks during one echo-interval AB, then it knows its acceleration and those of the two clocks along *all* their trajectories

This theorem is very interesting because of the following fact: the algorithm that proves it in absence of a gravitational field may be as well applied in presence of it. Of course, in this case it will provide a wrong acceleration along all the three trajectories excepting on the starting echo-interval of the clocks. But the importance of this fact is that the difference between these wrong accelerations (calculated as if the gravitational field were absent) and the true accelerations (measured by accelerometers along the world-lines of clocks and user in the gravitational field), is *all the dynamical effect that the gravitational field produces on the auto-locating system and the user*. Thus, it is a *complete*, although *relative* (to the choice of echo-interval), gravimetric measure.

Unfortunately, this fundamental Theorem 3.4.1 is only know in two-dimensional space-times. It is of crucial importance for the theory of relativistic positioning systems and for all its near-future applications (Earth surface, RNGSS, Deep Space Navigation) to have its four-dimensional generalization. But this problem remains open. As in the two-dimensional case, the starting point is

- the explicit knowledge of the coordinate transformation between the emission coordinates of four arbitrarily accelerated clocks and inertial coordinates, and
- the explicit knowledge of the metric in emission coordinates.

Fortunately, the solution to these problems is already known. If the world-lines $\gamma_A(\tau^A)$, $A = 1, \ldots, 4$, of four emitters with respect to inertial coordinates are known, we can evaluate the quantities

$$e_a \equiv \gamma_a - \gamma_4 \quad , \quad \Omega_a \equiv \frac{1}{2}(e_a)^2 \quad , \quad \chi \equiv *(e_1 \wedge e_2 \wedge e_3)$$

$$H \equiv *(\Omega_1\, e_2 \wedge e_3 + \Omega_2\, e_3 \wedge e_1 + \Omega_3\, e_1 \wedge e_2) \quad , \quad y_* \equiv \frac{1}{\xi \cdot \chi}\, i(\xi) H$$

where ξ is an arbitrary transversal vector, $\xi.\chi \neq 0$, and $a = 1, 2, 3$. Then, the answer to the first item is (see [2] and [6]):

Theorem 3.4.2 *Suppose known the world-lines $x = \gamma_A(\tau^A)$ of the four emitters of a positioning system with respect to an inertial coordinate system $\{x\}$, and let $\{\tau^A\}$ be their emission coordinates. In term of the quantities χ and y_* evaluated from the $\gamma_A(\tau^A)$'s, the coordinate transformation $x = \kappa(\tau^A)$ between emission and inertial coordinates is given by:*

$$x \equiv \kappa(\tau^R) = \gamma_4 + y_* + \frac{y_*^2\,\chi}{(y_* \cdot \chi) + \hat{\varepsilon}\sqrt{(y_* \cdot \chi)^2 - y_*^2\,\chi^2}}$$

where $\hat{\varepsilon}$ is the orientation of the position system with respect to the user.

And the answer to the second item is:

Theorem 3.4.3 *In terms of the above transformation $x = \kappa(\tau^R)$ and the world-lines $x = \gamma_A(\tau^A)$, the contravariant components of the metric, $g^{AB}(\tau^R)$, in the emission coordinates $\{\tau^R\}$ are given by:*

$$g^{AB}(\tau^R) = \frac{\Omega_{AB}}{\mu_A\,\mu_B} \quad ,$$

where

$$\Omega_{AB} \equiv \frac{1}{2}(\gamma_A - \gamma_B)^2 \quad , \quad \mu_A \equiv (\kappa - \gamma_A)\,.\,\dot{\gamma}_A \quad .$$

3.5 *Autonomous Positioning Systems*

Auto-locating systems contain *all* what a coordinate system needs in order to be drawn in a space in which a field of light-cones is given.[20] But if the light-cones are specifically related to a geometry (here the gravitational field represented by a metric), and we want to identify individually the coordinate system making in it gravimetry (measure of the metric), we need additional information. This information has to be also broadcast by the clocks and consists of:

[20]The field of light-cones defines all the possible coordinate hypersurfaces of the coordinate systems associated to all the possible relativistic positioning systems. Accordingly, the particular trajectories of the clocks select the specific ones for the auto-locating system and, consequently, its region of validity or domain of definition in the grid.

Fig. 5 An observer, determined by its unit velocity, projects its past light-cone on its celestial sphere

- dynamical data of the satellites (acceleration, gradiometry),
- observational data from them (e.g. particular masses, directions of reference quasars or pulsars, etc.), and
- gravitational knowledge of the coordinate region (theoretical, experimental or mixed).

This information constitutes the *autonomous data. Autonomous positioning systems* are auto-locating systems that broadcast autonomous data.

Autonomous positioning systems are, generically, the best locating systems we are able to construct. Their notion was already proposed in the first paper on positioning systems (see [3]).

4 Relativistic Stereometric Systems

4.1 The Notion

We already know that, to transform a finite region of the space-time in a laboratory, we need to endow it with, on one hand, a relativistic positioning system, to locate in the best form the significant parts of the physical system in question, and on the other hand, a relativistic stereometric system, to obtain their intrinsic physical properties.

Similarly to a relativistic positioning system, that consists essentially of four clocks broadcasting their times,[21] *a relativistic stereometric system consists of a system of four observers receiving the signals broadcast from a physical system.*

In relativity, the word 'observer' is used in different contexts with different meanings. Here an *observer* means an *eye* able to record and to analyze the input, and an eye[22] is a small (physically *local*, mathematically *differential*) object, determined by its space-time worldline, able to project its past light-cone on its celestial sphere (Fig. 5).

---

[21] The time broadcasted by every clock may be any convenient time, not necessarily its proper time, although proper time simplifies in general the theoretical analysis.

[22] More precisely, a 4π-wide hypergon eye or 4π steradian eye.

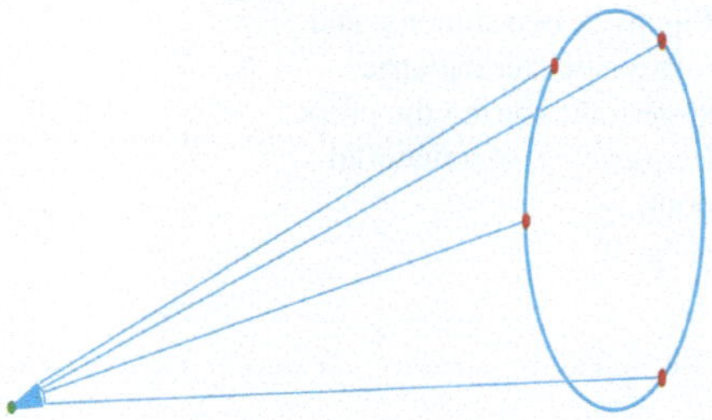

Fig. 6 Four events that are seen in a circle on the celestial sphere of an observer are also seen in a circle on the celestial sphere of any other observer at the same event, whatever its relative velocity

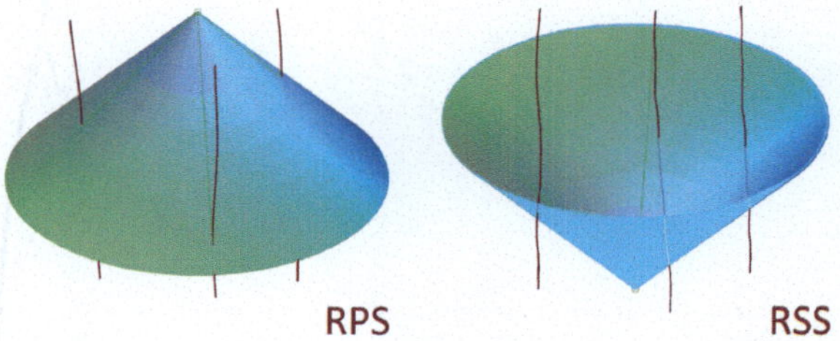

Fig. 7 Relativistic stereometric systems are the causal duals of relativistic positioning systems

There exists interesting papers on relativistic vision but they are very different in strategy, starting hypothesis and definition of 'eye', and their number is very small. It would be worthwhile to analyze and classify them, and specially to select those related to hypergon eyes.

But almost none of them consider the invariants of the observed figures. The study of these invariants is important for positioning as well as for stereometry. For example, any set of four events in the space-time (four stars in the sky or four material points of a system), that are seen on a circle of the celestial sphere of an observer, will be seen also on a circle, by any other observer at the same event, whatever be their relative velocity at the event (Fig. 6). In addition, the relative velocity of such two observers may be obtained by comparison of the relative positions on the circle of the four points as seen by them.[23]

From the point of view of the space-time geometry, *relativistic stereometric systems are the causal duals of relativistic positioning systems*: the four observers (*resp.* four clocks) of the relativistic stereometric system (*resp.* relativistic positioning system) receive (*resp.* broadcast) the signals emitted by the event to be analyzed (*resp.* the signals received by the event to be located). As a consequence, every geometric result on one of these systems generates another geometric result for the other system, by a simple change of time orientation (Fig. 7).

Relativistic stereometric systems are conceived to obtain the intrinsic properties of physical systems but, in their basic structure, they are also locating systems. As locating systems, the coordinates of every event are the (four) times of reception by the (four) observers of the signal emitted by the physical system at this event. According to the properties considered in Sect. 3.2, relativistic stereometric systems are *active*,

[23]In principle, a suitable choice of stars, in e.g. Hipparcos star catalogue, could help a spacecraft in the Solar system to estimate its velocity with respect to the Barycentric Celestial Reference System.

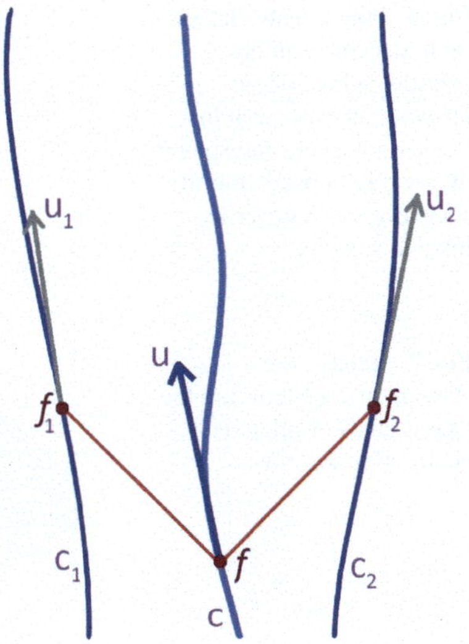

Fig. 8 A two-dimensional relativistic stereometric system measuring the proper frequency f of a material point

retarded and *virtual* locating systems, in contrast with the *passive*, *immediate* and *real* properties of relativistic positioning systems. As we have seen in Sect. 3.2, the active, retarded and virtual properties of relativistic stereometric systems are not the best as locating systems but, nevertheless, they are well adapted to the principal function of relativistic stereometric systems, because the intrinsic properties of a physical system cannot be obtained but by active, retarded and virtual ways. Thus, relativistic stereometric systems not only describe the intrinsic properties of events but also locate them.[24]

4.2 First Results in Relativistic Stereometry

Every observer of a relativistic stereometric system receives a (relativistic) perspective of the properties broadcast by the events. These four *observer-dependent* perspectives have to be combined in order to obtain the *observer-invariant* (*proper* or *intrinsic*) properties that generate them.

To show how to proceed, we shall present here the simplest case of relativistic stereometry, that of a material point broadcasting a signal in a flat two-dimensional space-time. In spite of its simplicity: (i) it leads to a simple but interesting extension of the unidirectional Doppler formula and (ii) it shows the difference between an ontic law and an epistemic one (Fig. 8).

[24]Of course, by broadcasting their proper times, a relativistic stereometric system may also act as a relativistic positioning system. But because of their dual ways of working, I believe it is clearer, for the moment, to study them separately.

Thus, let C, C_1 and C_2 be respectively the worldlines of a material point and of the two observers of the relativistic stereometric system in Minkowski space M_2. Suppose that at a particular instant[25] C broadcasts a signal of proper frequency f that is received by C_1 and C_2 at the Doppler frequencies f_1 and f_2 respectively. Let v_1 and v_2 be the scalar relative velocities of C with respect to C_1 and C_2 evaluated by them at the instants of reception of the frequencies f_1 and f_2 respectively. As it is well known, f_1 and f_2 are related to f by the Doppler expressions:

$$f_1 = f\sqrt{\frac{1-v_1}{1+v_1}} \quad , \quad f_2 = f\sqrt{\frac{1-v_2}{1+v_2}} \quad . \tag{1}$$

Denoting by v_{12} the relative scalar velocity of C_1 and C_2 at the instants of reception of the Doppler frequencies, one can prove:

Theorem 4.2.1 *In terms of the relative scalar velocity v_{12} of the observers C_1 and C_2 of a relativistic stereometric system at the instant of reception of the Doppler frequencies f_1 and f_2 from a material point C, the proper frequency f of the material point at the instant of emission is given by:*

$$f^2 = f_1 f_2 \sqrt{\frac{1+v_{12}}{1-v_{12}}} \quad . \tag{2}$$

Of course, if the material point is dragged by one of the observers of the system, say $f = f_2$, expression (2) coincides with the first of (1).

The Doppler frequencies also allow to know the relative velocities of the material point C at the instant of emission of the signal with respect to the observers C_1 and C_2 at the instants of its reception. By elimination of f from every one of the expressions (1) and (2), one has:

Theorem 4.2.2 *The relative velocities v_1 and v_2 of the material point C at the instant of emission of the signal with respect to the observers C_1 and C_2 at the instants of its reception are given by:*

$$v_1 = \frac{f_2\sqrt{1+v_{12}} - f_1\sqrt{1-v_{12}}}{f_2\sqrt{1+v_{12}} + f_1\sqrt{1-v_{12}}} \quad , \tag{3}$$

$$v_2 = \frac{f_1\sqrt{1+v_{12}} - f_2\sqrt{1-v_{12}}}{f_1\sqrt{1+v_{12}} + f_2\sqrt{1-v_{12}}} \quad . \tag{4}$$

The above results depend on the Doppler frequencies f_1 and f_2 but also on the quantity v_{12}, the scalar velocity of one of the observers of the relativistic stereometric system with respect to the other at the instants at which they receive the Doppler

[25] There is no matter here what instant-identifier is used: a clock associate to the point, measuring any time, non-necessarily proper, a flash, or any other pertinent one.

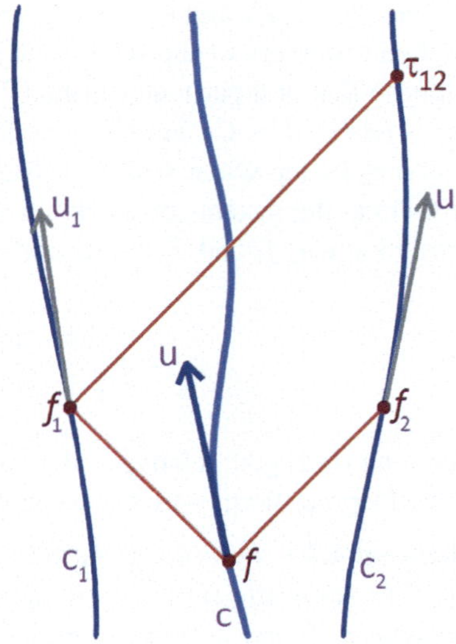

Fig. 9 The observer C_2 has been chosen as the 'physicist'. From the instant τ_{12} on, it may know all the quantities of the problem

frequencies. This quantity cannot be measured directly, but has to be calculated in terms of other physical quantities. This reason is sufficient to show that the above two theorems, in their present form, do not constitute results of epistemic relativity. In fact, these two theorems do not fulfill any of the three conditions of Sect. 2.1 characterizing epistemic relativity. For that, we must specify (i) what physicist we have chosen to make the experiment, (ii) when and where it will know the quantities needed to solve the problem and (iii) how can it know or measure these quantities. In the present case, we make the following choices:

- the physicist is one of the observers of the relativistic stereometric system, say C_2, as shown in Fig. 9,
- it will be informed of the pertinent quantities at the instant τ_{12} of reception of the information coming from the observer C_1,
- at that instant τ_{12}, it already knows the quantity f_2 measured and recorded by itself, it is being informed of the quantity f_1, and then may know the quantity v_{12} by computation.

At the instant of reception of the signal f_1, the observer C_1, in addition to send its measure, must send also its reflection or its proper frequency so as to allow C_2 to know its relative velocity at its instant τ_{12}. On the other hand, C_2 is supposed to know its proper acceleration, so that it is able to know its relative velocity between its instant of reception of the signal f_2 and the instant τ_{12}. The knowledge of these two velocities, allow it to calculate the velocity v_{12} of the theorems. For simplicity, suppose here that C_2 is geodesic. Then, the relative frequency between the proper time of C_1 at the instant of reception of the frequency f_1 and the proper time of C_2 at any instant is constant. Denoting it by ν_{12} and inverting the Doppler formula we have:

$$v_{12} = \frac{1 - \nu_{12}^2}{1 + \nu_{12}^2} \,. \tag{5}$$

Now if, for short, we call an epistemic theorem an ‘epistem’, the above two theorems in this geodesic case become respectively:

Epistem 1 *In terms of the frequencies f_1 and f_2 received by a relativistic stereometric system $\{C_1, C_2\}$ from a material point C and of the relative frequency ν_{12} of the proper time of the observer C_1 with respect to the geodesic observer C_2, the proper frequency f of the material point C is given by:*

$$f^2 = \frac{f_1 f_2}{\nu_{12}} \,. \tag{6}$$

Epistem 2 *In terms of the frequencies f_1 and f_2 received by a relativistic stereometric system $\{C_1, C_2\}$ from a material point emitter C and of the relative frequency ν_{12} of the proper time of the observer C_1 with respect to the geodesic observer C_2, the relative velocities v_1 and v_2 of the material point C with respect to the observers C_1 and C_2 at the instants of reception of the signals f_1 and f_2 are given by:*

$$v_1 = \frac{f_2 - f_1\nu_{12}}{f_2 + f_1\nu_{12}} \,, \quad v_2 = \frac{f_1 - f_2\nu_{12}}{f_1 + f_2\nu_{12}} \,. \tag{7}$$

These results are very simple, but show roughly the way of working with stereometric systems in epistemic relativity. Of course, the problem of determining the proper distances to its neighboring elements of a material point cannot be considered in a two-dimensional space-time, because there the celestial sphere of an observer reduces to two opposite points. It is evident that this work has to be extended and generalized in three and four dimensions. It remains an open problem.

5 Intrinsic Characterization of Gravitational Fields

In epistemic relativity, events are located in emission coordinates generated by relativistic positioning systems.

A complete set of gravimetric measurements (whatever be the methods) will lead to the experimental values of the components of the metric in emission coordinates.

It rest to identify this metric or, equivalently, its sources. For example, suppose we suspect that it corresponds either to a Kerr or to a Schwarzschild gravitational field, but that it can be neither one nor the other. How can we discern between these three possibilities?

The most part of the known gravitational fields (exact or approximate solutions to Einstein equations), and in particular Kerr and Schwarzschild, are known in very particular coordinate systems which have no simple relation with emission coordinates.

For example, for the Schwarzschild case, the direct procedure would be

- to calculate in Schwarzschild coordinates the equations of the field of light-cones,
- to model in these coordinates the world-lines of the clocks of the positioning system, parameterized with their proper time,
- to select with both results the four families of light-cones emitted by the clocks or, in other words, to obtain the transformation from emission coordinates to the Schwarzschild ones,
- to solve for the inverse transformation,
- to transform the Schwarzschild metric components in emission metric components,
- to compare them with the experimental values obtained in these emission coordinates.

Every one of these items presents non-negligible difficulties and can introduce specific uncertainties. It results in a hard and long procedure.

For this reason, to compare the experimental value of the gravitational field obtained in emission coordinates with a given solution of Einstein equations, it may be better to characterize intrinsically the solution and to check if the experimental data verify this characterization.

We understand here by *intrinsic characterization of a metric*, a set of necessary and sufficient local conditions on some of the concomitants of the metric that characterize it, its sources an their position, regardless of the coordinate system.

The idea is that an observer, from the sole measure of the gravitational field and its variations in a local region, is able to know the masses that produce it as well as its positions with respect to it.

The intrinsic characterization of individual metrics is, apart from some trivial cases, a relatively recent problem. I would like to draw attention on the pertinence not only of its development for all exact solutions of Einstein equations of physical interest, but also of its extension to the usual approximate solutions in experimental applications. By the reason above indicated, I believe this characterization to be the simplest and shortest way to identify gravitational fields in epistemic relativity.

The idea to characterize individual metrics intrinsically originates some years ago as part of a general project of IDEAL solution of problems, somehow similar in spirit to that of classic Greek mathematicians but of vivid actuality for the problems we were concerned, where the acronym stands for

- **I**ntrinsic (depending only of the concepts mentioned in the statement of the problem),
- **D**eductive (not involving inductive or inferential methods or arguments),
- **E**xplicit (expressing the elements of the solution non implicitly) and
- **Al**gorithmic (giving the solution as a flow chart with a finite number of steps).

The first solution intentionally obtained under the IDEAL spirit was that of the point particle in Newtonian gravity ([7]). I cite it here because the solution to its relativistic analog, the Schwarzschild gravitational field, in contrast with an extended opinion but like many other generic problems,[26] admits simpler expressions than that of Newtonian physics.

The intrinsic characterization of the Schwarzschild gravitational field is due to J. Ferrando and J.A. Sáez ([8]). Denoting by $Riem$ the Riemann tensor, by tr and $i(.)$ the trace operator and the interior product respectively, and by $\wedge$ the exterior product (considering the metric g as a double 1-form), they introduce the two scalars:

$$\sigma \equiv \left(\frac{1}{12} tr^2 Riem^3\right)^{\frac{1}{3}} \quad , \quad \alpha \equiv \frac{1}{9\sigma^2} g(d\sigma, d\sigma) + 2\sigma \quad , \tag{8}$$

allowing to construct the two tensors:

$$S \equiv -\frac{1}{3\sigma}\left(Riem + \frac{1}{2}\sigma g \wedge g\right) \quad , \quad Q \equiv i^2(d\sigma)S \quad . \tag{9}$$

and, with the help of an arbitrary direction x, generate a third, direction-dependent scalar θ:

$$\theta(x) \equiv 2Q(x,x) + trQ \quad . \tag{10}$$

These quantities may be evaluated for any gravitational field, and involve the gradient of a function of $Riem$, a third variation of the metric.

Their first result is:

Theorem 5.1 *A vacuum metric g, $Ric(g) = 0$, is the Schwarzschild metric if, and only if, the three scalars σ, α and θ are strictly positive,*

$$\sigma > 0 \quad , \quad \alpha > 0 \quad , \quad \theta(x) > 0 \quad , \tag{11}$$

and the two following tensors vanish:

$$i^2(d\sigma) * Riem = 0 \quad , \quad S^2 + S = 0 \quad , \tag{12}$$

where $*$ is the Hodge duality operator.

In the circumstances suggested by Fig. 10 this result assures the experimental physicist, whatever the coordinates used, that it is immersed in the field of a spherically symmetric mass. But what mass and where? The answer is also given in [8]:

Theorem 5.2 *The mass m and the radial coordinate r of a Schwarz-schild metric are given by:*

$$m = \sigma\alpha^{-\frac{3}{2}} \quad , \quad r = \alpha^{-\frac{1}{2}} \quad . \tag{13}$$

In fact, any other intrinsic characteristic of Schwarzschild metric may be obtained in an IDEAL way. For example, differentiating the second of relations (13) one

---

[26]Formulation of Maxwell equations, Cauchy problem for the permanence of electromagnetic waves, shock, detonation and deflagration waves in hydrodynamics, and some others.

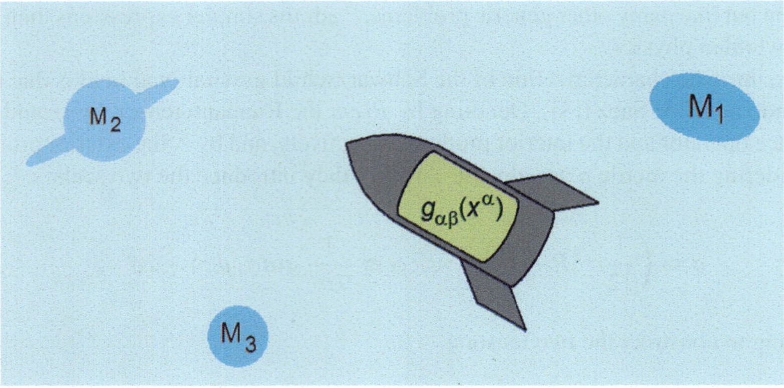

Fig. 10 An observer that measures the gravitational field in a local space is able to know the masses that produce them as well as their positions

obtains the intrinsic expression in terms of the invariant α of the radial codirection $-dr$ of the mass, allowing the physicist to know its direction, distance and value. Also, on the event horizon $r = 2m$ one would find $\alpha = 2\sigma$ and, as (8) shows, also $g(d\sigma, d\sigma)$ would vanish. Furthermore, as proven also in [8], the time-like direction of static evolution of the gravitational field, i.e. its integrable Killing vector field ξ, is given by:

$$\xi = -\sigma^{-\frac{4}{3}} \frac{Q(x)}{\sqrt{Q(x, x)}} \quad . \tag{14}$$

These results already allow us to appreciate the interest of local intrinsic characterizations of gravitational fields. But they are far from being *only* the characterization of the simplest gravitational field. The scalars σ, α and θ, in (8) and (10), and the tensors S and Q, in (9), may be evaluated in *any* gravitational field and their comparison with relations (11) and (12) will give us an information that we would study in detail, in particular if the gravitational field in question may be well described as a Schwarzschild perturbation.

At present, the intrinsic characterization of some other classes of vacuum exact solutions of Einstein equations are known, in particular Kerr metric (see [9] and references therein). We need to extend these results and, specially, develop these techniques for approximate solutions.

The above expressions, simple for a theoretical physicist, may seem complicated for an experimental one. They need to be broken down in measurable terms so as to conceive the devices and procedures to measure them. But there is no doubt that the questions that local intrinsic characterizations of metrics allow to decipher, deserve a deeper study.

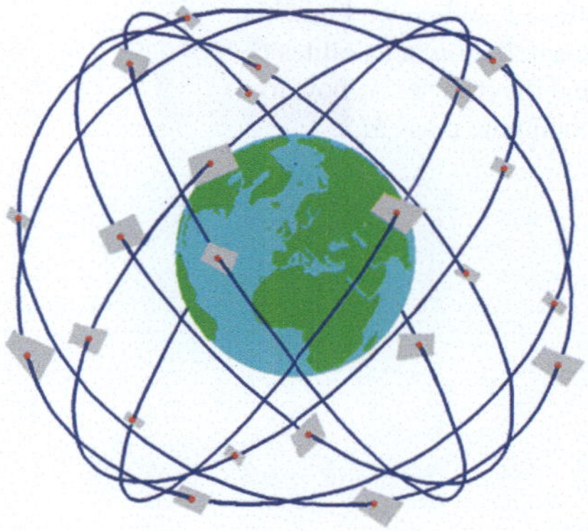

Fig. 11 A constellation $\mathcal{C}$ of satellized clocks around a mass M, every tetrad of them working as an auto-locating system, recording the data set Λ of all the signals emitted and received between them

6 Finite-Differential Geometry

As it is well known, the mathematical substratum of general relativity is Lorentzian differential geometry. It thus follows that differential geometric methods are *unavoidable* in relativity. But there exists many situations in which these differential methods appear to be manifestly *insufficient*.

Among these situations there are specific epistemic ones, involving the physical description in real time and by arbitrary observers of general variable gravitational fields, but also situations much more near of our usual ones in Earth's relativity (Fig. 11).

Thus, in a vacuum space-time, consider a mass M, not necessarily spherical, and a constellation $\mathcal{C}$ of satellized clocks around M and such that every tetrad of them works as an auto-locating system. Let Λ be the data set of all the signals that every clock emits and receives both, as an element of all the auto-locating systems at which it belongs and as a user of the auto-locating systems constituted by all the other clocks.

Suppose we know the mass M and the world-lines of the constellation $\mathcal{C}$. We can model the system as follows. Start from the space-time metric around M, integrate the geodesic equations and specify them for every one of the clocks so as to model the whole constellation $\mathcal{C}$. Then, integrate the equations of the light-cones, either starting from the general solution of the null geodesic equations, or integrating the geodesic distance function and considering its vanishing, so as to model any signal able to be emitted or received by any of the clocks of the constellation $\mathcal{C}$. With these two modeled ingredients, we are able to predict any of the data of the set Λ received or emitted by any of the clocks in terms respectively of the data emitted or received by all the other clocks.

The physical interest of such a mathematical model is not, however, very intense. On the contrary, the following scenario could be the prelude for a fine gravimetry of the Earth ... if we were able to develop the adequate mathematical instrument (Fig. 12).

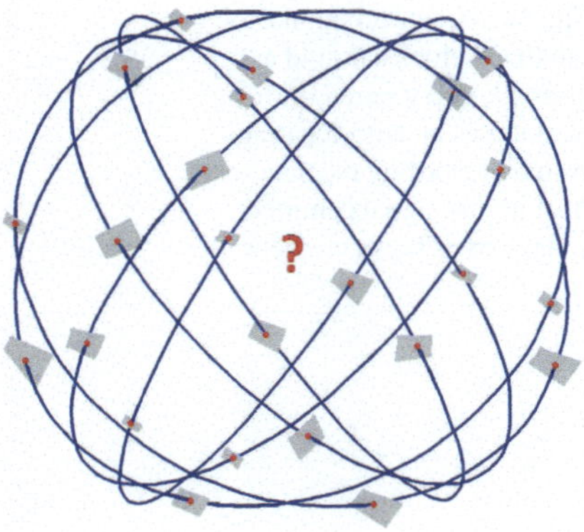

Fig. 12 When we know the constellation of satellites C and the data set Λ, how to obtain the mass M?

Suppose now that we know the constellation C and the data set Λ, and that we want to obtain the mass M. At present the only way we know to tackle the problem is the above model, but for this scenario it is heavy and unadapted to the starting point, the data set Λ. Indeed, Λ consists of time-like distances (geodesic time intervals between any two events on the world-line of every clock) and null distances (light links between every pair of clocks), that is to say, of values of the *distance function* between pairs of events causally separated. For this reason, any *direct* method to model this scenario would start with the obtainment, from the data set Λ, of the metric distance function by means of a suitable interpolation method. This is part of the adequate mathematical instrument that we would be able to develop for obtaining the mass M. But it is still insufficient.

The objective of *finite-differential geometry* is to study *finite and interchangeable* versions of the ingredients of differential geometry (metric, connection, curvature). Because for general relativity the basic ingredient is the metric g, and that its finite version, the distance function, already poses problems, we shall consider it here.[27]

It is well known that the finite version of the metric g is the *distance function* $D(x, y)$, or its half-square $\Omega(x, y)$, the Synge's *world-function*,

$$\Omega(x, y) \equiv \frac{1}{2} D(x, y)^2 \quad , \tag{15}$$

given by

$$\Omega(x, y) = \frac{1}{2} \int_0^1 g\left(\frac{d\gamma}{d\lambda}, \frac{d\gamma}{d\lambda}\right) d\lambda \quad , \tag{16}$$

and verifying its *fundamental equations*:

$$g^{\alpha\beta} \partial_\alpha \Omega \partial_\beta \Omega = 2\Omega \quad , \quad g^{ab} \partial_a \Omega \partial_b \Omega = 2\Omega \quad , \tag{17}$$

[27] This notion of finite-differential geometry was first presented as part of a lesson at the *International School on Relativistic Coordinates, Reference and Positioning Systems*, Salamanca, 2005. The mathematical results also appeared in [1].

where Greek indices correspond to coordinates at the point x and Latin ones at the point y.

Mathematically, *distance spaces*, i.e. manifolds endowed with a distance function, are well known,[28] but their link with differential geometry have not been sufficiently explored.

An important obstruction for the interchangeability between the differential and the finite concept is that the most part of distance functions *are not* geodesic distance functions of any metric.

It is known that, when a distance function $D(x, y)$ is the geodesic distance function of a metric, this metric may be obtained as minus the limit $y \to x$ of the mixed first derivatives of the half square (Synge's world function) of the distance function.[29] If this limit is applied to an arbitrary distance function, it may give rise to a zero, degenerate or regular metric. But, even when this metric is regular, the distance function $D(x, y)$ will not be generically the geodesic distance function of it.

Suppose, in our physical case, that the interpolated distance function from the data set Λ is at least of differentiability class 2. We can obtain a metric from the above limiting process, but this metric will depend strongly[30] on the interpolation method used, for which we have no control. For this and other applications, to discern if a distance function is a geodesic distance function of a metric, it would be convenient to have an IDEAL[31] criterion involving solely the distance function, without any limiting process.

I solved this problem some years ago. Let us introduce the following quantities of the first and second derivatives of the distance function $D(x, y)$:

$$V^\alpha_{\ell mn} \equiv \epsilon^{\alpha\lambda\mu\nu} D_{\ell\lambda} D_{m\mu} D_{n\nu} \quad , \quad V^{a\alpha} \equiv \epsilon^{a\ell mn} \epsilon^{\alpha\lambda\mu\nu} D_\ell D_\lambda D_{m\mu} D_{n\nu} \quad , \tag{18}$$

as well as the quantity:

$$V^\alpha \equiv V^\alpha_{\ell mn} x^\ell y^m z^n \; , \tag{19}$$

where x^ℓ, y^m, z^n are arbitrary independent directions. Define the two scalars

$$\Phi \equiv D_\lambda V^\lambda \quad , \quad \Psi \equiv \epsilon^{r\ell mn} V^\rho_{\ell mn} D_r D_\rho \quad , \tag{20}$$

and form the two quantities:

$$D^\alpha \equiv \frac{V^\alpha}{\Phi} \quad , \quad D^{a\alpha} \equiv 3\frac{V^{a\alpha}}{\Psi} \; . \tag{21}$$

Then, we have:

[28]The concept is due to Fréchet. Haussdorff named them 'metric spaces' ('metrischer Raum') but in our context it is better to call them 'distance spaces'.

[29]I am grateful to Abraham Harte for a pertinent observation on this fact.

[30]At points out of the data of Λ.

[31]See Sect. 5 for this notion.

Theorem 6.1 (Structure theorem for geodesic distance functions) *The necessary and sufficient condition for a distance function $D(x, y)$ to be the geodesic distance function of a metric, is that its derivatives verify:*

$$D^\rho D_{abc\rho} + D^\rho D^{m\sigma} \oint_{abc} (D_{am\rho} - D^n D_{mn\rho} D_a) D_{bc\sigma} = 0 \tag{22}$$

where the subscripts denote partial derivatives and D^a and $D^{a\alpha}$ are the quantities just defined.

In our physical case, these are the constraints to be imposed directly to any interpolated distance function on the data set Λ. Once these equations verified at the suitable degree of precision, one has to extract the metric from this distance function without any limiting process. I solved this problem together with the above one. The result is:

Theorem 6.2 (Metric of a geodesic distance function) *In terms of the partial derivatives D_α, $D_{a\alpha}$ and $D_{ab\alpha}$ of a geodesic distance function, the contravariant components $g^{\alpha\beta}$ of the metric at the point x, are given by*

$$g^{\alpha\beta} = D^\alpha D^\beta + D^{a\alpha} D^{b\beta} D_{ab\gamma} D^\gamma \quad , \tag{23}$$

where D^α and $D^{a\alpha}$ are the quantities above defined.

Coming back to our physical case, (23) would give us the metric in the region of the constellation. Then, the intrinsic characterization method of Sect. 5, once generalized to perturbations of Schwarzschild gravitational field, would give us the mass M that satellizes the constellation $\mathcal{C}$ and generates the data set Λ.

Theorems 6.1 and 6.2 give to the distance function $D(x, y)$ the interchangeable character with its differential homologue, the metric $g(x)$, that our finite-differential geometry requires.[32] But finite-differential geometry has to be developed in many other directions. Among them, perhaps the more urgent ones are the development of methods of interpolation and approximation of distance functions.

Acknowledgements This work has been supported by the Spanish "Ministerio de Economía y Competitividad", MICINN-FEDER project FIS2015-64552-P.

[32]On a metric manifold a distance function is generically *local* (normal geodesic domain), so that a (global) metric cannot be generically *interchanged* by a sole distance function, but by a suitable atlas of normal geodesic domains with distance functions submitted, in the intersection of charts, to compatibility conditions. In our case, nevertheless, the constellation $\mathcal{C}$ is contained clearly in a normal geodesic domain. Anyway, what we want here is to show the interest of the concept.

References

1. B. Coll, Relativistic positioning systems: perspectives and prospects. Acta Futura **7**, 35–47 (2013)
2. B. Coll, J.J. Ferrando, J.A. Morales-Lladosa, Positioning systems in Minkowski space-time: from emission to inertial coordinates. Class. Quantum Grav. **27**, 065013 (2010)
3. B. Coll, Elements for a theory of relativistic coordinate systems. Formal and physical aspects, in *Proceedings of the Spanish Relativity Meeting 2000 on Reference Frames and Gravitomagnetism (Valladolid, Spain)* (World Scientific, Singapore, 2001), pp. 53–65
4. B. Coll, J.M. Pozo, Relativistic positioning systems: the emission coordinates. Class. Quantum Grav. **23**, 7395 (2006)
5. B. Coll, J.J. Ferrando, J.A. Morales-Lladosa, Positioning in a flat two-dimensional space-time: the delay master equation. Phys. Rev. D **82**, 084038 (2010)
6. B. Coll, J.J. Ferrando, J.A. Morales-Lladosa, Positioning in Minkowski space-times: Bifurcation problem and observational data. Phys. Rev. D **86**, 084036 (2012)
7. B. Coll, J.J. Ferrando, The Newtonian point particle, in *Proceedings Spanish Relativity Meeting 1997, Palma de Mallorca, Spain* (Pub. Univ. Illes Balears, 1997), pp. 184–190
8. J.J. Ferrando, J.A. Sáez, An intrinsic characterization of the Schwarzschild metric. Class. Quantum Grav. **15**, 1323 (1998)
9. J.J. Ferrando, J.A. Sáez, An intrinsic characterization of the Kerr metric. Class. Quantum Grav. **26**, 075013 (2009)

Use of Geodesy and Geophysics Measurements to Probe the Gravitational Interaction

Aurélien Hees, Adrien Bourgoin, Pacome Delva, Christophe Le Poncin-Lafitte and Peter Wolf

Abstract Despite its extraordinary successes, the theory of General Relativity is likely not the ultimate theory of the gravitational interaction. Indeed, General Relativity as such is a classical theory and is therefore incomplete since it does not include any quantum effects. Moreover, most physicists agree that GR and the Standard Model are only effective field theories that are low-energy approximation of a more fundamental and more general theory that would provide a unified description of all the fundamental interactions. On the observational side, Dark Matter and Dark Energy are required to explain most of astrophysical and cosmological observations and very few is known and this two Dark components, which is sometimes interpreted as an hint that our theory of gravitation is incomplete. For these reasons, General Relativity is confronted to an increasing number of measurements, searching for deviations in more and more frameworks that extend General Relativity. Amongst all the measurements used to search for and to constrain deviations from General Relativity, a observations developed in the context of geodesy and geophysics are playing an important role like for example atomic clocks comparison, gravimetry measurements, satellite and lunar laser ranging, very long baseline interferometry, etc …In this communication, we present briefly each of these geodesic/geophysics measurements and show how they have recently been used to constrain extensions of General Relativity, model of Dark Matter or of Dark Energy.

A. Hees (✉) · P. Delva · C. Le Poncin-Lafitte · P. Wolf
SYRTE, Observatoire de Paris, Université PSL, CNRS, Sorbonne Université, LNE,
61 avenue de l'Observatoire, 75014 Paris, France
e-mail: aurelien.hees@obspm.fr

A. Bourgoin
Dipartimento di Ingegneria Industriale, University of Bologna,
via fontanelle 40, Bologna, Italy

D. Puetzfeld and C. Lämmerzahl (eds.), *Relativistic Geodesy*,
Fundamental Theories of Physics 196, https://doi.org/10.1007/978-3-030-11500-5_9

1 Introduction

The current paradigm to describe the gravitational interaction is the General theory of Relativity (GR) published by Einstein in 1915. Since 1915, this theory has been extremely successful making several predictions that were all verified observationally. Nowadays, the domain of application of GR is extremely large: (i) at small scales like for clocks comparison, time transfer, construction of time references, (ii) at the scale of the Solar System where GR needs to be taken into account to describe the motion of planets, of the Moon and of satellites, (iii) at the galactic scale to describe the galactic dynamics and also the motion of stars orbiting our Galactic Center, and (iv) at the cosmological scales to describe the behavior of the Universe and to explain large scales observations. In spite of the overwhelming success of these theory in describing much of the macroscopic observed universe, a number of open issues, both theoretical and experimental remain.

First of all, on the theoretical side, GR is a classical field theory while the other interactions from Nature are described within the framework of the Standard Model of particles physics, which is based on a quantum field theory. As such, GR is fundamentally incomplete since it does not include any quantum effects. Moreover, most physicists agree that GR and the Standard Model are only effective field theories that are low-energy approximation of a more fundamental and more general theory that would provide a unified description of all the fundamental interactions. Most attempts to derive such a theory that would unify the Standard Model with GR leads to deviations either in the Standard Model, either in the gravitational sector, either in both. These deviations occur at a level that is in general not predicted and not predictable.

On the observational side, several galactic observations cannot be explained within the frameworks of GR and of the Standard Model. In the most adopted paradigm, a new type of cold matter called Dark Matter (DM) is introduced. So far, DM has not been directly detected despite a large effort like e.g. with particles accelerator. The nature of this elusive type of matter is therefore still completely unknown. Note that while the introduction of new particles to explain DM correspond to a modification of the Standard Model, it has also been suggested that DM could be the result of a modification from GR (this is the case for the MOND theory for example, see e.g. [1]). In addition, for some model of DM, the occupation number of the corresponding field will be relatively high and the particle will actually behave as a classical field. This is the case for some model of ultra light bosonic DM. In that case, DM can also be interpreted as a classical modification from GR.

In addition to DM, large scale observations like the temperature fluctuations of the cosmic microwave background, the redshift-distance measurements from Supernova Ia or baryonic acoustic oscillations have shown that our Universe undergoes a phase of accelerated expansion that is attributed to Dark Energy (DE). In the simplest scenario, DE is attributed to a cosmological constant whose fine-tuned value is completely unexplained. A lot of modifications of GR have been suggested to provide a more elegant solution to DE by introducing at least a new dynamical field that would cause our Universe to accelerate (see e.g. [2]).

All these considerations have pushed the scientific community to confront GR to a wide number of experiments searching for tiny deviations that would be the hint of a new physics and that would guide the construction of new models of unification theory, of DM and of DE. So far, despite measurements that are more and more accurate, Einstein's theory has passed all the observational tests with flying color (see e.g. [3]). As already mentioned above, none of the alternative theory of gravitation developed is making a clear prediction for the actual level at which a deviation will become detectable. This property makes the search for new physics particularly difficult but also very exciting. These last decades, searches for deviations from GR have become a very active field of research. In particular, outstanding performance provided by the development of technology developed for geodesy or geophysics have been used to develop new high-accuracy fundamental physics experiments.

The goal of this communication is to give several successful examples of measurements and experiments that have been developed in the context of geodesy and geophysics and that have been used in order to search for new physics. The range of measurements that will be considered is relatively large. Amongst other, we will consider several laboratory experiments like the comparison of very accurate atomic clocks or atomic gravimetry. Best present day clocks reach relative uncertainties in the low 10^{18} range and progress is rapid with no obvious hard limit in sight. Amongst the applications of the comparison of high stable clocks is geodesy and the measure of the gravitational potential (see e.g. [4]) the experimental study of fundamental physics, and in particular the two fundamental theories mentioned above (GR and standard model). Gravimeters on the other hand are mainly used to measure the gravitational acceleration for geophysics purposes but have also found an application in testing GR.

At larger scales, GNSS satellites but also satellite laser ranging have been widely used in geodesy. In particular, they are used to derive high-precision models of Earth's gravity field. Both these techniques have also been used successfully to test the theory of GR with an impressive accuracy. In addition, Very Long Baseline Interferometry, used in particular to monitor the Earth's rotation has also been successfully used to probe the deflection of electromagnetic signals from quasar by the Sun.

Section 2 from this communication is devoted to the description of the different measurements and experiments that have been developed in the context of geodesy or geophysics and that have found an application in fundamental physics. Section 3 will present the theoretical background of GR, several alternative frameworks that allow scientists to search for deviation from GR and how measurements made using geodesic techniques have led to constraints on these alternative frameworks. We will also focus on the theoretical implications of such constraints. It is impossible to review all the alternative theories of gravitation and their constraints. In this communication, we will therefore select a couple of them that are very important and/or quite recent to illustrate this interplay between geodesy/geophysics and fundamental physics. We hope this will motivate more interactions between these two communities to develop more and more accurate experiments that would potentially lead to the discovery of a tiny deviation from GR that would be the sign of new physics.

2 Brief Description of Several Geodetic/Geophysical Measurements used for Fundamental Physics

The goal of this section is to give a brief overview of several different measurements that have been developed in the context of geodesy or geophysics and that have found some application in fundamental physics.

2.1 Gravimetry

Gravimetry is the measurement of the local gravitational (plus inertial due to the Earth's rotation) acceleration g experienced by an observer on the Earth. More generally, accelerometry measures the local acceleration experienced by an observer, whatever its origin (gravitational or inertial). Simply speaking a gravimetric measurement consists of "dropping" an object and measuring its acceleration using a ruler and a clock. In modern gravimetry the ruler is replaced by a laser whose wavelength is ultimately referenced to an atomic clock and to the SI definition of the meter and the second, thus providing an absolute measurement of local acceleration is SI units.

Classical absolute gravimeters use falling corner cubes and laser interferometry to determine the local acceleration [5]. More recently these instruments have become rivaled by gravimeters based on atom interferometry, that use falling atoms in quantum superpositions with lasers as the read-out of their acceleration [6]. Best absolute gravimeters reach uncertainties in the low 10^{-8} m/s^2. Additionally relative gravimeters and accelerometers measure time-variations of local acceleration, but not its absolute value, using in general electrostatic devices and readout (possibly superconducting). Such relative accelerometers are widely used is space missions like GRACE, GOCE or MICROSCOPE.

Gravimeters and accelerometers, whether relative or absolute, have been widely used in geodesy and geophysics (on ground and in space) for many decades. They provide the bulk of data that is used for the determination of the geopotential and numerous other applications. As described below, they have also been widely used in fundamental physics, in particular for tests of the universality of free fall, which is one of the foundations of general relativity.

2.2 Atomic Clock Comparison

Quite generally, an atomic clock is a device that delivers an electromagnetic signal (typically in the microwave or optical domain) whose frequency ν is locked to the unperturbed transition frequency between two quantum states of an atom or ion

$$\nu = \frac{2\pi(E_e - E_g)}{\hbar}, \tag{1}$$

where E_e and E_g are the exited and ground state energies of the atom/ion, and $\hbar$ is the reduced Planck constant. As all atoms/ions of a given isotope are identical, an atomic clock is thus a universal time piece, delivering the same proper time up to technical imperfections. The best atomic clocks are presently approaching impressive uncertainties of one part in 10^{18} [7–11] in fractional frequency which makes them some of the most accurate measurement devices ever built. They can be compared inside the same laboratory or over large distances using the exchange of electromagnetic signals in free space and via satellites, or through cables or optical fibers. The realization of such links, especially over large distances and with an uncertainty compatible with the best clocks is a major challenge today and a subject of active research [12].

Because of their exceptional stability and accuracy, and the fact that they are the basic building block of any space-time measurement, atomic clocks have been used in fundamental physics for many decades and some of those applications will be discussed in detail below. But more recently they have started to also be studied in the context of what has become known as "chronometric geodesy", the determination of geopotential differences between two distant locations using the comparison of two clocks [4, 13, 14]. Their measured frequency difference then provides a measurement of the geopotential difference at the two locations via the gravitational redshift of general relativity. A fractional uncertainty of 1×10^{-18} on such a measurement corresponds to an uncertainty of 0.1 m^2/s^2 in geopotential difference or 1 cm in height, which is comparable to the best present geopotential models obtained from gravimetry and satellite geodesy.

2.3 Satellite Laser Ranging

In 1964, NASA carried out the first laser ranging to a near-Earth satellite. Since that time, ranging precision has improved by a factor of a thousand from a few meters to a few millimeters, and more satellites equipped with corner cubes have been launched. During the subsequent decades, the first objectives of dedicated Satellite Laser Ranging (SLR) experiments (for example, STARLETTE, STELLA, LAGEOS) were to produce high accuracy orbitography of those satellites and, consequently, to study in details the Earth gravity field, i.e. the solid Earth. SLR technique is also used in altimeter missions mapping the ocean surface, for mapping changes in continental ice masses, for topography.

With a measurement accuracy better than the part-per-billion level, GR must be considered by adding additional perturbations to the orbit dynamics, corrections to the light-time computation and more fundamental aspects of the definition of the geocentric reference frame. While these effects are significant, they are generally not large enough to provide competitive tests of GR with those available from lunar laser ranging and other Solar System tests. An important exception, however, is the relativistic prediction of the Lense-Thirring orbit precession, i.e. the effect of frame-dragging on the satellite orbit due to the spinning Earth mass. Using the two LAGEOS satellites, [15, 16], Ciufolini and collaborators were able to detect at the level of 10% this effect. In 2012, a new mission, completely passive with 92 corner cubes, called LARES has been launched to improve this determination[17].

2.4 Lunar Laser Ranging

On August, 20th 1969, after ranging to the lunar retro-reflector placed during the Apollo 11 mission, the first LLR echo was detected at the McDonald Observatory in Texas. Currently, there are five stations spread over the world which have realized laser shots on five lunar retro-reflectors. Among these stations four are still operating: Mc Donald Observatory in Texas, Observatoire de la Côte d'Azur in France, Apache point Observatory in New Mexico and Matera in Italy while one on Maui, Hawaii has stopped lunar ranging since 1990. Concerning the lunar retro-reflectors three are located at sites of the Apollo missions 11, 14 and 15 and two are French-built array operating on the Soviet roving vehicle Lunakhod 1 and 2.

LLR is used to conduct high precision measurements of the light travel time of short laser pulses emitted at time t_1 by a LLR station, reflected by a lunar retro-reflector and finally received at time t_3 at a station receiver. The data are presented as normal points which combine time series of measured light travel time of photons, averaged over several minutes to achieve a higher signal-to-noise ratio measurement of the lunar range at some characteristic epoch. Each normal-point is characterized by one emission time (t_1 in universal time coordinate – UTC), one time delay (Δt_c in international atomic time – TAI) and some additional observational parameters as laser wavelength, atmospheric temperature and pressure etc.

LLR measurements are used to produce the Lunar ephemeris but also provide a unique opportunity to study the Moon's rotation, the Moon's tidal acceleration, the lunar rotational dissipation, etc [18]. In addition, LLR measurements have turn the Earth-Moon system into a laboratory to study fundamental physics and to conduct tests of the gravitation theory.

2.5 Interplanetary Spacecraft Tracking and Planetary Ephemerides

Modeling and understanding the motion of planets and minor bodies in the Solar System has been a long active field of research. This is important to understand the formation of our Solar System [REF NEEDED?], to successfully develop interplanetary space-missions, to predict an hypothetical "rendez-vous" between the Earth and an asteroid, to detect and quantify the presence of Dark Matter in our Solar System [19], etc...In addition, the analysis of the motion of the planet Mercury around the Sun was historically the first evidence in favor of GR with the explanation of the famous advance of the perihelion in 1915. From there, planetary ephemerides have always been a very powerful tool to constrain GR and alternative theories of gravitation.

Currently, three groups in the world are producing planetary ephemerides: the NASA Jet Propulsion Laboratory with the DE ephemerides [20–26], the French INPOP (Intégrateur Numérique Planétaire de l'Observatoire de Paris) ephemerides [27–32] and the Russian EPM ephemerides [33–37]. These analyses use an impressive number of different observations to produce high accurate planetary and asteroid trajectories. The observations used to produce ephemerides comprise radioscience observations of spacecraft that orbited around Mercury, Venus, Mars and Saturn, flyby tracking of spacecraft close to Mercury, Jupiter, Uranus and Neptune and optical observations of all planets.

2.6 Very Long Baseline Interferometry

VLBI is a geometric technique measuring the time difference in the arrival of a radio wavefront, emitted by a distant quasar, between at least two Earth-based radiotelescopes. VLBI observations are done daily since 1979 and the database contains nowadays almost 6000 24 h sessions, corresponding to 10 millions group-delay observations, with a present precision of a few picoseconds. One of the principal goals of VLBI observations is the kinematical monitoring of Earth rotation with respect to a global inertial frame realized by a set of defining quasars, the International Celestial Reference Frame [38], as defined by the International Astronomical Union [39]. The International VLBI Service for Geodesy and Astrometry (IVS) organizes sessions of observation, storage of data and distribution of products, in particular the Earth Orientation parameters. Because of this precision, VLBI is also a very interesting tool to test gravitation in the Solar System. Indeed, the gravitational fields of the Sun and the planets are responsible of relativistic effects on the quasar light beam through the propagation of the signal to the observing station and VLBI is able to detect these effects very accurately.

3 Theoretical Background and Tests of General Relativity

The classical theory of General Relativity provides a geometrical description of the gravitational interaction. It is based on two fundamental principles: (i) the Einstein Equivalence Principle (EEP) and (ii) the Einstein field equations that can be derived from the Einstein–Hilbert action. All GR extensions or alternative theories of gravitation will break at least one of these principles.

3.1 The Einstein Equivalence Principle

The first part of GR, the EEP, published in 1911 by Einstein [40], gives gravitation a geometric nature. This principle implies that gravity can be identified to space-time geometry which is described mathematically by a symmetric order 2 tensor, the space-time metric $g_{\mu\nu}$. More precisely, the EEP implies that there exists only one space-time metric to which all matter minimally couples to (see [41]). In practice, this means that the equations of motion for matter can be derived from the action of the Standard Model of particles in which the Minkowski metric $\eta_{\mu\nu}$ is replaced by the space-time metric $g_{\mu\nu}$

$$S_{\rm mat} = \int d^4x\sqrt{-g}\,\mathcal{L}_{\rm mat}\left(\Psi, g_{\mu\nu}\right) , \tag{2}$$

where g is the determinant of the space-time metric and $\mathcal{L}_{\rm mat}\left(\Psi, g_{\mu\nu}\right)$ is the standard Lagrangian for matter depending on the matter fields Ψ.

From a theoretical point of view, this part of Einstein theory allows one to derive the effects of gravitation from the space-time curvature. In particular, this implies that test bodies follow geodesic from this space-time. Furthermore, ideal clocks will measure the quadratic invariant of the space-time metric $d\tau^2 = -g_{\mu\nu}dx^\mu dx^\nu$. Similarly, the propagation of electromagnetic waves is governed by Maxwell equations in which standard derivatives are replaced by covariant derivatives [42]. At the geometric optic approximation, this implies that light rays are described by null geodesics.

From a phenomenological point of view, three aspects of the EEP can be tested (see [3, 43]): (i) the Universality of Free Fall (UFF), (ii) the Local Lorentz Invariance (LLI) and (iii) the Local Position Invariance (LPI). Furthermore, the Schiff conjecture stipulates that any complete, self-consistent theory of gravity that embodies the UFF necessarily embodies the two other parts of the EEP: the LPI and the LLI (see e.g. [3]). Nevertheless, no complete proof of this conjecture has been proposed and some counter-examples are known (see the discussion in Sect. 2.1.1 of [3]).

3.2 Einstein Field Equations

The EEP allows one to derive the effects of gravitation from the space-time metric $g_{\mu\nu}$. In order to have a complete theory, one needs to give a prescription to determine the form of this metric. In GR, the form of the metric is determined by solving the Einstein field equations

$$R_{\mu\nu} - \frac{1}{2} g_{\mu\nu} R + \Lambda g_{\mu\nu} = \frac{8\pi G}{c^4} T_{\mu\nu} \,, \tag{3}$$

where $R_{\mu\nu}$ and R are the Ricci tensor and scalar curvature from the metric, G the gravitational constant, c the speed of light in vacuum, $T_{\mu\nu}$ the stress-energy tensor and Λ the cosmological constant. These equations characterize the dynamic of space-time geometry essentially described in the left part of the equation and the matter/energy content in this space-time essentially described by the stress-energy tensor. In other words, this set of equations describes how space-time is curved by the presence of matter and energy.

The Einstein field equations can be derived by means of a variational principle from the Einstein–Hilbert action (see [44])

$$S_{\rm grav} = \frac{c^4}{16\pi G} \int d^4x \sqrt{-g} \, (R - 2\Lambda) \,. \tag{4}$$

The total action which describes completely the GR theory is the sum $S_{\rm grav} + S_{\rm mat}$ which makes GR an extremely simple theory from a conceptual point of view. It is also interesting to mention that due to the Lovelock theorem, the Einstein field equations are the only second order equations that can be derived from a least-principle action based on the space-time metric (and its derivatives) in a 4 dimensional Riemannian space (see [45, 46]). Following this theorem, alternative (metric) theories of gravitation will always imply one of the following (see [2]):

- The existence of new fields in addition to (or instead of) the space-time metric.
- The existence of higher order derivatives of the metric in the field equations.
- To work in a space-time with higher dimension than 4.
- To give up locality.

If the EEP specifies how different types of mass-energy react to gravitation, the Einstein field equations govern how gravitation is generated by matter and energy. A modification of the Einstein–Hilbert action will lead to different field equations that can be reflected in differences in space-time geometry. From an experimental point of view, it is of prime importance to search for metric deviations that would be produced by any deviations from GR. The class of metric theories of gravitation is huge and it is particularly difficult to analyze observational data in a general framework. So far, two frameworks have been widely used to analyze data: the parametrized post-Newtonian (PPN) formalism (see [3, 43]) and the fifth force formalism (see [47–51]). As we shall see in Sect. 4.2, these two formalisms have been very precisely constrained with observations at Solar System scales.

Nevertheless, it is useful to analyze experimental results in new extended frameworks. Indeed, even if observations lie very close to GR when analyzed within the PPN or fifth force framework, this does not mean that this has to be true in any other framework. The existing frameworks indeed cover a limited set of alternative theories of gravity (see [52, 53]). For example, a formalism aiming at considering violation of Lorentz invariance in the gravity sector has been develop within the SME framework (see [54–57]). In this formalism, an expansion at the level of the action is performed which naturally leads to a post-Newtonian metric that differs from the PPN one. Other examples of theories not entering the PPN or fifth force frameworks are given by the MOND (MOdified Newtonian Dynamics) phenomenology which produces a quadrupolar deviation from the Newton potential in the Solar System [58, 59], by the post-Einsteinian Gravity (PEG, see [60, 61]), by the parametrized post-Newtonian-Vainshteinian formalism [62], by Horndeski's gravity [63], by massive tensor-scalar theories, ... Currently, only few data analysis are exploring these formalisms or theories.

4 Frameworks to Search for New Physics and Constraints on Their Parameters

The task of confronting each of the modified gravitational theory to observations is huge. In practice, several frameworks have been used to analyze observations in order to search for new physics in the gravitational sector. In this section, we will present several of these frameworks and show how geodesy or geophysics measurements developed in Sect. 2 have been used to constrain their parameters. The list of the frameworks presented in this communication is obviously not exhaustive but demonstrates clearly the impact of geodesic measurements in fundamental physics.

The different formalisms presented in this section are motivated at different level from a theoretical perspective: some of them are complete theory characterized by a full action with new fields in addition to the standard space-time metric, some are phenomenology developed at the level of the action (the SME for example), some are phenomenology developed at the level of field equations (MOND for example), some are phenomenology developed at the level of the action (the PPN formalism is a good example) and some are only a phenomenological parametrization at the level of the observable (the redshift test or the UFF measurement are good example). They all have advantages and drawbacks but are aiming at facilitating the search for new physics with observations.

4.1 *Tests of the Einstein Equivalence Principle*

4.1.1 The Universality of Free Fall

The UFF states that the motion of a test body is independent of its composition. It has been constrained by various experiments. Precise tests of the UFF compare the free fall accelerations, a_1 and a_2, of two different test bodies 1 and 2 falling in the gravitational field sourced by a body S. A succinct phenomenological parametrization for the test of the UFF takes the form (see [3, 43]).

$$\eta = \left(\frac{\Delta a}{a}\right)_{S;1-2} = 2\frac{a_1 - a_2}{a_1 + a_2} \approx \left(\frac{m_P}{m_A}\right)_1 - \left(\frac{m_P}{m_A}\right)_2 , \tag{5}$$

where m_P and m_A are the passive and active masses of each body. This empirical formulation is quite generic and it is interesting to test the UFF by using different types of test bodies. The most thoroughly tested and known version of this principle is attributed to Galileo Galilei and consider two macroscopic test bodies of different compositions falling in the same gravitational field. This version is currently tested at the impressive level of 10^{-13} with torsion balances [51, 64–66]. In addition, LLR observations also provide a wonderful tool to test the UFF by searching for a differential acceleration between the Moon and the Earth with respect to the Sun. Formally, search for a breaking of the UFF using extended objects like the Earth and the Moon tests a combination of the Einstein Equivalence Principle and of the Strong Equivalence Principle (see e.g. [67]). Such a violation of the UFF will lead to a polarization of the Moon's orbit as notice by Nordtvedt [68]. Analyzes of LLR observations described in Sect. 2.4 has also provided a constraint on the UFF at the level of 10^{-13} on η [69, 70]. Very recently, new infrared observations from the Grasse station [71] has allowed to measure the range to the Moon during the new and the full Moon periods, when the signature from a breaking of the UFF is maximal. This lead to the best LLR constraint on the UFF between the Moon and the Earth (in the field of the Sun) given by: $\eta = (-3.8 \pm 7.1) \times 10^{-14}$ [67].

Finally, very recently, the first results of the MICROSCOPE space mission has been released. The MICROSCOPE project consists in performing a test of the UFF between two macroscopic bodies in space. The main advantage of performing this test in space comes from the fact that the free fall time baseline can be increased significantly compared to what is achievable on Earth. An analysis of the first dataset from this mission gave $\eta = (-1 \pm 9\,\text{(stat)} \pm 9\,\text{(syst)}) \times 10^{-15}$ [72]. This result, which has been interpreted for different theoretical models (see [73, 74]), is expected to be improved by roughly an order of magnitude with the full data of the mission.

More recently, the UFF has been tested by comparing the acceleration measured by a macroscopic mass with respect to the acceleration measured by a microscopic, quantum system and also by comparing the accelerations measured by two different microscopic quantum systems. This has been achieved using atom interferometry which provided a measurement of the local gravitational acceleration g (see

Sect. 2.1). The current constraints on the UFF between macroscopic and microscopic systems are at the level of 10^{-9} [75]. Similar constraints using two different types of atom interferometers have provided constraints at the level of 10^{-9} as well [76–78]. Furthermore, some modified theories of gravitation predict a violation of the UFF between bodies with different spins (see as examples [79–84], and references therein). Therefore, the UFF has also been tested by considering atoms with different spins and the related constraints are at the level of 10^{-7} [85]. One can wonder if anti-matter falls any differently than standard matter. Tests of the UFF using anti-hydrogen atoms are currently on-going at CERN with the AEgIS [86] and GBAR [87] experiments.

Predictions of a violation of the UFF between a black hole and standard matter has also been recently predicted and tested for a very particular case of Galileon theory by using astrophysical observations [88, 89].

Finally, the question of the validity of the UFF in the Dark sector remains completely open. Since dark matter has not been directly detected so far, tests of the EEP in the dark sector are model dependent. A lot of models introduced a long range interaction in the dark sector usually leading to a fifth force experienced either by standard matter or by Dark Matter (see [90–93]). This hypothetical fifth force has been constrained using various astrophysical observations (see for examples [92–98], and references therein). More recently, a model of Dark Matter coupled to a scalar field has been proposed by Damour et al. [99] and predicts a violation of the EEP between Dark Matter and standard matter. Initially introduced as a model leading to a time variation of the gravitational constant G, it has been shown that the violation of the EEP in the Dark sector can naturally produce Dark Energy [100–102]. This violation of the EEP has recently been constrained using galactic observations by Mohapi et al. [103].

The UFF is certainly the first observational principle upon which the gravitation theories (Newton's theory and later GR) have been built. As described in this section, testing the UFF can be performed using a wide various of test bodies and take various forms. In particular, observations that are used for geodesic purposes (LLR, atomic interferometry) gives some of the best current constraints. Considering that this principle is at the heart of the gravitation theory, testing its experimental validity is crucial and several projects aimed at improving the current searches for a UFF violations.

4.1.2 Local Lorentz Invariance

is a feature of relativity stating that the outcome of any local non gravitational experiment is independent of the velocity and orientation of the apparatus (see e.g. [3]). While completely integrated into special relativity, the effort to test LLI has recently increased. Indeed, it is often believed that models of quantum gravity will produce violations of Lorentz invariance. These models usually introduce a fundamental length (the Planck length). Since this length is not Lorentz invariant, in most scenarios, a breaking of Lorentz symmetry at some level is produced (see the discussion in

Sect. 2.1.2 from [3] or [104]). Several formalisms have been used to test Lorentz symmetry. The simplest one predicts anisotropies in the speed of light, often parametrized by the c^2-formalism (see e.g. [3]). Such a violation requires the existence of a privileged reference frame.

A violation of the LLI would produce a shift in the energy levels of a particle, which depends on the orientation of the quantization axis and the quantum number of the state. The "clock anisotropy" experiments verify that this shift is null. The first experiments of this type were done by [105] and [106]. The cooling of atoms and trapped ions then permitted to limit collision effects, which further increased the accuracy of the experiments [107–109]. Other methods exist to test the LLI. For example, Michelson–Morley type experiments and its numerous variants (see e.g. [110] for a review): let's cite the famous [111] experiment, which used a Fabry–Perot interferometer, and the experiments comparing the frequencies of electromagnetic cavities with each other, or with atomic clocks [112, 113].

Another widely used formalism to test Lorentz Invariance (LI) is called the Robertson-Mansouri-Sexl (RMS) framework [114–117] in which a modification of the kinematic Lorentz transformation is parametrized by functions which depend on the relative velocity **w** between a preferred frame – usually taken as the cosmic microwave background – and the observer frame. The three classical LI tests are the Michelson–Morley, Kennedy–Thorndike, and Ives–Stillwell experiments (see e.g. [114]); they are second-order tests as the LI violating signal depends on w^2/c^2, where $w = |\mathbf{w}|$ and c is the velocity of light in vacuum [116]. With the advent of heavy-ion storage rings, Ives–Stillwell type experiments gave until recently the best constraint on α, the time dilation parameter of the RMS parametrization. A limit of $|\alpha| \lesssim 8.4 \times 10^{-8}$ was found using ^{7}Li+ ions prepared in a storage ring to 6.4% and 3.0% of the speed of light [118]. The experiment described in [119] uses ^{7}Li+ ions confined at a velocity of 33.8% of the speed of light. When neglecting higher order RMS parameters, the constraint on the LI violating parameter is $|\alpha| \lesssim 2.0 \times 10{-}8$.

First-order tests in w/c are based on the comparison of clocks [115, 120]. Comparing atomic clocks on-board GPS satellites with ground atomic clocks, [121] obtained the constraint $|\alpha| \lesssim 10^{-6}$.

A recent test within the RMS framework is based on comparison of four optical lattice clocks using Sr atoms, two located at SYRTE, Observatoire de Paris, France [122, 123], one at PTB, Braunschweig, Germany [124, 125], and one at NPL, Teddington, UK [126]. These clocks are connected by two fibre links, one running from SYRTE to PTB operated in June 2015 [127], and one from SYRTE to NPL operated in June 2016. In a simplified set-up, an optical clock comparison using a phase noise compensated fibre link can be described as a two-way frequency transfer between two observers A and B (see e.g. [128–132]). By searching for a daily variation of the frequency difference these four strontium optical lattice clocks, [133] improve upon all previous tests of RMS time dilation parameter with $|\alpha| \lesssim 1.1 \times 10^{-8}$.

More recently, a very wide formalism has been developed to consider hypothetical violation of Lorentz symmetry in all fields of physics. Contrarily to the RMS framework, this formalism is dynamical. This framework has been named Standard Model Extension (SME) and contains an impressive number of parameters encoding hypothetical deviations from the Standard Model of particles or from GR (see e.g [134–137], and references therein). A number of these SME coefficients have been constrained by various experiments (for a review of the current constraints on SME parameters, see [138]). It is worth mentioning that in the matter-sectors, atomic clocks and gravimetry has been used to constrain some SME coefficients (see as examples [113, 139, 139–142]). The SME formalism has also been developed in the gravitational sector to model deviations from Einstein's theory at the level of the gravitational part of the action. This part of the SME formalism will be developed in Sect. 4.2.4.

4.1.3 Local Position Invariance

Local Position Invariance (LPI) stipulates that the outcome of any local non-gravitational experiment is independent of the space-time position of the freely-falling reference frame in which it is performed [3]. This principle is mainly tested by two types of experiments: (i) search for variations in the constants of Nature and (ii) redshift tests.

The question of the constancy of the constants of Nature was first addressed by Dirac. This question is driven by the principle of reason: what could be the reasons behind the specific values of the constants of Physics? (see the discussion in Sect. 2 of Damour [143]). This argument led to many developments of new theories where the constants of physics become dynamical entities. In parallel, many observational investigations try and search for any space/time evolution of the constants of Physics [144].

Amongst all the observations performed, atomic clocks have an important role leading to currently some of the best constraints currently available. In particular, linear drifts in the evolution of the fine structure constant α, in the ratio μ between the mass of the electron and the mass of the proton and in the ratio between the mass of the light quarks (up and down) and the quantum chromodynamics (QCD) energy scale Λ_3. Several groups in the world have pursued effort to constrain such hypothetical linear drifts: at SYRTE, Observatoire de Paris [145], NIST [146], Berkeley [147], NPL [148], PTB [149], …The current constraints on a linear drift variation of the three constants (fine structure constant, ratio between the mass of the electron and the mass of the proton and ration between the mass of the light quarks and the QCD energy scale) are at the level of 10^{-16} per year. Note that astrophysical observations also constrain variations of the constants of Nature [144] and furthermore, these observables can be related to other cosmological observables like the evolution of the cosmic microwave background temperature, spectral distortion of the cosmic microwave background and violation of the cosmic distance-duality relation (see for example [150–152]).

In addition to temporal variations of the constants of Nature, one can search for spatial variations. Regarding this, atomic clocks have also been widely used to search for a variation of the constants of Nature with respect to the gravitational potential of the Sun. The idea is to compare two clocks working on different atomic transitions and therefore sensitive differently to the different constants of Physics, located at the same place and search for periodic variations in their frequencies comparison. Several groups in the world have been measuring this effect which is now constrained at the level of 10^{-6}: the SYRTE [145], the USNO [153], Berkeley [147], NIST [154],...

The second way to test the LPI is to measure the gravitational redshift. The gravitational redshift is a consequence of the EEP predicted by A. Einstein, 1911. It was observed for the first time by Pound and Rebka in 1959 [155]. A simple and convenient formalism to test the gravitational redshift is to introduce a new parameter α_{red} defined through [3]

$$\frac{\Delta\nu}{\nu} = (1 + \alpha_{\text{red}})\frac{\Delta U}{c^2} \tag{6}$$

where $\Delta\nu$ is the difference between the observed frequency of the same signal measured in different locations in the gravitational potential U. The parameter α_{red} vanishes when the EEP is valid. The best constraint on α_{red} is at the level of 10^{-4} and has been obtained in 1976 by comparing the frequency of two clocks: one onboard a rocket and the other one on Earth [156]. An improvement on this constraint is expected in the near future by using observations of GNSS satellites Galileo V and VI. These two satellites were launched on August, 30th 2014. Because of a technical problem, the launcher brought them on a wrong, elliptic orbit which makes them difficult to use for GNSS purposed. Nevertheless, the eccentricity of the orbit induces a periodic modulation of the gravitational redshift, which combined with the good stability of recent GNSS clocks can be used to test the gravitational redshift to a very good level of accuracy [157]. Contrary to the GP-A experiment, it is possible to integrate the signal on a long duration, therefore improving the statistics. A thorough study of the statistical and systematic uncertainties has shown that a constraint on α_{red} at the level of 10^{-5} can be achieved with a year of data [157]. The main limitation for this experiment comes from systematic effects that have the same signature than the redshift signal. These systematics are mainly due to orbital mismodelling and in particular due to mismodelling of the solar radiation pressure. Nevertheless, Delva et al. [157] have shown that a year of data leads to a decorrelation between the two signals.

On the mid-term, the Atomic Clock Ensemble in Space (ACES) project [158] that aims at comparing atomic clocks on the international space station (ISS) with atomic clocks on Earth will provide a test of the redshift test at the level of 10^{-6}. The ACES payload is expected to be launched on board of the ISS in 2018 for a duration between 18 months up to 3 years. The payload includes the first cold atom clock in space, PHARAO, consisting of a Cs clock and of a H-maser. Another key element is the microwave link (MWL) that uses radio in a two-way configuration to compare space to ground clocks at unprecedented stability and accuracy. The test of the EEP is only one of the scientific objective of ACES. In addition, the measurement of the gravitational redshift can be used to measure gravitational potential differences between different clock locations, which is a new type of geodetic measurements using clocks called chronometric geodesy [4, 159–162].

In the longer term, several future space missions have been proposed to further improve the test of the gravitational redshift. One example is the SpaceTime Explorer and QUantum Equivalence Space Test (STE-QUEST) space mission [163], a M-class mission proposal that was pre-selected by the European Space Agency in 2010 together with four other missions for the cosmic vision M3 launch opportunity. It carries out tests of different aspects of the Einstein Equivalence Principle using atomic clocks, matter wave interferometry and long distance time/frequency links, providing fascinating science at the interface between quantum mechanics and gravitation that cannot be achieved, at that level of precision, in ground experiments. STE-QUEST will test the UFF at a level of 10^{-15}, two orders of magnitude better than the best present result using macroscopic test-objects, and at least 6 orders better than present UFF tests in the quantum regime. It will also improve best present results on redshift tests (in the field of the Earth, Sun, and Moon) by up to 4 orders of magnitude.

Finally, let us mention that all these redshift tests are performed in a low gravitational field in the Solar System. A "strong" field version of this test was recently performed by measuring the gravitational redshift from the star S0-2 orbiting the supermassive black hole in our Galactic Center, Sagittarius A* [98, 164–166].

4.1.4 Models of Ultralight Dark Matter

Motivated by the unsuccessful searches for a Dark Matter particle at high energy, models of light scalar DM have recently gained a lot of attention in the scientific community (see e.g. [167–189] and references therein). In those models, a light scalar field is introduced in addition to the standard space-time metric and to the standard model fields. Such scalar fields are also ubiquitous in theories with more than 4 dimensions, and in particular in string theory with the dilaton and the moduli fields [190–194].

Such models have been shown to produce nice galactic and cosmological predictions for very low masses of the scalar field ranging from 10^{-24} to 10^{-22} eV [169, 173, 174, 182–184, 186, 187, 189]. Because of the high occupation numbers in galactic halos, the scalar field can be treated as a classical field for masses $\ll$eV [169, 181] and this model of DM is actually a particular tensor-scalar modification of GR which can break the Einstein Equivalence Principle through the coupling between the scalar field and standard matter. The full action for this model writes

$$S = \frac{c^3}{16\pi G}\int d^4x\sqrt{-g}\left[R - 2g^{\mu\nu}\partial_\mu\varphi\partial_\nu\varphi - 2\frac{c^2 m_\varphi^2}{\hbar^2}\varphi\right] + S_{\mathrm{mat}}\left[g_{\mu\nu}, \Psi, \varphi\right] , \tag{7}$$

where φ is the scalar field of mass m_φ and S_{mat} is the matter action including the coupling between the scalar field and standard matter. A widely used parametrization of the interaction between regular matter and the scalar field is introduced in [195] and is given in terms of the following Lagrangian

$$\mathcal{L}_{\text{int}}^{(j)} = \varphi^j \left[\frac{d_e^{(j)}}{4\mu_0} F^2 - \frac{d_g^{(j)}\beta_3}{2g_3} \left(F^A\right)^2 - \sum_{i=e,u,d} \left(d_{m_i}^{(j)} + \gamma_{m_i} d_g^{(1)} \right) m_i \bar{\psi}_i \psi_i \right] . \quad (8)$$

The superscripts $^{(j)}$ indicate the type of coupling considered: a coupling linear in φ or a coupling proportional to φ^2. In this Lagrangian, $F_{\mu\nu}$ is the standard electromagnetic Faraday tensor, μ_0 the magnetic permeability, $F_{\mu\nu}^A$ the gluon strength tensor, g_3 the QCD gauge coupling, β_3 the β function for the running of g_3, m_j the mass of the fermions (electron and light quarks), γ_{m_j} the anomalous dimension giving the energy running of the masses of the QCD coupled fermions and ψ_j the fermion spinors. The constants $d_j^{(i)}$ characterize the interaction between the scalar field φ and the different matter sectors.

This interaction Lagrangian leads to the following effective dependency of five constants of Nature

$$\alpha_{\text{EM}}(\varphi) = \alpha \left(1 + d_e^{(i)} \varphi^i\right) , \quad (9a)$$

$$m_j(\varphi) = m_j \left(1 + d_{m_j}^{(i)} \varphi^i\right) \quad \text{for } j = e, u, d \quad (9b)$$

$$\Lambda_3(\varphi) = \Lambda_3 \left(1 + d_g^{(i)} \varphi^i\right) , \quad (9c)$$

where α_{EM} is the electromagnetic fine structure constant, m_j are the fermions (electron and quarks up, down and strange) masses, Λ_3 is the QCD mass scale Λ_3 and the superscripts $^{(i)}$ indicate the type of coupling considered (linear for $i = 1$ and quadratic for $i = 2$).

At the cosmological level, this scalar field will oscillate at a frequency ω that is directly related to its mass m_φ through $\omega = c^2 m_\varphi / \hbar$

$$\varphi(t) = \varphi_0 \cos(\omega t + \delta) . \quad (10)$$

In addition, at the cosmological level, it can be shown that this scalar field can be interpreted as a perfect fluid whose mean pressure vanishes, making it a good DM candidate. Under the assumption that the DM is made completely by one mode of this scalar field, the amplitude of the scalar oscillations are directly determined by the DM energy density through

$$\rho_\varphi = \frac{c^6}{4\pi G \hbar^2} \frac{m_\varphi^2 \varphi_0^2}{2} . \quad (11)$$

In addition, the coupling of the scalar field with the standard model fields will exhibit signatures that can be searched for using different types of measurements. These signatures depend highly on the type of coupling considered.

Linear coupling: When the coupling between standard matter and the scalar field is linear in the Lagrangian from Eq. (8), the scalar field will be the sum of two distinct contributions: (i) a static Yukawa contribution, which is characteristic of massive gauge boson and which is independent of the identification of the scalar field to dark matter and (ii) an oscillating contribution which is identified as dark matter.

The Yukawa interaction is a generic feature appearing in most theories that introduce a new massive gauge boson. Measurements of the UFF are particularly sensitive to such a modification of the 2-body interaction. The best current measurements of the UFF on Earth have been done by the Eöt-Wash laboratory [65, 66]. This consists of a measurement of the differential acceleration between two test masses at the Earth's surface. Two types of pairs of tests masses have been used: (i) Be versus Ti and (ii) Be versus Al. For each of these pairs of test masses, a violation of the UFF in the field of the Earth, in the field of the Sun and in the field generated by the galactic dark matter distribution has been searched for. It is possible to reinterpret these constraints in terms of the coefficients that appear in the Lagrangian from Eq. (8) [196]. Note that in order to probe the UFF at very short distances, the Eöt-Wash group also performed an experiment where they made a body of Uranium rotate around the test masses to search for a violation of the UFF in the gravitational field of that body [64]. This measurement also has the advantage of being sensitive to different linear combinations of the matter-scalar coupling coefficients. For this particular experiment, the test masses are made of Cu and Pb and the ^{238}U source is located 10.2 cm from the test masses. Figure 1 shows the upper limit on the various coefficients that parametrize the interaction between the scalar field and the standard model fields.

In addition to the Yukawa 2-body interaction, if the scalar field is identified as DM, it will oscillate and these oscillations will be reflected in the time evolution of the constants of Nature because of the scalar/matter coupling. This is a signature of the Einstein Equivalence Principle that can be searched by comparing atomic transitions frequencies using atomic clocks as described in Sect. 2.2. Recently, this model has been constrained by three different teams: (i) using Dysprosium (Dy) atomic transitions at Standford [198], (ii) using the dual Cs/Rb atomic fountain from SYRTE [197]. These three searches constrain the scalar/matter linear coupling constant for a range of scalar mass m_φ between 10^{-24} and 10^{-12} eV as shown on Fig. 1.

The clocks are most powerful for low scalar field masses while UFF experiments become more interesting at larger masses. The UFF measurements start to deter when the interaction length scale of the Yukawa interaction $\lambda_\varphi \sim 1/m_\varphi$ is of the same order of the distance between the body that generate gravitation and the test masses. That is why, although the measurements from MICROSCOPE are one order of magnitude better than the ones on Earth (which is reflected for low masses), for larger masses, the corresponding upper limit falls quickly because of the altitude of the satellite while Earth experiments are more powerful to probe short Yukawa interaction length.

Note that the so-called *natural* couplings — usually defined as coupling of the order of unity — are excluded for scalar field masses m_φ up to $\sim 10^{-5}$ eV for $d_e^{(1)}$, up to $\sim 10^{-4}$ eV for $d_{\hat{m}}^{(1)} - d_g^{(1)}$ and up to 10^{-5} eV for $d_{m_e}^{(1)} - d_g^{(1)}$.

Quadratic coupling: The phenomenology arising in the case of a quadratic coupling between the scalar field and the standard model fields is quite different from the one

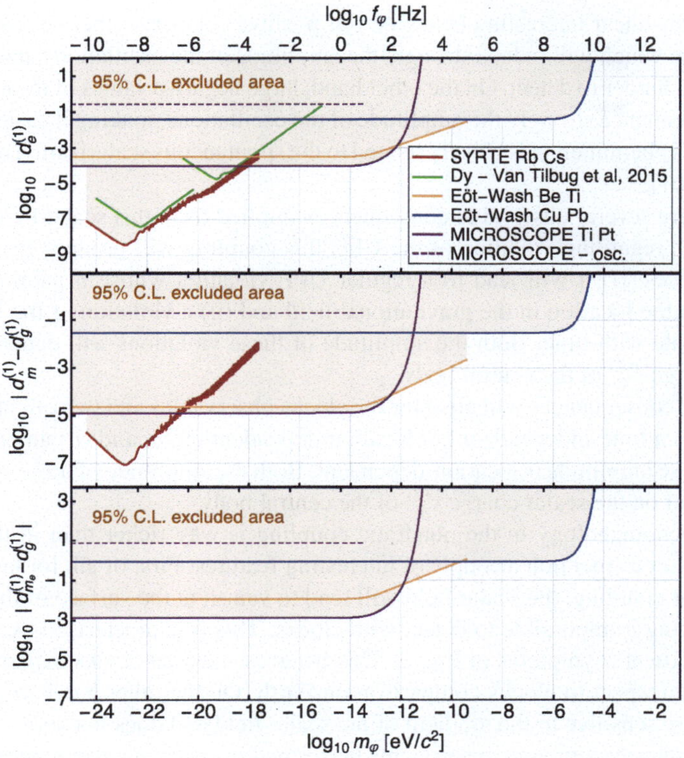

Fig. 1 Upper limit (at 95% confidence level) on the various scalar/matter coupling coefficients in the case of a linear coupling between matter and the scalar field. The SYRTE Cs/Rb analysis is from [197], the Dy analysis is presented in [198], the UFF measurement around Earth between Be and Ti is from [65], the UFF measurement between Cu and Pb in the gravitational field of a ^{238}U body is from [64], MICROSCOPE's result is presented in [72, 73]. The constraints derived from clock measurements assumed that the scalar field comprises all local DM while the UFF constraints do not rely on this assumption

that arises in the case of the linear coupling. In particular, it can be shown that no Yukawa interaction is present in that case, making it a big difference with respect to the linear interaction [199]. There is still an oscillatory mode that can be identified as DM but the amplitude of the oscillation will be impacted by the presence of other bodies and is now dependent on the distance to bodies, leading to a very rich phenomenology. In particular, the mode of the scalar field that can be identified as DM writes

$$\varphi^{(2)}(t, \boldsymbol{x}) = \varphi_0 \cos\left(\frac{m_\varphi c^2}{\hbar} t + \delta\right)\left[1 - s_A^{(2)} \frac{G M_A}{c^2 r}\right], \tag{12}$$

where $s_A^{(2)}$ depends on the scalar/matter couplings and on the compactness of the body A. It has been shown that the case of large couplings or large compactness leads to non-linear interesting behavior. For positive couplings, there is a screening mechanism which will reduce strongly the amplitude of the oscillations, making this scalar field harder to detect. On the other hand, large negative values of the couplings constants enhance strongly the amplitude of the oscillations, making it easier to constrain. This phenomenon is closely related to the spontaneous scalarization described in [200, 201].

There are several observable consequences implied from this scalar field profile. First of all, regarding violation of the UFF, this coupling will produce two distinct consequences: (i) it will lead to a regular UFF violation whose η parameter will depend on the location in the gravitational field and (ii) a violation of the UFF that will oscillate with time. Both the amplitude of these violations will depend on the scalar charge $s_A^{(2)}$ of the central body.

Similar consequences will arise for the clocks observables and two effects will be present: (i) a time-independent but location-dependent effect and (ii) an oscillating effect whose amplitude is location-dependent. Both the amplitude of these violations will depend on the scalar charge $s_A^{(2)}$ of the central body.

The phenomenology in the quadratic coupling is way richer than in the linear coupling. Let us mention a couple of interesting features. First of all, for large compactness or coupling, the scalar field will tend to vanish at the surface of the central body, making it impossible to detect with clocks. This is characterized by the presence of vertical asymptotes in Fig. 2. This behavior favored clocks experiments in space with respect to clocks comparison on Earth. On the other hand, violation of the UFF are sensitive to the gradient of the scalar field and does not suffer from the same divergences although experiments performed in space are also favored.

On Fig. 2 is presented the current constraints on the quadratic coupling constant $d_e^{(2)}$. Similar figures for the other coupling coefficients can be found in [196].

It is worth mentioning that, contrary to the linear coupling case, values corresponding to so-called "natural couplings" (i.e. $d_i^{(2)}$ of the order of unity) are either not constrained at all, or only very marginally constrained for extremely small DM masses. This leaves a lot of space for so-called "natural" models to exist in the context of quadratic couplings.

This type of search for Dark Matter is a remarkable example of interdisciplinary use of experiments built for metrology purposes that end up to be useful in a totally different context: to search for Dark Matter.

4.1.5 Cosmological Implications of Laboratory Measurements

While the previous section is devoted to searches of Dark Matter with laboratory measurements, these have also some implications at the cosmological level. For example, a coupling between a scalar field and the electromagnetic part of the Lagrangian such that the electromagnetic part of the action writes

$$S_{\text{EM}} = -\frac{1}{4\mu_0}\int d^4x\sqrt{-g}\left(1 - d_e^{(1)}\varphi\right) F_{\mu\nu}F^{\mu\nu} + q_p \int A_\mu dx^\mu\,, \tag{13}$$

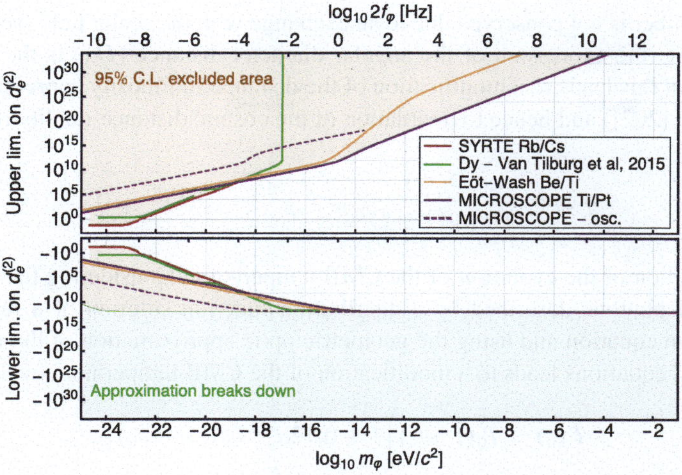

Fig. 2 Upper and lower (MRA) limits (at 95% confidence level) on the scalar/electromagnetic coupling coefficients $d_e^{(2)}$ in the case of a quadratic coupling between matter and the scalar field. Similar figures for the other coupling coefficients can be found in [196]/ The constraints have been derived using the following measurements: the SYRTE Cs/Rb data from [197], the Dy measurements from [198], the UFF measurement around Earth between Be and Ti from [65] and the MICROSCOPE's result presented in [72]. Note that the dashed line is not an actual constraint but an estimate of the potential sensitivity that would be obtained by searching for an oscillating violation of the UFF within MICROSCOPE data

where q_p is the electric charge of a particle interacting with the EM field. As mentioned in the previous section, this type of coupling has implications on atomic clocks and on the universality of free fall. In addition to these effects, four cosmological observables are modified (with respect to GR) and are intimately related to each other in this class of theories (see [151]): (i) temporal variation of the fine structure constant, (ii) violation of the distance-duality relation, (iii) modification of the evolution of the Cosmic Microwave Background (CMB) temperature and (iv) CMB spectral distortions. It is worth to insist on the fact that the derivation relies only on the matter part of the action and not on the gravitational part. This means that our results apply to a very wide class of gravitation theories. In a Friedman–Lemaître–Robertson–Walker space-time, the expressions of the four observables are given by the following expressions:

- temporal variation of the fine structure constant. A straightforward identification in the action leads to

$$\frac{\Delta\alpha}{\alpha} = \frac{\alpha(z) - \alpha_0}{\alpha_0} = d_e^{(1)} \left(\varphi(z) - \varphi_0\right) \tag{14}$$

  where z is the redshift and the subscripts 0 refer to $z = 0$.
- violation of the cosmic distance-duality relation. The optic geometric limit of the modified Maxwell equations shows that photons propagate on null geodesics but

their number is not conserved due to an exchange with the scalar field (see [202]). Therefore, the expression of the angular diameter distance (D_A) is the same as in GR but this leads to a modification of the distance-luminosity expression (D_L) (see also [202]) and hence to a violation of the cosmic distance-duality relation:

$$\eta(z) = \frac{D_L(z)}{D_A(z)(1+z)^2} = 1 + \frac{d_e^{(1)}}{2}\left(\varphi(z) - \varphi_0\right) . \tag{15}$$

- modification of the evolution of the CMB temperature. Considering the CMB as a gaz of photons described by a distribution function solution of a relativistic Boltzman equation and using the geometric optic approximation of the modified Maxwell equations leads to a modification of the CMB temperature evolution:

$$T(z) = T_0(1+z)\left[1 + 0.12 d_e^{(1)}\left(\varphi(z) - \varphi_0\right)\right] . \tag{16}$$

- spectral distortion of the CMB. Using the same approach as the one sketched in the last item, one can show that the evolution of the CMB radiation leads to deviations from its black body spectrum. When this deviation is parametrized by a chemical potential μ, one can show that its expression at current epoch is given by

$$\mu = 0.47 d_e^{(1)}\left[\varphi(z_{\rm CMB}) - \varphi_0\right] . \tag{17}$$

To summarize, a coupling between the scalar field and EM implies that the four observables are intimately linked to each other through the relations

$$d_e^{(1)}\left[\varphi(z) - \varphi_0\right] = \frac{\Delta\alpha}{\alpha} = 2\left[\eta(z) - 1\right] = 8.33\left[\frac{T(z)}{T_0(1+z)} - 1\right] , \tag{18a}$$

$$\mu = 0.47\frac{\Delta\alpha(z_{\rm CMB})}{\alpha} = 0.94\left[\eta(z_{\rm CMB}) - 1\right] = 3.92\left[\frac{T(z_{\rm CMB})}{T_0(1+z_{\rm CMB})} - 1\right] . \tag{18b}$$

These relations hold for all theories whose matter part of the action can be cast in the form of the action (13). This class of theories is very large and includes all metric theories. These relations imply also that if a deviation of the EEP is observed around Earth, it is likely to have cosmological counterparts as well, which would be an excellent check to confirm any observed deviations. On the other hand, it also shows that local measurements can have cosmological interpretation.

As an example, several parametrizations of the distance-duality relationship have been considered in the literature: $\eta(z) = \eta_0$, $\eta(z) = 1 + \eta_1 z$, $\eta(z) = 1 + \eta_2\frac{z}{1+z}$, $\eta(z) = 1 + \eta_3 \ln(1+z)$ and for the evolution of the CMB temperature $T(z) = (1+z)^{1-\beta}$. Assuming that the theory of gravitation is described by the multiplicative coupling introduced in Eq. (13) (which is a large class of theories including GR), we can use the relations from Eq. (18) to transform observational constraints on one type of observations into constraints on another type. As an example, we use the constraint from optical clocks from [146] ($\dot\alpha/\alpha = (1.6 \pm 2.3) \times 10^{-17}\,\mathrm{yr}^{-1}$ to constrain the cosmic distance-duality parameters: $\eta_1 = \eta_2 = \eta_3 = (1 \pm 1.4) \times 10^{-7}$ and the evolution of the CMB temperature $\beta = -0.3 \pm 0.3$ (see [151]). These estimations are a factor 5 better than similar estimations based on measurements of variations of α on astrophysical distances and order of magnitudes better than direct measurements of the cosmic distance-duality. This illustrates very clearly the interplay between small scales measurements and cosmological interpretation.

4.1.6 Violations of the Equivalence Principle and Dark Energy

The action from Eq. (7) with a vanishing potential (i.e. a massless scalar field) has also been studied in the context of the acceleration of the cosmic expansion (see e.g. [203, 204] and references therein). In this case, it is useful to work with the cosmological energy density and pressure associated with the scalar field and it can be shown that the Dark Energy equation of state is given by (see[204]) $w_\varphi = -1 + \frac{2}{3}\frac{\varphi'^2}{\Omega_\varphi}$ where $\Omega_\varphi = \frac{8\pi G \rho_\varphi}{3H^2c^2}$. This equation allows one to derive an expression that makes the connection between the temporal evolution of α to cosmological variables [205]

$$\left.\frac{\dot{\alpha}}{\alpha}\right|_0 = -d_e^{(1)} H_0 \sqrt{\frac{3}{2}\Omega_{\varphi 0}(1 + w_{\varphi 0})}\,. \tag{19}$$

This relation is very useful in order to combine data in a global analysis. For example, [205] uses atomic clocks constraints on α and cosmological observations (Supernovae Ia and Hubble parameter measurement) in order to constrain the parameters for this Dark Energy model. In particular, it is shown that one type of observations only would not constrain this model at all but that it's really the combination of local measurements with astrophysical observations that is useful. It is also remarkable to see in that context the importance of local measurements that have been developed in the context of metrology and geodesy and to appreciate the impact of such measurements into a completely different field: cosmology.

4.1.7 The Matter-Gravity Sector of the Standard Model Extension (SME)

The SME framework is a wide framework that aims at parametrizing all possible violation of Lorentz symmetry in all sectors of physics. A part of this framework aims at parametrizing violations of the EEP by modifying the matter action as

$$S_{\text{Matter}} = \int d\lambda\, c\left(-m\sqrt{-(g_{\mu\nu} + 2c_{\mu\nu})u^\mu u^\nu} - (a_{\text{eff}})_\mu\, u^\mu\right), \tag{20}$$

where the particle's worldline tangent is $u^\mu = dx^\mu/d\lambda$ [55]. The fields $c_{\mu\nu}$ and $(a_{\text{eff}})_\mu$ are new dynamical fields whose kinematic terms are unspecified. In the linearized gravity limit, the motion of test masses depend only on the background value (or vacuum expectation value) of these fields (the background values will be denoted by a bar). The coefficients $\bar{c}_{\mu\nu}$ and $(\bar{a}_{\rm eff})_\mu$ are species dependent, therefore leading to the three facets of the EEP.

The $\bar{c}_{\mu\nu}$ coefficients have been widely constrained by different laboratory experiments including atomic clock measurements [139–142, 206] and optical cavity [207, 208].

On the opposite, the $(\bar{a}_{\rm eff})_\mu$ components are less constrained. The temporal component is currently only constrained through redshift test [141]. Superconducting gravimeters have recently been used to measure the spatial components [209] at the level of 10^{-5---6} GeV. The idea is to search for periodic variation in the gravitational acceleration. The main difficulty from such measurements come from tidal effects that produce similar signatures and that needs to be removed carefully. A recent reanalysis of LLR observations have provided constraints two orders of magnitude more stringent [210] and are currently the best constraints on these parameters. Note that a reanalysis of the MICROSCOPE measurements within the context of the SME would provide an improvement on these constraints [211]. Such an analysis is currently on-going.

4.1.8 Conclusion

In conclusion, the EEP is an essential part of Einstein's theory that is shared by a number of alternative theories of gravitation. All the theories of gravitation satisfying the EEP are called metric theories. In these theories, it is sufficient to know the space-time metric (to which matter is minimally coupled) to infer all effects produced by gravitation (motion of bodies, light propagation, behavior of clocks, ...). It is interesting to mention that the equivalence principle is a generalization of the fact that all bodies seem to fall with the same acceleration in a gravitational field. Nevertheless, as mentioned by T. Damour [143]: "Despite its name, the equivalence principle is not one of the basic principles of physics. There is nothing taboo about having an observable violation of the EP. In contrast, one can argue (notably on the basis of the central message of Einsteins theory of general relativity) that the historical tendency of physics is to discard any, a priori given, absolute structure (principle of absence of absolute structures)." It is therefore highly important to pursue our quest to test the various facets of this principle. Several on-going projects or proposals aim at improving our current searches for deviations from the EEP in the standard phenomenological framework used to test the UFF, LLI and LPI.

In all the different kind of tests, experiments to measure geodetic/geophysical properties like clocks, gravimeter, LLR have an important role regarding tests of the Einstein Equivalence Principle. It is also amazing that such measurements have implications in several different fields related to the search for Dark Matter and even at the cosmological level with the search for Dark Energy.

4.2 Tests of the Einstein Field Equations

There exist several frameworks to test the form of the space-time metric, the second building block of General Relativity. The two most famous ones are the parametrized post-Newtonian formalism (PPN) and the fifth force formalism. Nevertheless, these two formalisms do not cover all the extensions from GR (see e.g. [60, 61]) and it is useful to analyze experimental results in new extended frameworks. Indeed, even if observations lie very close to GR when analyzed within the PPN or fifth force framework, this does not mean that this has to be true in any other framework. In addition, a framework that would allow a direct comparison between local tests of GR and cosmological observation has not been developed so far and still needs to be developed.

In this section, we will review recent constraints obtained for different frameworks used to probe the form of space-time geometry.

4.2.1 The Parametrized Post-Newtonian Formalism

The PPN formalism is providing an interface between theoretical developments and data analyses. It is fully described in [43]. In this formalism, a phenomenological expansion at the level of the space-time metric is performed by introducing 10 dimensionless parameters. From an observational point of view, these parameters can be constrained regardless of any considerations about the hypothetical underlying theory. From a theoretical point of view, if the post-Newtonian metric of a theory can be matched to the PPN metric, it can automatically be constrained by all observations used to constrain the PPN parameters.

In the simplest case, only two PPN parameters are considered and the spacetime metric for a spherically symmetric configuration is given by

$$ds^2 = -\left(1 + 2\frac{\phi_N}{c^2} + 2\beta\frac{\phi_N^2}{c^4} + \dots\right)c^2dt^2 + \left(1 - 2\gamma\frac{\phi_N}{c^2} + \dots\right)dx^2\,, \quad (21)$$

where ϕ_N is the Newtonian potential and γ and β are the PPN parameters. These parameters take the value of 1 in GR and my deviate from 1 in alternative theories.

The PPN formalism was historically the first one developed and therefore was extensively constrained by several different types of observations. The best constraint on the γ parameter was obtained by measuring precisely the Shapiro time delay with the Cassini spacecraft when it was cruising between Jupiter and Saturn in 2003 and is given by (see [212])

$$\gamma - 1 = (2.1 \pm 2.3) \times 10^{-5}\,. \quad (22)$$

This constraint has been confirmed by measuring the light deflection around the Sun using Very Long Baseline Interferometry (see [213, 214]), by using radioscience observations of spacecraft orbiting Mars (see [215]) or Mercury (see [31]).

The β PPN parameter is essentially constrained through orbital dynamics. Lunar Laser Ranging analysis provides a constraint at the level of 10^4 (see [67, 69, 70]). Furthermore, planetary ephemerides analysis currently provides constraints at the level of 10^{-5} (see [31, 32, 35, 37, 215, 216]). The other PPN parameters are also very well constrained. An updated list of constraints can be found in [3].

4.2.2 The Fifth Force Formalism

In the fifth force formalism, a Yukawa-type deviation from Newtonian gravity is considered. The deviation from the Newtonian potential is given by

$$\phi = -\frac{GM}{r}\left(1 + \alpha e^{-r/\lambda}\right) , \tag{23}$$

where α is the strength of the interaction and λ a characteristic length scale. The idea is to constrain the couple of parameters (α, λ). An impressive number of experiments and observations have also been used to constrain this formalism at various scales: at Solar System scales by using planetary ephemerides analyses [215], Lunar Laser Ranging observations [217], ranging measurements of LAGEOS satellite around Earth [50]. At lower distances, several geophysical observations have been used to constrain a hypothetical fifth force. A short distances, torsion balances provide the best constraints. For $\lambda \leq \mu$m, it becomes extremely hard to constrain the strength of the interaction because of several quantum noise sources (like e.g. the Casimir effect). The current limit on the Yukawa parameters are summarized in Fig. 3. In conclusion, the fifth force formalism is very well constrained except at very low and very large distances where deviations can still be searched for.

4.2.3 A Temporal Evolution of the Gravitational Constant *G*

As soon as a new field is introduced in addition to the standard space-time metric to describe the gravitational interaction, it is possible that the gravitational constant G will become space-time dependent. This is always the case when a coupling between the scalar field and the scalar curvature is introduced like e.g. in Brans–Dicke theory [218–220]. This feature also appears in several model of Dark Matter [99] or of Dark Energy [101, 202].

A very interesting aspect is related to the fact that a temporal evolution of G might still be present even when screening mechanisms are acting. Screening mechanisms are non-linear effects that produce a strong reduction of the deviations from General Relativity in a specific regime. Several mechanisms are know like e.g. the chameleon mechanism [221–223] where the deviation is reduced in region of high matter density, the symmetron mechanism [224, 225] where a scalar field is screened through a symmetry restoring mechanism in region of high density or the Vanshtein mechanism [226, 227] appearing in massive gravity and in Galileons where deviations are hidden

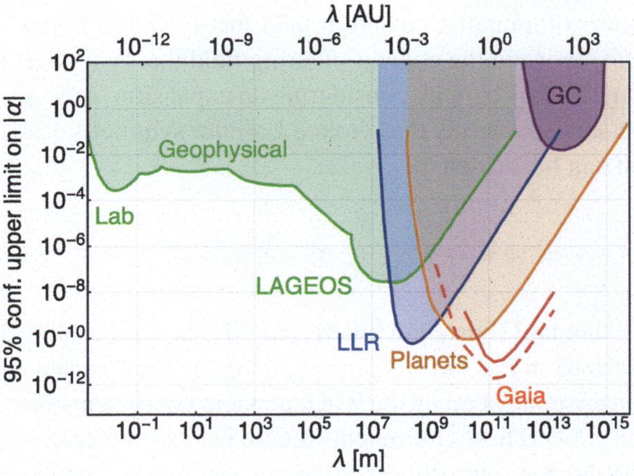

Fig. 3 95% upper confidence limit on the strength $|\alpha|$ of the Yukawa interaction as a function of the interaction length scale λ. Several geodesic techniques have provided the best current limit on the Yukawa interaction. The red curves are limits that are expected from the Gaia mission. The purple curve is a limit obtained around the supermassive black hole from our Galactic Center [98]

below a certain length scale or the k-mouflage mechanism [228] which is due to non-linearity in the scalar field kinetic term in the action. While these mechanisms strongly reduced the deviations from GR in the static limit, allowing these theories to pass the standard Solar System tests of gravity, it has been shown that in some cases, the temporal variation of the gravitational constant will not be screened and will be a perfect signature to search for and to constrain such theories [229]. That argument shows that constraining an hypothetical $\dot{G}$ is highly important.

So far, only linear variation of the gravitational constant has been searched for. The best measurements of $\dot{G}/G$ are at the level of 10^{-13} yr^{-1} from Lunar Laser Ranging [230] and from planetary ephemerides [32, 215], both these important measurements relying on techniques developed for geodesy.

4.2.4 The Gravitational Sector of the Standard Model Extension (SME)

As mentioned previously, the SME is a very wide framework aiming at considering systematically Lorentz violations in any sector of Physics. In the pure gravity sector, SME can be interpreted as a metric extension of GR. Any action-based model that breaks local Lorentz symmetry either explicitly or spontaneously can be matched to a subset of the SME coefficients. Therefore, constraints on SME coefficients can directly constrain these models. Matches between various toy models and coefficients in the SME have been achieved for models that produce effective $\bar{s}^{\mu\nu}$, $\bar{c}_{\mu\nu}$, $\bar{a}_{\mu}$, and other coefficients. This includes vector and tensor field models of spontaneous Lorentz-symmetry breaking [54, 55, 231–234], models of quantum gravity

[235, 236] and noncommutative quantum field theory [237]. Furthermore, Lorentz violations may also arise in the context of string field theory models [238].

The SME framework is built by considering an expansion at the level of the action by introducing a series of terms that violate Lorentz symmetry [54]. At the lowest order, this action can be written as

$$\mathcal{L} = \mathcal{L}_{\rm EH} + \frac{c^3}{32\pi G} h^{\mu\nu} \bar{s}^{\alpha\beta} \mathcal{G}_{\alpha\mu\nu\beta} + \dots , \tag{24}$$

where $\mathcal{L}_{\rm EH}$ is the standard Einstein–Hilbert term, $\mathcal{G}_{\alpha\mu\nu\beta}$ is the double dual of the Einstein tensor linearized in $h_{\mu\nu}$ (with $h_{\mu\nu} = g_{\mu\nu} - \eta_{\mu\nu}$). The Lorentz-violating effects in this expression are controlled by the 9 independent coefficients in the traceless and dimensionless $\bar{s}^{\mu\nu}$ [54]. These coefficients are treated as constants in asymptotically flat cartesian coordinates. The ellipses represent additional terms in a series including terms that break CPT symmetry for gravity; such terms are detailed elsewhere [239–242] and are part of the so-called nonminimal SME expansion. Note that the process by which one arrives at the effective quadratic Lagrangian (24) is consistent with the assumption of the spontaneous breaking of local Lorentz symmetry.

Several measurements have been used to search for a breaking of the Lorentz invariance and to constrain the SME parameters. Amongst them, measurements dedicated to geodesy and geophysics are the most important ones like e.g.: atomic interferometry [243, 244], gravimetry [209, 245], planetary ephemerides [246], very long baseline interferometry [247], lunar laser ranging [210, 248, 249]. Table 1 summarizes the order of magnitude of the constraints on the SME coefficients from the pure gravity sector that have been obtained using the techniques mentioned in Sect. 2. More informations about these constraints and about other constraints can be found in the reviews [56, 57].

Table 1 Order of magnitude of the constraints on the pure gravitational SME coefficients derived from measurements that are related to geodesy and geophysics

Experiments	$\bar{s}^{TT}$	$\bar{s}^{TJ}$	$\bar{s}^{JK}$
Atom interferometry [244]	–	$\sim 10^{-5}$	$\sim 10^{-9}$
Gravimeters [209]	–	$\sim 10^{-7}$	$\sim 10^{-9}$
Gravimeters [245]	–	–	$\sim 10^{-10}$
Planetary ephemerides [246]	–	$\sim 10^{-8}$	$\sim 10^{-10}$
Lunar laser ranging [249]	–	$\sim 10^{-9}$	$\sim 10^{-11}$
VLBI [247]	10^{-4}	–	–

4.2.5 Modified Newtonian Dynamics (MOND)

The MOND phenomenology was introduced in 1983 by M. Milgrom [250–252] in order to explain observations at the galactic scale (in particular galactic rotation curves) without the introduction of Dark Matter. The main idea behind the MOND phenomenology is to replace the standard Newton acceleration g_N valid in "high" gravitational field by $\sqrt{a_0 g_N}$ for regions of the Universe where the gravitational field is very low. More precisely, the MOND phenomenology depends on a MOND interpolating function ν and on a MOND acceleration scale a_0. The interpolating function is making the transition between the Newtonian regime and the MONDian regime appearing in low gravitational fields. While developed initially as a purely phenomenological Newtonian model, the MOND phenomenology has later been extended as relativistic theories. A review of different relativistic MOND theories can be found in [253] Bruneton and Esposito-Farse or in [1]. A review of the different interpolating functions used in the literature can also be found in [1].

Recently, it has been shown that contrarily to what was previously expected, the MOND phenomenology would produce detectable effects in the Solar System. This effect is due to the non-linearity of the MOND phenomenology which implies that the external galactic gravitational field will play a role in this theory. This External Field Effect was first discovered by [58] and was studied later in [59]. The main contribution from this External Field Effect in the Solar System takes the form of a quadrupolar deviation from the Newton potential

$$\phi_N = -\frac{GM}{r} - \frac{Q_2}{2} x^i x^j \left(e_i e_j - \frac{1}{3}\delta_{ij} \right) , \tag{25}$$

where the unit vector e_i points toward the galactic center and Q_2 is a parameter that depends directly on the MOND interpolating function ν and on the MOND acceleration scale a_0. Values of this parameter for different interpolating functions are given in [58, 59, 254].

The interesting point is that this modification of gravitation does not enter the fifth force or the PPN formalisms and it can be used efficiently to detect or constrain the MOND phenomenology. A recent analysis based on the Saturn ephemeris (see Sect. 2.5) obtained mainly from the radio tracking data of the Cassini spacecraft (between 2004 and 2013) led to a constraint on Q_2 given by [26]

$$Q_2 = (3 \pm 3) \times 10^{-27}\ \mathrm{s}^{-2} . \tag{26}$$

This constraint on MOND theory is the best currently available on Solar System scales. When combined with galactic rotation curves observations, this analysis narrows down the possible interpolating functions allowed to only one class of function as shown in [254].

As such, this local constraint does not constraint the MOND theory. What is interesting is to combine this local constraint with astrophysical measurements of galactic rotation curves. Indeed, several classes of interpolating function ν have been used to explain the behavior of galactic rotation curves. Using observations of 27 galactic rotation curves (using data from [255]), fitting them with different MOND phenomenology and combining this analysis from the Solar System leads to a very stringent constraint on the MOND phenomenology. Indeed, all but one class of interpolating function is excluded by this combined analysis [254]. This result shows the power of combining multi-scales measurements and once again how geodesic measurements can help to improve our understanding of fundamental physics, even at a totally different scale in our Universe.

4.2.6 Conclusion

In conclusion, any modification of GR will at least lead to a modification of the form of the space-time metric. This will be induced by new fields in addition to the regular space-time metric, by the consideration of higher dimensions, the inclusion of derivatives of the space-time metric of the order higher than 2, ...Measuring the curvature of space-time around different body in different location in space and time in our Universe is therefore one of the best way to search for new physics beyond GR. Two formalisms have been widely used so far to search for deviations from GR at the level of the space-time metric: the PPN formalism and the fifth force formalism. In this section, we have shown that geodesic measurements are very important to constrain these formalisms. In addition, not every alternative theory of gravitation enters these two frameworks. In the previous sections, we give examples of other modifications from GR that do not enter the PPN or fifth force frameworks. It is important to reanalyze existing measurements in the context of these GR extensions and in that context, geodetic measurements are also very important.

5 Conclusion and Outlook

In this communication, we highlight how experiments and measurements that have been developed in the context of geodesy and geophysics have found very interesting applications in fundamental physics. These two fields are very highly active field of research that thrives on technological developments, in particular of clocks, long distance radio and optical links, and inertial sensors (accelerometers, gyroscopes). It is impossible to describe all the constraints on the modified theory of gravitation that have been derived using such measurements and in this paper, we focused only on several important and/or recent ones. This shows the vitality of the field and many more exciting results are expected in the near and more distant future, which will hopefully pave the way towards the new physics beyond general relativity and the standard model of particle physics. And it is likely that measurements or experiments developed in the context of geodesy will continue to play a major role in that process.

References

1. B. Famaey, S.S. McGaugh, Modified Newtonian dynamics (MOND): observational phenomenology and relativistic. Living Rev Relativ **15**, 10 (2012)
2. T. Clifton, P.G. Ferreira, A. Padilla, C. Skordis, Modified gravity and cosmology. Phys. Rep. **513**, 1–189 (2012)
3. C.M. Will, The confrontation between general relativity and experiment. Living Rev. Relativ. **17**, 4 (2014)
4. G. Lion, I. Panet, P. Wolf, C. Guerlin, S. Bize, P. Delva, Determination of a high spatial resolution geopotential model using atomic clock comparisons. J. Geodesy **91**, 597–611 (2017)
5. Microg Lacoste. FG5-X and FGL absolute gravity meters
6. μ QUANS. Absolute quantum gravimeter
7. C.W. Chou, D.B. Hume, J.C.J. Koelemeij, D.J. Wineland, T. Rosenband, Frequency comparison of two high-accuracy Al^+ optical clocks. Phys. Rev. Lett. **104**(7), 070802 (2010)
8. K. Beloy, N. Hinkley, N.B. Phillips, J.A. Sherman, M. Schioppo, J. Lehman, A. Feldman, L.M. Hanssen, C.W. Oates, A.D. Ludlow, Atomic clock with 1×10^{-18} room-temperature blackbody stark uncertainty. Phys. Rev. Lett. **113**(26), 260801 (2014)
9. T.L. Nicholson, S.L. Campbell, R.B. Hutson, G.E. Marti, B.J. Bloom, R.L. McNally, W. Zhang, M.D. Barrett, M.S. Safronova, G.F. Strouse, W.L. Tew, J. Ye, Systematic evaluation of an atomic clock at 2×10^{-18} total uncertainty. Nat. Commun. **6**, 6896 (2015)
10. I. Ushijima, M. Takamoto, M. Das, T. Ohkubo, H. Katori, Cryogenic optical lattice clocks. Nat. Photon. **9**, 185–189 (2015)
11. N. Huntemann, C. Sanner, B. Lipphardt, C. Tamm, E. Peik, Single-ion atomic clock with 3×10^{-18} systematic uncertainty. Phys. Rev. Lett. **116**(6), 063001 (2016)
12. Peter Wolf. Viewpoint: Next generation clock networks. *Physical Review: Viewpoint*, Physics 9, 51, May 11, 2016 2016
13. H. Denker, L. Timmen, C. Voigt, S. Weyers, E. Peik, H.S. Margolis, P. Delva, P. Wolf, G. Petit, Geodetic methods to determine the relativistic redshift at the level of 10 (-18) - 18 in the context of international timescales: a review and practical results. J. Geodesy **92**, 487–516 (2018)
14. T.E. Mehlstäubler, G. Grosche, C. Lisdat, P.O. Schmidt, H. Denker, Atomic clocks for geodesy. Rep. Prog. Phys. **81**(6), 064401 (2018)
15. I. Ciufolini, E. Pavlis, F. Chieppa, E. Fernandes-Vieira, J. Perez-Mercader, Test of general relativity and measurement of the lense-thirring effect with two earth satellites. Science **279**, 2100 (1998)
16. I. Ciufolini, E.C. Pavlis, A confirmation of the general relativistic prediction of the Lense-Thirring effect. Nature **431**, 958–960 (2004)
17. I. Ciufolini, A. Paolozzi, E.C. Pavlis, R. Koenig, J. Ries, V. Gurzadyan, R. Matzner, R. Penrose, G. Sindoni, C. Paris, H. Khachatryan, S. Mirzoyan, A test of general relativity using the LARES and LAGEOS satellites and a GRACE Earth gravity model. Measurement of Earth's dragging of inertial frames. Eur. Phys. J. C **76**, 120 (2016)
18. J.O. Dickey, P.L. Bender, J.E. Faller, X.X. Newhall, R.L. Ricklefs, J.G. Ries, P.J. Shelus, C. Veillet, A.L. Whipple, J.R. Wiant, J.G. Williams, C.F. Yoder, Lunar laser ranging: a continuing legacy of the apollo program. Science **265**, 482–490 (1994)
19. N.P. Pitjev, E.V. Pitjeva, Constraints on dark matter in the solar system. Astron. Lett. **39**, 141–149 (2013)
20. E.M. Standish, The JPL planetary ephemerides. Celest. Mech. **26**, 181–186 (1982)
21. X.X. Newhall, E.M. Standish, J.G. Williams, DE 102 - A numerically integrated ephemeris of the moon and planets spanning forty-four centuries. A&A **125**, 150–167 (1983)
22. E.M. Standish Jr., The observational basis for JPL's DE 200, the planetary ephemerides of the Astronomical Almanac. A&A **233**, 252–271 (1990)
23. E.M. Standish, Testing alternate gravitational theories, in *IAU Symposium*, ed. by S.A. Klioner, P.K. Seidelmann, M.H. Soffel, vol. 261 (2010), pp. 179–182

24. E.M. Standish, J.G. Williams, Orbital ephemerides of the Sun, Moon, and Planets, in *Explanatory Supplement to the Astronomical Almanac*, ed. by S.E. Urban, P.K. Seidelmann, 3rd edn. (Univeristy Science Books, 2012), pp. 305–346
25. W.M. Folkner, J.G. Williams, D.H. Boggs, R. Park, P. Kuchynka, The planetary and lunar ephemeris DE 430 and DE431. IPN Prog. Rep. **42**(196) (2014)
26. A. Hees, W.M. Folkner, R.A. Jacobson, R.S. Park, Constraints on modified Newtonian dynamics theories from radio tracking data of the Cassini spacecraft. Phys. Rev. D **89**(10), 102002 (2014)
27. A. Fienga, H. Manche, J. Laskar, M. Gastineau, INPOP06: a new numerical planetary ephemeris. A&A **477**, 315–327 (2008)
28. A. Fienga, J. Laskar, T. Morley, H. Manche, P. Kuchynka, C. Le Poncin-Lafitte, F. Budnik, M. Gastineau, L. Somenzi, INPOP08, a 4-D planetary ephemeris: from asteroid and time-scale computations to ESA Mars Express and Venus Express contributions. A&A **507**, 1675–1686 (2009)
29. A. Fienga, J. Laskar, P. Kuchynka, C. Le Poncin-Lafitte, H. Manche, M. Gastineau, Gravity tests with INPOP planetary ephemerides, in *IAU Symposium*, ed. by S.A. Klioner, P.K. Seidelmann, M.H. Soffel, vol. 261 (2010), pp. 159–169
30. A. Fienga, J. Laskar, P. Kuchynka, H. Manche, G. Desvignes, M. Gastineau, I. Cognard, G. Theureau, The INPOP10a planetary ephemeris and its applications in fundamental physics. Celest. Mech. Dyn. Astrono. **111**(3), 363–385 (2011)
31. A.K. Verma, A. Fienga, J. Laskar, H. Manche, M. Gastineau, Use of MESSENGER radioscience data to improve planetary ephemeris and to test general relativity. A&A **561**, A115 (2014)
32. A. Fienga, J. Laskar, P. Exertier, H. Manche, M. Gastineau, Numerical estimation of the sensitivity of INPOP planetary ephemerides to general relativity parameters. Celest. Mech. Dyn. Astron. **123**, 325–349 (2015)
33. E.V. Pitjeva, High-precision ephemerides of planets EPM and determination of some astronomical constants. Solar Syst. Res. **39**, 176–186 (2005)
34. E.V. Pitjeva, EPM ephemerides and relativity, in *Proceedings of IAU Symposium 261*, ed. by S.A. Klioner, P.K. Seidelmann, M.H. Soffel (2010), pp. 170–178
35. E.V. Pitjeva, N.P. Pitjev, Relativistic effects and dark matter in the Solar system from observations of planets and spacecraft. MNRAS **432**, 3431–3437 (2013)
36. E.V. Pitjeva, Updated IAA RAS planetary ephemerides-EPM2011 and their use in scientific research. Solar Syst. Res. **47**, 386–402 (2013)
37. E.V. Pitjeva, N.P. Pitjev, Development of planetary ephemerides EPM and their applications. Celest. Mech. Dyn. Astron. **119**, 237–256 (2014)
38. A.L. Fey, D. Gordon, C.S. Jacobs, C. Ma, R.A. Gaume, E.F. Arias, G. Bianco, D.A. Boboltz, S. Böckmann, S. Bolotin, P. Charlot, A. Collioud, G. Engelhardt, J. Gipson, A.-M. Gontier, R. Heinkelmann, S. Kurdubov, S. Lambert, S. Lytvyn, D.S. MacMillan, Z. Malkin, A. Nothnagel, R. Ojha, E. Skurikhina, J. Sokolova, J. Souchay, O.J. Sovers, V. Tesmer, O. Titov, G. Wang, V. Zharov, The second realization of the international celestial reference frame by very long baseline interferometry. AJ **150**, 58 (2015)
39. M. Soffel, S.A. Klioner, G. Petit, P. Wolf, S.M. Kopeikin, P. Bretagnon, V.A. Brumberg, N. Capitaine, T. Damour, T. Fukushima, B. Guinot, T.-Y. Huang, L. Lindegren, C. Ma, K. Nordtvedt, J.C. Ries, P.K. Seidelmann, D. Vokrouhlický, C.M. Will, C. Xu, The IAU 2000 resolutions for astrometry, celestial mechanics, and metrology in the relativistic framework: explanatory supplement. Astron. J. **126**, 2687–2706 (2003)
40. A. Einstein, Über den Einfluß der Schwerkraft auf die Ausbreitung des Lichtes. Annalen der Physik **340**, 898–908 (1911). Traduction anglaise dans [256]
41. K.S. Thorne, C.M. Will, Theoretical frameworks for testing relativistic gravity I. Foundations. ApJ **163**, 595 (1971)
42. C.W. Misner, K.S. Thorne, J.A. Wheeler, *Gravitation*. Physics Series (W. H. Freeman, 1973)
43. C.M. Will, *Theory and Experiment in Gravitational Physics* (1993)

44. A. Einstein, Die Grundlage der allgemeinen Relativitätstheorie. Annalen der Physik **354**, 769–822 (1916). Traduction anglaise dans [256]
45. D. Lovelock, The Einstein tensor and its generalizations. J. Math. Phys. **12**, 498–501 (1971)
46. D. Lovelock, The Four-Dimensionality of Space and the Einstein Tensor. J. Math. Phys. **13**, 874–876 (1972)
47. E. Fischbach, D. Sudarsky, A. Szafer, C. Talmadge, S.H. Aronson, Reanalysis of the Eotvos experiment. Phys. Rev. Lett. **56**, 3–6 (1986)
48. C. Talmadge, J.-P. Berthias, R.W. Hellings, E.M. Standish, Model-independent constraints on possible modifications of Newtonian gravity. Phys. Rev. Lett. **61**, 1159–1162 (1988)
49. E. Fischbach, C. Talmadge, Six years of the fifth force. Nature **356**, 207–215 (1992)
50. E. Fischbach, C.L. Talmadge, *The Search for Non-Newtonian Gravity*, Aip-Press Series (Springer, Berlin, 1999)
51. E.G. Adelberger, J.H. Gundlach, B.R. Heckel, S. Hoedl, S. Schlamminger, Torsion balance experiments: a low-energy frontier of particle physics. Prog. Part. Nucl. Phys. **62**, 102–134 (2009)
52. A. Hees, B. Lamine, S. Reynaud, M.-T. Jaekel, C. Le Poncin-Lafitte, V. Lainey, A. Füzfa, J.-M. Courty, V. Dehant, P. Wolf, Radioscience simulations in general relativity and in alternative theories of gravity. Class. Quantum Gravity **29**(23), 235027 (2012)
53. A. Hees, W. Folkner, R. Jacobson, R. Park, B. Lamine, C. Le Poncin-Lafitte, P. Wolf, Tests of gravitation at Solar System scales beyond the PPN formalism, in *Journées 2013 Systèmes de référence spatio-temporels* ed. by N. Capitaine (2014), pp. 241–244
54. Q.G. Bailey, V.A. Kostelecký, Signals for Lorentz violation in post-Newtonian gravity. Phys. Rev. D **74**(4), 045001 (2006)
55. V.A. Kostelecký, J.D. Tasson, Matter-gravity couplings and Lorentz violation. Phys. Rev. D **83**(1), 016013 (2011)
56. A. Hees, Q. Bailey, A. Bourgoin, H. Pihan-Le Bars, C. Guerlin, C. Le Poncin-Lafitte, Tests of Lorentz symmetry in the gravitational sector. Universe **2**, 30 (2016)
57. J.D. Tasson, The standard-model extension and gravitational tests. Symmetry **8**, 111 (2016)
58. M. Milgrom, MOND effects in the inner Solar system. MNRAS **399**, 474–486 (2009)
59. L. Blanchet, J. Novak, External field effect of modified Newtonian dynamics in the Solar system. MNRAS **412**, 2530–2542 (2011)
60. M.-T. Jaekel, S. Reynaud, Post-Einsteinian tests of linearized gravitation. Class. Quantum Gravity **22**, 2135–2157 (2005)
61. M.-T. Jaekel, S. Reynaud, Post-Einsteinian tests of gravitation. Class. Quantum Gravity **23**, 777–798 (2006)
62. A. Avilez-Lopez, A. Padilla, P.M. Saffin, C. Skordis, The parametrized post-Newtonian-vainshteinian formalism. J. Cosmol. Astropart. Phys. **6**, 044 (2015)
63. M. Hohmann, Parametrized post-Newtonian limit of Horndeski's gravity theory. Phys. Rev. D **92**(6), 064019 (2015)
64. G.L. Smith, C.D. Hoyle, J.H. Gundlach, E.G. Adelberger, B.R. Heckel, H.E. Swanson, Short-range tests of the equivalence principle. Phys. Rev. D **61**(2), 022001 (1999)
65. S. Schlamminger, K.-Y. Choi, T.A. Wagner, J.H. Gundlach, E.G. Adelberger, Test of the equivalence principle using a rotating torsion balance. Phys. Rev. Lett. **100**(4), 041101 (2008)
66. T.A. Wagner, S. Schlamminger, J.H. Gundlach, E.G. Adelberger, Torsion-balance tests of the weak equivalence principle. Class. Quantum Gravity **29**(18), 184002 (2012)
67. V. Viswanathan, A. Fienga, O. Minazzoli, L. Bernus, J. Laskar, M. Gastineau, The new lunar ephemeris INPOP17a and its application to fundamental physics. MNRAS **476**, 1877–1888 (2018)
68. K. Nordtvedt, Testing relativity with laser ranging to the Moon. Phys. Rev. **170**, 1186–1187 (1968)
69. J.G. Williams, S.G. Turyshev, D.H. Boggs, Lunar laser ranging tests of the equivalence principle with the Earth and Moon. Inte. J. Mod. Phys. D **18**, 1129–1175 (2009)
70. J.G. Williams, S.G. Turyshev, D. Boggs, Lunar laser ranging tests of the equivalence principle. Class. Quantum Gravity **29**(18), 184004 (2012)

71. C. Courde, J.M. Torre, E. Samain, G. Martinot-Lagarde, M. Aimar, D. Albanese, P. Exertier, A. Fienga, H. Mariey, G. Metris, H. Viot, V. Viswanathan, Lunar laser ranging in infrared at the Grasse laser station. A&A **602**, A90 (2017)
72. P. Touboul, G. Métris, M. Rodrigues, Y. André, Q. Baghi, J. Bergé, D. Boulanger, S. Bremer, P. Carle, R. Chhun, B. Christophe, V. Cipolla, T. Damour, P. Danto, H. Dittus, P. Fayet, B. Foulon, C. Gageant, P.-Y. Guidotti, D. Hagedorn, E. Hardy, P.-A. Huynh, H. Inchauspe, P. Kayser, S. Lala, C. Lämmerzahl, V. Lebat, P. Leseur, F. Liorzou, M. List, F. Löffler, I. Panet, B. Pouilloux, P. Prieur, A. Rebray, S. Reynaud, B. Rievers, A. Robert, H. Selig, L. Serron, T. Sumner, N. Tanguy, P. Visser, MICROSCOPE mission: first results of a space test of the equivalence principle. Phys. Rev. Lett. **119**(23), 231101 (2017)
73. J. Bergé, P. Brax, G. Métris, M. Pernot-Borràs, P. Touboul, J.-P. Uzan, MICROSCOPE mission: first constraints on the violation of the weak equivalence principle by a light scalar dilaton. Phys. Rev. Lett. **120**(14), 141101 (2018)
74. P. Fayet. MICROSCOPE limits for new long-range forces and implications for unified theories. ArXiv e-prints (2017)
75. A. Peters, K.Y. Chung, S. Chu, Measurement of gravitational acceleration by dropping atoms. Nature **400**, 849–852 (1999)
76. A. Peters, K.Y. Chung, S. Chu, High-precision gravity measurements using atom interferometry. Metrologia **38**, 25–61 (2001)
77. S. Merlet, Q. Bodart, N. Malossi, A. Landragin, F. Pereira Dos Santos, O. Gitlein, L. Timmen, SHORT COMMUNICATION: comparison between two mobile absolute gravimeters: optical versus atomic interferometers. Metrologia **47**, L9–L11 (2010)
78. L. Zhou, S. Long, B. Tang, X. Chen, F. Gao, W. Peng, W. Duan, J. Zhong, Z. Xiong, J. Wang, Y. Zhang, M. Zhan, Test of equivalence principle at 1 0^{-8} level by a dual-species double-diffraction Raman atom interferometer. Phys. Rev. Lett. **115**(1), 013004 (2015)
79. F.W. Hehl, P. von der Heyde, G.D. Kerlick, J.M. Nester, General relativity with spin and torsion: foundations and prospects. Rev. Mod. Phys. **48**, 393–416 (1976)
80. A. Peres, Test of equivalence principle for particles with spin. Phys. Rev. D **18**, 2739–2740 (1978)
81. B. Mashhoon, Gravitational couplings of intrinsic spin. Class. Quantum Gravity **17**, 2399–2409 (2000)
82. Y.N. Obukhov, Spin, gravity, and inertia. Phys. Rev. Lett. **86**, 192–195 (2001)
83. W.-T. Ni, Searches for the role of spin and polarization in gravity. Rep. Prog. Phys. **73**(5), 056901 (2010)
84. W.-T. Ni, Searches for the role of spin and polarization in gravity: a five-year update, in *International Journal of Modern Physics Conference Series*, vol. 40 (2016), pp. 1660010–146
85. M.G. Tarallo, T. Mazzoni, N. Poli, D.V. Sutyrin, X. Zhang, G.M. Tino, Test of einstein equivalence principle for 0-spin and half-integer-spin atoms: search for spin-gravity coupling effects. Phys. Rev. Lett. **113**(2), 023005 (2014)
86. S. Aghion, O. Ahlén, C. Amsler, A. Ariga, T. Ariga, A. S. Belov, G. Bonomi, P. Bräunig, J. Bremer, R. S. Brusa, L. Cabaret, C. Canali, R. Caravita, F. Castelli, G. Cerchiari, S. Cialdi, D. Comparat, G. Consolati, J. H. Derking, S. Di Domizio, L. Di Noto, M. Doser, A. Dudarev, A. Ereditato, R. Ferragut, A. Fontana, P. Genova, M. Giammarchi, A. Gligorova, S. N. Gninenko, S. Haider, J. Harasimovicz, S. D. Hogan, T. Huse, E. Jordan, L. V. Jørgensen, T. Kaltenbacher, J. Kawada, A. Kellerbauer, M. Kimura, A. Knecht, D. Krasnický, V. Lagomarsino, A. Magnani, S. Mariazzi, V. A. Matveev, F. Moia, G. Nebbia, P. Nédélec, M. K. Oberthaler, N. Pacifico, V. Petráček, C. Pistillo, F. Prelz, M. Prevedelli, C. Regenfus, C. Riccardi, O. Røhne, A. Rotondi, H. Sandaker, P. Scampoli, A. Sosa, J. Storey, M. A. Subieta Vasquez, M. Špaček, G. Testera, D. Trezzi, R. Vaccarone, C. P. Welsch, and S. Zavatarelli. Prospects for measuring the gravitational free-fall of antihydrogen with emulsion detectors. J. Instrum. **8**, 8013P (2013)
87. P. Perez, Y. Sacquin, The GBAR experiment: gravitational behaviour of antihydrogen at rest. Class. Quantum Gravity **29**(18), 184008 (2012)

88. L. Hui, A. Nicolis, Proposal for an observational test of the vainshtein mechanism. Phys. Rev. Lett. **109**(5), 051304 (2012)
89. J. Sakstein, B. Jain, J.S. Heyl, L. Hui, Tests of gravity theories using supermassive black holes. ApJl **844**, L14 (2017)
90. J.A. Frieman, B.-A. Gradwohl, Dark matter and the equivalence principle. Phys. Rev. Lett. **67**, 2926–2929 (1991)
91. B.-A. Gradwohl, J.A. Frieman, Dark matter, long-range forces, and large-scale structure. ApJ **398**, 407–424 (1992)
92. S.M. Carroll, S. Mantry, M.J. Ramsey-Musolf, C.W. Stubbs, Dark-matter-induced violation of the weak equivalence principle. Phys. Rev. Lett. **103**(1), 011301 (2009)
93. S.M. Carroll, S. Mantry, M.J. Ramsey-Musolf, Implications of a scalar dark force for terrestrial experiments. Phys. Rev. D **81**(6), 063507 (2010)
94. M. Kesden, M. Kamionkowski, Tidal tails test the equivalence principle in the dark-matter sector. Phys. Rev. D **74**(8), 083007 (2006)
95. M. Kesden, M. Kamionkowski, Galilean equivalence for galactic dark matter. Phys. Rev. Lett. **97**(13), 131303 (2006)
96. C.W. Stubbs, Experimental limits on any long range nongravitational interaction between dark matter and ordinary matter. Phys. Rev. Lett. **70**, 119–122 (1993)
97. Y. Bai, J. Salvado, B.A. Stefanek, Cosmological constraints on the gravitational interactions of matter and dark matter. J. Cosmol. Astropart. Phys. **10**, 029 (2015)
98. A. Hees, T. Do, A.M. Ghez, G.D. Martinez, S. Naoz, E.E. Becklin, A. Boehle, S. Chappell, D. Chu, A. Dehghanfar, K. Kosmo, J.R. Lu, K. Matthews, M.R. Morris, S. Sakai, R. Schödel, G. Witzel, Testing general relativity with stellar orbits around the supermassive black hole in our galactic center. Phys. Rev. Lett. **118**(21), 211101 (2017)
99. T. Damour, G.W. Gibbons, C. Gundlach, Dark matter, time-varying G, and a dilaton field. Phys. Rev. Lett. **64**, 123–126 (1990)
100. J.-M. Alimi, A. Füzfa, Is dark energy abnormally weighting? Int. J. Mod. Phys. D **16**, 2587–2592 (2007)
101. J.-M. Alimi, A. Füzfa, The abnormally weighting energy hypothesis: the missing link between dark matter and dark energy. J. Cosmology Astropart. Phys. **9**, 14 (2008)
102. A. Füzfa, J.-M. Alimi, Toward a unified description of dark energy and dark matter from the abnormally weighting energy hypothesis. Phys. Rev. D **75**(12), 123007 (2007)
103. N. Mohapi, A. Hees, J. Larena, Test of the equivalence principle in the dark sector on galactic scales. J. Cosmol. Astropart. Phys. **3**, 032 (2016)
104. D. Mattingly, Modern tests of lorentz invariance. Living Rev. Relativ. **8**, 5 (2005)
105. V.W. Hughes, H.G. Robinson, V. Beltran-Lopez, Upper limit for the anisotropy of inertial mass from nuclear resonance experiments. Phys. Rev. Lett. **4**, 342–344 (1960)
106. R.W.P. Drever, A search for anisotropy of inertial mass using a free precession technique. Philos. Mag. **6**, 683–687 (1961)
107. T.E. Chupp, R.J. Hoare, R.A. Loveman, E.R. Oteiza, J.M. Richardson, M.E. Wagshul, A.K. Thompson, Results of a new test of local Lorentz invariance: a search for mass anisotropy in ^{21}Ne. Phys. Rev. Lett. **63**, 1541–1545 (1989)
108. S.K. Lamoreaux, J.P. Jacobs, B.R. Heckel, F.J. Raab, E.N. Fortson, New limits on spatial anisotropy from optically-pumped sup201Hg and ^{199}Hg. Phys. Rev. Lett. **57**, 3125–3128 (1986)
109. J.D. Prestage, J.J. Bollinger, W.M. Itano, D.J. Wineland, Limits for spatial anisotropy by use of nuclear-spin-polarized Be-9(+) ions. Phys. Rev. Lett. **54**, 2387–2390 (1985)
110. M.P. Haugan, C.M. Will, Modern tests of special relativity. Phys. Today **40**, 69–86 (1987)
111. A. Brillet, J.L. Hall, Improved laser test of the isotropy of space. Phys. Rev. Lett. **42**, 549–552 (1979)
112. P.L. Stanwix, M.E. Tobar, P. Wolf, C.R. Locke, E.N. Ivanov, Improved test of Lorentz invariance in electrodynamics using rotating cryogenic sapphire oscillators. Phys. Rev. D **74**(8), 081101 (2006)

113. P. Wolf, S. Bize, A. Clairon, A.N. Luiten, G. Santarelli, M.E. Tobar, Tests of Lorentz invariance using a microwave resonator. Phys. Rev. Lett. **90**(6), 060402 (2003)
114. H.P. Robertson, Postulate versus observation in the special theory of relativity. Rev. Mod. Phys. **21**, 378–382 (1949)
115. R. Mansouri, R.U. Sexl, A test theory of special relativity. I - simultaneity and clock synchronization. II - first order tests. Gen. Relativ. Gravitat. **8**, 497–513 (1977)
116. R. Mansouri, R.U. Sexl. A test theory of special relativity: II. first order tests. Gen. Relativ. Gravit. **8**, 515–524 (1977)
117. R. Mansouri, R.U. Sexl, A test theory of special relativity: III Second-order tests. Gen. Relativ. Gravit. **8**, 809–814 (1977)
118. S. Reinhardt, G. Saathoff, H. Buhr, L.A. Carlson, A. Wolf, D. Schwalm, S. Karpuk, C. Novotny, G. Huber, M. Zimmermann, R. Holzwarth, T. Udem, T.W. Hänsch, G. Gwinner, Test of relativistic time dilation with fast optical atomic clocks at different velocities. Nat. Phys. **3**, 861–864 (2007)
119. B. Botermann, D. Bing, C. Geppert, G. Gwinner, T.W. Hänsch, G. Huber, S. Karpuk, A. Krieger, T. Kühl, W. Nörtershäuser, C. Novotny, S. Reinhardt, R. Sánchez, D. Schwalm, T. Stöhlker, A. Wolf, G. Saathoff, Test of time dilation using stored Li^+ ions as clocks at relativistic speed. Phys. Rev. Lett. **113**(12), 120405 (2014)
120. C.M. Will, Clock synchronization and isotropy of the one-way speed of light. Phys. Rev. D **45**, 403–411 (1992)
121. P. Wolf, G. Petit, Satellite test of special relativity using the global positioning system. Phys. Rev. A **56**, 4405–4409 (1997)
122. J. Lodewyck, S. Bilicki, E. Bookjans, J.-L. Robyr, C. Shi, G. Vallet, R. Le Targat, D. Nicolodi, Y. Le Coq, J. Guéna, M. Abgrall, P. Rosenbusch, S. Bize, Optical to microwave clock frequency ratios with a nearly continuous strontium optical lattice clock. Metrologia **53**, 1123 (2016)
123. R. Le Targat, L. Lorini, Y. Le Coq, M. Zawada, J. Guéna, M. Abgrall, M. Gurov, P. Rosenbusch, D.G. Rovera, B. Nagórny, R. Gartman, P.G. Westergaard, M.E. Tobar, M. Lours, G. Santarelli, A. Clairon, S. Bize, P. Laurent, P. Lemonde, J. Lodewyck, Experimental realization of an optical second with strontium lattice clocks. Nat. Commun. **4**, 2109 (2013)
124. S. Falke, N. Lemke, C. Grebing, B. Lipphardt, S. Weyers, V. Gerginov, N. Huntemann, C. Hagemann, A. Al-Masoudi, S. Häfner, S. Vogt, U. Sterr, C. Lisdat, A strontium lattice clock with 3×10^{-17} inaccuracy and its frequency. New J. Phys. **16**(7), 073023 (2014)
125. C. Grebing, A. Al-Masoudi, D. Sören, H. Sebastian, G. Vladislav, W. Stefan, L. Burghard, R. Fritz, S. Uwe, L. Christian, Realization of a timescale with an accurate optical lattice clock. Optica **3**(6), 563–569 (2016)
126. I.R. Hill, R. Hobson, W. Bowden, E.M. Bridge, S. Donnellan, E.A. Curtis, P. Gill, A low maintenance Sr optical lattice clock. In *Journal of Physics Conference Series*, volume 723 of *Journal of Physics Conference Series*, page 012019, June 2016
127. C. Lisdat, G. Grosche, N. Quintin, C. Shi, S.M.F. Raupach, C. Grebing, D. Nicolodi, F. Stefani, A. Al-Masoudi, S. Dörscher, S. Häfner, J.-L. Robyr, N. Chiodo, S. Bilicki, E. Bookjans, A. Koczwara, S. Koke, A. Kuhl, F. Wiotte, F. Meynadier, E. Camisard, M. Abgrall, M. Lours, T. Legero, H. Schnatz, U. Sterr, H. Denker, C. Chardonnet, Y. Le Coq, G. Santarelli, A. Amy-Klein, R. Le Targat, J. Lodewyck, O. Lopez, P.-E. Pottie, A clock network for geodesy and fundamental science. Nat. Commun. **7**, 12443 (2016)
128. P.A. Williams, W.C. Swann, N.R. Newbury, High-stability transfer of an optical frequency over long fiber-optic links. J. Opt. Soc. Am. B Opt. Phys. **25**, 1284 (2008)
129. G. Grosche, O. Terra, K. Predehl, R. Holzwarth, B. Lipphardt, F. Vogt, U. Sterr, H. Schnatz, Optical frequency transfer via 146 km fiber link with 10^-19 relative accuracy. Opt. Lett. **34**, 2270 (2009)
130. G. Grosche, Eavesdropping time and frequency: phase noise cancellation along a time-varying path, such as an optical fiber. Opt. Lett. **39**, 2545 (2014)
131. F. Stefani, O. Lopez, A. Bercy, W.-K. Lee, C. Chardonnet, G. Santarelli, P.-E. Pottie, A. Amy-Klein, Tackling the limits of optical fiber links. J. Opt. Soc. Am. B Opt. Phys. **32**, 787 (2015)

132. J. Geršl, P. Delva, P. Wolf, Relativistic corrections for time and frequency transfer in optical fibres. Metrologia **52**, 552 (2015)
133. P. Delva, J. Lodewyck, S. Bilicki, E. Bookjans, G. Vallet, R. Le Targat, P.-E. Pottie, C. Guerlin, F. Meynadier, C. Le Poncin-Lafitte, O. Lopez, A. Amy-Klein, W.-K. Lee, N. Quintin, C. Lisdat, A. Al-Masoudi, S. Dörscher, C. Grebing, G. Grosche, A. Kuhl, S. Raupach, U. Sterr, I.R. Hill, R. Hobson, W. Bowden, J. Kronjäger, G. Marra, A. Rolland, F.N. Baynes, H.S. Margolis, P. Gill, Test of special relativity using a fiber network of optical clocks. Phys. Rev. Lett. **118**(22), 221102 (2017)
134. D. Colladay, V.A. Kostelecký, CPT violation and the standard model. Phys. Rev. D **55**, 6760–6774 (1997)
135. D. Colladay, V.A. Kostelecký, Lorentz-violating extension of the standard model. Phys. Rev. D **58**(11), 116002 (1998)
136. V.A. Kostelecký, M. Mewes, Signals for Lorentz violation in electrodynamics. Phys. Rev. D **66**(5), 056005 (2002)
137. J.D. Tasson, What do we know about Lorentz invariance? Rep. Prog. Phys. **77**(6), 062901 (2014)
138. V.A. Kostelecký, N. Russell, Data tables for Lorentz and CPT violation. Rev. Mod. Phys. **83**, 11–32 (2011)
139. H. Pihan-Le Bars, C. Guerlin, R.-D. Lasseri, J.-P. Ebran, Q.G. Bailey, S. Bize, E. Khan, P. Wolf, Lorentz-symmetry test at Planck-scale suppression with nucleons in a spin-polarized ^{133}Cs cold atom clock. Phys. Rev. D **95**(7), 075026 (2017)
140. P. Wolf, F. Chapelet, S. Bize, A. Clairon, Cold atom clock test of Lorentz invariance in the matter sector. Phys. Rev. Lett. **96**(6), 060801 (2006)
141. M.A. Hohensee, S. Chu, A. Peters, H. Müller, Equivalence principle and gravitational redshift. Phys. Rev. Lett. **106**(15), 151102 (2011)
142. M.A. Hohensee, N. Leefer, D. Budker, C. Harabati, V.A. Dzuba, V.V. Flambaum, Limits on violations of Lorentz symmetry and the Einstein equivalence principle using radio-frequency spectroscopy of atomic dysprosium. Phys. Rev. Lett. **111**(5), 050401 (2013)
143. T. Damour, Theoretical aspects of the equivalence principle. Class. Quantum Gravity **29**(18), 184001 (2012)
144. J.-P. Uzan, Varying constants, gravitation and cosmology. Living Rev. Relativ. **14**, 2 (2011)
145. J. Guéna, M. Abgrall, D. Rovera, P. Rosenbusch, M.E. Tobar, P. Laurent, A. Clairon, S. Bize, Improved tests of local position invariance using Rb87 and Cs133 fountains. Phys. Rev. Lett. **109**(8), 080801 (2012)
146. T. Rosenband, D.B. Hume, P.O. Schmidt, C.W. Chou, A. Brusch, L. Lorini, W.H. Oskay, R.E. Drullinger, T.M. Fortier, J.E. Stalnaker, S.A. Diddams, W.C. Swann, N.R. Newbury, W.M. Itano, D.J. Wineland, J.C. Bergquist, Frequency ratio of Al^+ and Hg^+ single-ion optical clocks; metrology at the 17th decimal place. Science **319**, 1808– (2008)
147. N. Leefer, C.T.M. Weber, A. Cingöz, J.R. Torgerson, D. Budker, New limits on variation of the fine-structure constant using atomic dysprosium. Phys. Rev. Lett. **111**(6), 060801 (2013)
148. R.M. Godun, P.B.R. Nisbet-Jones, J.M. Jones, S.A. King, L.A.M. Johnson, H.S. Margolis, K. Szymaniec, S.N. Lea, K. Bongs, P. Gill, Frequency ratio of two optical clock transitions in $^{171}Yb^+$ and constraints on the time variation of fundamental constants. Phys. Rev. Lett. **113**(21), 210801 (2014)
149. N. Huntemann, B. Lipphardt, C. Tamm, V. Gerginov, S. Weyers, E. Peik, Improved limit on a temporal variation of m_p/m_e from comparisons of Yb^+ and Cs atomic clocks. Phys. Rev. Lett. **113**(21), 210802 (2014)
150. P. Brax, C. Burrage, A.-C. Davis, G. Gubitosi, Cosmological tests of the disformal coupling to radiation. J. Cosmol. Astropart. Phys. **11**, 1 (2013)
151. A. Hees, O. Minazzoli, J. Larena, Breaking of the equivalence principle in the electromagnetic sector and its cosmological signatures. Phys. Rev. D **90**(12), 124064 (2014)
152. R.F.L. Holanda, K.N.N.O. Barros, Searching for cosmological signatures of the Einstein equivalence principle breaking. Phys. Rev. D **94**(2), 023524 (2016)

153. S. Peil, S. Crane, J.L. Hanssen, T.B. Swanson, C.R. Ekstrom, Tests of local position invariance using continuously running atomic clocks. Phys. Rev. A **87**(1), 010102 (2013)
154. N. Ashby, T.P. Heavner, S.R. Jefferts, T.E. Parker, A.G. Radnaev, Y.O. Dudin, Testing local position invariance with four cesium-fountain primary frequency standards and four NIST hydrogen masers. Phys. Rev. Lett. **98**(7), 070802 (2007)
155. R.V. Pound, G.A. Rebka, Gravitational red-shift in nuclear resonance. Phys. Rev. Lett. **3**, 439–441 (1959)
156. R.F.C. Vessot, M.W. Levine, E.M. Mattison, E.L. Blomberg, T.E. Hoffman, G.U. Nystrom, B.F. Farrel, R. Decher, P.B. Eby, C.R. Baugher, Test of relativistic gravitation with a space-borne hydrogen maser. Phys. Rev. Lett. **45**, 2081–2084 (1980)
157. P. Delva, A. Hees, S. Bertone, E. Richard, P. Wolf, Test of the gravitational redshift with stable clocks in eccentric orbits: application to Galileo satellites 5 and 6. Class. Quantum Gravity **32**(23), 232003 (2015)
158. L. Cacciapuoti, C. Salomon, Atomic clock ensemble in space. J. Phys. Conf. Ser. **327**(1), 012049 (2011)
159. M. Soffel, H. Herold, H. Ruder, M. Schneider, Relativistic theory of gravimetric measurements and definition of thegeoid. Manuscr. Geod. **13**, 143–146 (1988)
160. S.M. Kopejkin, Relativistic Manifestations of gravitational fields in gravimetry and geodesy. Manuscripta Geodaetica **16** (1991)
161. J. Müller, M. Soffel, S.A. Klioner, Geodesy and relativity. J. Geodesy **82**, 133–145 (2008)
162. P. Delva, J. Lodewyck, Atomic clocks: new prospects in metrology and geodesy. Acta Futura, (7), 67–78, **7**:67–78, November 2013
163. B. Altschul, Q.G. Bailey, L. Blanchet, K. Bongs, P. Bouyer, L. Cacciapuoti, S. Capozziello, N. Gaaloul, D. Giulini, J. Hartwig, L. Iess, P. Jetzer, A. Landragin, E. Rasel, S. Reynaud, S. Schiller, C. Schubert, F. Sorrentino, U. Sterr, J.D. Tasson, G.M. Tino, P. Tuckey, P. Wolf, Quantum tests of the Einstein equivalence principle with the STE-QUEST space mission. Adv. Space Res. **55**, 501–524 (2015)
164. S. Zucker, T. Alexander, S. Gillessen, F. Eisenhauer, R. Genzel, Probing post-Newtonian physics near the galactic black hole with stellar redshift measurements. ApJl **639**, L21–L24 (2006)
165. M. Grould, F.H. Vincent, T. Paumard, G. Perrin, General relativistic effects on the orbit of the S2 star with GRAVITY. ArXiv e-prints (2017)
166. Gravity Collaboration, R. Abuter, A. Amorim, N. Anugu, M. Bauböck, M. Benisty, J. P. Berger, N. Blind, H. Bonnet, W. Brandner, A. Buron, C. Collin, F. Chapron, Y. Clénet, V. Coudé Du Foresto, P. T. de Zeeuw, C. Deen, F. Delplancke-Ströbele, R. Dembet, J. Dexter, G. Duvert, A. Eckart, F. Eisenhauer, G. Finger, N. M. Förster Schreiber, P. Fédou, P. Garcia, R. Garcia Lopez, F. Gao, E. Gendron, R. Genzel, S. Gillessen, P. Gordo, M. Habibi, X. Haubois, M. Haug, F. Haußmann, T. Henning, S. Hippler, M. Horrobin, Z. Hubert, N. Hubin, A. Jimenez Rosales, L. Jochum, K. Jocou, A. Kaufer, S. Kellner, S. Kendrew, P. Kervella, Y. Kok, M. Kulas, S. Lacour, V. Lapeyrère, B. Lazareff, J.-B. Le Bouquin, P. Léna, M. Lippa, R. Lenzen, A. Mérand, E. Müler, U. Neumann, T. Ott, L. Palanca, T. Paumard, L. Pasquini, K. Perraut, G. Perrin, O. Pfuhl, P. M. Plewa, S. Rabien, A. Ramírez, J. Ramos, C. Rau, G. Rodríguez-Coira, R.-R. Rohloff, G. Rousset, J. Sanchez-Bermudez, S. Scheithauer, M. Schöller, N. Schuler, J. Spyromilio, O. Straub, C. Straubmeier, E. Sturm, L. J. Tacconi, K. R. W. Tristram, F. Vincent, S. von Fellenberg, I. Wank, I. Waisberg, F. Widmann, E. Wieprecht, M. Wiest, E. Wiezorrek, J. Woillez, S. Yazici, D. Ziegler, and G. Zins. Detection of the gravitational redshift in the orbit of the star S2 near the Galactic centre massive black hole. A&A **615**, L15 (2018)
167. S. Weinberg, A new light boson? Phys. Rev. Lett. **40**, 223–226 (1978)
168. J. Preskill, M.B. Wise, F. Wilczek, Cosmology of the invisible axion. Phys. Lett. B **120**, 127–132 (1983)
169. W. Hu, R. Barkana, A. Gruzinov, Fuzzy cold dark matter: the wave properties of ultralight particles. Phys. Rev. Lett. **85**, 1158–1161 (2000)
170. F. Piazza, M. Pospelov, Sub-eV scalar dark matter through the super-renormalizable Higgs portal. Phys. Rev. D **82**(4), 043533 (2010)

171. A. Khmelnitsky, V. Rubakov, Pulsar timing signal from ultralight scalar dark matter. J. Cosmol. Astropart. Phys. **2**, 019 (2014)
172. N.K. Porayko, K.A. Postnov, Constraints on ultralight scalar dark matter from pulsar timing. Phys. Rev. D **90**(6), 062008 (2014)
173. H.-Y. Schive, T. Chiueh, T. Broadhurst, Cosmic structure as the quantum interference of a coherent dark wave. Nat. Phys. **10**, 496–499 (2014)
174. J. Beyer, C. Wetterich, Small scale structures in coupled scalar field dark matter. Phys. Lett. B **738**, 418–423 (2014)
175. A. Arvanitaki, J. Huang, K. Van Tilburg, Searching for dilaton dark matter with atomic clocks. Phys. Rev. D **91**(1), 015015 (2015)
176. Y.V. Stadnik, V.V. Flambaum, Searching for dark matter and variation of fundamental constants with laser and maser interferometry. Phys. Rev. Lett. **114**(16), 161301 (2015)
177. Y.V. Stadnik, V.V. Flambaum, Can dark matter induce cosmological evolution of the fundamental constants of nature? Phys. Rev. Lett. **115**(20), 201301 (2015)
178. P.W. Graham, D.E. Kaplan, J. Mardon, S. Rajendran, W.A. Terrano, Dark matter direct detection with accelerometers. Phys. Rev. D **93**(7), 075029 (2016)
179. A. Arvanitaki, S. Dimopoulos, K. Van Tilburg, Sound of dark matter: searching for light scalars with resonant-mass detectors. Phys. Rev. Lett. **116**(3), 031102 (2016)
180. Y.V. Stadnik, V.V. Flambaum, Enhanced effects of variation of the fundamental constants in laser interferometers and application to dark-matter detection. Phys. Rev. A **93**(6), 063630 (2016)
181. Y.V. Stadnik, V.V. Flambaum, Improved limits on interactions of low-mass spin-0 dark matter from atomic clock spectroscopy. Phys. Rev. A **94**(2), 022111 (2016)
182. D.J.E. Marsh, Axion cosmology. Phys. Rep. **643**, 1–79 (2016)
183. L.A. Ureña-López, A.X. Gonzalez-Morales, Towards accurate cosmological predictions for rapidly oscillating scalar fields as dark matter. J. Cosmol. Astropart. Phys. **7**, 048 (2016)
184. E. Calabrese, D.N. Spergel, Ultra-light dark matter in ultra-faint dwarf galaxies. MNRAS **460**, 4397–4402 (2016)
185. D. Blas, D.L. Nacir, S. Sibiryakov, Ultralight dark matter resonates with binary pulsars. Phys. Rev. Lett. **118**(26), 261102 (2017)
186. T. Bernal, V.H. Robles, T. Matos, Scalar field dark matter in clusters of galaxies. MNRAS **468**, 3135–3149 (2017)
187. L. Hui, J.P. Ostriker, S. Tremaine, E. Witten, Ultralight scalars as cosmological dark matter. Phys. Rev. D **95**(4), 043541 (2017)
188. C. Abel, N.J. Ayres, G. Ban, G. Bison, K. Bodek, V. Bondar, M. Daum, M. Fairbairn, V.V. Flambaum, P. Geltenbort, K. Green, W.C. Griffith, M. van der Grinten, Z.D. Grujić, P.G. Harris, N. Hild, P. Iaydjiev, S.N. Ivanov, M. Kasprzak, Y. Kermaidic, K. Kirch, H.-C. Koch, S. Komposch, P.A. Koss, A. Kozela, J. Krempel, B. Lauss, T. Lefort, Y. Lemière, D.J.E. Marsh, P. Mohanmurthy, A. Mtchedlishvili, M. Musgrave, F.M. Piegsa, G. Pignol, M. Rawlik, D. Rebreyend, D. Ries, S. Roccia, D. Rozpędzik, P. Schmidt-Wellenburg, N. Severijns, D. Shiers, Y.V. Stadnik, A. Weis, E. Wursten, J. Zejma, G. Zsigmond, Search for axion-like dark matter through nuclear spin precession in electric and magnetic fields. Phys. Rev. X **7**(4), 041034 (2017)
189. T. Bernal, L.M. Fernández-Hernández, T. Matos, M.A. Rodríguez-Meza, Rotation curves of high-resolution LSB and SPARC galaxies with fuzzy and multistate (ultralight boson) scalar field dark matter. MNRAS **475**, 1447–1468 (2018)
190. M.B. Green, J.H. Schwarz, E. Witten, *Superstring Theory* (1988)
191. T. Damour, A.M. Polyakov, The string dilation and a least coupling principle. Nucl. Phys. B **423**, 532–558 (1994)
192. T. Damour, A.M. Polyakov, String theory and gravity. Gen. Relativ. Gravit. **26**, 1171–1176 (1994)
193. M. Gasperini, F. Piazza, G. Veneziano, Quintessence as a runaway dilaton. Phys. Rev. D **65**(2), 023508 (2001)

194. T. Damour, F. Piazza, G. Veneziano, Runaway dilaton and equivalence principle violations. Phys. Rev. Lett. **89**(8), 081601 (2002)
195. T. Damour, J.F. Donoghue, Equivalence principle violations and couplings of a light dilaton. Phys. Rev. D **82**(8), 084033 (2010)
196. A. Hees, O. Minazzoli, E. Savalle, Y.V. Stadnik, P. Wolf, Violation of the equivalence principle from light scalar dark matter. Phys. Rev. D **98**(6), 064051 (2018)
197. A. Hees, J. Guéna, M. Abgrall, S. Bize, P. Wolf, Searching for an oscillating massive scalar field as a dark matter candidate using atomic hyperfine frequency comparisons. Phys. Rev. Lett. **117**(6), 061301 (2016)
198. K. Van Tilburg, N. Leefer, L. Bougas, D. Budker, Search for ultralight scalar dark matter with atomic spectroscopy. Phys. Rev. Lett. **115**(1), 011802 (2015)
199. T.A. de Pirey Saint Alby, N. Yunes, Cosmological evolution and Solar System consistency of massive scalar-tensor gravity. Phys. Rev. D **96**(6), 064040 (2017)
200. T. Damour, G. Esposito-Farese, Nonperturbative strong-field effects in tensor-scalar theories of gravitation. Phys. Rev. Lett. **70**, 2220–2223 (1993)
201. T. Damour, G. Esposito-Farèse, Tensor-scalar gravity and binary-pulsar experiments. Phys. Rev. D **54**, 1474–1491 (1996)
202. O. Minazzoli, A. Hees, Late-time cosmology of a scalar-tensor theory with a universal multiplicative coupling between the scalar field and the matter Lagrangian. Phys. Rev. D **90**(2), 023017 (2014)
203. S.M. Carroll, Quintessence and the rest of the world: suppressing long-range interactions. Phys. Rev. Lett. **81**, 3067–3070 (1998)
204. N.J. Nunes, J.E. Lidsey, Reconstructing the dark energy equation of state with varying alpha. Phys. Rev. D **69**(12), 123511 (2004)
205. C.J.A.P. Martins, A.M.M. Pinho, Fine-structure constant constraints on dark energy. Phys. Rev. D **91**(10), 103501 (2015)
206. M.A. Hohensee, H. Müller, R.B. Wiringa, Equivalence principle and bound kinetic energy. Phys. Rev. Lett. **111**(15), 151102 (2013)
207. H. Müller, S. Herrmann, A. Saenz, A. Peters, C. Lämmerzahl, Optical cavity tests of Lorentz invariance for the electron. Phys. Rev. D **68**(11), 116006 (2003)
208. H. Müller, Testing Lorentz invariance by the use of vacuum and matter filled cavity resonators. Phys. Rev. D **71**(4), 045004 (2005)
209. N.A. Flowers, C. Goodge, J.D. Tasson, Superconducting-gravimeter tests of local lorentz invariance. Phys. Rev. Lett. **119**(20), 201101 (2017)
210. A. Bourgoin, C. Le Poncin-Lafitte, A. Hees, S. Bouquillon, G. Francou, M.-C. Angonin, Lorentz symmetry violations from matter-gravity couplings with lunar laser ranging. Phys. Rev. Lett. **119**(20), 201102 (2017)
211. H. Pihan-Le Bars, C. Guerlin, P. Wolf, Progress on testing Lorentz symmetry with MICROSCOPE. ArXiv e-prints (2017)
212. B. Bertotti, L. Iess, P. Tortora, A test of general relativity using radio links with the Cassini spacecraft. Nature **425**, 374–376 (2003)
213. S.B. Lambert, C. Le Poncin-Lafitte, Determining the relativistic parameter γ using very long baseline interferometry. A&A **499**, 331–335 (2009)
214. S.B. Lambert, C. Le Poncin-Lafitte, Improved determination of γ by VLBI. A&A **529**, A70 (2011)
215. A.S. Konopliv, S.W. Asmar, W.M. Folkner, Ö. Karatekin, D.C. Nunes, S.E. Smrekar, C.F. Yoder, M.T. Zuber, Mars high resolution gravity fields from mro, mars seasonal gravity, and other dynamical parameters. Icarus **211**(1), 401–428 (2011)
216. R.S. Park, W.M. Folkner, A.S. Konopliv, J.G. Williams, D.E. Smith, M.T. Zuber, Precession of mercury's perihelion from ranging to the MESSENGER spacecraft. AJ **153**, 121 (2017)
217. E.G. Adelberger, B.R. Heckel, A.E. Nelson, Tests of the gravitational inverse-square law. Ann. Rev. Nucl. Part. Sci. **53**, 77–121 (2003)
218. P. Jordan, Formation of the stars and development of the universe. Nature **164**, 637–640 (1949)

219. C. Brans, R.H. Dicke, Mach's principle and a relativistic theory of gravitation. Phys. Rev. **124**, 925–935 (1961)
220. T. Damour, G. Esposito-Farese, Tensor-multi-scalar theories of gravitation. Class. Quantum Gravity **9**, 2093–2176 (1992)
221. J. Khoury, A. Weltman, Chameleon cosmology. Phys. Rev. D **69**(4), 044026 (2004)
222. J. Khoury, A. Weltman, Chameleon fields: awaiting surprises for tests of gravity in space. Phys. Rev. Lett. **93**(17), 171104 (2004)
223. A. Hees, A. Füzfa, Combined cosmological and solar system constraints on chameleon mechanism. Phys. Rev. D **85**(10), 103005 (2012)
224. K. Hinterbichler, J. Khoury, Screening long-range forces through local symmetry restoration. Phys. Rev. Lett. **104**(23), 231301 (2010)
225. K. Hinterbichler, J. Khoury, A. Levy, A. Matas, Symmetron cosmology. Phys. Rev. D **84**(10), 103521 (2011)
226. A.I. Vainshtein, To the problem of nonvanishing gravitation mass. Phys. Lett. B **39**, 393–394 (1972)
227. E. Babichev, C. Deffayet, R. Ziour, The vainshtein mechanism in the decoupling limit of massive gravity. Jo. High Energy Phys. **5**, 98 (2009)
228. E. Babichev, C. Deffayet, R. Ziour, k-MOUFLAGE gravity. Int. J. Mod. Phys. D **18**, 2147–2154 (2009)
229. E. Babichev, C. Deffayet, G. Esposito-Farèse, Constraints on shift-symmetric scalar-tensor theories with a vainshtein mechanism from bounds on the time variation of G. Phys. Rev. Lett. **107**(25), 251102 (2011)
230. F. Hofmann, J. Müller, Relativistic tests with lunar laser ranging. Class. Quantum Gravity **35**(3), 035015 (2018)
231. M.D. Seifert, Vector models of gravitational Lorentz symmetry breaking. Phys. Rev. D **79**(12), 124012 (2009)
232. V.A. Kostelecký, R. Potting, Gravity from local Lorentz violation. Gen. Relativ. Gravit. **37**, 1675–1679 (2005)
233. V.A. Kostelecký, R. Potting, Gravity from spontaneous Lorentz violation. Phys. Rev. D **79**(6), 065018 (2009)
234. B. Altschul, Q.G. Bailey, V.A. Kostelecký, Lorentz violation with an antisymmetric tensor. Phys. Rev. D **81**(6), 065028 (2010)
235. R. Gambini, J. Pullin, Nonstandard optics from quantum space-time. Phys. Rev. D **59**(12), 124021 (1999)
236. V.A. Kostelecký, M. Mewes, Electrodynamics with Lorentz-violating operators of arbitrary dimension. Phys. Rev. D **80**(1), 015020 (2009)
237. S.M. Carroll, J.A. Harvey, V.A. Kostelecký, C.D. Lane, T. Okamoto, Noncommutative field theory and Lorentz violation. Phys. Rev. Lett. **87**(14), 141601 (2001)
238. V.A. Kostelecký, R. Lehnert, Stability, causality, and Lorentz and CPT violation. Phys. Rev. D **63**(6), 065008 (2001)
239. Q.G. Bailey, V.A. Kostelecký, R. Xu, Short-range gravity and Lorentz violation. Phys. Rev. D **91**(2), 022006 (2015)
240. V.A. Kostelecký, J.D. Tasson, Constraints on Lorentz violation from gravitational Čerenkov radiation. Phys. Lett. B **749**, 551–559 (2015)
241. V.A. Kostelecký, M. Mewes, Testing local Lorentz invariance with gravitational waves. Phys. Lett. B **757**, 510–514 (2016)
242. Q.G. Bailey, D. Havert, Velocity-dependent inverse cubic force and solar system gravity tests. Phys. Rev. D **96**(6), 064035 (2017)
243. H. Müller, S.-W. Chiow, S. Herrmann, S. Chu, K.-Y. Chung, Atom-interferometry tests of the isotropy of post-Newtonian gravity. Phys. Rev. Lett. **100**(3), 031101 (2008)
244. K.-Y. Chung, S.-W. Chiow, S. Herrmann, S. Chu, H. Müller, Atom interferometry tests of local Lorentz invariance in gravity and electrodynamics. Phys. Rev. D **80**(1), 016002 (2009)
245. C.-G. Shao, Y.-F. Chen, R. Sun, L.-S. Cao, M.-K. Zhou, Z.-K. Hu, C. Yu, H. Müller, Limits on Lorentz violation in gravity from worldwide superconducting gravimeters. Phys. Rev. D **97**(2), 024019 (2018)

246. A. Hees, Q.G. Bailey, C. Le Poncin-Lafitte, A. Bourgoin, A. Rivoldini, B. Lamine, F. Meynadier, C. Guerlin, P. Wolf, Testing Lorentz symmetry with planetary orbital dynamics. Phys. Rev. D **92**, 064049 (2015)
247. C. Le Poncin-Lafitte, A. Hees, S. lambert, Lorentz symmetry and very long baseline interferometry. Phys. Rev. D **94**(12), 125030 (2016)
248. J.B.R. Battat, J.F. Chandler, C.W. Stubbs, Testing for Lorentz violation: constraints on standard-model-extension parameters via lunar laser ranging. Phys. Rev. Lett. **99**(24), 241103 (2007)
249. A. Bourgoin, A. Hees, S. Bouquillon, C. Le Poncin-Lafitte, G. Francou, M.-C. Angonin, Testing Lorentz symmetry with lunar laser ranging. Phys. Rev. Lett. **117**(24), 241301 (2016)
250. M. Milgrom, A modification of the Newtonian dynamics as a possible alternative to the hidden mass hypothesis. ApJ **270**, 365–370 (1983)
251. M. Milgrom, A modification of the Newtonian dynamics - implications for galaxy systems. ApJ **270**, 384 (1983)
252. M. Milgrom, A modification of the Newtonian dynamics - implications for galaxies. ApJ **270**, 371–389 (1983)
253. J.-P. Bruneton, G. Esposito-Farèse, Field-theoretical formulations of MOND-like gravity. Phys. Rev. D **76**(12), 124012 (2007)
254. A. Hees, B. Famaey, G.W. Angus, G. Gentile, Combined solar system and rotation curve constraints on MOND. MNRAS **455**, 449–461 (2016)
255. R.A. Swaters, R.H. Sanders, S.S. McGaugh, Testing modified Newtonian dynamics with rotation curves of dwarf and low surface brightness galaxies. ApJ **718**, 380–391 (2010)

Operationalization of Basic Relativistic Measurements

Bruno Hartmann

Abstract We present a novel phenomenological foundation of relativistic physics. That means, we focus on the observable entities and make no mathematical preassumptions. Like Einstein for relativistic kinematics we start from vivid measurement operations and simple natural principles. Seeking, formulating and refining operational definitions reveals the physical meaning. We grasp the basic observables (length, duration, inertial mass, momentum, energy) in a physical way. We define an order of energy and impulse from a physical comparison. Each step (the construction of "sufficiently constant" reference devices and of a machinery, which "functions" for a basic measurement) follows from practical requirements. One can directly count the tangible measurement units and ultimately derive the fundamental equations (e.g. the kinetic energy-velocity relation or the mass-energy equivalence).

1 Introduction

In relativistic geodesy physicists develop the measurement practice. One manufactures increasingly precise reference devices, operates with them, and fits the data with equations of relativity theory. We present a foundation of these basic equations. We define all mathematical variables and operations strictly from the underlying physical operations. We review that approach for relativistic kinematics and dynamics.

We all use the terms length, duration, mass or energy in our everyday lives and come to an intuition about them. The challenge in learning about physics is how to reconcile the common understanding with the scientific concepts that are more precise [1]. Among physics education researchers broad consensus shows, that one "constructs" new knowledge from prior knowledge [2]. Our goal is to help understand learning as "the refinement of everyday thinking". The common everyday experiences and language serves as a starting point to understand space and time and energy. We develop the preconceptions into a physicists view of measurements so that one can describe these observables more precisely. We reveal a novel aspect of physics,

B. Hartmann (✉)
Humboldt University, Berlin, Germany
e-mail: brunohartmannjr@gmail.com

D. Puetzfeld and C. Lämmerzahl (eds.), *Relativistic Geodesy*,
Fundamental Theories of Physics 196, https://doi.org/10.1007/978-3-030-11500-5_10

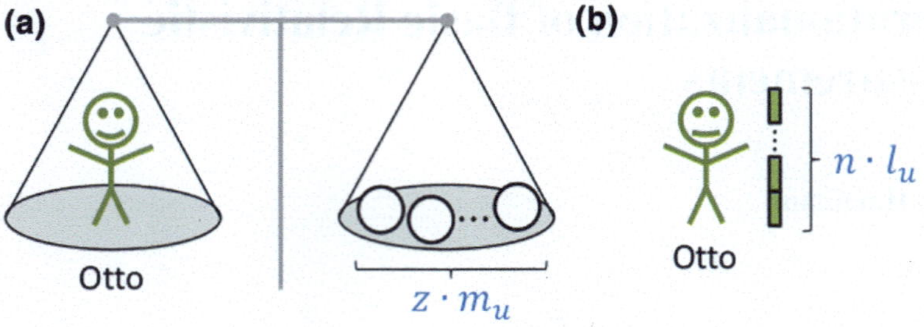

Fig. 1 **a** Count unit weights m_u on balance scale **b** count rulers along straight line

the methodological knowledge of a quantification, i.e. of a direct measurement. We construct precise physical quantities of space, time and energy hands-on and then derive the basic equations. We develop the theory from an empirical basis and not retrospectively from non-vivid expressions and formal equations. For the resulting formalism one can understand scope and limitations.

Before our approach is outlined we specify what we understand as a basic measurement. According to Gerthsen [3] "Physics is a measuring science". One defines "measuring a quantity" by the direct or indirect comparison with a measurement unit. Mostly this is not done directly - one does not use an energymeter - but indirectly by measuring other primary quantities and then using an agreed formula. This risks to be circular. For physics Gerthsen demands solely operational definitions, that specify the quantity to be measured, e.g. inertial mass or energy, by its effect in a measurement process.

We are long familiar with basic measurements as in the case of weights and lengths. We order the basic observables through a practical comparison, for example "heavier than" if for example a person called Otto outweighs a reference body on a balance scale (Fig. 1a). One wants to specify the values also numerically: "how many times" heavier. Hence a procedure is needed that leads to a quantification. The procedure for finding these values is the measurement. In a weight measurement one successively piles up weight units until the pile is sufficiently precise neither heavier nor lighter than Otto. One manufactures standard reference objects, units which are all of the same weight, and counts them. If one needs z weight units then the weight of Otto is equal to z times the weight of the reference object. Formally we write a measurement result $m[\mathcal{O}] = z \cdot m_u$. One also knows the basic observable length. One defines: Otto is "longer than" a measuring stick, if he towers over the latter. For the measurement one manufactures equally long measuring sticks and successively places them along a straight line until they cover Otto (Fig. 1b). If one needs n sticks, then the length of Otto is equal to n times the length of the reference stick $l[\mathcal{O}] = n \cdot l_u$.

Helmholtz [4] specifies the basic measurement process. Essentially one begins from counting *same* objects. The basic observable (weight, length etc.) is ordered by a practical comparison. A measurement quantifies the basic observable. One obtains a quantified observable expressed as a number (of reference devices) times its dimension (standard weight, standard length etc.). A basic measurement requires knowledge of the method of comparison (of a particular attribute of both bodies), we write ">", "=" or "<", and of the physical concatenation method. Obviously the way of concatenation (piling weight units, placing meter sticks along straight line, coupling impulse units etc.) generally depends on the type of measure. One simply writes the formal symbol "+". Our physical foundation will specify the underlying tangible operations. We will describe an approach how space-time in Sect. 2, inertial mass and momentum in Sect. 3, and energy in Sect. 4 can be introduced as an observable and measurable concept. We will derive the exact mathematical relations between these physical quantities in the end.

2 Kinematics

First we outline Helmholtz program for measuring relativistic motion [5]. We define a spatiotemporal order "longer than" by the practical comparison, whether one object or process covers the other. Without one word of mathematics Alice can manufacture uniform running light clocks "$\mathcal{L}$" and place them literally one after the other or side by side "$\mathcal{L} + \mathcal{L}$" and then count, how many building blocks one needs to assemble a regular grid which covers the measurement object (Fig. 2a). These are the elementary measurement operations. From their structure and overlap we derive mathematical relations, e.g. for two different observers Alice and Bob the formal Lorentz transformation (Fig. 2b).

2.1 Reference Process

The construction of a basic measurement process for space-time intervals makes it necessary to define a suitable reference process. Traditionally one manufactures rulers and clocks. An arbitrary chunk of matter does not function as a measurement instrument. Lorenzen, Janich [6] explain the actions that lead to these artifacts. How does one manufacture increasingly precise clocks if one did not have them before? By testing procedures one can examine the straight form of a constructed ruler and the uniform running of a clock. Such test norms developed historically from grinding practice, from intuitively controlled actions in everyday technical work. With Lorenzen, Janich [6] we can presuppose (circularity free and without mathematical presuppositions) that every observer can produce "straight", "rigid" rulers and "uniform running" clocks. Provided the classical measurement practice one gets the Euclidean metric. We will demonstrate the step to relativistic kinematics.

Einstein did analyze the principles for measuring moving objects. To make any sense of the concept of time Einstein [7] required an additional process that establishes a connection between clocks at distant locations. In principle one could use any type of process. Most favorable for the theory one chooses a process about which

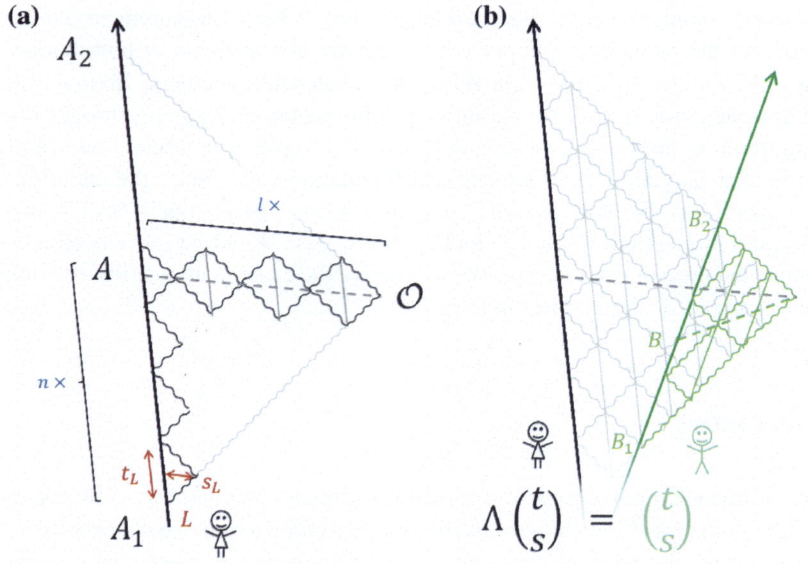

Fig. 2 **a** Count regular layout of light clocks $\mathcal{L}$ **b** intrinsic operations by Alice and Bob

we know something certain. For the motion of light this holds much more than for any other process. By local comparison with classical rulers and clocks the light propagates in a uniform, isotropic and straight way. The motion of light provides a universal reference for any intrinsic observer.

Let Alice construct a light clock $\mathcal{L}$ with two nearby mirrors in a rigid frame. Between both mirrors the light constantly oscillates back and forth (vertical sequence in Fig. 2a). Each tick of her reference device covers the same standard distance $s_{\mathcal{L}}$ and because of the light principle each successive tick takes the same standard time $t_{\mathcal{L}}$.[1] We use the light clock as a new measurement unit. The width of the light clock represents unit length $s_{\mathcal{L}}$ and each tick a unit time $t_{\mathcal{L}}$.

[1]One replaces the traditional rulers and clocks by new light clocks because they approximate the aspired ideal of uniform running more precisely. Contemporary metrology regards the speed of light c as an invariant natural constant and introduces optical clocks as a frequency standard. The definition of the standard length $s_{\mathcal{L}}$ is based on a *given* standard duration $t_{\mathcal{L}}$ and the universal speed of light c.

Basic dimensions (unit length, unit time etc.) are "arbitrarily chosen constant reference measures" [8]. Contemporary metrology refers to the standard duration t_{Cs} of a Cesium period and the speed of light c. On paper one defines the atomic second $\mathrm{sec_{SI}} := 9192631770 \cdot t_{\mathrm{Cs}}$ as a multiple of that standard duration (to match the calendrical second). This defines a standard distance, which light propagates in one atomic second of flight. Then one defines the standard meter, again on paper, as a certain fraction of that distance $299792458 \cdot \mathrm{m_{SI}} := (c \cdot \mathrm{sec_{SI}})$. These new SI units are more practical. Now everyone can reproduce the SI second and meter from invariant natural processes (Cs-period and speed of light) and fixed numerical factors (to match the traditional prototypes, that refer to a tropical year and to a platinum-iridium standard bar from the international bureau of weights and standards, that are not accessible and change in time!).

2.2 *Quantification*

In the measurement Alice joins together light clock ticks one after another until the sequence covers e.g. the waiting interval from a starting moment A_1 to an end in moment A (Fig. 2a). Consider a swarm of identically constituted light clocks. Alice can place ticking light clocks literally side by side. By letting their inner light rays overlap all light clocks tick synchronized. This connection (dashed line in Fig. 2a) represents Alice *simultaneous straight measurement path* $\overline{A\mathcal{O}}$ towards the distant event $\mathcal{O}$. From solely congruent building blocks Alice assembles a grid which covers larger objects or processes, e.g. the relative motion of Bob or the light ray from A_1 to $\mathcal{O}$. In both segments of her material model Alice counts the number n of congruent ticks and the number l of congruent clocks. Alice measures the spatiotemporal interval $\overline{A_1\mathcal{O}}$

$$\begin{aligned} t_{\overline{A_1\mathcal{O}}} &= n \cdot t_{\mathcal{L}} \\ s_{\overline{A_1\mathcal{O}}} &= l \cdot s_{\mathcal{L}} \end{aligned} \tag{1}$$

by the number of measurement units and their dimension (standard duration $t_{\mathcal{L}}$ and width $s_{\mathcal{L}}$).

By inspection of the regular pattern of light clocks (Fig. 2a) one counts for the outgoing light ray $(t, s)_{\overline{A_1\mathcal{O}}} = (n \cdot t_{\mathcal{L}}, 2n \cdot s_{\mathcal{L}})$ and for the returning light ray $(t, s)_{\overline{\mathcal{O}A_2}} = (n \cdot t_{\mathcal{L}}, -2n \cdot s_{\mathcal{L}})$ the *same* distance units and duration units (n the number of ticks from A_1 to A resp. from A to A_2, $2n$ the number of clocks sitting side by side along $\overline{A\mathcal{O}}$). Both measured values are proportional. In standard light clock dimensions the velocity of light has the value $c^{(\mathcal{L})} := \frac{2n}{n} = 2$. For any given duration t_c the light ray travels a distance

$$s_c = \underbrace{2 \cdot \frac{s_{\mathcal{L}}}{t_{\mathcal{L}}}}_{\equiv c} \cdot t_c \; . \tag{2}$$

For a light clock built so that $t_{\mathcal{L}} := \mathrm{sec}_{\mathrm{SI}}$ and $(2 \cdot s_{\mathcal{L}}) =: 299792458 \cdot \mathrm{m}_{\mathrm{SI}}$ (definition SI-meter) we express the speed of light in conventional SI-units $c = 2 \cdot \frac{s_{\mathcal{L}}}{t_{\mathcal{L}}} = 299792458 \cdot \frac{\mathrm{m}}{\mathrm{sec}}$. For the outgoing and the returning light ray we measure the same velocity $c_{out} = c_{in}$ because our measurement unit, the light clock is build from two-way light cycles on equal footing.

Alice can express her *direct measurement* of the spatiotemporal interval $\overline{A_1\mathcal{O}}$ (by literally covering the object with a layout of measurement units) with the help of $t_{\overline{A_1 A}} = \frac{1}{2} \cdot t_{\overline{A_1 A_2}}$

$$(t_{\overline{A_1\mathcal{O}}}\, , \, s_{\overline{A_1\mathcal{O}}}) \overset{(2)}{=} \left(\frac{1}{2} \cdot t_{\overline{A_1 A_2}} \, , \, \frac{c}{2} \cdot t_{\overline{A_1 A_2}} \right) \tag{3}$$

in terms of measuring just the round trip waiting times $t_{\overline{A_1 A_2}}$. That corresponds to the familiar principle of laser ranging measurements, by sending a light ray in moment A_1 to a distant object $\mathcal{O}$ and waiting until the reflection returns in A_2 (Fig. 2a). We derive that basic equation from direct measurements (1) of the outgoing and returning light ray.

2.3 Lorentz Transformation

We utilize a reference process, which represents two dimensions (the length and the duration of the light cycle between two mirrors). Alice and Bob manufacture their own light clocks and place them intrinsically until the measurement object is covered. Along the measurement paths everyone counts the number of ticks and of clocks. The relation between these numbers that is the Lorentz transformation.

Let one more observer Bob measure the distant object $\mathcal{O}$. By connecting his light clocks in an intrinsically timelike resp. spacelike way Bob constructs his straight simultaneous measurement path to the object $\overline{B\mathcal{O}}$ (dashed line in Fig. 2b). He can also determine his simultaneity lines from symmetric round trip signaling times. For the moments of departure B_1 and arrival B_2 of a light signal reflected from an event (anywhere on line $\overline{B\mathcal{O}}$) that is simultaneous to the event B one counts the same number of light clock ticks $t_{\overline{B_1 B}} = t_{\overline{B B_2}}$ in Fig. 2b. That corresponds to Einstein's convention of synchronizing distant clocks. Alice and Bob conduct the same intrinsic operations with light clocks. Though the results are not identical. Alice constructs simultaneity lines $\overline{A\mathcal{O}}$ with different orientation than Bob's simultaneity lines $\overline{B\mathcal{O}}$ (dashed lines in Fig. 2b).

Let Alice and Bob measure the same spacetime interval $\overline{\mathcal{P}\mathcal{O}}$. Without restricting generality both measure a segment of the moving object $\mathcal{O}$, let all start from the joint moment $\mathcal{P}$. Figure 3 shows the interrelation of their laser ranging measurements:

a the clock ticks until Alice A_1 sends out the light signal
$b + b$ the ticks until the reflection from Bob B_1 returns in A_2
$d + d$ the ticks until the reflection from the object $\mathcal{O}$ returns in A_3
m the clock ticks until the coinciding light signal passes Bob B_1
$n + n$ the ticks until the reflection from the object O returns to Bob B_2.

From his round trip signaling durations m, n Bob measures for the moving object

$$t^{(\mathcal{B})}_{\overline{\mathcal{P}\mathcal{O}}} \overset{(3)}{=} m + n$$
$$s^{(\mathcal{B})}_{\overline{\mathcal{P}\mathcal{O}}} = c \cdot n \quad .$$

Alice measures for the same object from her laser ranging durations a, d

$$t^{(\mathcal{A})}_{\overline{\mathcal{P}\mathcal{O}}} = a + d$$
$$s^{(\mathcal{A})}_{\overline{\mathcal{P}\mathcal{O}}} = c \cdot d \quad .$$

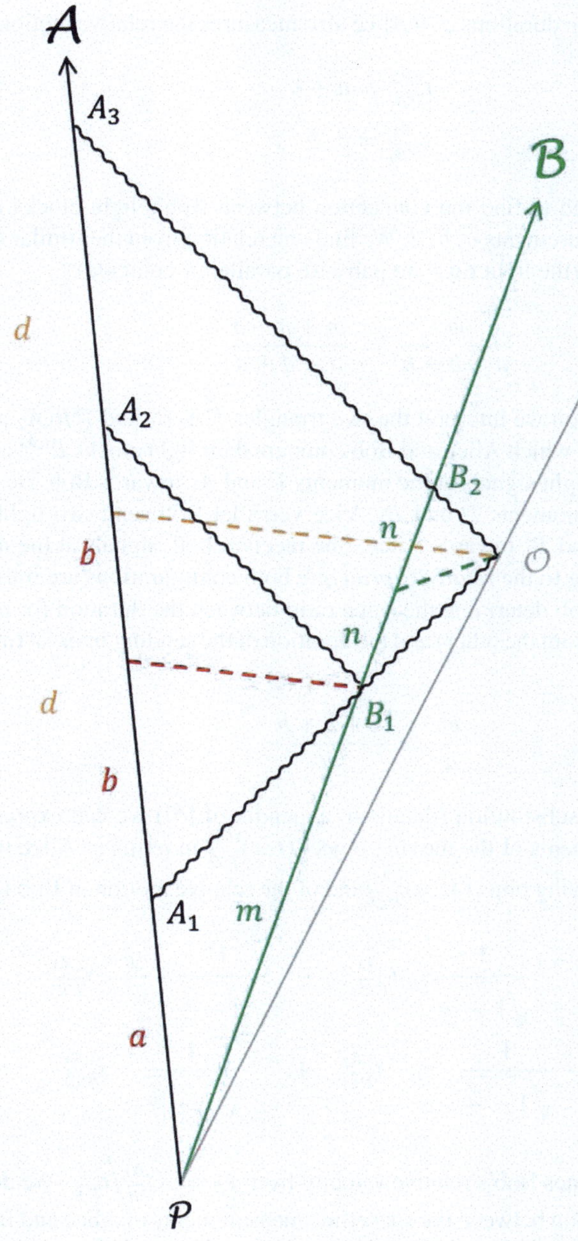

Fig. 3 coinciding light rays in the laser ranging measurements by $\mathcal{A}$lice and $\mathcal{B}$ob

With the laser ranging durations a, b Alice also measures the relative motion of Bob

$$t^{(\mathcal{A})}_{\overline{\mathcal{P}B_1}} = a + b$$
$$s^{(\mathcal{A})}_{\overline{\mathcal{P}B_1}} = c \cdot b \quad .$$

Now we only need to find the connection between Bob's light clocks readings m, n and Alice measurements a, b, d. We find one relation from the similar triangles $\mathcal{P}B_1A_2$ and $\mathcal{P}B_2A_3$ (the light rays are pairwise parallel or coincide)

$$\frac{m}{a+b+b} = \frac{m+n+n}{a+d+d} \quad .$$

For the second relation we interpret the two triangles $\mathcal{P}A_1B_1$ and $\mathcal{P}B_1A_2$ as a calibration procedure by which Alice and Bob compare their light clocks $\mathcal{L}^{(\mathcal{A})}$ and $\mathcal{L}^{(\mathcal{B})}$: Let Alice send two light signals at the moments $\mathcal{P}$ and A_1 towards Bob. He receives both signals at the moments $\mathcal{P}$ and B_1. Vice versa let Bob send two light signals at the moments $\mathcal{P}$ and B_1 towards Alice. She receives both signals at the moments $\mathcal{P}$ and A_2. According to the *relativity principle* both configurations are intrinsically similar. Alice and Bob determine the same ratio between the duration for receiving both signals (heard from the other) and the duration of the sending interval (measured by themselves)

$$\frac{m}{a} \stackrel{!}{=} \frac{a+b+b}{m} \quad .$$

After successive substitution (details in appendix of [5]) we can express Bob's spacetime measurements of the moving object $(t,s)^{(\mathcal{B})}_{\overline{\mathcal{P}\mathcal{O}}}$ in terms of Alice two measurements of the moving object $(t,s)^{(\mathcal{A})}_{\overline{\mathcal{P}\mathcal{O}}}$ and of the relative motion of Bob $(t,s)^{(\mathcal{A})}_{\overline{\mathcal{P}B_1}}$

$$\begin{aligned} t^{(\mathcal{B})}_{\overline{\mathcal{P}\mathcal{O}}} &= \frac{1}{\sqrt{1-\frac{v_B^2}{c^2}}} \cdot t^{(\mathcal{A})}_{\overline{\mathcal{P}\mathcal{O}}} - \frac{1}{\sqrt{1-\frac{v_B^2}{c^2}}} \cdot \frac{v_B}{c^2} \cdot s^{(\mathcal{A})}_{\overline{\mathcal{P}\mathcal{O}}} \\ s^{(\mathcal{B})}_{\overline{\mathcal{P}\mathcal{O}}} &= -\frac{1}{\sqrt{1-\frac{v_B^2}{c^2}}} \cdot v_B \cdot t^{(\mathcal{A})}_{\overline{\mathcal{P}\mathcal{O}}} + \frac{1}{\sqrt{1-\frac{v_B^2}{c^2}}} \cdot s^{(\mathcal{A})}_{\overline{\mathcal{P}\mathcal{O}}} \end{aligned} \tag{4}$$

where Alice determines Bob's relative velocity from $v_B := s^{(\mathcal{A})}_{\overline{\mathcal{P}B}} \big/ t^{(\mathcal{A})}_{\overline{\mathcal{P}B}}$. We derive the Lorentz transformation between the spacetime measurements of Alice and Bob. That is a strictly physical approach to physics where initially mathematics must remain outside - and then every step where mathematics is introduced requires an extra justification. In the same way we develop Helmholtz program for direct measurements of interactions.

Provided the classical measurement practice we get Galilei kinematics. In analyzing the measuring of moving objects Einstein discovered the additional need to establish a physical connection between clocks at different locations and speeds. Our intrinsic operations with light clocks represent the classical metric locally (Euclidean geometry). For the connection Einstein chose the universal motion of light. By including the (local) light principle and the relativity principle we derive Poincare kinematics. Our locally regular grid of light clocks can potentially spread out into every direction. It represents our *metric connection* between distant measurements.

In local laser ranging practice the accelerations of an object are negligible compared to the enormous speed of light. For larger configurations though effects accumulate for example in the twin paradox [5]. One ought to keep the clocks and rulers in an inertial reference frame. We conclude the discussion of kinematics with a brief remark on the *principle of inertia* that also connects to the following discussion of dynamics. In practical experimentation one probes the effects of certain sources on a given object. When those actions become reproducible one can utilize the knowledge for building and steering machines [9]. We identify the action of external forces on a given body by changes in its state of motion - relative to other objects that are not effected by systematic changes of the source. One can examine a given process with the reference to the "relatively fixed" stars or one can mount the reference instruments to the rigid ground of Earth, whose enormous mass will practically not be effected by systematic changes of the source. One can also provide local reference devices on a microscopic scale. In portable electronic devices one can increasingly find microelectromechanical systems with integrated accelerometers, i.e. an intrinsic detector that measures proper acceleration resp. inertial motion. Thus, for all practical purposes the selection of a (sufficiently) inertial reference is well defined. Then one can apply the principle of inertia: "Every body with no forces acting on it remains - as judged from the lab - in a state of rest or of uniform rectilinear motion" [10].

3 Momentum and Inertial Mass

Our goal is to also define impulse, inertial mass (and later energy) as a basic observable that can be measured directly. Momentum can be seen as a power to overrun other objects in collisions. We define a direct comparison method. Let two generic bodies a and b run into each other with initial velocities v_a resp. v_b. According to Galilei (in [11]) object a has "more momentum than" object b if in an head-on collision test a overruns b (they collide, stick together and move in a's initial direction). We symbolize this case with $p\,[a] > p\,[b]$. In a special case they can come to rest; then their momentum is the same $p\,[a] = p\,[b]$. We remark that the comparison of inertial mass is a special case where the initial velocities of the colliding bodies must be the same. Sommerfeld [12] gives essentially the same definition by the reverse process: "Impulse means (with regards to direction and magnitude) that kick, which is capable of generating a given state of motion from the initial state of rest."

For quantification we want to generate that kick from a certain number k of congruent standard kicks. In physics didactics one uses the example of a cannon which is shooting snowballs at an object. With every shot the velocity of the object decreases. One counts the momentum units until the object stops. This gives an intuitive idea of a direct impulse measurement. We define suitable reference objects for these observables. We provide standard bodies with a standard velocity. Then we construct such a collision machinery: The generic particle collides with a swarm of standard momentum carriers. We count how many impulse units one must connect in order to reproduce the same impact. Then we can quantify the impact of any given object $\mathcal{O}$ as a number k times the impulse of our reference bodies. Formally we write the measurement result $p[\mathcal{O}] = k \cdot p_u$. Similarly we determine the mass and velocity by independent measurement procedures. A quantitative description of the collision machinery leads to the basic equation between momentum, inertial mass and velocity. We derive the momentum-velocity relation.

3.1 Reference Process

We define standard bodies and springs as sufficiently constant reference objects for "impact" and "capability to do work". Consider a reservoir of standard bodies with the same mass (Note that we can test this through head-on collisions.) and standard springs $\mathcal{S}$ with the same capability to do work (We can test this by catapulting standard bodies). For our measurements in entire mechanics we refer to two elementary standard processes:

1. Let a particle pair of two standard bodies with the same mass $m_1 = m_2$ and standard velocities $\pm \mathbf{v}_S$ in opposing directions compress a spring that stays at rest (Fig. 4 t_1).
2. Let the compressed spring that remains at rest catapult two initially resting standard objects into opposite directions (Fig. 4 t_2).

Hence standard particles with a standard momentum equal in magnitude but opposite in direction compress a standard spring. Vice versa, the standard spring $\mathcal{S}$ turns standard particles with a standard mass into standard momentum carriers. The initial and final state of that inelastic collision (of irrelevant internal structure) are well-defined.

If Alice couples the compression and decompression of her spring then the particle pair, which initially flew towards each other, will instantly be catapulted apart (Fig. 4 from t_0 to t_2). Alice generates an eccentric elastic collision between bodies of the same mass. If Bob drives by with the same horizontal velocity, he will see the same process as an elastic transversal collision (Fig. 4 Bob). One particle kicks in from below and rebounds antiparallel. The other particle moves on with the same velocity v into a slightly deflected direction. The kinematics is well-defined by the symmetry (collision of two equivalent bodies) and by the relativity principle (the view from a moving observer). For any given initial velocity v and transversal impact velocity w one determines, with Feynman's trick [13], the deflection angle α

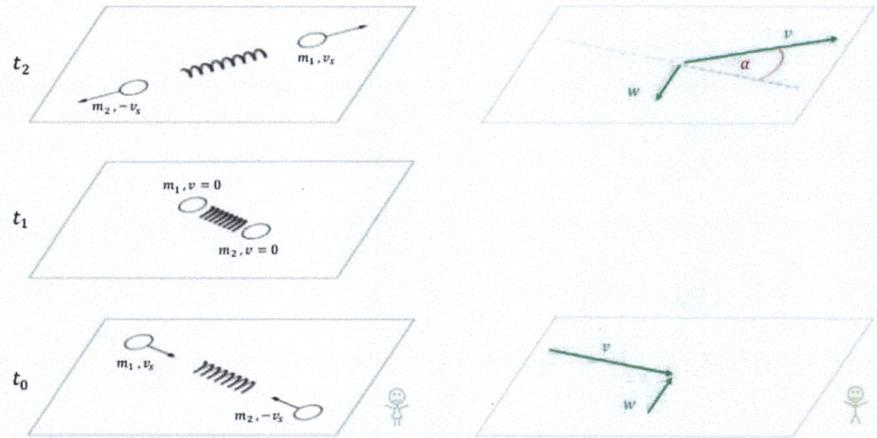

Fig. 4 t_0: A spring is compressed by particles moving in opposite directions. t_1: After the process the compressed spring has capability to do work. t_2: Decompression of the charged spring. Alice observes a symmetric collision. The same process appears as a transversal kick when Bob drives by with the same horizontal velocity to the left

$$\sin\left(\frac{\alpha}{2}\right) = \frac{\sqrt{1 - \frac{v_x^2}{c^2}}}{v} \cdot w \ . \tag{5}$$

Vice versa for a given angle one can determine the necessary velocity v.

3.2 *Assemble Collision Models*

From these standard kicks we assemble increasingly complex collision models. In Fig. 5 we construct a reversion process, where 10 radial kicks with the same impact reverse the motion of a particle, that comes in from the lower left side with a velocity v_2. Similarly we construct a reversion process for another particle that comes in from the opposite side (upper right) with velocity v_1. Let this particle be slower and require only half the number of standard kicks. We choose the velocity v_1 so that after one kick with the same impact its motion is deflected by the same angle $\alpha_1 = 2 \cdot \alpha_2$ as before after two kicks for the faster particle.

We align the reversion processes in the depicted way so that all the temporary activated recoil particles in the center can be captured again and recycled. In the net result only the motion of the three incident particles (one from left and two from right) is exactly reversed. We determine the relation between their velocities v_1 and v_2 from matching the two building blocks, so that the total machinery functions. Depending on the velocities v_1 resp v_2 we steer two types of radial kicks from the outside (highlighted in Fig. 5). To leave no trace in the external reservoir all pairs of

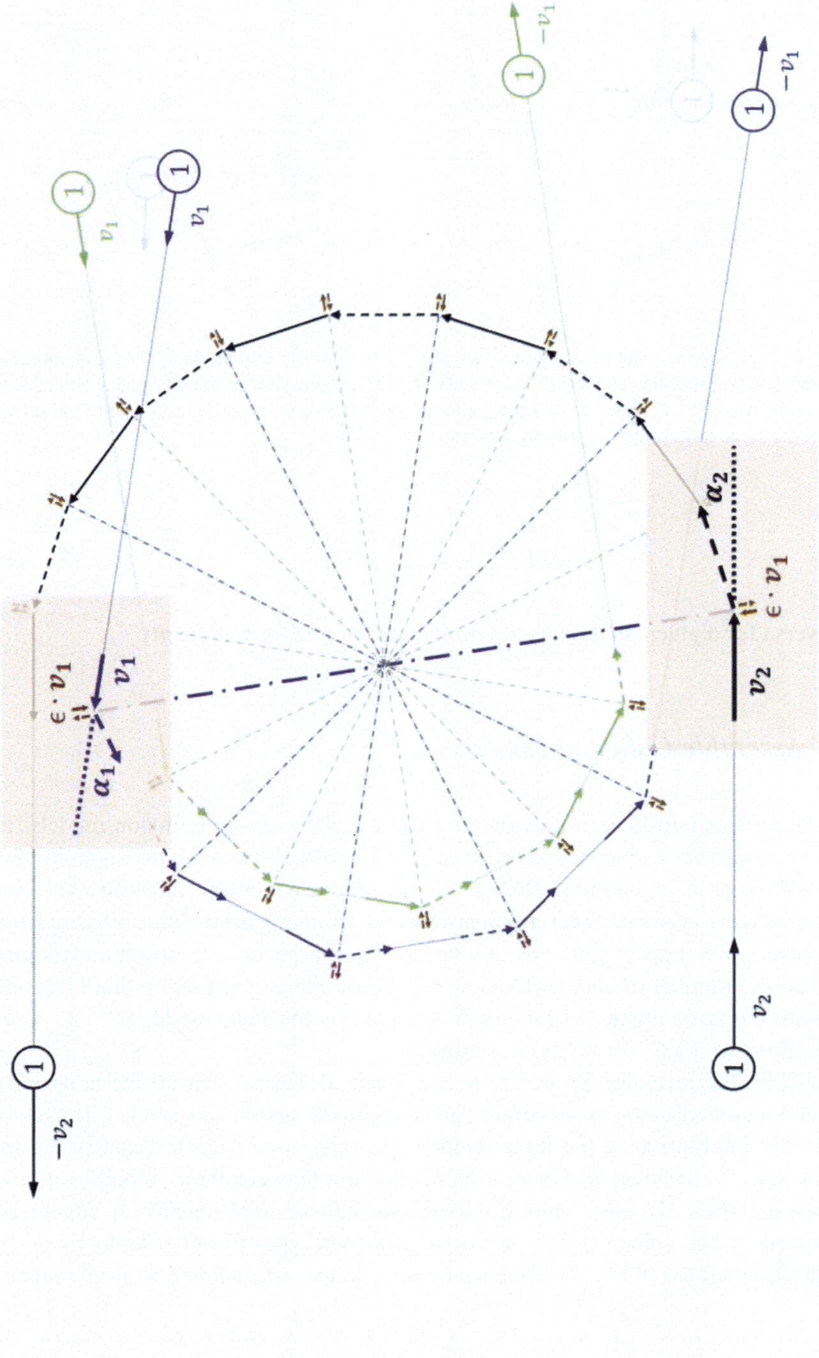

Fig. 5 Align standard collisions

recoil particles must have the same radial velocities $w := \epsilon \cdot v_1$ and are aligned at diametrically opposed locations. Finally, in order to link the collision of one particle from left and two particles from right we match the deflection angles $\alpha_1 = 2 \cdot \alpha_2$. This leads, after substitution (5), to a relation between the initial velocities v_1 and v_2. One can generalize the construction for the collision of $1 + z$ particles in the same way.

By refinement of the radial kicks one can construct a similar model for the elastic head-on collision between two generic bodies: Let one standard body with mass m (from left) and a rigid composite of z standard objects with equal masses $m_1 = m_2 = \cdots = m_z$ (from right) collide and rebound off of one another with reversed velocities. We symbolize only the velocity changes of the standard particles

$$v(m) = \mathbf{u}\,,\ v_1 = \cdots = v_z = \mathbf{v} \quad \Rightarrow \quad v(m) = -\mathbf{u}\,,\ v_1 = \cdots = v_z = -\mathbf{v}\ .$$

Then in Poincare kinematics one can derive [14], that the initial velocities satisfy the relation

$$\frac{\mathbf{u}}{\sqrt{1-\frac{u^2}{c^2}}} \overset{(5)}{=} -z \cdot \frac{\mathbf{v}}{\sqrt{1-\frac{v^2}{c^2}}}\ . \tag{6}$$

We do not presuppose how the velocities of two generic objects change in an elastic collision. The trick is to mediate their direct interaction by steering a process with an external reservoir. The model is built from one elementary collision process between standard elements which must behave in a symmetrical way. From the geometric layout we derive the relation (6) between the amount of matter and the impact velocities.

3.3 *Quantification*

This gedanken experiment allows a direct impulse measurement. From everyday experience one knows, what is meant by "impact" in a collision. With our collision test we can specify the concept more precisely. We measure the momentum of particle a directly with an aggregate of l standard impulse carriers that has the same impact. In an inelastic head-on collision test this means that neither body overruns the other. For our version of the collision test, which is reversible, the corresponding criterion is that we catapult all incoming particles back into the reversed direction. In that aggregate each impulse carrier represents the same standard impulse p_1. According to the congruence principle

$$p\,[a] \overset{\text{(direct)}}{=} p_1 + \cdots + p_1 \overset{\text{(Congr.)}}{=:} l \cdot p_1$$

we measure "how many times" larger the momentum of particle a is than the impulse p_1 of one reference body.

In the same way we measure the inertia m of the body a with an equally massive aggregate of standard elements m_i and the latter according to the congruence principle

$$m\,[a] \overset{\text{(direct)}}{=} m_1 + \cdots + m_1 \overset{\text{(Congr.)}}{=:} z \cdot m_1$$

by the number z of standard elements and their unit mass m_1.

3.4 General Collision Law

In the collision model we can count the coinciding numbers of standard elements. When we build the model in Galilei kinematics we probe one standard particle with the mass m and the velocity $n \cdot v_1$. Then we find the same number n of impulse units. Here we measure the momentum $p[m] = n \cdot p_1$. For an aggregate of z standard particles $m_i, i = 1, \ldots, z$ with the same velocity one can conduct the collision model z times. Thus we count that z times more impulse units have the same impact. We measure the momentum $p[m_1, \ldots, m_z] = z \cdot n \cdot p_1$.

This leads to the general theorem: The particle a with mass $m_a = z \cdot m_1$ (m_1 a standard mass) and velocity $v_a = n \cdot v_1$ (v_1 the standard velocity) has a momentum

$$p\,[a] \;=\; z \cdot n \,\cdot\, p_1 \;.$$

The equation between these physical quantities looks more familiar in the numerical form

$$\frac{p\,[a]}{p_1} = \frac{m_a}{m_1} \cdot \frac{v_a}{v_1} \;.$$

In Galilei kinematics the magnitude of the impulse equals the magnitude of the mass times the magnitude of the velocity.

In Poincare kinematics we conduct the same direct impulse measurement. We assemble the collision test model from well-defined radial kicks (Fig. 5). In the relation (6) between the impact velocities w, v_i and the particle number z appear additional Lorentz terms γ. For a given particle a with the velocity v one counts the number of impulse units, that generate the same impact. This leads to the relativistic relation between momentum and velocity [14]

$$p\,[a] \;=\; \{\gamma \cdot m \cdot v\} \,\cdot\, p_1 \;.$$

4 Energy

We define energy as a basic observable which can be measured directly. In the familiar context of working with machines one can identify the "source" that drives

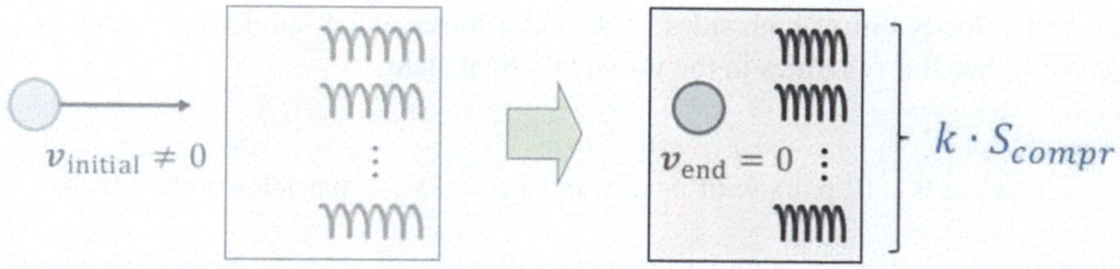

Fig. 6 Direct measurement of kinetic energy: count compressed standard springs

the process and the "raw material" on which work is done. We regard energy as a capability of one source or system to do work against another system. One also knows what is meant if one source has more capability to do work (lift more weights, grind more corn, generate more heat etc.) than another. Leibniz and Helmholtz give a more precise physical definition. One source $\mathcal{S}$ has more energy than another source $\mathcal{T}$ if, until complete exhaustion, the work of $\mathcal{S}$ *exceeds* the work of $\mathcal{T}$ on the same test system [15]. Formally we write $E[\mathcal{S}] > E[\mathcal{T}]$.

We focus on the kinetic form of energy which is associated with slowing down the motion of a body. When the body stops this form of energy is exhausted. According to Papadouris, Constantinou [1] (p. 209) we measure the kinetic energy possessed by a moving object by exploring "the extent of damage that it could cause" by colliding with certain other objects. We construct a machinery which functions for a direct measurement, a calorimeter model. The measurement process works as follows: A calorimeter absorbs a given object, stops it and generates a swarm of standard energy sources (Fig. 6). We will count how many standard springs $\mathcal{S}$ an incoming object with v_{initial} can compress before it stops ($v_{\text{end}} = 0$). Finally we derive the kinetic energy-velocity equation.

4.1 Steering a Calorimeter Model

We illustrate the measurement process with an example. Consider the elastic head-on collision (3.2) between one fast standard particle with the mass m and an aggregate of 9 elements with the standard mass m_i, $i = 1, \ldots, 9$. For a drive-by observer the incident particle kicks the resting aggregate into motion and rebounds with reduced velocity to the left (Fig. 7 t_1). From those deceleration kicks we build our calorimeter model. On the left we place again a suitable number of 7 reservoir elements into the way, such that they get kicked out with the same standard velocity. The incident particle successively rebounds with reduced velocity (Fig. 7 t_1 to t_5) until it stops inside the calorimeter. On the left side of the calorimeter we kick $7 + 3$ reservoir elements into motion. On the right side we generate $9 + 5 + 1$ impulse carriers with the same standard velocity. Thus, for a particle with velocity $5 \cdot v$ we mobilize a total of 25 initially resting reservoir elements. We kick 10 particle pairs with the same standard velocity out of both sides of the calorimeter and 5 single recoil particles. We symbolize the velocities in the initial and final state

$$\begin{aligned} &v(m) = 5 \cdot \mathbf{v}\ ,\ v_1 = \cdots = v_{25} = \mathbf{0} \quad \Rightarrow \\ &\qquad v(m) = \mathbf{0}\ ,\ 10 \text{ pairs with } v_i = \mathbf{v} \text{ and } v_j = -\mathbf{v}\ ,\ 5 \text{ particles with } v_k = \mathbf{v}\ . \end{aligned} \tag{7}$$

When we absorb the same standard particle with a higher velocity, we have to couple more suitable deceleration kicks, that happen before the five collisions t_1 to t_5 in Fig. 7. We mobilize more reservoir elements and create more standard particle pairs and impulse carriers.

4.2 Quantification

Now we interpret the model from a physical perspective. The direct comparison methods specify the physical meaning of our reference objects as units for energy and momentum. The standard impulse carrier m with $v(m) = v_S$ represents the unit of momentum p_1. It also has an energy, namely one half of the energy of the standard spring, because one particle pair can compress one standard spring Sect. 3.1. The calorimeter output (7) has the same capability to do work as the incident particle, because our calorimeter model is reversible. This is plausible because each building block is essentially an elastic head-on collision and because one can steer the complete process in both ways.

Hence the kinetic energy $E[a]$ of the incident particle a is completely transformed into potential energy of the absorber material; therein every spring $\mathcal{S}$ carries the same standard energy. According to the congruence principle

$$E\,[a] \overset{\text{(direct)}}{=} E[\underbrace{\mathcal{S} + \cdots + \mathcal{S}}_{k \text{ times}}] \overset{\text{(Congr.)}}{=:} k \cdot E\,[\mathcal{S}]$$

we measure "how many times" larger its kinetic energy is than the potential energy of one standard spring $\mathcal{S}$. We quantify the energy as a certain number k of reference units times the dimension of the reference objects.

4.3 Basic Equations

In the calorimeter model we can count the coinciding numbers of activated standard springs, reservoir elements, and velocity units. We will now derive the relation between these physical quantities. This step leads to the basic dynamical equations.

In the example in Fig. 7 one can count the absorption effect for one standard particle with the mass m and the velocity $n = 5$ times the standard velocity $\mathbf{v}$. In

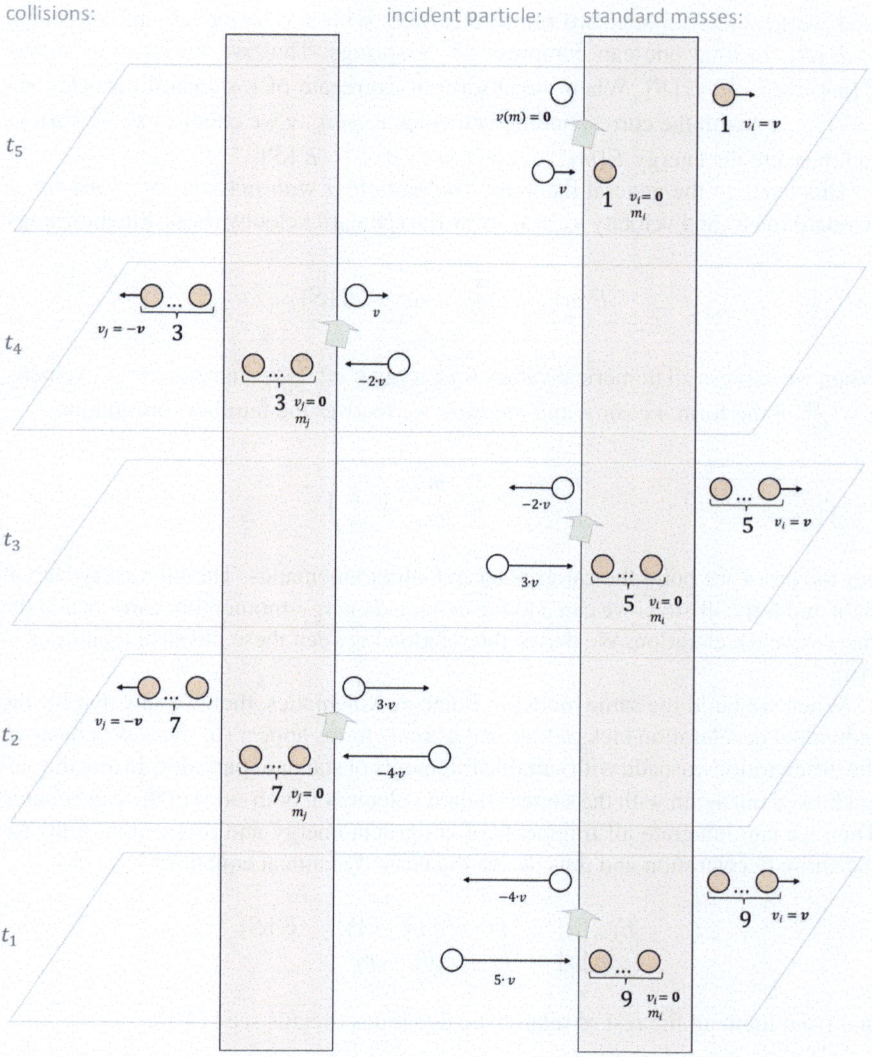

Fig. 7 An incident particle with the mass m comes to rest by a series of collisions with initially resting elements with standard masses m_i on the left and right side of a calorimeter. The deceleration kicks at t_1 to t_5 bring particle m with velocity $5 \cdot \mathbf{v}$ to rest. We generate $7 + 3$ particle pairs with $v_i = \mathbf{v}$ and $v_j = -\mathbf{v}$ and $2 + 2 + 1$ impulse carriers with $v_k = \mathbf{v}$

total we generate 25 standard impulse carriers with a velocity $\pm\mathbf{v}$ and an energy $\frac{1}{2} \cdot E[\mathcal{S}]$. In total one can compress $25 \cdot \frac{1}{2}$ springs. Thus we measure an energy $E[m] = 25 \cdot \frac{1}{2} \cdot E[\mathcal{S}]$. When we absorb an aggregate of z standard particles m_i, $i = 1, \ldots, z$ with the corresponding velocities $v_i = n \cdot \mathbf{v}$ we count $z \cdot n^2 \cdot \frac{1}{2}$ springs and measure the energy $E[m_1, \ldots, m_z] = z \cdot n^2 \cdot \frac{1}{2} \cdot E[\mathcal{S}]$.

This leads to the general theorem: The particle a with mass $m_a = z \cdot m_1$ (m_1 a standard mass) and velocity $\mathbf{v}_a = n \cdot \mathbf{v}$ ($\mathbf{v}$ the standard velocity) has a kinetic energy

$$E[a] = \frac{1}{2} \cdot z \cdot n^2 \cdot E[\mathcal{S}] .$$

When we express all numerical values for energy $k =: \frac{E[a]}{E[\mathcal{S}]}$, mass $z =: \frac{m_a}{m_1}$, velocity $n =: \frac{v_a}{v_1}$ in the form *measure/unit measure* we recover the familiar formulation

$$\frac{E[a]}{E[\mathcal{S}]} = \frac{1}{2} \cdot \frac{m_a}{m_1} \cdot \left(\frac{v_a}{v_1}\right)^2 .$$

For the proof we build the calorimeter in Galilei kinematics. Then for every step of right and left collisions we can add the extracted energy-momentum carriers and the successive deceleration. We derive the relation between these physical quantities in [16].

When we build the same model in Poincare kinematics, then we saw that for the individual deceleration kicks additional Lorentz terms appear (6). Now we construct the deceleration-cascade with suitable fragments of standard particles, so that they all get kicked out again with the same standard velocity on both sides of the calorimeter. Then we can integrate all fragments of congruent energy and momentum units for the entire deceleration and thus derive the basic dynamical equations

$$\begin{aligned} E_{\text{kin}}[a] &= \left\{m \cdot c^2 \cdot (\gamma - 1)\right\} \cdot E[\mathcal{S}] \\ p[a] &= \{m \cdot \gamma \cdot v\} \cdot p_1 \end{aligned}$$

and from them all the rest of relativistic dynamics (details see [14]).

5 Discussion

We introduce a novel strictly physical foundation of the measure of energy, momentum and inertial mass without taking equations of motion etc. as a basis. This approach draws on measurement principles of Galilei, Leibniz and Helmholtz and more recent developments: Einstein's relativity principle, light principle and the measurement theoretical foundation of kinematics as well as the protophysical understanding of the origin of reference devices and procedures. In the debate on the status of Einstein's clock postulate, of providing clocks and rods as *unstructured entities*, to give empirical meaning to the assertions of the theory, Weyl represented the axiomatic view (that measurement instruments ought to be regarded as solutions to differential equations) while Pauli "denied the very possibility of applying a single conceptual model to the theory and to the measurement instruments that verify it"; Einstein took a stance in between [17]. The main idea of protophysics [6] is that physics does not at the same time provide a theory of its measurement units. A theory of measurements in the foundation of physics therefore also implies a systematization of measurement conventions and norms, which the instruments must obey [15].

References

1. R. Duit. Teaching and learning of energy in K-12 education, ed. by R.F. Chen, A. Eisenkraft, D. Fortus, J. Krajcik, K. Neumann, J. Nordine, A. Scheff (Springer, Heidelberg, 2014), pp. 153
2. D. Hammer, Student ressources for learning introductory physics. Am. J. Phys. **68**, 52 (2000)
3. C. Gerthsen, H. Vogel, *Gerthsen Physik* (Springer, Berlin, 1995)
4. H.V. Helmholtz, Zählen und Messen, erkenntnistheoretisch betrachtet. *Philosophische Vorträge und Aufsätze*, ed. by H. Hörz, S. Wollgast (Akademie Verlag, Berlin, 1971), pp. 109
5. B. Hartmann, Operationalization of Relativistic Motion (Kinematics) (2012), arXiv:1205.2680
6. P. Janich, *Das Maß der Dinge: Protophysik von Raum Zeit und Materie* (Suhrkamp, 1997)
7. A. Einstein, *Grundzüge der Allgemeinen Relativitätstheorie* (Springer, Berlin, 2002)
8. J. Wallot, *Grössengleichungen Einheiten und Dimensionen* (Johann Ambrosius Barth, Leipzig, 1952)
9. H. Hertz, Einleitung zur Mechanik. *Zur Grundlegung der theoretischen Physik*, ed. by R. Rompe, H.-J. Treder (Akademie Verlag, Berlin, 1984), pp. 82
10. R.U. Sexl, H.K. Urbantke, *Relativity, Groups, Particles - Special Relativity and Relativistic Symmetry in Field and Particle Physics* (Springer, Berlin, 2001)
11. H. Weyl, *Philosophy of Mathematics and Natural Science* (Princeton University Press, Princeton, 1949), p. 139
12. A. Sommerfeld, *Mechanik* (Verlag Harry Deutsch, Thun, 1994), pp. 4
13. R.P. Feynman, R.B. Leighton, M. Sands, *The Feynman Lectures on Physics - Mainly Mechanics, Radiation and Heat* (Addison-Wesley Publishing Company, Boston, 1977), pp. 4–1–4–8
14. B. Hartmann, Operationalization of Relativistic Energy-Momentum. Dissertation, Humboldt-University, urn:nbn:de:kobv:11-100233941 (2015), pp. 115–138
15. O. Schlaudt, *Messung als konkrete Handlung - Eine kritische Untersuchung über die Grundlagen der Bildung quantitativer Begriffe in den Naturwissenschaften* (Verlag Königshausen & Neumann, Würzburg, 2009)
16. B. Hartmann, Operationalization of Basic Observables in Mechanics (2015). arXiv:1504.03571
17. M. Giovanelli, But one must not legalize the mentioned sin: phenomenological vs. dynamical treatments of rods and clocks in Einstein's thought. Stud. Hist. Philos. Mod. Phys. **48**, 20–44 (2014)

Can Spacetime Curvature be Used in Future Navigation Systems?

Hernando Quevedo

Abstract We argue that the curvature generated by a gravitational field can be used to calculate the corresponding metric which determines the trajectories of freely falling test particles. To this end, we present a method to compute the metric from a given curvature tensor. We use Petrov's classification to handle the structure and properties of the curvature tensor, and Cartan's structure equations in an orthonormal tetrad to investigate the differential equations that relate the curvature with the metric. The second structure equation is integrated to obtain the explicit expression for the connection 1−form from which the components of the orthonormal tetrad are obtained by using the first structure equation. This opens the possibility of using the curvature of astrophysical objects like the Earth to determine the position of freely falling satellites that are used in modern navigation systems.

1 Introduction

One of the most important practical applications of general relativity is the Global Positioning System (GPS), the most advanced navigation system known today. It consists essentially in a set of artificial satellites freely falling in the gravitational field of the Earth. To determine the location of any point on the Earth by using the method of triangulation, it is necessary to know the exact position of several satellites at a given moment of time. This means that the path of each satellite must be determined as exact as possible. In fact, due to the accuracy expected from the GPS, specially

---

H. Quevedo (✉)
Instituto de Ciencias Nucleares, Universidad Nacional Autónoma de México,
AP 70543, 04510 Mexico, DF, Mexico
e-mail: quevedo@nucleares.unam.mx

H. Quevedo
Dipartimento di Fisica and ICRANet, Università di Roma "La Sapienza",
00185 Rome, Italy

H. Quevedo
Department of Theoretical and Nuclear Physics, Kazakh National University,
Almaty 050040, Kazakhstan

D. Puetzfeld and C. Lämmerzahl (eds.), *Relativistic Geodesy*,
Fundamental Theories of Physics 196, https://doi.org/10.1007/978-3-030-11500-5_11

for navigation purposes, it is necessary to take into account relativistic effects for the determination of the satellites trajectories and the gravitational field of the Earth. This method is therefore essentially based upon the use of the geodesic equations of motion for each satellite. Moreover, it is necessary to consider the fact that according to special and general relativity clocks inside the satellites run differently than clocks on the Earth surface. Indeed, it is known that not taking relativistic effects into account would lead to an error in the determination of the position which could grow up to 10 kilometers per day.

The curvature seems to be an alternative way to determine the position of any point on the surface of the Earth. Indeed, if we could measure the curvature of the spacetime around the Earth, and from it the corresponding metric, one could imagine that the determination of the position of the satellites could be carried out in a different way. Maybe this method could be more efficient and more accurate. To this end, it is necessary to measure the curvature of spacetime. Several devices have been proposed for this purpose. The five-point curvature detector [1] consists of four mirrors and a light source. By measuring the distances between all the components of the detector, it is possible to determine the curvature. Another method uses a local orthonormal frame which is Fermi-Walker propagated along a geodesic [2]. A gyroscope is directed along each vector of the frame so that the relative acceleration will allow the determination of the curvature components. The gravitational compass [3] is a tetrahedral arrangement of springs with test particles on each vertex. Using the geodesic deviation equation, from the strains in the springs it is possible to infer the components of the curvature. More recently, a generalized geodesic deviation equation was derived which, when applied to a set of test particles, can be used to measure the components of the curvature tensor [4].

It seems therefore to be now well established that the curvature can be measured by using different devices that are within the reach of modern technology. The question arises whether it is possible to obtain the metric from a given curvature tensor. This is the problem we will address in this work. In Sect. 2, we study a particular matrix representation of the curvature tensor which allows us to calculate its eigenvalues in a particularly simple way. Petrov's classification is used to represent the curvature matrix in terms of its eigenvalues. In Sect. 3, we use Cartan's formalism to derive all the algebraic and differential equations which must be combined and integrated to determine the components of the metric from the components of the curvature. As particular examples, we present the Schwarzschild, Taub-NUT and Kasner metrics with cosmological constant. All the components of the metric are found explicitly in terms of the components of the curvature tensor. It turns out that for a given vacuum solution it is possible to find several generalizations which include the cosmological constant.

2 Matrix Representation of the Curvature Tensor

There are several ways to represent and study the properties of the curvature tensor. Here, we will use a method which is based upon the formalism of differential forms and the matrix representation of the curvature tensor. The reason is simple. Imagine an observer in a gravitational field. Locally, the observer can introduce a set of four vectors e_a to perform measurements and experiments. Although it is possible to choose the direction of each vector arbitrarily, the most natural choice would be to construct an orthonormal system, i.e., $e_a \otimes e_b = \eta_{ab} = \mathrm{diag}(+1, -1, -1, -1)$. Of course, the observer could also choose a local metric which depends on the point. Nevertheless, the choice of a constant local metric facilitates the process of carrying out measurements in space and time. This choice is also in the spirit of the equivalence principle which states that locally it is always possible to introduce a system in which the laws of special relativity are valid. The set of vectors e_a can be used to introduce a local frame ϑ^a by using the orthonormality condition $e_a \rfloor \vartheta^b = \delta^b_a$, where $\rfloor$ is the internal product. The set of 1-forms ϑ^a determines a local orthonormal tetrad that is the starting point for the construction of the formalism of differential forms which is widely used in general relativity.

There is an additional advantage in choosing a local orthonormal frame. General relativity is a theory constructed upon the assumption of diffeomorphism invariance, i.e, it is invariant with respect to arbitrary changes of coordinates $x^\mu \to x^{\mu'}$ such that $J = \det\left(\frac{\partial x^{\mu'}}{\partial x^\mu}\right) \neq 0$. Once a local orthonormal frame ϑ^a is chosen, the only freedom which remains is the transformation $\vartheta^a \to \vartheta^{a'} = \Lambda^{a'}_{\ a}\vartheta^a$, where $\Lambda^{a'}_{\ a}$ is a Lorentz transformation, satisfying the condition $\Lambda^{a'}_{\ a}\Lambda_{a'b} = \eta_{ab}$. This means that the diffeomorphism invariance reduces locally to the Lorentz invariance, which is easier to be handled.

In the local orthonormal frame, the line element can be written as

$$ds^2 = g_{\mu\nu}dx^\mu \otimes dx^\nu = \eta_{ab}\vartheta^a \otimes \vartheta^b \,, \tag{1}$$

with

$$\vartheta^a = e^a_{\ \mu}dx^\mu \,. \tag{2}$$

The components $e^a_{\ \mu}$ are called tetrad vectors, and can be used to relate tetrad components with coordinate components. For instance, the components of the metric are given in terms of the tetrad vectors by $g_{\mu\nu} = e^a_{\ \mu}e^b_{\ \nu}\eta_{ab}$. The exterior derivative of the local tetrad is given in terms of the connection $1-$form ω_{ab} as [5]

$$d\eta_{ab} = \omega_{ab} + \omega_{ba} \,. \tag{3}$$

Since the local metric is constant, the above expression vanishes, indicating that the connection $1-$form is antisymmetric. Furthermore, the first structure equation

$$d\vartheta^a = -\omega^a_{\ b} \wedge d\vartheta^b \,, \tag{4}$$

can be used to calculate all the components of the connection 1−form. Finally, the curvature 2−form is defined as

$$\Omega^a{}_b = d\omega^a{}_b + \omega^a{}_c \wedge \omega^c{}_b \tag{5}$$

in terms of a connection. In this differential form representation, the Ricci and Bianchi identities can be expressed as

$$\Omega^a{}_b \wedge \vartheta^b = 0 \,, \qquad d\Omega^a{}_b + \omega^a{}_c \wedge \Omega^c{}_b - \Omega^a{}_c \wedge \omega^c{}_b = 0 \,, \tag{6}$$

respectively.

The curvature 2−form can be decomposed in terms of the canonical basis $\vartheta^a \wedge \vartheta^b$ as

$$\Omega^a{}_b = \frac{1}{2} R^a{}_{bcd} \vartheta^c \wedge \vartheta^d \,, \tag{7}$$

where $R^a{}_{bcd}$ are the components of the Riemann curvature tensor in the tetrad representation.

It is well known that the curvature tensor can be decomposed in terms of its irreducible parts which are the Weyl tensor [6]

$$W_{abcd} = R_{abcd} + 2\eta_{[a|[c} R_{d]|b]} + \frac{1}{6} R \eta_{a[d} \eta_{c]b} \,, \tag{8}$$

the trace-free Ricci tensor

$$E_{abcd} = 2\eta_{[b|[c} R_{d]|a]} - \frac{1}{2} R \eta_{a[d} \eta_{c]b} \,, \tag{9}$$

and the curvature scalar

$$S_{abcd} = -\frac{1}{6} R \eta_{a[d} \eta_{c]b} \,, \tag{10}$$

where we use the following convention for the components of the Ricci tensor:

$$R_{ab} = \eta^{cd} R_{cabd} \,. \tag{11}$$

Due to the symmetry properties of the components of the curvature tensor, it is possible to represent it as a (6×6)-matrix by introducing the bivector indices A, B, ... which encode the information of two different tetrad indices, i.e., $ab \rightarrow A$. We follow the convention used in [5] which establishes the following correspondence between tetrad and bivector indices

$$01 \rightarrow 1 \,, \quad 02 \rightarrow 2 \,, \quad 03 \rightarrow 3 \,, \quad 23 \rightarrow 4 \,, \quad 31 \rightarrow 5 \,, \quad 12 \rightarrow 6 \,. \tag{12}$$

This correspondence can be applied to all the irreducible components of the Riemann tensor given in Eqs. (8)–(10). Then, the bivector representation of the Riemann tensor reads

$$R_{AB} = W_{AB} + E_{AB} + S_{AB} \,, \tag{13}$$

with

$$W_{AB} = \begin{pmatrix} M & N \\ N & -M \end{pmatrix}, \tag{14}$$

$$E_{AB} = \begin{pmatrix} P & Q \\ Q & -P \end{pmatrix}, \tag{15}$$

$$S_{AB} = -\tfrac{R}{12} \begin{pmatrix} I_3 & 0 \\ 0 & -I_3 \end{pmatrix}. \tag{16}$$

Here M, N and P are (3×3) real symmetric matrices, whereas Q is antisymmetric.

We see that the bivector representation of the curvature is in fact given in terms of the (3×3)-matrices M, N, P, Q and the scalar R, suggesting an equivalent representation in terms of only (3×3)-matrices. Indeed, since (13) represents the irreducible pieces of the curvature with respect to the Lorentz group $SO(3, 1)$ and, in turn, this group is isomorphic to the group $SO(3, C)$, it is possible to introduce a local complex basis where the curvature is given as a (3×3)-matrix. This is the $SO(3, C)$-representation of the Riemann tensor [6, 7]:

$$\mathcal{R} = W + E + S \,, \tag{17}$$

$$W = M + iN \,, \tag{18}$$

$$E = P + iQ \,, \tag{19}$$

$$S = \tfrac{1}{12} R\, I_3 \,. \tag{20}$$

In this representation, Einstein's equations can be written as algebraic equations. Consider, for instance, a vacuum spacetime for which $E = 0$ and $S = 0$. Then, the vanishing of the Ricci tensor in terms of the components of the Riemann tensor corresponds to the algebraic condition

$$\mathrm{Tr}(W) = 0 \,, \quad W^T = W \,. \tag{21}$$

In general, from Einstein's equations in the presence of matter

$$R_{ab} - \frac{1}{2} R \eta_{ab} + \Lambda \eta_{ab} = -\kappa T_{ab} \,, \tag{22}$$

we find that

$$R = 4\Lambda + \kappa T \,, \quad T = \eta^{ab} T_{ab} \,, \tag{23}$$

and the components of the curvature tensor satisfy the relationships

$$\eta^{cd} R_{cabd} = \kappa T_{ab} + \left(\Lambda + \frac{\kappa}{2} T\right) \eta_{ab} \; . \tag{24}$$

It is then easy to see that the following curvature tensor

$$S = \frac{1}{12}(4\Lambda + \kappa T) \, \text{diag}(1, 1, 1) \; , \tag{25}$$

$$E = \frac{\kappa}{2} \begin{pmatrix} T_{11} - T_{00} + \frac{1}{2}T & T_{12} - iT_{03} & T_{13} + iT_{02} \\ T_{12} - iT_{03} & T_{22} - T_{00} + \frac{1}{2}T & T_{23} - iT_{01} \\ T_{13} - iT_{02} & T_{23} + iT_{01} & T_{33} - T_{00} + \frac{1}{2}T \end{pmatrix} \; , \tag{26}$$

$$W \text{ arbitrary } (3 \times 3) - \text{ matrix with } \text{Tr}(W) = 0 \; , \quad W^T = W \; , \tag{27}$$

satisfies Einstein's equations identically. Thus, we see that the matrix W has only ten independent components, the matrix E is hermitian with nine independent components and the scalar piece S has only one component.

The energy-momentum tensor determines completely only the trace-free Ricci tensor and the scalar curvature. The Weyl tensor contains in general ten independent components. However, since the local tetrad ϑ^a is defined modulo transformations of the Lorentz group $SO(3, 1)$, we can use the six independent parameters of the Lorentz group to fix six components of the Weyl tensor. Accordingly, we can use the eigenvalues of the matrix W to write the four remaining parameters in the form

$$W^{'} = \begin{pmatrix} a_1 + ib_1 & & \\ & a_2 + ib_2 & \\ & & -a_1 - a_2 - i(b_1 + b_2) \end{pmatrix} \; . \tag{28}$$

In fact, this is the most general case of a Weyl tensor, and corresponds to a type I curvature tensor in Petrov's classification. If the eigenvalues of the matrix W are degenerate, then $a_2 = a_1 = a$ and $b_2 = b_1 = b$ and therefore

$$W^{^D} = \begin{pmatrix} a + ib & & \\ & a + ib & \\ & & -2a - 2ib \end{pmatrix} \; , \tag{29}$$

which represents a type D curvature tensor.

In general, all the eigenvalues can depend on the coordinates x^μ of the spacetime. The real part of the eigenvalues a_1 and a_2 represent the gravitoelectric part of the curvature, whereas the imaginary part b_1 and b_2 correspond to the gravitomagnetic field, i.e., the gravitational field generated by the motion of the source.

3 Integration of Cartan's Structure Equations

Our aim now is to show that for a given curvature tensor it is possible to integrate Cartan's equation in order to compute the components of the metric. To this end, it is necessary to rewrite Cartan's equations so that the dependence on the spacetime coordinates becomes explicit. First, let us introduce the components of the anholonomic connection $\Gamma^a{}_{bc}$ by means of the relationship

$$\omega^a{}_b = \Gamma^a{}_{bc}\vartheta^c \,, \tag{30}$$

and the condition $\Gamma_{abc} = -\Gamma_{bac}$. Then, from the definition of the connection 1−form, we obtain

$$e^a{}_{[\mu,\nu]} = \Gamma^a{}_{bc} e^b_{[\nu} e^c_{\mu]} \,, \tag{31}$$

which represents a differential equation for the components of the tetrad vectors $e^a{}_\mu$. Here, the square brackets denote antisymmetrization. On the other hand, the exterior derivative of the curvature 2−form yields

$$d\Omega^a{}_b = \frac{1}{2}\left(R^a{}_{bcd,\mu} e^\mu_e + 2R^a{}_{bfd}\Gamma^f{}_{ec}\right)\vartheta^e \wedge \vartheta^c \wedge \vartheta^d \,, \tag{32}$$

which together with

$$d\Omega^a{}_b = \frac{1}{2}\left(R^f{}_{bed}\Gamma^a{}_{fc}\right)\vartheta^e \wedge \vartheta^c \wedge \vartheta^d \,, \tag{33}$$

leads to the following equation

$$R^a{}_{b[cd,|\mu|} e^\mu_{e]} = R^a{}_{f[cd}\Gamma^f{}_{|b|e]} - R^a{}_{b[cd}\Gamma^f{}_{|f|e]} - 2R^a{}_{bf[c}\Gamma^f{}_{de]} \,. \tag{34}$$

This equation represents an algebraic relationship between the components of the tetrad vectors $e^a{}_\mu$ and the components of the connection 1−form $\Gamma^a{}_{bc}$.

Finally, the components of the curvature tensor can be expressed in terms of the anholonomic components of the connection as

$$\frac{1}{2}R^a{}_{bcd} = \Gamma^a{}_{b[d,|\mu|} e^\mu_{c]} + \Gamma^a{}_{be}\Gamma^e{}_{[cd]} + \Gamma^a{}_{e[c}\Gamma^e{}_{|b|d]} \,, \tag{35}$$

which can be considered as a system of partial differential equations for the components of the connection with the components of the curvature and the tetrad vectors as variable coefficients.

To integrate Cartan's equations we proceed as follows. First, we consider the 20 particular independent equations (35) together with the 18 equations which follow from Eq. (34). The idea is to obtain from here all the 24 anholonomic components of the connection $\Gamma^a{}_{bc}$. Then, this result is used as input to solve the 24 independent equations which follow from Eq. (31). This procedure leads to a large number of equations which are complicated to be handled. They have been analyzed with some detail in [7]. Here, we will limit ourselves to quoting the some of the final results obtained previously.

4 Type *D* Metrics

Consider a type *D* curvature tensor with eigenvalue $a + ib$, and suppose that

$$a = a(x^3)\,, \quad b = b(x^3)\,, \tag{36}$$

i.e., we assume that the curvature depends on only one spatial coordinate. Furthermore, it is well known that type *D* spacetimes can have a maximum of four Killing vector fields. Then, we will consider spacetimes with two Killing vector fields which can be taken along the coordinates x^0 and x^1; consequently,

$$g_{\mu\nu,0} = g_{\mu\nu,1} = 0\,, \quad g_{\mu\nu,0} = \frac{\partial g_{\mu\nu}}{\partial x^0}\,. \tag{37}$$

This means that the only relevant spatial direction should be x^3. Therefore, we can use the diffeomorphism invariance of general relativity in order to bring four metric components into any desired form. We then assume that

$$g_{30} = g_{31} = g_{32} = 0\,, \quad g_{33} = g_{33}(x^3)\,. \tag{38}$$

In terms of the local tetrad, the above assumption implies that

$$\vartheta^3 = \sqrt{g_{33}}dx^3 = e^3_{\ \dot{3}}dx^3\,, \tag{39}$$

where the dot denotes coordinate indices. As a consequence we have that

$$d\vartheta^3 = 0\,, \tag{40}$$

which implies that six components of the tetrad vectors vanish, namely,

$$e^0_{\ \dot{3}} = e^1_{\ \dot{3}} = e^2_{\ \dot{3}} = e^3_{\ \dot{0}} = e^3_{\ \dot{1}} = e^3_{\ \dot{2}}\,. \tag{41}$$

This means that we now have a system of only ten components of $e^a_{\ \mu}$ that are unknown. On the other hand, the vanishing of the exterior derivative of ϑ^3 implies that

$$\Gamma^3_{\ [ab]} = 0\,, \tag{42}$$

which drastically simplifies the set of differential equations for the components of the connection. A detailed analysis of the resulting equations shows that it is convenient to consider particular cases which are obtained for different choices of some components of the connection. In fact, it turns out that the choices

$$\Gamma^{1}{}_{21} = 0\,, \quad \Gamma^{1}{}_{23} \neq 0 \tag{43}$$

and

$$\Gamma^{1}{}_{21} \neq 0\,, \quad \Gamma^{1}{}_{23} = 0 \tag{44}$$

lead to completely different solutions which we will analyze in the following subsections.

It is then possible to show that with these simplifying assumptions, we can integrate the set of partial differential equations. Several arbitrary functions arise in the tetrad vectors which can then be absorbed by means of coordinate transformations.

4.1 Schwarzschild and Taub-NUT Metrics

The particular choice

$$\Gamma^{1}{}_{21} \neq 0\,, \quad \Gamma^{1}{}_{23} = 0 \tag{45}$$

leads to a compatible set of algebraic and differential equations which allow us to calculate all the components of the tetrad vectors. We present the final results without the details of calculations which can be consulted in [7].

Consider, for instance, the following curvature tensor in the $SO(3, C)$ representation:

$$\mathcal{R} = -\frac{M}{r^3}\,\mathrm{diag}(1, 1, -2) + \frac{\Lambda}{3}\,\mathrm{diag}(1, 1, 1)\,, \tag{46}$$

where $r = x^3$. Then, the integration of all the differential equations yields

$$e^{3}{}_{\dot{3}}\left(\alpha - \frac{2M}{r} - \frac{\Lambda}{3}r^2\right)^{-1/2}, \quad e^{2}{}_{\dot{2}} = r\,, \quad e^{1}{}_{\dot{2}} = rF^{1}_{\dot{2}}\,, \tag{47}$$

$$e^{0}{}_{\dot{m}} = C^{0}{}_{\dot{m}}\left(\alpha - \frac{2M}{r} - \frac{\Lambda}{3}r^2\right)^{1/2}, \quad e^{0}_{\dot{2}} = F^{0}_{\dot{2}}\left(\alpha - \frac{2M}{r} - \frac{\Lambda}{3}r^2\right)^{1/2}, \tag{48}$$

where $m = 0, 1$, α and $C^{0}{}_{\dot{m}}$ are arbitrary real constants and $F^{0}_{\dot{2}}$ and $F^{1}_{\dot{2}}$ are non-zero functions of the coordinate x^2. It is then possible to find a coordinate system in which the above tetrad vector components lead to the line element

$$ds^2 = \left(\alpha - \frac{2M}{r} - \frac{\Lambda}{3}r^2\right)dt^2 - \frac{dr^2}{\alpha - \frac{2M}{r} - \frac{\Lambda}{3}r^2} - r^2(d\theta^2 + \sin^2\theta d\phi^2) , \quad (49)$$

which represents the Schwarzschild-de-Sitter spacetime.

Consider now a curvature tensor with gravitoelectric and gravitomagnetic components:

$$\mathcal{R} = -\frac{M + iP}{(r + iC)^3}\,\mathrm{diag}(1, 1, -2) + \frac{\Lambda}{3}\,\mathrm{diag}(1, 1, 1) , \quad (50)$$

where P and C are arbitrary real constants. It is then possible to show that the result of the integration leads to a line element of the form

$$ds^2 = \Delta_1(dt + 2C\cos\theta d\phi)^2 - \frac{dr^2}{\Delta_1} - (r^2 + C^2)\left(\Delta_2 \sin^2\theta d\phi^2 + \frac{d\theta^2}{\Delta_2}\right) , \quad (51)$$

with

$$\Delta_1 = (r^2 + C^2)\left[\frac{P}{C}(r^2 - C^2) - 2Mr - \frac{\Lambda}{3}(r^2 + C^2)^2\right]^{-1} , \quad (52)$$

$$\Delta_2 = \frac{P}{C} + \frac{4}{3}\Lambda C^2 . \quad (53)$$

Different choices of the parameters P and C lead to different particular solutions of Einstein's equations. For instance, the choice

$$P = l\left(1 - \frac{4}{3}\Lambda l^2\right) , \quad C = l \quad (54)$$

corresponds to the Taub-NUT metric with cosmological constant [8], where l is the NUT parameter. Furthermore, the choice

$$P = kl\left(1 - \frac{4}{3}\Lambda l^2\right) , \quad C = l , \quad k = -1, 0, +1 \quad (55)$$

is known as the Cahen-Defrise spacetime [9].

The Taub-NUT metric is obtained for the choice $P = l$ and $C = l$ with $\Lambda = 0$. It is then possible to obtain several different generalizations which include the cosmological constant. In fact, the simplest choice corresponds to $P = C = l$ and the cosmological constant entering only the scalar part of the curvature. Other generalizations are obtained by choosing the free parameter P as a polynomial in Λ, for instance,

$$C = l , \quad P = l\left(c_1 + c_2\Lambda l^2 + c_3\Lambda^2 l^4 + \cdots\right) , \quad (56)$$

where c_1, c_2, etc. are dimensionless constants. Another example is obtained for the choice

$$P = l\,, \quad C = l\left(c_1 + c_2\Lambda l^2 + c_3\Lambda^2 l^4 + \cdots\right)\,. \tag{57}$$

All these examples generalize the Taub-NUT metric to include the cosmological constant. In principle, all of them should represent different physical configurations since they all differ in the behavior of the Weyl tensor. This opens the possibility of analyzing anti-de-Sitter spacetimes which are equivalent from the point of view of the scalar curvature, but different from the point of view of the Weyl curvature.

We conclude that in the particular case analyzed here the method presented above can be used to generate new solutions of Einstein's equations with cosmological constant.

4.2 Generalized Kasner Metrics

Another particular choice of the connection components given by

$$\Gamma^1{}_{21} = 0\,, \quad \Gamma^1{}_{23} \neq 0 \tag{58}$$

leads to a set of algebraic and differential equations which can be integrated completely for a curvature tensor with only gravitomagnetic components, i.e.,

$$\mathcal{R} = a(x^3)\,\mathrm{diag}(1, 1, -2) + \frac{\Lambda}{3}\,\mathrm{diag}(1, 1, 1)\,. \tag{59}$$

Indeed, after applying a series of coordinate transformations, the corresponding line element can be expressed as

$$\begin{aligned} ds^2 = & \frac{|a_0|}{|3a_0 - \frac{\Lambda}{2}|^{2/3}} dt^2 - \frac{(a_0')^2}{2|a_0|(3a_0 - \frac{\Lambda}{2})^2} dr^2 \\ & - \frac{1}{|3a_0 - \frac{\Lambda}{2}|^{2/3}} (dX^2 + dY^2)\,, \end{aligned} \tag{60}$$

with

$$a_0 = a + \frac{\Lambda}{2} < 0\,, \tag{61}$$

and the prime represents derivation with respect to $r = x^3$. Here we see that the metric can be calculated immediately from the gravitoelectric component of the curvature $a(r)$. Several particular metrics can be written down. We quote only the metric that follows from the eigenvalue

$$a = -\frac{\gamma}{r^{2\beta}} - \frac{\Lambda}{3}\,, \tag{62}$$

where γ and β are real constants. For this case, we obtain

$$ds^2 = \left(\gamma - \tfrac{\Lambda}{6} r^{2\beta}\right) r^{-2\beta/3} dt^2 - \tfrac{2}{9}\beta^2 \left(\gamma - \tfrac{\Lambda}{6} r^{2\beta}\right)^{-1} r^{2(\beta-1)} dr^2 \\ - r^{4\beta/3}(dX^2 + dY^2) \ . \tag{63}$$

In the limiting case $\Lambda = 0$, we obtain for each value of β a particular case of the Kasner metric [10]. In general, the above line element represents a generalization of the Kasner space which includes the cosmological constant. We see that in this particular case we have chosen a curvature eigenvalue which contains the cosmological constant explicitly. This has been done in order to obtain a simple expression for the Kasner metric with Λ. However, one can always change in the function $a(r)$ the term containing the cosmological constant, in order to obtain different solutions. The simplest spacetime would correspond to the one in which the Weyl tensor does not depend on the cosmological constant, i.e.,

$$a = -\frac{\gamma}{r^{2\beta}} \ , \tag{64}$$

for which we obtain the generalized Kasner metric

$$ds^2 = \frac{\left|-\frac{\gamma}{r^{2\beta}} + \frac{\Lambda}{2}\right|}{\left|-\frac{3\gamma}{r^{2\beta}} + \Lambda\right|^{2/3}} dt^2 - \frac{2\beta^2\gamma^2}{r^{2(2\beta+1)}\left|-\frac{\gamma}{r^{2\beta}} + \frac{\Lambda}{2}\right|\left(-\frac{3\gamma}{r^{2\beta}} + \Lambda\right)^2} dr^2 \\ - \frac{1}{\left|-\frac{3\gamma}{r^{2\beta}} + \Lambda\right|^{2/3}} \left(dX^2 + dY^2\right) \ . \tag{65}$$

This particular choice seems to be more complicated than the solution (63); however, from a physical point of view it corresponds to the simplest choice in which the Weyl tensor is not affected by the presence of the cosmological constant.

We see that it is possible to obtain several generalizations of the Kasner metric with cosmological constant and, in principle, each of them should correspond to a different physical configuration.

5 Conclusions

In this work, we presented a method to calculate the components of the metric tensor from the components of the Riemann curvature tensor. We use the formalism of differential forms and Cartan's structure equations in order to calculate explicitly the algebraic and differential equations that relate the components of the local tetrad vectors with the components of the connection 1−form and the curvature 2−form.

We integrate the differential equations for the case of a type D curvature tensor in Petrov's classification which is characterized by only one complex eigenvalue. We found that for a given curvature eigenvalue, it is possible to obtain different metrics, depending on some assumptions made for the components of the connection 1−form. For the computation of explicit examples, we assume that the curvature eigenvalue depends on only one spatial coordinate. This simplifies the set of differential equations and allows us to carry out the integration completely. We obtain as concrete examples two classes of spacetimes. The first class contains the Schwarzschild metric, the Taub-NUT metric and several generalizations which include the cosmological constant. The second class contains a family of particular Kasner spacetimes with cosmological constant.

The main result of the present work is that is possible to obtain the metric from the curvature. Furthermore, we found that for any given vacuum spacetime, we can apply the procedure presented in this work to obtain different generalizations which include the cosmological constant. This means that solutions of Einstein's equations with cosmological constant are not unique. The main physical difference between different spacetimes with cosmological constant is reflected in the Weyl tensor which behaves differently for each metric.

The concrete examples of curvature analyzed in this work involve terms with gravitoelectric monopole (mass parameter) and gravitomagnetic monopole (NUT parameter) only. In the case of an astrophysical gravitational source, a more realistic situation involves higher mass and angular momentum multipole moments. It is then easy to see that if we consider the Weyl tensor in the form

$$W = -\sum_{n=1}^{\infty} \frac{m_n}{r^{2n+1}} \operatorname{diag}(1, 1, -2) , \tag{66}$$

the integration of the structure equations can be performed in a way similar to the one used to obtain the Schwarzschild and the Taub-NUT metrics. The explicit metric components can be computed by using the general formula presented here and in [7]. The resulting metric will contain the parameters m_n which correspond to higher mass multipole moments. In this way, one could generate exact solutions with a prescribed set of multipoles. In a realistic situation, for instance in the case of the Earth, one would need only a limited number of moments m_n, whose values can be from the measurement of the curvature components.

To take into account higher gravitomagnetic moments, it will be probably necessary to generalize the method presented here. Indeed, the presence of rotational moments implies that the curvature must depend on at least two spatial coordinates (a radial and an angular coordinate). In addition, it will probably necessary to consider not only type D, but also type I Weyl tensors. In this case, we need to construct a more general method than the one presented here. However, if we fix the angular coordinate and consider, for instance, the equatorial plane of the gravitational source, the curvature will depend only on the radial distance and it will be possible to consider a Weyl tensor of the form

$$W = -\sum_{n=1}^{\infty} \frac{m_n + i j_n}{r^{2n+1}} \operatorname{diag}(1, 1, -2) , \tag{67}$$

where the parameters j_n represent the multipoles of the curvature generated by the rotation of the source. Of course, this would be only an approximation of a realistic compact object since the dependence on the angular coordinate is completely neglected. However, since in the case of an object like the Earth, the deviations from spherical symmetry due to rotation are very small, one could expect that this equatorial plane approximation would lead results with a good degree of accuracy.

We conclude that the method presented in this work can be used, in principle, to generate particular metrics, describing the gravitational field of realistic compact objects. It would be interesting to investigate this problem in detail in the case of the Earth, to study the possibility of developing new navigation systems by using as input the curvature of the spacetime around our planet.

Acknowledgements This work has been supported by the UNAM-DGAPA-PAPIIT, Grant No. IN111617.

References

1. J.L. Synge, *Relativity: The General Theory* (North-Holland, Amsterdam, 1960)
2. F.A.E. Pirani, Acta Phys. Pol. **15**, 389 (1956)
3. P. Szekeres, J. Math. Phys. **6**, 1387 (1965)
4. D. Puetzfeld, Y.N. Obukhov, Phys. Rev. D **93**, 044073 (2016)
5. K.S. Thorne, C.W. Misner, J.A. Wheeler, *Gravitation* (W. H. Freeman, San Francisco, 1973)
6. R. Debever, J. Geheniau, Bull. Acad. R. Soc. Belg. **42**, 114 (1956)
7. H. Quevedo, Gen. Relativ. Gravit. **24**, 693 (1992)
8. M. Demiański, Phys. Lett. A **42**, 157 (1972)
9. M. Cahen, L. Defrise, Commun. Math. Phys. **11**, 16 (1968)
10. M. MacCallum, C. Hoenselaers, H. Stephani, D. Kramer, E. Herlt, *Exact Solutions of Einstein's Field Equations* (Cambridge University Press, Cambridge, 2003)

World-Line Perturbation Theory

Jan-Willem van Holten

Abstract The motion of a compact body in space and time is commonly described by the world line of a point representing the instantaneous position of the body. In General Relativity such a world-line formalism is not quite straightforward because of the strict impossibility to accommodate point masses and rigid bodies. In many situations of practical interest it can still be made to work using an effective hamiltonian or energy-momentum tensor for a finite number of collective degrees of freedom of the compact object. Even so exact solutions of the equations of motion are often not available. In such cases families of world lines of compact bodies in curved space-times can be constructed by a perturbative procedure based on generalized geodesic deviation equations. Examples for simple test masses and for spinning test bodies are presented.

1 Test Bodies in General Relativity

The newtonian theory of gravity is a theory of instantaneous action at a distance, which is consistent with the concept of absolute time and absolute simultaneity. This allows for the existence of rigid bodies. Taking that for granted Newton proved in the *Principia* that the orbital motion of a homogeneous spherical rigid body is correctly represented by that of a point mass located at the center of gravity. Thus he was able to explain the motion of the moon orbiting the earth. Also in more general situations the motion of a rigid body can be represented by a single curve: its world line, identified as the orbit in space and time of the center of gravity, whilst the remaining kinematical degrees of freedom are restricted to rotational motion of the body about the center of gravity and specify the orientation of the body at every point of the world line. In this frame work there is no obstacle to consider the limit of a very small rigid body of which the gravitational influence on the motions of other bodies is negligeable. Such a body which probes the gravitational field without disturbing it is commonly referred to as a test body. It is characterized by a finite number of

J.-W. van Holten (✉)
Nikhef, Science Park 105, 1098 XG Amsterdam, The Netherlands
e-mail: v.holten@nikhef.nl

D. Puetzfeld and C. Lämmerzahl (eds.), *Relativistic Geodesy*,
Fundamental Theories of Physics 196, https://doi.org/10.1007/978-3-030-11500-5_12

physical degrees of freedom as its motion is completely specified in terms of its world line plus orientational degrees of freedom.

In General Relativity (GR) the situation is more subtle, as strictly point-like particles cannot be accommodated in the theory. Most importantly any object of finite non-zero mass m has an associated Schwarzschild radius

$$\rho_S = \frac{2Gm}{c^2},$$

such that if the size of the object shrinks below this scale it becomes a black hole with a finite surface area equal to $A = 4\pi\rho_S^2$ for spherical bodies. Thus any massive body is an extended body, at least in classical GR. In situations where quantum effects become relevant this may change, but then there are other limitations like the Compton wave length opposing complete localization of objects.

If a classical object also possesses an internal angular momentum (spin) there are additional complications. First of all in GR rigid bodies can not exist and there is no unique, observer-independent center of mass. Indeed as there is no absolute simultaneity the relative positions of different particles composing the body at any fixed time depend on the state of motion of the body with respect to the observer. Some elementary considerations showing this state of affairs are discussed for a simple two-body system in Appendix A. In addition, in a relativistic context for a composite system it is more appropriate to discuss the motion of a center of energy or *centroid* rather than a center of mass, although such a concept is still observer-dependent. Actually it has been shown [1–3] that under reasonable assumptions the world lines of all possible centroids of an object with mass m and spin s fill a time-like oriented tube of radius

$$\rho_M = \frac{s}{mc}.$$

The upshot of this discussion is that strictly speaking in GR no unique world line can be associated with the motion of massive bodies and any particular choice of representative world line is at least in part a matter of convenience and requires careful specification.

Nevertheless there are circumstances in which the localization of a compact body is possible with sufficient accuracy that for practical purposes it may be regarded as a mass point moving along a world line. Moreover if its mass is small enough that one can neglect the associated space-time curvature and its influence on other bodies, one can still regard such an approximate point mass as a test body probing the space-time geometry in which it moves.

In the context of GR the space-time geometry and the motion of compact bodies are linked by the Einstein equations[1]

$$G_{\mu\nu} + 8\pi G\, T_{\mu\nu} = 0, \tag{1}$$

[1]Here and in the following we use natural units in which the speed of light $c = 1$.

where the Einstein tensor $G_{\mu\nu}$ is specified by the space-time geometry, and the energy-momentum tensor $T_{\mu\nu}$ describes the physical degrees of matter. Irrespective of the precise background geometry the Bianchi identities for the Einstein tensor require the energy-momentum tensor to be divergence-free:

$$\nabla_\mu T^{\mu\nu} = 0. \tag{2}$$

For a compact body with mass m moving as an approximate point mass along a world line $\xi^\mu(\tau)$ parametrized by the proper time τ the effective energy-momentum tensor is [4]

$$T^{\mu\nu}(x) = \frac{m}{\sqrt{-g}} \int d\tau\, \dot{\xi}^\mu \dot{\xi}^\nu \delta^4(x - \xi(\tau)), \tag{3}$$

with the overdot denoting a derivative w.r.t. proper time.[2] Its divergence vanishes if the world line is a geodesic:

$$\nabla_\mu T^{\mu\nu} = \frac{m}{\sqrt{-g}} \int d\tau \left[\ddot{\xi}^\nu + \Gamma_{\mu\lambda}^{\;\;\nu}(\xi)\dot{\xi}^\mu \dot{\xi}^\lambda\right] \delta^4(x - \xi(\tau)) = 0. \tag{4}$$

Neglecting the back reaction of the compact body is allowed if the contributions of the test body to the geometry of space time are too small to be of interest. As an illustration take the local background geometry to be that of flat Minkowski space-time:

$$g^{(0)}_{\mu\nu} = \eta_{\mu\nu} \quad \text{and} \quad \Gamma^{(0)\;\nu}_{\mu\lambda} = 0. \tag{5}$$

In particular we can then choose to work in the local inertial frame in which the body is at rest:

$$T^{(0)}_{00} = m\delta^3(\mathbf{x}), \quad T^{(0)}_{i0} = T^{(0)}_{ij} = 0 \qquad i, j = (1, 2, 3). \tag{6}$$

Taking into account these energy-momentum source terms, the solution of the Einstein equation is modified to first order to read

$$g^{(1)}_{\mu\nu} = \eta_{\mu\nu} + h_{\mu\nu}, \tag{7}$$

where the correction term satisfies the linearized Einstein equation

$$\begin{aligned} &\Box h_{\mu\nu} - \partial_\mu \partial^\lambda h_{\lambda\nu} - \partial_\nu \partial^\lambda h_{\lambda\mu} + \partial_\mu \partial_\nu h_\lambda^{\;\lambda} - \eta_{\mu\nu} \left(\Box h_\lambda^{\;\lambda} - \partial^\kappa \partial^\lambda h_{\kappa\lambda}\right) \\ &= -8\pi G\, T^{(0)}_{\mu\nu}. \end{aligned} \tag{8}$$

Removing gauge degrees of freedom by the De Donder conditions

---

[2] In this paper the delta function is defined as a scalar density of weight 1/2 such that $\int d^4y\, f(y)\, \delta^4(y - x) = f(x)$.

$$\partial^\mu h_{\mu\nu} = \frac{1}{2}\,\partial_\nu h_\mu^{\ \mu},$$

the Einstein equation simplifies further to

$$\Box\left(h_{\mu\nu} - \frac{1}{2}\,\eta_{\mu\nu} h_\lambda^{\ \lambda}\right) = -16\pi G\, T^{(0)}_{\mu\nu}, \tag{9}$$

which has the solution

$$h_{i0} = 0, \quad h_{ij} = \delta_{ij} h_{00}, \quad h_{00} = \frac{2Gm}{r}. \tag{10}$$

With this correction we obtain the modified line element

$$g^{(1)}_{\mu\nu} dx^\mu dx^\nu = -\left(1 - \frac{2Gm}{r}\right) dt^2 + \left(1 + \frac{2Gm}{r}\right) d\mathbf{r}^2. \tag{11}$$

It follows that the geometry near the test body deviates strongly from flat space on the scale of its Schwarzschild radius, and it can be considered as a near point-like object only as long as the external curvature is comparatively small on this scale.

Actually to first order in G the geometry specified by the line element (11) coincides with that of Schwarzschild space-time in isotropic co-ordinates:

$$g_{\mu\nu} dx^\mu dx^\nu = -\left(\frac{2r - Gm}{2r + Gm}\right)^2 dt^2 + \left(1 + \frac{Gm}{2r}\right)^4 d\mathbf{r}^2. \tag{12}$$

In hindsight it is not such a surprise that the test-particle approximation is the limit of a black-hole geometry, as it is well-known that the standard black-hole space-times are exact vacuum solutions of the Einstein equations characterized by a finite number of parameters like mass, spin and electric charge, in which respect they resemble test bodies. The difference is of course that their action on the space-time geometry has been fully taken into account to the extent that –in contrast to test particles– their energy-momentum tensor has been absorbed completely in the space-time curvature, i.e. it has become part of the Einstein tensor in Eq. (1).

2 The Motion of Test Bodies

Restricting our considerations to situations where the internal degrees of freedom of a compact body do not take part in its gravitational interaction and its size and gravitational back reaction can be neglected, its motion can be represented adequately by a world line which is a time-like geodesic of the external space-time geometry. This is the simplest case of the test-body approximation. The more elaborate test-body limit of a compact body with spin will be discussed in Sect. 4 and later.

This description of a test body as an object with a finite numer of degrees of freedom to which one can assign at any moment a representative position has considerable mathematical advantage: to such a body we can associate a finite-dimensional phase space in which the evolution of the system is described by a simple curve. This orbit is generated by a hamiltonian H depending on a finite number of phase-space variables including position and momentum. As by construction it neglects the finite size and gravitational back reaction of the body, it is of course an *effective hamiltonian*, its validity restricted by the test-body limit.

The effective hamiltonian dynamics of a massive test body with no other degrees of freedom, like spin or charge, and an energy-momentum tensor of the form (1) is straightforward to construct. The phase space is spanned by the position variables $\xi^{\mu}(\tau)$ and the momentum variables $\pi_{\mu}(\tau)$, with the usual equal proper-time canonical Poisson brackets

$$\{\xi^{\mu}, \pi_{\nu}\} = \delta_{\nu}^{\mu}. \tag{13}$$

The free hamiltonian

$$H = \frac{1}{2m}\, g^{\mu\nu}[\xi]\, \pi_{\mu}\pi_{\nu} \tag{14}$$

then generates the equations of motion

$$\dot{\xi}^{\mu} = \{\xi^{\mu}, H\} = \frac{1}{m}\, g^{\mu\nu}\pi_{\nu}, \qquad \dot{\pi}_{\mu} = \left\{\pi_{\mu}, H\right\} = \frac{1}{2m}\, \partial_{\mu}g^{\nu\lambda}\, \pi_{\nu}\pi_{\lambda}. \tag{15}$$

By simple algebra these equations can be rewritten in the form

$$\pi_{\mu} = m g_{\mu\nu}\dot{\xi}^{\nu}, \qquad \ddot{\xi}^{\mu} + \Gamma_{\lambda\nu}^{\;\;\mu}\dot{\xi}^{\lambda}\dot{\xi}^{\nu} = 0, \tag{16}$$

reproducing the geodesic equation derived earlier from the energy-momentum tensor (1).

In this hamiltonian frame work it is easy to find constants of motion, simplifying the solution of the geodesic equation. There is at least one universal constant independent of the specific metric $g_{\mu\nu}$, the hamiltonian itself:

$$H = -\frac{m}{2} \quad \Leftrightarrow \quad g_{\mu\nu}\, \dot{\xi}^{\mu}\dot{\xi}^{\nu} = -1. \tag{17}$$

It establishes the usual relation between proper time and co-ordinate time. Other constants of motion depend on the symmetries of the background space-time, as implied by Noether's theorem. For example a quantity

$$J[\xi, \pi] = \alpha^{\mu}[\xi]\, \pi_{\mu} \tag{18}$$

is a constant of motion if α^{μ} is a Killing vector, representing an isometry of the metric:

$$\dot{J} = \{J, H\} = 0 \quad \Leftrightarrow \quad \nabla_{\mu}\alpha_{\nu} + \nabla_{\nu}\alpha_{\mu} = 0. \tag{19}$$

As an example in static space-times the kinetic energy E of the test body is a constant of motion:

$$\alpha^\mu = (-1, 0, 0, 0) \quad \Rightarrow \quad E = -\pi_t. \tag{20}$$

In particular in Minksowki space-time:

$$E = -m\, \eta_{tt}\, \dot{\xi}^t = m\, \frac{dt}{d\tau} = \gamma m, \tag{21}$$

where γ is the time-dilation factor.

In spherically symmetric geometries, such as Minkowski, Schwarzschild and Freedmann–Lemaitre type cosmological space-times, all three components of angular momentum are conserved:

$$\begin{aligned} J_1 &= -\sin\varphi\, \pi_\theta - \operatorname{cotan}\theta\, \cos\varphi\, \pi_\varphi, \\ J_2 &= \cos\varphi\, \pi_\theta - \operatorname{cotan}\theta\, \sin\varphi\, \pi_\varphi, \\ J_3 &= \pi_\varphi. \end{aligned} \tag{22}$$

Here the momenta $(\pi_r, \pi_\theta, \pi_\varphi)$ are defined w.r.t. a polar co-ordinate frame (r, θ, φ) such that the hamiltonian is given by

$$2mH = -g^{tt}(t,r)\pi_t^2 + g^{rr}(t,r)\pi_r^2 + g^{\theta\theta}(t,r)\left(\pi_\theta^2 + \frac{\pi_\varphi^2}{\sin^2\theta}\right). \tag{23}$$

Clearly if the metric components $g_{\mu\nu}$ are time independent the system is both static and spherically symmetric, and energy and angular momentum are all conserved. This holds in particular for Minkowski and Schwarzschild space-time.

The standard metric of Schwarzschild space-time in Droste co-ordinates for an object with mass M can be obtained from the isotropic co-ordinate representation (12) by the co-ordinate transformation

$$\bar{r} = r\left(1 + \frac{GM}{2r}\right)^2, \tag{24}$$

reproducing the static, spherically symmetric line element

$$g_{\mu\nu}dx^\mu dx^\nu = -\left(1 - \frac{2GM}{\bar{r}}\right)dt^2 + \frac{d\bar{r}^2}{1 - \frac{2GM}{\bar{r}}} + \bar{r}^2\left(d\theta^2 + \sin^2\theta\, d\varphi^2\right). \tag{25}$$

Replacing $\bar{r} \rightarrow r$ we can then write the hamiltonian for a test body in Schwarzschild space-time as

$$2mH = -\frac{\pi_t^2}{1 - \frac{2GM}{r}} + \left(1 - \frac{2GM}{r}\right)\pi_r^2 + \frac{1}{r^2}\left(\pi_\theta^2 + \frac{\pi_\varphi^2}{\sin^2\theta}\right), \tag{26}$$

which is a special instance of a static hamiltonian of the type (23). The conservation of kinetic energy then holds in the form

$$\varepsilon \equiv \frac{E}{m} = \left(1 - \frac{2GM}{r}\right)\frac{dt}{d\tau}. \tag{27}$$

Rotating the co-ordinate system such that the plane of the orbit is at constant $\theta = \pi/2$, we have $J_1 = J_2 = \pi_\theta = 0$ and

$$\ell \equiv \frac{J_3}{m} = r^2\frac{d\varphi}{d\tau}. \tag{28}$$

In addition the universal constant of motion $H = -m/2$ implies that

$$\left(\frac{dr}{d\tau}\right)^2 = \varepsilon^2 - \left(1 - \frac{2GM}{r}\right)\left(1 + \frac{\ell^2}{r^2}\right). \tag{29}$$

In particular there are circular orbits $r = R =$ constant for which

$$\begin{aligned} \frac{d\varphi}{d\tau} &= \frac{\ell}{R^2} = \sqrt{\frac{GM}{R^3}}\,\frac{1}{\sqrt{1 - \frac{3GM}{R}}}, \\ \frac{dt}{d\tau} &= \frac{\varepsilon}{1 - \frac{2GM}{R}} = \frac{1}{\sqrt{1 - \frac{3GM}{R}}}. \end{aligned} \tag{30}$$

Observe that these equations reproduce Keplers third law for circular orbits:

$$\frac{d\varphi}{dt} = \sqrt{\frac{GM}{R^3}} \quad \Rightarrow \quad T^2 = \frac{4\pi^2}{GM}R^3,$$

where T is the orbital period measured by the clock of a distant observer keeping co-ordinate time t, and the radial co-ordinate R is determined by the orbital circumference through the relation $L = 2\pi R$.

3 Geodesic Deviations

In most space-time geometries general exact solutions of the geodesic equation are difficult to obtain, and when they are available they are often expressed in terms of non-elementary transcendental functions [5]. However given one particular geodesic curve it is possible to find approximate solutions for arbitrary nearby geodesics using the geodesic deviation equation and its higher-order generalizations [6, 7]. This procedure can also be explained in terms of the Schild ladder construction [8, 9], for which next-to-leading order corrections can be obtained using the results reviewed in this section.

In terms of the tangent vector $u^{\mu} = \dot{\xi}^{\mu}$ the geodesic equation (16) reads

$$D_{\tau} u^{\mu} = \dot{u}^{\mu} + \Gamma_{\lambda\nu}^{\;\;\;\mu} u^{\lambda} u^{\nu} = 0. \tag{31}$$

Now consider a family of world lines

$$\xi^{\mu}(\tau, \sigma) = \xi^{\mu}(\tau) + \sigma n^{\mu}(\tau), \tag{32}$$

obtained by a displacement of the geodesic in the direction of a vector field $n^{\mu}(\tau)$ scaled by the real parameter σ. The corresponding changes in the world line and its tangent vector are described covariantly by the vectors

$$\Delta\xi^{\mu} = \sigma n^{\mu}, \quad \Delta u^{\mu} = D_{\tau}\Delta\xi^{\nu} = \sigma D_{\tau} n^{\mu}. \tag{33}$$

For this displacement to respect the geodesic equation (31) to first order in σ we require

$$\Delta D_{\tau} u^{\mu} = D_{\tau}\Delta u^{\mu} + [\Delta, D_{\tau}]\, u^{\mu} = \sigma \left(D_{\tau}^{2} n^{\mu} - R_{\lambda\nu\kappa}^{\;\;\;\;\;\mu} u^{\lambda} u^{\kappa} n^{\nu} \right) = 0. \tag{34}$$

For example, we know all circular orbits in Schwarzschild space-time in analytical form. Then we can construct approximate non-circular (eccentric) solutions in the same plane by adding a geodesic deviation σn^{μ}; generically this is an oscillating term as the test body moves periodically closer and farther from the central mass between periastron and apastron. As the period of this perturbation is in general different from the period of the circular orbit one starts from, the periastron and apastron will shift their azimuthal directions each turn. This phenomenon is well-known ever since it was calculated by Einstein for the planet Mercury in its orbit around the sun. It turns out that for Mercury the first-order deviation (33) from motion on a circle with the period of Mercury's orbit actually already gives full numerical agreement with the observed periastron shift of 43 arcseconds per century [6].

There is however no obstacle to include second and higher-order contributions in σ to the displaced geodesics $\xi^{\mu}(\sigma, \tau)$. As Δ creates a covariant displacement in the direction of n, but n is not necessarily displaced parallel to itself (it is not required to be a tangent vector field to a family of geodesics crossing the world line $\xi^{\mu}(\tau)$), it follows that in general

$$\Delta^{2}\xi^{\mu} = \sigma\Delta n^{\mu} \equiv \sigma^{2} m^{\mu} \neq 0. \tag{35}$$

The geodesic family $\xi^\mu(\tau, \sigma)$ is then parametrized to second order in σ by

$$\xi^\mu(\tau, \sigma) = \xi_1^\mu(\tau) + \sigma n^\mu(\tau) + \frac{1}{2}\sigma^2 \left(m^\mu - \Gamma_{\lambda\nu}^{\;\;\mu} n^\lambda n^\nu\right) + \cdots \tag{36}$$

Using the properties of the vector field n it is straightforward to generalize the derivation of Eq. (34) and show that $(n^\mu, m^\mu, ...)$ are solutions of a hierarchy of deviation equations [6]

$$\begin{aligned}
D_\tau^2 n^\mu - R_{\lambda\nu\kappa}^{\quad\;\mu} u^\kappa u^\lambda n^\nu &= 0, \\
D_\tau^2 m^\mu - R_{\lambda\nu\kappa}^{\quad\;\mu} u^\kappa u^\lambda m^\nu &= \left(\nabla_\rho R_{\kappa\nu\lambda}^{\quad\;\mu} - \nabla_\kappa R_{\rho\lambda\nu}^{\quad\;\mu}\right) u^\kappa u^\lambda n^\rho n^\nu \\
&\quad + 4 R_{\kappa\nu\lambda}^{\quad\;\mu} u^\kappa n^\nu D_\tau n^\lambda, \\
&\;\;\ldots
\end{aligned} \tag{37}$$

We note in passing, that whereas the first-order deviations n^μ provide information about the curvature of space-time through the Riemann tensor $R_{\kappa\nu\lambda\mu}$, the second-order deviations provide further information about the gradient $\nabla_\rho R_{\kappa\nu\lambda\mu}$ of the Riemann tensor. Thus with sufficient knowledge of families of geodesics in some domain of space-time one can reconstruct the Riemann tensor in the whole domain in terms of a Taylor series from the geodesic deviations w.r.t. a given geodesic [10–12].

The procedure described here has been worked out for Schwarzschild space-time up to and including the second-order deviations [6, 13, 14]. The results for orbits in the equatorial plane $\theta = \pi/2$ are summarized by the parametrized expansions

$$\begin{aligned}
t(\tau, \sigma) &= \xi_0^t(\sigma)\tau + \sum_{n=1}^{\infty} \xi_n^t(\sigma) \sin n\omega(\sigma)\tau, \\
r(\tau, \sigma) &= \xi_0^r(\sigma) + \sum_{n=1}^{\infty} \xi_n^r(\sigma) \cos n\omega(\sigma)\tau, \\
\varphi(\tau, \sigma) &= \xi_0^\varphi(\sigma)\tau + \sum_{n=1}^{\infty} \xi_n^\varphi(\sigma) \sin n\omega(\sigma)\tau.
\end{aligned} \tag{38}$$

The coefficients $\xi_n^\mu(\sigma)$ and the fundamental frequency $\omega(\sigma)$ are computed order by order in σ by solving the hierarchy of deviation equations (37). Of course the terms of order σ^0 represent the parameters of the circular parent orbit we encountered in Eq. (30):

$$\xi_0^t(0) = \frac{1}{\sqrt{1 - \frac{3GM}{R}}}, \quad \xi_0^r(0) = R, \quad \xi_0^\varphi(0) = \sqrt{\frac{GM}{R^3}} \frac{1}{\sqrt{1 - \frac{3GM}{R}}}. \tag{39}$$

All other terms have non-trivial dependence on the expansion parameter σ:

$$\xi_0^\mu(\sigma) = \xi_0^\mu(0) + \sigma\rho_1^\mu + \frac{1}{2}\sigma^2\rho_2^\mu + \cdots,$$
$$\xi_1^\mu(\sigma) = \sigma n_1^\mu + \frac{1}{2}\sigma^2 n_2^\mu + \cdots, \tag{40}$$
$$\xi_2^\mu(\sigma) = \frac{1}{2}\sigma^2 m_2^\mu + \cdots,$$

each $\xi_n^\mu(\sigma)$ contributing only terms of order σ^n and higher, whilst the angular frequency

$$\omega(\sigma) = \omega_0 + \sigma\omega_1 + \cdots \tag{41}$$

also depends on the order of approximation, reflecting the anharmonicity of the non-circular deviations. The explicit expressions for the coefficients ρ_n^μ, n_n^μ and m_n^μ for $n = 1$ and $n = 2$ and the expressions for the frequencies ω_0, ω_1 are summarized in Appendix B for the restricted case $\rho_n^r = 0$. This restriction implies, that we compare the non-circular orbits to a *fixed* circular orbit, although in general one might wish to adapt the circular reference orbit order by order in σ to improve convergence. The expressions for the unrestricted case can be found in Ref. [13].

Observe that the deviations are bound and periodic as long as the angular frequency ω is real. For $R < 6GM$ Eq. (77) in the appendix shows that it develops an imaginary part, indicating exponential run-away behaviour. Thus the circular orbit with $R = 6GM$ is the innermost stable circular orbit (ISCO) for a simple test body in Schwarzschild space-time.

Finally from the solutions of the geodesic deviations we can evaluate the constants of motion ε and ℓ for the non-circular orbits by substitution of the solutions for $dt/d\tau$ and $d\varphi/d\tau$ including first and second order deviations into the expressions (27) and (28). The resulting constants of motion are then also written as a series expansion in the deviation parameter σ. Up to and including second order deviations the result is

$$\varepsilon(\sigma) = \varepsilon_0 + \sigma\varepsilon_1 + \frac{1}{2}\sigma^2\varepsilon_2 + \cdots, \quad \ell(\sigma) = \ell_0 + \sigma\ell_1 + \frac{1}{2}\sigma^2\ell_2 + \cdots, \tag{42}$$

where ε_0 and ℓ_0 are the energy and angular momentum per unit of mass for circular orbits:

$$\varepsilon_0 = \frac{R - 2GM}{\sqrt{R(R - 3GM)}}, \quad \ell_0 = \sqrt{\frac{GMR^2}{R - 3GM}}. \tag{43}$$

In the restricted case $\rho_1^r = 0$ the first-order corrections vanish: $\varepsilon_1 = \ell_1 = 0$, but the second-order corrections do not:

$$\varepsilon_2 = -\frac{GM}{2R^{7/2}} \frac{R^2 - 9GMR + 6(GM)^2}{(R-3GM)^{3/2}},$$
$$\ell_2 = -\frac{3\sqrt{GM}}{2R^2} \frac{(R-2GM)(R-7GM)}{(R-3GM)^{3/2}}. \quad (44)$$

The most important aspect of these expressions is that all dependence on the proper time has disappeared and therefore they are true constants of motion indeed. This is a strong consistency check on the results listed in Appendix B.

One of the useful aspects of the perturbative construction of orbits by the method of geodesic deviations is that they provide explicit and accurate expressions for the position of the test body as a function of the evolution parameter (proper time). Most results in the literature describe the time-dependence of the orbits only in an implicit way. The explicit time-dependent representation of the orbits of a test body in Schwarzschild space-time is especially useful in the computation of the emission of gravitational waves by extreme mass ratio (EMR) binaries: systems consisting of a compact stellar object, like a white dwarf, neutron star or stellar-mass black hole orbiting a supermassive black hole such as found in the center of many galaxies. Gravitational wave emission by EMR binaries has been studied in detail in Ref. [14].

4 Spinning Test Bodies

The motion of electrically neutral, non-spinning test bodies is represented by geodesics in space-time. As soon as new degrees of freedom come into play the world line of a test body becomes non-geodesic. For example, an electrically charged particle in the presence of electric or magnetic fields is subject to a Lorentz force in addition to the effect of curvature. Its world line will differ from that of a similar neutral particle which follows a geodesic.

Similarly the effect of rotation represented by internal angular momentum (spin) will change the motion of test body in curved space-time even in the absence of other fields of force [15, 16]. One can think of this as a form of gravitational spin-orbit coupling. In this section we discuss the motion of spinning test bodies in more detail.

There is an extended literature on spinning test particles; for a review see e.g. Ref. [17]. The most common approaches are based on a multipole expansion requiring a dynamical constraint as an implicit choice of the center of mass fixing the monopole term. As this choice is not unique we have developed in Refs. [18–20] a different formalism which does not incorporate any constraints but only requires appropriate initial conditions to solve the equations of motion. In this formulation the spin degrees of freedom are described by a real anti-symmetric tensor field $\Sigma^{\mu\nu}(\tau)$, which has six components defined on the world line. Three of these components can be represented equivalently by a space-like axial vector, a magnetic-type dipole moment which we identify with the spin proper. Similarly the other three components are described equivalently by a space-like real vector, an electric-type dipole moment which in the gravitational context is interpreted as a mass dipole. Denoting the four-velocity of the test body by the time-like unit vector $u^\mu = \dot{\xi}^\mu$, we can decompose the spin tensor $\Sigma^{\mu\nu}$ accordingly as follows

$$\Sigma^{\mu\nu} = -\frac{1}{\sqrt{-g}}\,\varepsilon^{\mu\nu\kappa\lambda} u_\kappa S_\lambda + u^\mu Z^\nu - u^\nu Z^\mu, \tag{45}$$

with S_μ and Z^μ representing the proper spin and mass dipole components, respectively. The decomposition (45) can be inverted to get

$$S_\mu = \frac{1}{2}\sqrt{-g}\,\varepsilon_{\mu\nu\kappa\lambda} u^\nu \Sigma^{\kappa\lambda}, \quad Z^\mu = \Sigma^{\mu\nu} u_\nu. \tag{46}$$

As mentioned both vectors are space-like by construction:

$$S \cdot u = 0, \quad Z \cdot u = 0. \tag{47}$$

Hence each contains only three degrees of freedom, their time components vanishing in the rest frame $u = (1, 0, 0, 0)$. From these expressions it is inferred directly that under the group of three-dimensional spatial rotations and parity transformations in the rest frame Z is a real vector, whereas S actually is an axial pseudo-vector.

To obtain equations of motion for a spinning body we follow the same route as for a non-rotating body starting from an energy-momentum tensor and rederiving the results from an effective hamiltonian. Following [19] the energy-momentum tensor is taken to have support on a world line $\xi^\mu(\tau)$ with

$$\begin{aligned} T^{\mu\nu} &= m \int d\tau\, u^\mu u^\nu \frac{1}{\sqrt{-g}}\,\delta^4(x - \xi(\tau)) \\ &+ \frac{1}{2}\nabla_\lambda \int d\tau \left(u^\mu \Sigma^{\nu\lambda} + u^\nu \Sigma^{\mu\lambda}\right) \frac{1}{\sqrt{-g}}\,\delta^4(x - \xi(\tau)). \end{aligned} \tag{48}$$

Its covariant divergence vanishes: $\nabla_\mu T^{\mu\nu} = 0$, if

$$\begin{aligned} m D_\tau u^\mu &= \dot{u}^\mu + \Gamma_{\lambda\nu}^{\;\;\mu} u^\lambda u^\nu = \frac{1}{2}\,\Sigma^{\kappa\lambda} R_{\kappa\lambda\;\;\nu}^{\;\;\;\mu} u^\nu, \\ D_\tau \Sigma^{\mu\nu} &= \dot{\Sigma}^{\mu\nu} + u^\lambda \Gamma_{\lambda\kappa}^{\;\;\mu} \Sigma^{\kappa\nu} + u^\lambda \Gamma_{\lambda\kappa}^{\;\;\nu} \Sigma^{\mu\kappa} = 0. \end{aligned} \tag{49}$$

Alternatively in the hamiltonian formulation we introduce a set of classical Poisson-Dirac brackets on the particle phase space defined by [18, 20]

$$\{\xi^\mu, \pi_\nu\} = \delta^\mu_\nu, \qquad \{\pi_\mu, \pi_\nu\} = \frac{1}{2}\, \Sigma^{\kappa\lambda} R_{\kappa\lambda\mu\nu},$$

$$\{\Sigma^{\mu\nu}, \pi_\lambda\} = \Gamma_{\lambda\kappa}^{\;\;\mu} \Sigma^{\nu\kappa} - \Gamma_{\lambda\kappa}^{\;\;\nu} \Sigma^{\mu\kappa}, \tag{50}$$

$$\{\Sigma^{\mu\nu}, \Sigma^{\kappa\lambda}\} = g^{\mu\kappa}\Sigma^{\nu\lambda} - g^{\mu\lambda}\Sigma^{\nu\kappa} - g^{\nu\kappa}\Sigma^{\mu\lambda} + g^{\nu\lambda}\Sigma^{\mu\kappa}.$$

These brackets have several important features. First, they define an algebra with structure functions rather than structure constants. Moreover these structure functions encode all of the usual space-time geometry: the metric, the connection and the Riemann curvature tensor. In addition the properties of these geometric objects guarantee that the Jacobi identities for cyclic triple brackets are satisfied by this Poisson-Dirac algebra without specifying the dynamics of the system. Therefore the bracket algebra is consistent independently of the choice of hamiltonian. In particular the free hamiltonian (14):

$$H = \frac{1}{2m}\, g^{\mu\nu}[\xi]\, \pi_\mu \pi_\nu,$$

generates the same equations of motion (49) as derived from the energy-momentum tensor, providing in addition the identification of π_μ with the kinetic momentum (16):

$$\pi_\mu = m g_{\mu\nu} u^\nu.$$

To solve the equations of motion (49) it is again convenient to first identify constants of motion. In this case there are 3 universal constants of motion, independent of the specific geometry: the hamiltonian itself:

$$H = -\frac{m}{2},$$

as long as it is proper-time independent; and two spin invariants:

$$I = \frac{1}{2}\, g_{\mu\kappa} g_{\nu\lambda} \Sigma^{\mu\nu} \Sigma^{\kappa\lambda} = S^2 - Z^2,$$

$$D = \frac{1}{8}\, \sqrt{-g}\, \varepsilon_{\mu\nu\kappa\lambda} \Sigma^{\mu\nu} \Sigma^{\kappa\lambda} = S \cdot Z. \tag{51}$$

Clearly I is a real scalar whilst D is a pseudoscalar under three-dimensional spatial rotations and parity transformations.

In addition there may exist other constants of motion connected with symmetries of the background space time. The construction in Eqs. (18), (19) based on the presence of Killing vectors can be generalized immediately to include spin [21, 22], as follows: a quantity

$$J[\xi, \pi, \Sigma] = \alpha^\mu[\xi]\, \pi_\mu + \frac{1}{2}\, \beta_{\mu\nu}[\xi]\, \Sigma^{\mu\nu} \tag{52}$$

is a constant of motion if α^{μ} is a Killing vector and $\beta_{\mu\nu}$ its gradient:

$$\nabla_{\mu}\alpha_{\nu} + \nabla_{\nu}\alpha_{\mu} = 0, \qquad \beta_{\mu\nu} = -\beta_{\nu\mu} = \nabla_{\mu}\alpha_{\nu}. \tag{53}$$

Observe that the symmetric part of $\beta_{\mu\nu}$ vanishes by construction as a result of the Killing condition; this property also implies the identity

$$\nabla_{\lambda}\beta_{\mu\nu} = R_{\mu\nu\lambda}^{\;\;\;\;\;\kappa}\, \alpha_{\kappa}, \tag{54}$$

which is necessary to show that J is a constant of motion.

5 Spinning Test Bodies in Schwarzschild Space-Time

Applying the general formalism above to the case of Schwarzschild space-time we first construct the generalization of the constants of motion (20) and (22), to wit the kinetic energy based on the Killing vector of time-translation invariance:

$$E = -\pi_t - \frac{GM}{r^2}\,\Sigma^{tr} = m\left(1 - \frac{2GM}{r}\right)u^t - \frac{GM}{r^2}\,\Sigma^{tr}, \tag{55}$$

and the total angular momentum based on the Killing vectors of invariance under rotations:

$$\begin{aligned} J_1 &= -\sin\varphi\,\pi_{\theta} - \operatorname{cotan}\theta\cos\varphi\,\pi_{\varphi} \\ &\quad -r\sin\varphi\,\Sigma^{r\theta} - r\sin\theta\cos\theta\cos\varphi\,\Sigma^{r\varphi} + r^2\sin^2\theta\cos\varphi\,\Sigma^{\theta\varphi}, \\ J_2 &= \cos\varphi\,\pi_{\theta} - \operatorname{cotan}\theta\sin\varphi\,\pi_{\varphi} \\ &\quad +r\cos\varphi\,\Sigma^{r\theta} - r\sin\theta\cos\theta\sin\varphi\,\Sigma^{r\varphi} + r^2\sin^2\theta\sin\varphi\,\Sigma^{\theta\varphi}, \\ J_3 &= \pi_{\varphi} + r\sin^2\theta\,\Sigma^{r\varphi} + r^2\sin\theta\cos\theta\,\Sigma^{\theta\varphi}. \end{aligned} \tag{56}$$

As for the spinless test bodies we can orient the co-ordinate system such that the z-axis coincides with the direction of the total angular momentum:

$$\mathbf{J} = (0.0, J), \qquad J = mr^2u^{\varphi} + r\Sigma^{r\varphi}. \tag{57}$$

The total spin J is now composed of the contributions from orbital angular momentum and from spin in the z-direction. The contributions in the perpendicular directions must then cancel. This is expressed by the conditions

$$\Sigma^{r\theta} = -mru^{\theta}, \qquad \Sigma^{\theta\varphi} = \frac{J}{r^2} \cot\theta. \tag{58}$$

As a transverse component of spin must be compensated by a transverse component of orbital angular momentum, the spin proper can precess only if the orbital angular momentum precesses as well. Orbits can therefore be strictly planar if and only if the spin and orbital angular momentum are permanently aligned.

With that restriction it is still possible to find planar and even circular orbits. They necessarily require all θ-components of the spin tensor to vanish:

$$\Sigma^{t\theta} = \Sigma^{r\theta} = \Sigma^{\theta\varphi} = 0, \tag{59}$$

and therefore $D = S \cdot Z = 0$ identically. Moreover for circular orbits $r = R$ is a constant, and $u^r = \dot{u}^r = 0$. This is sufficient to fix the motion of a spinning test body, in the sense that we can derive equations for the time dilation u^t and the angular velocity u^{φ} in terms of the energy per unit of mass $\varepsilon = E/m$ and the total angular momentum per unit of mass $\eta = J/m$:

$$\begin{aligned} \varepsilon R^2 \left(1 - u^{t\,2}\right) &= R\,(R - 3GM)\, u^t - \left(R^2 - 3GMR + 3(GM)^2\right) u^{t\,3}, \\ \eta \left(2GM + R^3 u^{\varphi 2}\right) &= R^3 u^{\varphi} \left[1 - \frac{R^3 u^{\varphi 2}}{GM}\left(1 - \frac{6GM}{R} + \frac{6(GM)^2}{R^2}\right)\right]. \end{aligned} \tag{60}$$

Finally the solutions also have to satisfy the hamiltonian constraint

$$\left(1 - \frac{2GM}{R}\right) u^{t\,2} = 1 + R^2 u^{\varphi 2} \geq 1. \tag{61}$$

Therefore of the five parameters $(\varepsilon, \eta, R, u^t, u^{\varphi})$ characterizing a circular obit only two are independent. Implicitly by Eqs. (55) and (57) they also determine the non-vanishing components of the spin tensor; these are related to the velocity components by

$$\begin{aligned} \frac{1}{m}\, \Sigma^{tr} \left(u^{t\,2} - 1\right) &= u^t \left[(R - 3GM) u^{t\,2} - R\right], \\ \frac{1}{m}\, \Sigma^{r\varphi} \left(2GM + R^3 u^{\varphi 2}\right) &= (R - 2GM) R u^{\varphi} \left[1 - \left(\frac{R}{GM} - 3\right) R^2 u^{\varphi 2}\right]. \end{aligned} \tag{62}$$

Having a complete parametrization of circular orbits for spinning test bodies it is straightforward to construct non-circular orbits by the method of world-line deviations, a direct generalization of the geodesic-deviation procedure explained in Sect. 3. First the deviations $(\delta\xi, \delta u, \delta\Sigma)$ near a given reference world line can be recombined in covariant expressions

$$\Delta\xi^\mu = \delta\xi^\mu, \quad \Delta u^\mu = \delta u^\mu + \delta\xi^\lambda \Gamma_{\lambda\nu}^{\;\;\mu} u^\nu,$$
$$\Delta\Sigma^{\mu\nu} = \delta\Sigma^{\mu\nu} + \delta\xi^\lambda \Gamma_{\lambda\kappa}^{\;\;\mu} \Sigma^{\kappa\nu} + \delta\xi^\lambda \Gamma_{\lambda\kappa}^{\;\;\nu} \Sigma^{\mu\kappa}. \tag{63}$$

Note that there is no a priori relation between the variations $\Delta\xi^\mu$ and $\Delta\Sigma^{\mu\nu}$, but all variations are linked by the first-order world-line deviation equations

$$D_\tau \Delta\xi^\mu = \Delta u^\mu,$$
$$D_\tau \Delta u^\mu - R_{\nu\kappa\;\;\lambda}^{\;\;\;\mu} u^\kappa u^\lambda \Delta\xi^\nu = \frac{1}{2m} \Sigma^{\rho\sigma} R_{\rho\sigma\;\;\nu}^{\;\;\;\mu} \Delta u^\nu + \frac{1}{2m} \Delta\Sigma^{\rho\sigma} R_{\rho\sigma\;\;\nu}^{\;\;\;\mu} u^\nu$$
$$+ \frac{1}{2m} \Sigma^{\rho\sigma} \nabla_\lambda R_{\rho\sigma\;\;\nu}^{\;\;\;\mu} u^\nu \Delta\xi^\lambda, \tag{64}$$
$$D_\tau \Delta\Sigma^{\mu\nu} = \left(R_{\kappa\lambda\sigma}^{\;\;\;\;\;\mu} \Sigma^{\sigma\nu} + R_{\kappa\lambda\sigma}^{\;\;\;\;\;\nu} \Sigma^{\mu\sigma} \right) u^\kappa \Delta\xi^\lambda.$$

Here we will consider in particular planar non-circular orbits. The special conditions (59) then still apply. The only new degrees of freedom are the radial velocity u^r and the mass dipole moment $\Sigma^{t\varphi}$ which become non-zero. Above we observed that for circular orbits the parameters ε and η, representing energy and total angular momentum per unit of mass of the test body, can be varied independently even for circular orbits, e.g. by adjusting z-component of the spin. Therefore there is always a circular orbit with the same ε and η as the non-circular orbit we wish to construct. For simplicity we will take this circular orbit as the reference orbit. Then the covariant deviation equations (64) for near-circular orbits in Schwarzschild space-time reduce to a set of *homogeneous* linear differential equations for the deviations as functions of proper time τ. Now as the spin-tensor deviations are independent of the orbital deviations there are actually two independent types of solutions of the coupled deviation equations. Indeed, the first-order approximation for the world lines of spinning bodies takes the form

$$t(\tau) = u^t \tau + \sigma_+ n_+^t \sin\omega_+(\tau - \tau_+) + \sigma_- n_-^t \sin\omega_-(\tau - \tau_-),$$
$$\varphi(\tau) = u^\varphi \tau + \sigma_+ n_+^\varphi \sin\omega_+(\tau - \tau_+) + \sigma_- n_-^\varphi \sin\omega_-(\tau - \tau_-),$$
$$r(\tau) = R + \sigma_+ n_+^r \cos\omega_+(\tau - \tau_+) + \sigma_- n_-^r \cos\omega_-(\tau - \tau_-),$$
$$\Sigma^{tr}(\tau) = \Sigma_0^{tr} + \sigma_+ N_+^{tr} \cos\omega_+(\tau - \tau_+) + \sigma_- N_-^{tr} \cos\omega_-(\tau - \tau_-), \tag{65}$$
$$\Sigma^{r\varphi}(\tau) = \Sigma_0^{r\varphi} + \sigma_+ N_+^{r\varphi} \cos\omega_+(\tau - \tau_+) + \sigma_- N_-^{r\varphi} \cos\omega_-(\tau - \tau_-),$$
$$\Sigma^{t\varphi}(\tau) = \sigma_+ N_+^{t\varphi} \sin\omega_+(\tau - \tau_+) + \sigma_- N_-^{t\varphi} \sin\omega_-(\tau - \tau_-),$$

where $\sigma_\pm$ are the two independent expansion parameters for the different deviation modes, which are characterized by fundamental angular frequencies $\omega_\pm$. The corresponding amplitudes are denoted by $(n_\pm^\mu, N_\pm^{\mu\nu})$ for the orbital and spin-tensor deviations respectively. The explicit expressions for the frequencies and amplitudes are collected in Appendix C.

Finally the lowest order terms Σ_0^{tr} and $\Sigma_0^{r\varphi}$ for the spin tensor components are the circular-orbit values satisfying equations (62), and $\tau_\pm$ are constants of integration, one for each type of deviation, which fix their initial values. For a complete derivation of these results I refer to Ref. [20].

6 Stability of Orbits and the ISCO

In Sect. 3 it was shown that for radial co-ordinates $R < 6GM$ the deviations from circular orbits show exponential runaway behaviour. Therefore $R = 6GM$ is the innermost stable circular orbit. In the case of spinning particles we can similarly investigate the solutions (65) of the equations for deviations from circular orbits for instabilities and determine the existence of an ISCO for different values of the z-component of spin per unit of mass

$$\sigma = \frac{R\Sigma^{r\varphi}}{m},$$

which is the spin-contribution to η.

The stable deviations (65) are characterized by real-valued angular frequencies $\omega_\pm$. If any one of these frequencies develops an imaginary part runaway behaviour sets in and the circular orbits become unstable. The frequencies themselves are given by the expressions (80) in Appendix C:

$$\omega_\pm^2 = \frac{1}{2}\left(A \pm \sqrt{A^2 - 4B}\right),$$

where A and B represent long expressions in terms of the parameters of the circular reference orbit. For $\omega_\pm$ to be real the square root on the right-hand side must be real, and the whole expression must be nonnegative as it represents a real square. This results in the following inequalities

$$A \geq 0 \quad \text{and} \quad 0 \leq 4B \leq A^2. \tag{66}$$

These conditions are plotted in Fig. 1 as a function of the radial co-ordinate R and the orbital angular momentum

$$\ell = R^2 u^\varphi = \eta - \sigma,$$

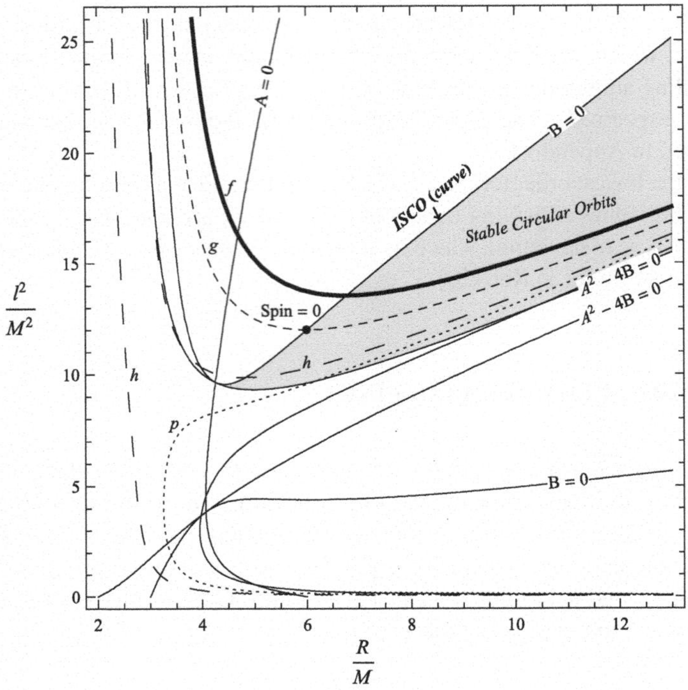

Fig. 1 Region of stability of circular orbits in the R-ℓ^2 plane

both measured in units of GM. The curves labeled *f*, *g*, *h* represent lines of constant σ. In particular *g* is the curve for spinless test bodies ($\sigma = 0$); it leaves the region of stability at $R = 6GM$, as expected. For retrograde spin values $\sigma < 0$ as represented by the curve *f* the ISCO is reached earlier, whilst prograde spin values $\sigma > 0$ as represented by the curve *h* stabilize circular orbits in a range of values $R < 6GM$. From these results one can infer the radial co-ordinate R of the ISCO as a function of the spin parameter σ, as plotted in Fig. 2.

The steep line for spin values $\sigma > 0.55$ has been included for completeness; here the upper limit on B in inequality (66) takes over from the condition $B > 0$ as the main stability criterion. However these large spin values are physically unrealistic as they can only be obtained in cases where the test-body limit is not applicable, such as binary black holes of comparable mass. Also plotted in Fig. 2 is the curve obtained by minimizing the orbital angular momentum ℓ as a function of R at fixed spin. Cleary the two curves largely coincide.

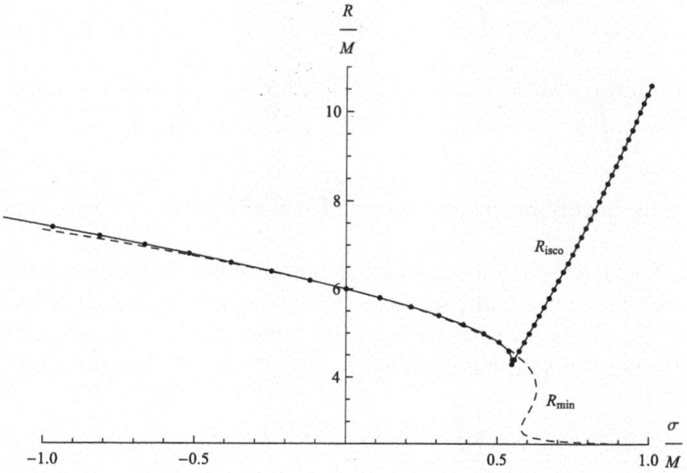

Fig. 2 ISCO radius R as function the spin parameter σ. The dashed line represents the values of R for which the orbital angular momentum reaches its minimum

7 Non-minimal Hamiltonian Dynamics of Spinning Test Bodies

The motion of test bodies has been modeled so far using the minimal hamiltonian (14). However, it is not difficult to construct more complicated hamiltonians to model test bodies with additional interactions such as spin-curvature couplings. As the Dirac-Poisson brackets (50) are closed and model independent the equations of motion can be derived in straightforward fashion for any such extended hamiltonian. For example, one can include Stern–Gerlach type of interactions as discussed in Refs. [18, 20, 23]. In this case the extended test-body hamiltonian is

$$H = \frac{1}{2m}\, g^{\mu\nu}[\xi]\, \pi_\mu \pi_\nu + \frac{\kappa}{4}\, R_{\mu\nu\kappa\lambda}[\xi]\, \Sigma^{\mu\nu} \Sigma^{\kappa\lambda}. \tag{67}$$

In terms of the four-velocity $u^\mu = \dot{\xi}^\mu$ the corresponding equations of motion read

$$\begin{aligned} \pi_\mu &= m g_{\mu\nu} u^\nu, \\ m D_\tau u^\mu &= \frac{1}{2}\, \Sigma^{\kappa\lambda} R_{\kappa\lambda\ \ \nu}^{\ \ \ \mu} u^\nu - \frac{\kappa}{4}\, \Sigma^{\kappa\lambda} \Sigma^{\rho\sigma} \nabla_\mu R_{\kappa\lambda\rho\sigma}, \\ D_\tau \Sigma^{\mu\nu} &= -\kappa \Sigma^{\rho\sigma} \left(R_{\rho\sigma\ \ \lambda}^{\ \ \ \mu} \Sigma^{\lambda\nu} + R_{\rho\sigma\ \ \lambda}^{\ \ \ \nu} \Sigma^{\mu\lambda} \right). \end{aligned} \tag{68}$$

As in the minimal case these equations can also be derived by requiring the vanishing of the covariant divergence of a suitable energy-momentum tensor [19]

$$T^{\mu\nu} = T^{\mu\nu}_{min} + \frac{\kappa}{4} \nabla_\kappa \nabla_\lambda \int d\tau \left(\Sigma^{\mu\lambda}\Sigma^{\kappa\nu} + \Sigma^{\nu\lambda}\Sigma^{\mu\kappa} \right) \frac{1}{\sqrt{-g}}\, \delta^4(x - \xi(\tau))$$
$$+ \frac{\kappa}{4} \int d\tau\, \Sigma^{\rho\sigma} \left(R_{\rho\sigma\lambda}^{\;\;\;\;\;\mu} \Sigma^{\lambda\nu} + R_{\rho\sigma\lambda}^{\;\;\;\;\;\nu} \Sigma^{\lambda\mu} \right) \frac{1}{\sqrt{-g}}\, \delta^4(x - \xi(\tau)). \tag{69}$$

Here $T^{\mu\nu}_{min}$ is the energy-momentum tensor (48) of a spinning test body with minimal dynamics.

Remarkably all conservation laws for spinning bodies we derived in the minimal case carry over to the case with Stern–Gerlach interactions. In particular any constant of motion (52), (53) associated with a Killing vector α^μ is also conserved by the Stern–Gerlach terms in the hamiltonian:

$$\frac{\kappa}{4} \left\{ J, R_{\mu\nu\kappa\lambda} \Sigma^{\mu\nu} \Sigma^{\kappa\lambda} \right\} = 0. \tag{70}$$

For example, in a static and spherically symmetric background like Schwarzschild or Reissner–Nordstrøm space-time the kinetic energy E and the angular momentum 3-vector $\mathbf{J}$ given by Eqs. (55) and (56) are again conserved.

This form of non-minimal hamiltonian dynamics predicts some interesting effects. In particular, as the hamiltonian is a constant of motion which by evaluation in a curvature-free region is seen to be expressed in terms of the inertial mass by $H = -m/2$, the hamiltonian constraint gets modified to read

$$g^{\mu\nu} \pi_\mu \pi^\nu + \frac{\kappa m}{2}\, R_{\mu\nu\kappa\lambda}\, \Sigma^{\mu\nu} \Sigma^{\kappa\lambda} + m^2 = 0. \tag{71}$$

Then the four-velocity is no longer normalized to be a time-like unit vector; instead the time-like unit vector tangent to the world line actually is

$$n^\mu = \frac{u^\mu}{\left(1 + \frac{\kappa}{2m}\, R_{\mu\nu\kappa\lambda}\, \Sigma^{\mu\nu} \Sigma^{\kappa\lambda} \right)^{1/2}}. \tag{72}$$

Considering a particle at rest:

$$g_{tt} \left(\frac{dt}{d\tau} \right)^2 = 1 + \frac{\kappa}{2m}\, R_{\mu\nu\kappa\lambda}\, \Sigma^{\mu\nu} \Sigma^{\kappa\lambda}, \tag{73}$$

this is seen to imply that the spin-curvature coupling represents an additional source of gravitational time-dilation. A similar effect related to spinning particles interacting with electromagnetic fields was conjectured in Refs. [24, 25].

8 Final Remarks

The motion of test bodies carrying a finite number of relevant degrees of freedom like momentum, spin or charge can be represented by world lines in space-time to the extent that we can assign them a well-defined position and that their back reaction on space-time geometry can be neglected. Convenient position co-ordinates are not necessarily those of a center of mass (or for that matter a center of charge) in the local rest frame, as the example of spinning test bodies shows. In that case we find it preferable to associate the world line of free particles with the line on which the spin tensor is covariantly constant. The mass dipole moment can then be taken to represent the effective position of the mass with respect to that world line.

This is also clear from the corresponding energy-momentum tensor which receives contributions from both the spin proper and the mass dipole. In a next step this can be used to compute the back reaction of the test body on the space-time geometry as discussed in the simple example in Sect. 1. In general this procedure also includes determining the self-force and the gravitational waves emitted by test bodies in the specific background under discussion [14, 26].

As another application it has been shown in the literature how the motion of test bodies can be used to reconstruct the geometry of space-time [11]. Simple geodesic motion of a sufficient number of test bodies allows one to determine the curvature at a point in space by measuring the geodesic deviations in its neighborhood. By including higher-order corrections as in Eqs. (36), (37) one could also determine the derivatives of the curvature to obtain the curvature in a region around the point of interest. As Eq. (64) show, an alternative method is to measure first-order world-line deviations of spinning test bodies, which also depend on the gradient of the Riemann curvature tensor.

Acknowledgements I am indebted to Richard Kerner, Roberto Collistete jr., Gideon Koekoek, Giuseppe d'Ambrosi, S. Satish Kumar and Jorinde van de Vis for pleasant and informative discussions and collaboration on various aspects of the topics discussed. This work is supported by the Foundation for Fundamental Research of Matter (FOM) in the Netherlands.

A Observer-Dependence of the Center of Mass in Relativity

To illustrate the observer-dependence of the center of mass of an extended body we consider a simple example: the motion of two equal test bodies revolving at constant angular velocity in Minkowski space on a circular orbit around an observer located in the origin of an inertial frame with cartesian co-ordinates (t, x, y, z). The plane of the orbit is taken to be the x-y-plane. In Fig. 3 we plot the projection of the world lines in the x-t-plane, represented by the two widely oscillating curves. At any moment the two test bodies are at equal distance to the observer and in opposite phase with respect to the origin. In this frame the center of mass is located at the origin $x = 0$ and moves in a straight line along the t-axis in space-time.

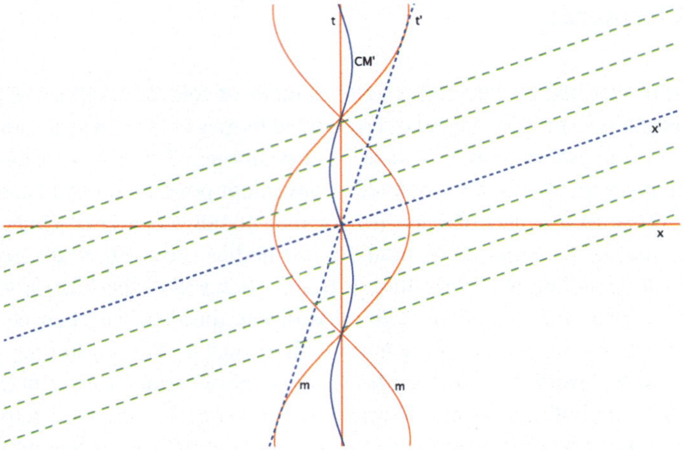

Fig. 3 The world line of the center of mass of two equal masses (e.g., a binary star system) in circular orbit with respect to a stationary observer on the axis of orbital angular momentum is represented by the line $x = 0$. The world line of the center of mass with respect to an observer in an inertial frame (t', x') moving at constant velocity along the x-axis is represented by the oscillating curve labeled CM′

A second observer in another inertial frame (t', x', y', z') moving with constant velocity v in the positive x-direction has a different notion of simultaneity, as defined by the appropriate Lorentz transformation. The lines $t' =$ constant are represented by the dashed slant lines parallel to the x'-axis. In the limit of large masses and slow rotation the center of mass CM′ with respect to this moving frame is located halfway between the masses at fixed time t'. The world line of CM′ is now represented by the single curve oscillating at smaller amplitude around the line $x = 0$ in the original frame. In fact for the observer in relative motion CM′ moves in the negative x'-direction while oscillating around the line $x' = -vt'$.

It is obvious that in curved space-time the notion of simultaneity is further complicated because of the non-existence of global inertial frames, resulting in additional distortions of the world line CM′ with respect to the world line in the local inertial frame (t, x, y, z) fixed to the center of rotation.

B Coefficients for Geodesic Deviations in Schwarzschild Geometry

The coefficients for the deviations of bound equatorial orbits w.r.t. parent circular orbits have been calculated for Schwarzschild space-time up to second order; with the restriction $\rho_1^r = \rho_2^r = 0$ explained in the main text one gets the following results [6]:

a. Secular terms:

$$\rho_1^t = 0, \; \rho_2^t = \frac{3GM}{2R^{5/2}} \frac{R+GM}{(R-3GM)^{3/2}},$$
$$\rho_1^\varphi = 0, \; \rho_2^\varphi = \frac{3}{2R^{7/2}} \sqrt{\frac{GM}{R}} \frac{(R-2GM)(R+GM)}{(R-3GM)^{3/2}}. \tag{74}$$

b. First-order periodic terms:

$$n_1^t = -\sqrt{\frac{4GMR}{(R-2GM)(R-6GM)}}, \; n_2^t = 0,$$
$$n_1^r = \sqrt{1-\frac{2GM}{R}}, \qquad n_2^r = 0, \tag{75}$$
$$n_1^\varphi = -\frac{2}{R}\sqrt{\frac{R-2GM}{R-6GM}}, \qquad n_2^\varphi = 0.$$

c. Second order periodic terms:

$$m_2^t = \sqrt{\frac{GM}{R^2}} \frac{2R^2 - 15GMR + 14(GM)^2}{(R-2GM)(R-6GM)^{3/2}},$$
$$m_2^r = -\frac{1}{R^2} \frac{(R-2GM)(R-7GM)}{R-6GM}, \tag{76}$$
$$m_2^\varphi = \frac{1}{2R^{5/2}} \frac{(R-2GM)(5R-32GM)}{(R-6GM)^{3/2}}.$$

d. Angular frequency:

$$\omega_0 = \sqrt{\frac{GM}{R^3} \frac{R-6GM}{R-3GM}}, \quad \omega_1 = 0. \tag{77}$$

In the non-restricted case with $\rho_1^r \neq 0$ also the coefficients ρ_1^t, ρ_1^φ, n_2^μ and ω_1 all become non-zero as well [13].

C Coefficients for Spinning World-Line Deviations in Schwarzschild Geometry

The first-order planar deviations of circular orbits of spinning particles for constant energy and total angular momentum in Schwarzschild space-time are expressed conveniently in terms of the following combinations of orbital and spin parameters [20]

$$\alpha = \frac{2(R-GM)}{R(R-2GM)}\,u^t - \frac{2\varepsilon}{R-2GM}, \qquad \beta = -\frac{GMu^\varphi}{mR},$$

$$\gamma = \frac{2R-5GM}{R(R-2GM)}\,u^\varphi + \frac{GM\eta}{R^3(R-2GM)}, \; \zeta = -\frac{GM(R-2GM)}{mR^4}\,u^t, \tag{78}$$

$$\kappa = -\frac{2(R-2GM)}{R^2}\,\varepsilon, \qquad \lambda = 2Ru^\varphi - \frac{GM\eta}{R^2},$$

and

$$\mu = -\frac{2(R-3GM)}{R^3} + \frac{2(R-4GM)}{R^3}\,\varepsilon u^t + u^{\varphi\,2} + \frac{2GM}{R^3}\,\eta u^\varphi,$$

$$\nu = \frac{(R-GM)(R-3GM)}{R-2GM}\,mu^\varphi + \frac{GMm\eta}{R^2(R-2GM)},$$

$$\sigma = \frac{(R-GM)(R-3GM)}{GM(R-2GM)}\,mu^t - \frac{mR\varepsilon}{GM},$$

$$\chi = \frac{(R^2-4GMR+5(GM)^2)}{GM(R-2GM)^2}\,mu^\varphi u^t - \frac{GM(3R-4GM)}{R^3(R-2GM)^2}\,m\eta u^t - \frac{m\varepsilon u^\varphi}{GM}. \tag{79}$$

With these definitions the frequencies of the first-order planar deviations are

$$\begin{aligned} \omega_\pm^2 &= \frac{1}{2}\left(A \pm \sqrt{A^2-4B}\right), \\ A &= \mu - \alpha\kappa - \beta\nu - \gamma\lambda - \zeta\sigma, \\ B &= \beta\left(\kappa\chi - \mu\nu + \gamma(\lambda\nu - \kappa\sigma)\right) + \zeta\left(\lambda\chi - \mu\sigma - \alpha(\lambda\nu - \kappa\sigma)\right), \end{aligned} \tag{80}$$

whilst the amplitudes are given by

$$\begin{aligned} n_\pm^t &= \lambda(\beta\gamma - \alpha\zeta) + \beta(\omega_\pm^2 - \mu), \\ n_\pm^\varphi &= -\kappa(\beta\gamma - \alpha\zeta) + \zeta(\omega_\pm^2 - \mu), \\ n_\pm^r &= \omega_\pm(\beta\kappa + \zeta\lambda), \end{aligned} \tag{81}$$

and

$$
\begin{aligned}
N_{\pm}^{tr} &= \frac{m\omega_{\pm} R^2}{GM}\left(1 - \frac{2GM}{R}\right) n_{\pm}^{t} + \frac{2mR}{GM}\left[\left(1 - \frac{GM}{R}\right) u^{t} - \varepsilon\right] n_{\pm}^{r}, \\
N_{\pm}^{r\varphi} &= -m\omega_{\pm} R n_{\pm}^{\varphi} - \frac{m}{R^2}\left(\eta + R^2 u^{\varphi}\right) n_{\pm}^{r}, \\
N_{\pm}^{t\varphi} &= \omega_{\pm}^2 (\omega_{\pm}^2 - \mu + \alpha\kappa + \gamma\lambda).
\end{aligned}
\tag{82}
$$

References

1. C. Møller, Sur la dynamique des systèmes ayant un moment angulaire interne. Ann. Inst. Henri Poincaré **11**, 251 (1949)
2. C. Møller, *The Theory of Relativity* (Clarendon Press, Oxford, 1952)
3. L.F. Costa, J. Natário, Center of mass, spin supplementary conditions, and the momentum of spinning particles. Fund. Theor. Phys. **179**, 215 (2015)
4. S. Weinberg, *Gravitation and Cosmology* (Wiley, New York, 1972)
5. E. Hackmann, Geodesic equations and algebro-geometric methods (2015), arXiv:1506.00804v1 [gr–qc]
6. R. Kerner, J.W. van Holten, R. Collistete jr., Relativistic epicycles: another approach to geodesic deviations. Class. Quantum Gravity **18**, 4725 (2001)
7. R. Colistete Jr., C. Leygnac, R. Kerner, Higher-order geodesic deviations applied to the Kerr metric. Class. Quantum Gravity **19**, 4573 (2002)
8. J. Ehlers, F.A.E. Pirani, A. Schild, The geometry of free fall and light propagation, in *General Relativity: Papers in Honnor of J.L. Synge*, ed. by L. O'Raifeartaigh (Oxford University Press, Oxford, 1972)
9. C.W. Misner, K.S. Thorne, J.A. Wheeler, *Gravitation* (Freeman and Co., San Francisco, 1970)
10. P. Szekeres, The gravitational compass. J. Math. Phys. **6**, 1387 (1965)
11. D. Puetzfeld, Y.N. Obukhov, Generalized deviation equation and determination of the curvature in general relativity. Phys. Rev. D **93**, 044073 (2016)
12. D. Philipp, D. Puetzfeld, C. Lämmerzahl, On the applicability of the geodesic deviation equation in general relativity (2016), arXiv:1604.07173 [gr–qc]
13. G. Koekoek, J.W. van Holten, Epicycles and Poincaré resonances in general relativity. Phys. Rev. D **83**, 064041 (2011)
14. G. Koekoek, J.W. van Holten, Geodesic deviations: modeling extreme mass-ratio systems and their gravitational waves. Class. Quantum Gravity **28**, 225022 (2011)
15. M. Mathison, Neue Mechanik materieller Systeme. Acta Phys. Pol. **6**, 163 (1937)
16. W. Tulczyjew, Motion of multipole particles in general relativity theory. Acta Phys. Pol. **18**, 393 (1959)
17. J. Steinhoff, Canonical formulation of spin in general relativity, Ph.D. thesis (Jena University) (2011), arXiv:1106.4203v1 [gr-qc]
18. G. d'Ambrosi, S.S. Kumar, J.W. van Holten, Covariant Hamiltonian spin dynamics in curved space-time. Phys. Lett. B **743**, 478 (2015)
19. J.W. van Holten, Spinning bodies in general relativity. Int. J. Geom. Methods Mod. Phys. **13**, 1640002 (2016)
20. G. d'Ambrosi, S.S. Kumar, J.W. van Holten, J. van de Vis, Spinning bodies in curved space-time. Phys. Rev. D **93**, 04451 (2016)
21. J. Ehlers, E. Rudolph, Dynamics of extended bodies in general relativity: center-of-mass description and quasi-rigidity. Gen. Relativ. Gravit. **8**, 197 (1977)

22. R. Ruediger, Conserved quantities of spinning test particles in general relativity. Proc. R. Soc. Lond. **A375**, 185 (1981)
23. I. Khriplovich, A. Pomeransky, Equations of motion for spinning relativistic particle in external fields. Surv. High Energy Phys. **14**, 145 (1999)
24. J.W. van Holten, On the electrodynamics of spinning particles. Nucl. Phys. B **356**, 3–26 (1991)
25. J.W. van Holten, Relativistic time dilation in an external field. Phys. A **182**, 279 (1992)
26. G. d'Ambrosi, J.W. van Holten, Ballistic orbits in Schwarzschild space-time and gravitational waves from EMR binary mergers. Class. Quantum Gravity **32**, 015012 (2015)

On the Applicability of the Geodesic Deviation Equation in General Relativity

Dennis Philipp, Dirk Puetzfeld and Claus Lämmerzahl

Abstract Within the theory of General Relativity, we study the solution and range of applicability of the standard geodesic deviation equation in highly symmetric spacetimes. In the Schwarzschild spacetime, the solution is used to model satellite orbit constellations and their deviations around a spherically symmetric Earth model. We investigate the spatial shape and orbital elements of perturbations of circular reference curves. In particular, we reconsider the deviation equation in Newtonian gravity and then determine relativistic effects within the theory of General Relativity by comparison. The deviation of nearby satellite orbits, as constructed from exact solutions of the underlying geodesic equation, is compared to the solution of the geodesic deviation equation to assess the accuracy of the latter. Furthermore, we comment on the so-called Shirokov effect in the Schwarzschild spacetime and limitations of the first order deviation approach.

1 Introduction

Applications in space based geodesy and gravimetry missions require the precise knowledge of satellite orbits and possible deviations of nearby ones. In such a context, one of the satellites may serve as the reference object and measurements are performed w.r.t. this master spacecraft. GRACE-FO, the successor of the long-lasting GRACE mission, aims at measuring the change of the separation between two spacecrafts with some 10 nm accuracy [1, 2]. The change of this distance is then used to obtain information about the gravitational field of the Earth, i.e. to measure the Newtonian multipole moments of the gravitational potential, and to deduce information about the mass distribution and its temporal variations.

D. Philipp (✉) · D. Puetzfeld · C. Lämmerzahl
Center of Applied Space Technology and Microgravity (ZARM), University of Bremen, Bremen, Germany
e-mail: dennis.philipp@zarm.uni-bremen.de

D. Puetzfeld
e-mail: dirk.puetzfeld@zarm.uni-bremen.de
URL: http://puetzfeld.org

D. Puetzfeld and C. Lämmerzahl (eds.), *Relativistic Geodesy*,
Fundamental Theories of Physics 196, https://doi.org/10.1007/978-3-030-11500-5_13

In this work, we investigate the geodesic deviation equation in General Relativity (GR) in the case of highly symmetric spacetimes. Our aim is to develop a measure for the quality of approximation that the deviation equation provides to model test bodies and their orbit deviations in different orbital configurations. To achieve this, we construct general solutions of the geodesic deviation equation and compare them to exact solutions of the underlying geodesic equation in a Schwarzschild spacetime. Our analysis allows us to reveal physical and artificial effects of such an approximative description. In particular, we comment on an effect that was reported for the first time in 1973 by Shirokov [3].

The structure of the paper is as follows: In Sect. 2, we reconsider the deviation equation in Newtonian gravity. This is followed by an investigation of the first order geodesic deviation equation in static, spherically symmetric spacetimes in GR in Sect. 3. A direct comparison between the solution for the Schwarzschild spacetime, which we focus on for the rest of this work, and the Newtonian results unveils relativistic effects. In Sect. 4, we describe the shape of perturbed orbits and the influence of six free integration parameters on the general solution. These parameters are connected to orbital elements of the orbit under consideration, and they determine how it is obtained from a perturbation of the reference curve. We assess the range of applicability of the deviation equation by comparing its solutions to deviations constructed "by hand" from exact solutions of the underlying geodesic equation. We study physical effects such as perigee precession and the redshift due to time dilation between the reference and perturbed orbit. Building on these results, we uncover some artificial effects previously reported in the literature. Our conclusions and a brief outlook on future applications are given in Sect. 5. Appendix A contains a summary of our notations and conventions.

2 Orbit Deviations in Newtonian Gravity

There are two basic equations that govern Newtonian gravitational physics; the field equation, also known as Poisson's equation

$$\Delta U(x) = \partial_\mu \partial^\mu U(x) = 4\pi G \rho(x)\,, \tag{1a}$$

and the equation of motion

$$\ddot{x}^\mu = -\partial^\mu U(x)\,, \tag{1b}$$

where the $(x^\mu) = (x, y, z)$ are Cartesian coordinates and the overdot denotes derivatives w.r.t. the Newtonian absolute time t. Here and in the following, Greek indices are spatial indices and take values $1, 2, 3$. The field Eq. (1a) relates the Newtonian gravitational potential U to the mass density ρ and introduces Newton's gravitational constant G as a factor of proportionality. Outside a spherically symmetric (and static) mass distribution, i.e. in the region where $\rho = 0$, we obtain as a solution of the Laplace equation $\Delta U = 0$:

$$U(r) = -GM/r\,, \tag{2}$$

where M is the mass of the central object, obtained by integrating the mass density over the three-volume of the source, and r is the distance to the center of the gravitating mass. The equation of motion (1b) describes how point particles move in the gravitational potential given by U.

2.1 Newtonian Deviation Equation

We now recall the derivation of the Newtonian deviation equation, see, e.g., Ref. [4] and references therein. For a given reference curve $Y^\mu(t)$ that fulfills the equation of motion we construct a second curve $X^\mu(t) = Y^\mu(t) + \eta^\mu(t)$ and introduce the deviation η. This second curve shall be a solution of the equation of motion as well (at least up to linear order, as we will see below). Hence, we get

$$\ddot{X}^\mu = \ddot{Y}^\mu + \ddot{\eta}^\mu = -\partial^\mu U(X) = -\partial^\mu U(Y+\eta)\,. \tag{3}$$

For small deviations we linearize the potential around the reference object with respect to the deviation,

$$U(X) = U(Y+\eta) = U(Y) + \eta^\nu \partial_\nu U(Y) + \mathcal{O}(\eta^2)\,. \tag{4}$$

Thereupon, the first order deviation equation in Newtonian gravity becomes (spatial indices are raised and lowered with the Kronecker delta δ^μ_ν)

$$\ddot{\eta}^\mu = -[\partial^\mu \partial_\nu U(Y)]\,\eta^\nu =: K^\mu{}_\nu \eta^\nu\,. \tag{5}$$

For a homogeneous ($\partial_\nu U \equiv 0$) or vanishing ($U \equiv 0$) gravitational potential, the deviation vector has the simple linear time dependence $\eta^\mu(t) = A^\mu t + B^\mu$, with constants A^μ and B^μ. A non-linear time dependence of the deviation is caused by second derivatives of the Newtonian gravitational potential, i.e. if $K^\mu{}_\nu \neq 0$.

Since we are interested in the deviation for highly symmetric situations, we now use the potential (2) outside a spherically symmetric mass distribution and introduce usual spherical coordinates by

$$(x, y, z) = (r \sin\vartheta \cos\varphi, r \sin\vartheta \sin\varphi, r \cos\vartheta)\,. \tag{6}$$

Due to the symmetry of the situation we can, without loss of generality, restrict the reference curve to lie within the equatorial plane that is defined by $\vartheta = \pi/2$. Applying the coordinate transformation $x^a \to \tilde{x}^a$ from Cartesian to spherical coordinates, Eq. (5) turns into

$$
\ddot{\tilde{\eta}}^\nu \tilde{\partial}_\nu x^\mu + 2\dot{\tilde{\eta}}^\nu \tilde{\partial}_\nu \dot{x}^\mu + \tilde{\eta}^\nu \tilde{\partial}_\nu \ddot{x}^\mu \\
= \left[\partial_r U(r)\, \partial^\mu \partial_\nu\, r + \partial_r^2 U(r)(\partial^\mu r)(\partial_\nu r)\right] \tilde{\eta}^\sigma \tilde{\partial}_\sigma x^\nu\,, \tag{7}
$$

where $(\tilde{\eta}^\mu) = (\eta^r, \eta^\vartheta, \eta^\varphi)$ are the components of the deviation in the new coordinates and $(\tilde{\partial}_\mu) = (\partial_r, \partial_\vartheta, \partial_\varphi)$. These are three equations for the three unknown components of the deviation. All angular terms can be eliminated by appropriate combinations of these equations and a straightforward but rather lengthy calculation yields the system of differential equations

$$
\ddot{\eta}^\vartheta = -\frac{\dot{R}}{R}\dot{\eta}^\vartheta - \left(\frac{GM}{R^3} + \frac{\ddot{R}}{R}\right)\eta^\vartheta\,, \tag{8a}
$$

$$
\ddot{\eta}^r = \left(\dot{\Phi}^2 + \frac{2GM}{R^3}\right)\eta^r + \left(2\dot{R}\dot{\Phi} + R\ddot{\Phi}\right)\eta^\varphi + 2R\dot{\Phi}\,\dot{\eta}^\varphi\,, \tag{8b}
$$

$$
\ddot{\eta}^\varphi = -\frac{2\dot{R}}{R}\dot{\eta}^\varphi - \frac{2\dot{\Phi}}{R}\dot{\eta}^r - \left(\frac{\ddot{R}}{R} + \frac{GM}{R^3} - \dot{\Phi}^2\right)\eta^\varphi - \frac{\ddot{\Phi}}{R}\eta^r\,, \tag{8c}
$$

where the quantities represented by capital letters (R, Φ) are in general functions of time t and describe the trajectory of the reference object, along which the system (8) must be solved. In geodesy, the quantity $K^\mu{}_\nu \neq 0$ and the Eq. (5) are known in the framework of gradiometry. However, we did not find the system of differential equations (8), describing the Newtonian deviations from a general reference curve, published elsewhere in this form.

2.2 Deviation from Circular Reference Curves

One particular case is the deviation from a circular reference orbit with constant radius R. This special situation was already considered by Greenberg [5], who derived (only) the oscillating solutions. However, in [4] the full solution for this case can be found. In the following we briefly summarize the results in a form that we will use later to compare to the relativistic results. The azimuthal motion of the reference orbit is described by

$$
\dot{\Phi} = \sqrt{\frac{GM}{R^3}} \;\Rightarrow\; \Phi(t) = \sqrt{\frac{GM}{R^3}}\,t =: \Omega_K\,t\,. \tag{9}
$$

The quantity Ω_K is the well known Keplerian frequency and leads to the Keplerian orbital period $2\pi/\Omega_K$. For the circular reference orbit, the conditions $\ddot{R} = \dot{R} = \ddot{\Phi} \equiv 0$ hold, and the system (8) yields three ordinary second order differential equations of which the last two are coupled (see Eqs. (29), (33) and (34) in [5] for comparison)

$$\ddot{\eta}^{\vartheta} = -\Omega_K^2\,\eta^{\vartheta}\,, \tag{10a}$$

$$\ddot{\eta}^{r} = 2R\,\Omega_K\,\dot{\eta}^{\varphi} + 3\Omega_K^2\,\eta^{r}\,, \tag{10b}$$

$$R\,\ddot{\eta}^{\varphi} = -2\Omega_K\,\dot{\eta}^{r}\,. \tag{10c}$$

The first equation for the deviation in the ϑ-direction describes a simple harmonic oscillation around the reference orbital plane and is decoupled from the remaining ones. The general real-valued solution is given by

$$\eta^{\vartheta}(t) = \frac{C_{(5)}}{R}\cos\Omega_K t + \frac{C_{(6)}}{R}\sin\Omega_K t\,. \tag{11a}$$

The parameters $C_{(5)}$ and $C_{(6)}$ are the amplitudes of the two fundamental solutions (normalized to the reference radius R) and the deviation component η^{ϑ} oscillates with the Keplerian frequency Ω_K. In [5], Greenberg derived the oscillating solutions for the remaining two equations. However, the general solution, cf. [4], is given by

$$\eta^{r}(t) = C_{(1)} + C_{(2)}\sin\Omega_K t + C_{(3)}\cos\Omega_K t\,, \tag{11b}$$

$$R\,\eta^{\varphi}(t) = 2\left(C_{(2)}\cos\Omega_K t - C_{(3)}\sin\Omega_K t\right) - \frac{3}{2}\,\Omega_K C_{(1)} t + C_{(4)}\,. \tag{11c}$$

Summarizing the results, the perturbed orbit is described by

$$r(t) = R + \eta^{r}(t) = R + C_{(1)} + C_{(2)}\sin\Omega_K t + C_{(3)}\cos\Omega_K t, \tag{12a}$$

$$\begin{aligned}\varphi(t) &= \Omega_K\,t + \eta^{\varphi}(t) = \Omega_K\left(1 - \frac{3}{2R}C_{(1)}\right)t + \frac{C_{(4)}}{2R}\\ &\quad + \frac{2}{R}\left(C_{(2)}\cos\Omega_K t - C_{(3)}\sin\Omega_K t\right),\end{aligned} \tag{12b}$$

$$\vartheta(t) = \frac{\pi}{2} + \eta^{\vartheta}(t) = \pi/2 + \frac{C_{(5)}}{R}\cos\Omega_K t + \frac{C_{(6)}}{R}\sin\Omega_K t\,. \tag{12c}$$

Obviously, there are several possibilities to perturb the reference orbit. The parameters $C_{(i)}, i = 1\ldots 6$ define the initial position and velocity (or the orbital elements) of the test body that follows the perturbed curve. The meaning of these parameters and their impact on the perturbed orbit was studied briefly in Ref. [4], and the analysis will be extended in Sect. 4 in the context of the general relativistic results. Note that the only frequency appearing in the solution so far is the Keplerian frequency Ω_K.

3 Geodesic Deviation in General Relativity

In GR, the equation of motion for structureless test bodies takes the form of the geodesic equation

$$\frac{d^2x^a}{ds^2} = -\Gamma_{bc}{}^a(x)\frac{dx^b}{ds}\frac{dx^c}{ds}\,. \tag{13}$$

See Refs. [6, 7] for reviews of methods to derive this, and higher order equations of motion, by means of multipolar techniques. Latin indices denote spacetime indices, taking values 0, 1, 2, 3, and $\Gamma_{bc}{}^a$ are the connection coefficients of the underlying spacetime (Christoffel symbols).

As in the Newtonian case, we consider two neighboring curves $Y^a(s)$ and $X^a(s)$, both of them are now assumed to be geodesics, and s is the proper time measured along the curve $Y^a(s)$. Choosing $Y^a(s)$ as the reference curve, we may introduce, in a coordinate representation, the deviation $\eta^a(s)$ w.r.t. the neighboring curve $X^a(s)$ as

$$\eta^a(s) := X^a(s) - Y^a(s)\,. \tag{14}$$

Denoting the normalized four-velocity along the reference curve by $\dot{Y}^a := \mathrm{d}Y^a/\mathrm{d}s$, it can be shown that the second covariant derivative of the deviation fulfills

$$\frac{\mathrm{D}^2\eta^a(s)}{\mathrm{d}s^2} = R^a{}_{bcd}(Y)\,\dot{Y}^b\eta^c\dot{Y}^d + \mathcal{O}(\eta^2)\,, \tag{15}$$

up to the linear order in the deviation and its first derivative, along the reference curve. This is the well-known geodesic deviation or Jacobi equation, in which $R^a{}_{bcd}$ denotes the curvature of spacetime. For more details on its systematic derivation, in particular its possible generalizations, and an overview of the literature see Ref. [8]. From Eq. (15), we infer that the deviation η^a will have a non-linear time dependence if and only if the spacetime is curved. For vanishing curvature, the deviation can only grow linearly in time as in the Newtonian situation for $K^\mu_{\ \nu} = 0$. The Newtonian quantity $K^\mu_{\ \nu}$ that measures second derivatives of the gravitational potential is replaced by the curvature tensor in GR.

In the following, we will focus on the solutions of Eq. (15) in the case of timelike geodesics that may correspond to satellite orbits around the Earth. In particular, we are going to assume an orthogonal parametrization, see Sect. III in [8], in which the deviation is orthogonal to the velocity along the reference curve $\eta_a\dot{Y}^a = 0$ – cf. Fig. 1 for a sketch. We will further assume the reference curve Y^a to be a circular geodesic and construct orbits out of its perturbation.

3.1 Deviation Equation in Spherically Symmetric and Static Spacetimes

In a spherically symmetric and static spacetime that is described by the metric

$$ds^2 = A(r)dt^2 - B(r)dr^2 - r^2(d\vartheta^2 + \sin^2\vartheta d\varphi^2)\,, \tag{16}$$

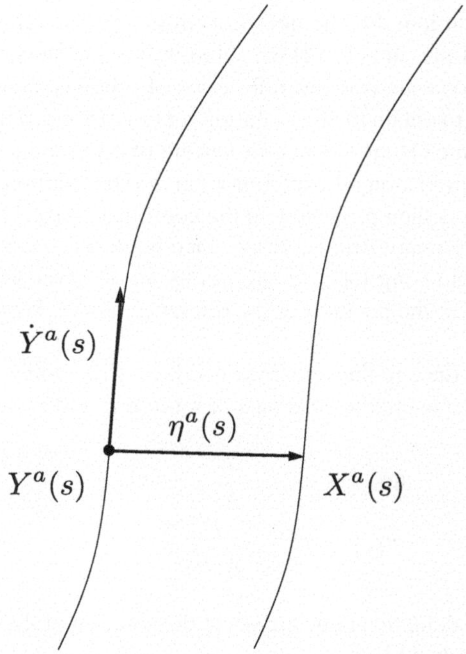

Fig. 1 Sketch of the deviation of two nearby geodesics Y^a and $X^a = Y^a + \eta^a$. Here we depict the case of the orthogonal correspondence – in which the deviation vector η^a is chosen to be orthogonal to the four-velocity $\dot{Y}^a$ along the reference geodesic

we use spherical coordinates $(x^a) = (t, r, \vartheta, \varphi)$ and choose units such that the speed of light c and Newton's gravitational constant G are equal to one. The angles ϑ and φ are the usual polar and azimuthal angles as in spherical coordinates and the radial coordinate r is defined such that spheres at a radius r have area $4\pi r^2$. In these coordinates, the reference geodesic shall be represented by $(Y^a) = (T, R, \Theta, \Phi)$. Due to the symmetry of the spacetime we can, without loss of generality, assume that the reference geodesic is confined to the equatorial plane. Thus, we have $\Theta \equiv \pi/2$ and $\dot{\Theta} = \ddot{\Theta} = 0$. For geodesics in the considered spacetime there exist constants of motion that correspond to the conservation of energy E and angular momentum L, see for example [9]. Since the metric (16) does neither depend on the time coordinate t nor on the angle φ, i.e. ∂_t and ∂_φ are Killing vector fields, the constants of motion are given by

$$E := A(r)\,\dot{t} = \text{const.}\,, \quad L := r^2\,\dot{\varphi} = \text{const.} \tag{17}$$

The general solution of the first order geodesic deviation equation (15) in the spacetime (16) was given by Fuchs [10] in terms of first integrals, which remain to be solved. Unfortunately, this solution is not applicable to the simplest case of the deviation from a circular reference geodesic. The condition $\dot{R} = 0$ causes singularities in terms $\sim 1/\dot{R}$ that appear in the equations. Shirokov [3] was the first to derive periodic solutions for the deviation from circular reference geodesics in Schwarzschild spacetime. In [11, 12] the solution for Schwarzschild spacetime and circular reference geodesics was given in terms of relativistic epicycles. However, another possible way to obtain the full solution for circular reference geodesics in the more general spacetime (16) is to refer the system of differential equations (15) to a parallel propagated tetrad along the reference curve. The solution of the equations in this reference system is then projected on the coordinate basis [13]. This method is of direct relevance for relativistic geodesy, since it allows to describe the deviation as observed in the comoving local tetrad, i.e. by an observer with an orthonormal frame who is located at the position of the reference object. We will use the results of this method here.

The motion along the circular reference geodesic with radius R in the equatorial plane can be described using the constants of motion E and L from Eq. (17):

$$\Phi(s) = \dot{\Phi}\, s = \frac{L}{R^2}\, s =: \Omega_\Phi\, s\,, \tag{18a}$$

$$T(s) = \dot{T}\, s = \frac{E}{A(R)}\, s\,. \tag{18b}$$

After some lengthy calculations one arrives at the solution of the deviation equation using the result of Fuchs [13]

$$\eta^t(s) = \frac{LE}{A\sqrt{L^2+R^2}}\, f(s)\,, \tag{19a}$$

$$\eta^r(s) = \frac{ER}{\sqrt{AB}\sqrt{L^2+R^2}}\, g(s)\,, \tag{19b}$$

$$\eta^\vartheta(s) = \frac{C_{(5)}}{R}\cos\Omega_\Phi s + \frac{C_{(6)}}{R}\sin\Omega_\Phi s\,, \tag{19c}$$

$$\eta^\varphi(s) = \frac{\sqrt{L^2+R^2}}{R^2}\, f(s)\,, \tag{19d}$$

where the two proper time dependent functions $f(s)$ and $g(s)$ are given by

$$f(s) = \sqrt{\frac{\Delta}{k^2}}\left(C_{(2)}\cos ks - C_{(3)}\sin ks\right) + \frac{(k^2-\Delta)}{\sqrt{\Delta}}C_{(1)}s + C_{(4)}\,, \tag{20a}$$

$$g(s) = C_{(1)} + C_{(2)}\sin ks + C_{(3)}\cos ks\,, \tag{20b}$$

$$k^2 := \frac{2A''A - A'^2}{2AB(2A - A'R)} + \frac{3A'}{2AB\,R}\,, \tag{20c}$$

$$\Delta := \frac{2A'}{AB\,R}\,. \tag{20d}$$

The prime denotes derivatives w.r.t. the radial coordinate and the metric functions $A = A(R)$, $B = B(R)$ are to be evaluated at the reference radius R. Furthermore, for a circular geodesic the constants of motion (17) can be expressed by

$$E^2 = \frac{2A^2}{2A - A'R}, \quad L^2 = \frac{R^3 A'}{2A - A'R}. \tag{21}$$

3.2 Deviation Equation in Schwarzschild Spacetime

In GR, the Schwarzschild spacetime serves as the simplest model of an isolated and spherically symmetric central object and might be used as a first order approximation of an astrophysical object like the Earth.[1] The metric functions in Eq. (16) are then given by

$$A(r) = 1 - \frac{2m}{r}, \quad B(r) = A(r)^{-1}. \tag{22}$$

The constants of motion E and L as well as the remaining quantities k and Δ are uniquely defined by the radius R of the circular reference geodesic

$$\begin{aligned} \Delta &= \frac{4m}{R^3}, & k^2 &= \frac{m(R-6m)}{R^3(R-3m)}, \\ E^2 &= \frac{(R-2m)^2}{R(R-3m)}, & L^2 &= \frac{mR^2}{R-3m}. \end{aligned} \tag{23}$$

We should mention that the mass of the Earth in the units that we use is $m \approx 0.5$ cm. The parameters $C_{(1,\ldots,6)}$ will be used to model different orbital scenarios. Using the constants in Eq. (23), we can simplify the solution (19) for the case of Schwarzschild spacetime. We find[2]:

$$\eta^t(s) = \frac{\sqrt{mR}}{R-2m} f(s), \tag{24a}$$

$$\eta^r(s) = g(s), \tag{24b}$$

$$\eta^\vartheta(s) = \frac{C_{(5)}}{R} \cos \Omega_\Phi s + \frac{C_{(6)}}{R} \sin \Omega_\Phi s, \tag{24c}$$

$$\eta^\varphi(s) = \frac{f(s)}{R}, \tag{24d}$$

where the two functions $f(s)$ and $g(s)$ are given by

---

[1] Planets do not possess any net charge, therefore we do not consider charged solutions like, for example, the Reissner–Nordstrøm spacetime.

[2] W.r.t. Eq. (19) we have slightly redefined the constant parameters $C_{(1,\ldots,6)}$ in a way such that $\eta^r(s) = g(s)$. This is always possible since all coefficients preceding the functions $f(s)$ and $g(s)$ in (19) are constant because the reference radius R is constant.

$$f(s) = 2\sqrt{\frac{R}{R-6m}}\left(C_{(2)}\cos ks - C_{(3)}\sin ks\right) - \frac{3}{2}\Omega_{\Phi}\frac{R-2m}{R-3m}C_{(1)}\,s + C_{(4)}\,, \tag{25a}$$

$$g(s) = C_{(1)} + C_{(2)}\sin ks + C_{(3)}\cos ks\,. \tag{25b}$$

Notice that the function $f(s)$ contains, besides periodic and constant parts, a term that grows linearly with the reference proper time s for $C_{(1)} \neq 0$. This contribution is not bounded and will, thus, limit the validity of the framework since we work with the first order deviation equation, i.e. the deviation η^a is assumed to be small and only contributions up to first order were considered. We observe that in the general relativistic solution of the first order deviation equation two distinct frequencies appear:

$$k = \sqrt{\frac{m}{R^3}}\sqrt{\frac{R-6m}{R-3m}} = \Omega_K\sqrt{\frac{R-6m}{R-3m}}\,, \tag{26a}$$

$$\Omega_{\Phi} = \sqrt{\frac{m}{R^3}}\sqrt{\frac{R}{R-3m}} = \Omega_K\sqrt{\frac{R}{R-3m}}\,. \tag{26b}$$

In the Newtonian limit these two frequencies coincide and yield the Keplerian frequency Ω_K. Figure 2 shows the difference $\Omega_{\Phi} - k$ between both frequencies for reference radii that correspond to satellite orbits from 100 km to $3.6 \cdot 10^4$ km above the surface of the Earth. It is worthwhile to note that the general relativistic solution

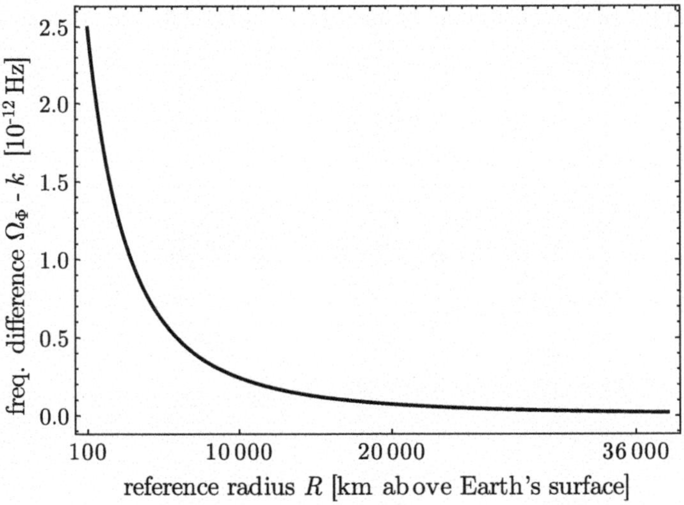

Fig. 2 The difference between the frequencies Ω_{Φ} and k, which appear in the solution of the deviation equation, is shown for reference radii that belong to satellite orbits around the Earth. The frequency difference is of the order of some 10^{-12} Hz, which yields a period difference in the range of 10–30 μs. We consider as the mean Earth radius $R_{\oplus} = 6.37 \cdot 10^3$ km

(24), (25) approaches the correct Newtonian limit (11) for $c \to \infty$. Studying the difference allows to uncover relativistic effects in the following. Observe that the normalization of the reference four-velocity yields

$$(1 - 2m/R)\,\dot{T}^2 - R^2\dot{\Phi}^2 = (1 - 2m/R)\dot{T}^2 - r^2\Omega_\Phi^2 = 1 \tag{27}$$

$$\Rightarrow \quad \dot{T} = \sqrt{\frac{R}{R - 3m}}\,. \tag{28}$$

When we parametrize the circular reference orbit by coordinate time we get

$$\tilde{\Omega}_\Phi := \frac{d\Phi}{ds}\left(\frac{dT}{ds}\right)^{-1} = \Omega_\Phi \dot{T}^{-1} = \sqrt{\frac{m}{R^3}} = \Omega_K\,. \tag{29}$$

Hence, Kepler's third law holds perfectly well for circular orbits in the Schwarzschild spacetime when the orbit is parametrized by coordinate time.

4 Applicability of the Geodesic Deviation Equation

In this section, we study the applicability of the first order deviation equation (15) in Schwarzschild spacetime to describe the motion of a test body that is close to a given circular reference geodesic. Its worldline is determined by a small initial perturbation of that reference curve, described by the solution (24). In the following, we investigate the shape of the perturbed orbits as well as physical and artificial effects, which are present in the solution.

To describe different orbital configurations we have to examine the impact of the parameters $C_{(1,\ldots,6)}$ on the perturbed orbit. A proper way to do this is to investigate the impact of each parameter separately since the different effects can be superimposed in this linearized framework. Here, we extend the brief analysis that was done in Ref. [4]. The connection between the parameters $C_{(i)}$ and the orbital elements of the perturbed orbit are summarized in Table 1. Hence, for a specific orbital configuration that is to be modeled we can determine the parameters that must be taken into account from the table and describe that satellite configuration within the framework of the geodesic deviation equation using the solution (24).

4.1 Shape of the Perturbed Orbits

The following sections are named after the geometric shape of the perturbed orbits, caused by the choice of the respectively considered parameter(s). All orbits that we discuss in the following are shown in Fig. 3.

Table 1 The initial position and velocity of the test body that follows the perturbed orbit X^a. Choose one parameter that shall be the only non-zero one, then the initial state can be read off from the table. For combinations of different parameters the effects can be superimposed. We list the orbital elements such as eccentricity e, semi-major axis a, the ascending node Ω_a (longitude), the inclination i, distance to the perigee d_p, distance to the apogee d_a and the argument of the perigee ω. For the definitions of these orbital elements see Fig. 5. When two values are given, the orbital element depends on the sign of the respective parameter

$\neq 0$:	$C_{(1)}$	$C_{(2)}$	$C_{(3)}$	$C_{(4)}$	$C_{(5)}$	$C_{(6)}$
$r(0)$	$R + C_{(1)}$	R	$R + C_{(3)}$	R	R	R
$\vartheta(0)$	$\pi/2$	$\pi/2$	$\pi/2$	$\pi/2$	$\pi/2 + C_{(5)}/R$	$\pi/2$
$\varphi(0)$	0	$\delta\omega_{(2)}$	0	$C_{(4)}/R$	0	0
$\dot{r}(0)$	0	$C_{(2)}\, k$	0	0	0	0
$\dot{\vartheta}(0)$	0	0	0	0	0	$\Omega_\Phi\, C_{(6)}/R$
$\dot{\varphi}(0)$	$\Omega_\Phi + \delta\omega_{(1)}$	Ω_Φ	$\Omega_\Phi +$ $k\, \delta\omega_{(3)}$	Ω_Φ	Ω_Φ	Ω_Φ
e	0	$C_{(2)}/R$	$C_{(3)}/R$	0	0	0
a	$R + C_{(1)}$	R	R	R	R	R
Ω_a	0	0	0	0	$\pi/2(2 - \text{sgn } C_{(5)})$	$\pi/2(1 + \text{sgn } C_{(6)})$
i	0	0	0	0	$C_{(5)}/R$	$C_{(6)}/R$
d_p	$R + C_{(1)}$	$R - C_{(2)}$	$R - C_{(3)}$	R	R	R
d_a	$R + C_{(1)}$	$R + C_{(2)}$	$R + C_{(3)}$	R	R	R
ω	0	$3\pi\Omega_\Phi/(2k)$; $\pi\Omega_\Phi/(2k)$	$\pi\Omega_\Phi/k$; 0	0	0	0
Shape:	Circular	Elliptical	Elliptical	Circular	Circular, inclined	Circular, inclined

4.1.1 Circular Perturbation

If we set all parameters but $C_{(1)}$ equal to zero, the perturbed orbit remains in the reference orbital plane and has still a circular shape. This perturbed orbit is given by

$$r = R + \eta^r = R + C_{(1)}\,, \tag{30a}$$

$$t(s) = \dot{T}s + \eta^t(s) = \dot{T}s - \frac{3}{2}\dot{T}\frac{m}{R - 3m}\frac{C_{(1)}}{R}s =: \left(\dot{T} + \delta t_{(1)}\right)s, \tag{30b}$$

$$\varphi(s) = \Omega_\Phi\, s + \eta^\varphi(s) = \Omega_\Phi\, s - \frac{3}{2}\Omega_\Phi\frac{R - 2m}{R - 3m}\frac{C_{(1)}}{R}s =: (\Omega_\Phi + \delta\omega_{(1)})\, s. \tag{30c}$$

The reference and the perturbed orbit are shown in Fig. 3 for one reference period and a chosen reference radius of 5000 km above the surface of the Earth. As one would expect, a positive radial perturbation $C_{(1)}$ yields a smaller azimuthal frequency, $\dot{\varphi} = \Omega_\Phi + \delta\omega_{(1)} < \Omega_\Phi$, as compared to the reference motion. The frequency and radial perturbations are related via

$$\frac{\delta\omega_{(1)}}{\Omega_\Phi} = -\frac{3}{2}\frac{R - 2m}{R - 3m}\frac{C_{(1)}}{R}\,, \tag{31}$$

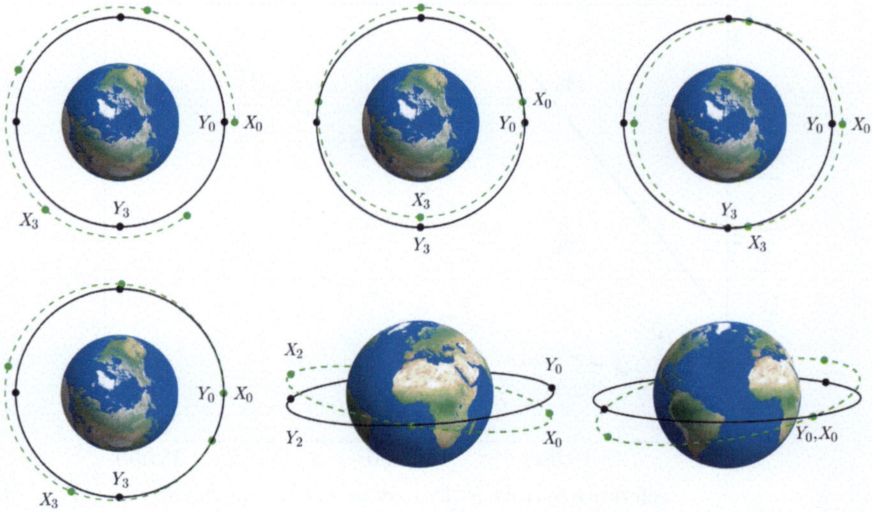

Fig. 3 The solid line (black) shows the circular reference orbit, whereas the dashed line (green) shows the perturbed orbit as calculated with the solution of the deviation equation for: only $C_{(1)} \neq 0$ (top, left), only $C_{(2)} \neq 0$ (top, middle), only $C_{(3)} \neq 0$ (top, right), a combination of both such that initially $r(0) = R$, $\dot{r}(0) = 0$ (bottom left) and a pendulum orbit as the result of an inclined perturbation using only $C_{(5)}$ (bottom middle) and $C_{(6)}$ (bottom right). We have marked the respective positions Y_n on the reference orbit and X_n on the perturbed orbit for reference proper time values $s = n/4 \cdot 2\pi/\Omega_\Phi$. We used a reference radius of $R = R_\oplus + 5000\,\text{km}$ and a mean Earth radius $R_\oplus = 6.37 \cdot 10^3$ km.eps

such that they are not independent. Since we work with the first order deviation equation, we have to ensure that the radial perturbation is indeed small, i.e. $C_{(1)}/R \ll 1$. This is related to upper bounds for $C_{(1)}$ that need to be chosen in a proper way. For a satellite orbit of about 10^4 km above the surface of the Earth, the normalized perturbation is $C_{(1)}/R \approx 10^{-7}\text{m}^{-1}\, C_{(1)}$. Hence, the allowed values for $C_{(1)}$ strongly depend on the chosen reference radius and given upper bounds for the radial perturbation. For various values of the reference radius - ranging from Low Earth Orbits (LEO) to geostationary ones - and the parameter $C_{(1)}$, we show the magnitude of the normalized radial perturbation $C_{(1)}/R$ in Fig. 4. To decide whether the description of a satellite configuration within the framework of the first order deviation equation is useful or not, one has to define the reference radius and the maximal radial perturbation for the desired scenario. The value $C_{(1)}/R$ can then be estimated from Fig. 4 and if it fulfills the condition $C_{(1)}/R \ll 1$ the solution may give a simple and useful description.

After a full azimuthal period on the reference orbit, $s = 2\pi/\Omega_\Phi$, and the perturbed orbit is not yet closed since $\varphi(2\pi/\Omega_\Phi) \neq 2\pi$. The deficit angle $\Delta\alpha$ is given by

$$\Delta\alpha = \varphi\left(2\pi/\Omega_\Phi\right) - 2\pi = 2\pi \frac{\delta\omega_{(1)}}{\Omega_\Phi} = -3\pi \frac{R-2m}{R-3m} \frac{C_{(1)}}{R}. \tag{32}$$

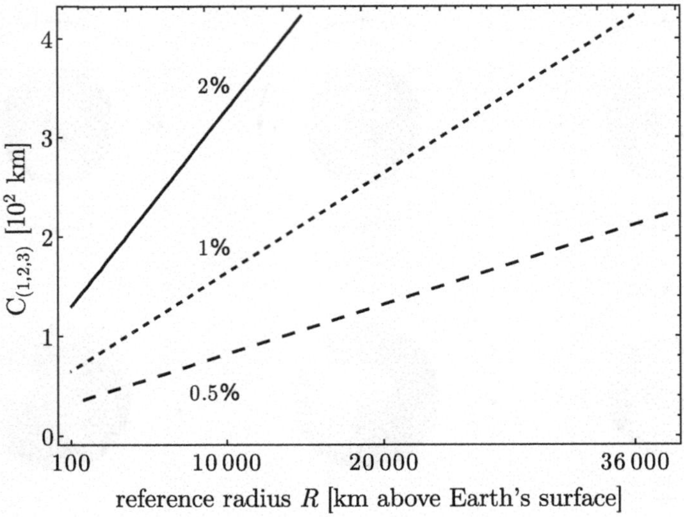

Fig. 4 The magnitude of the normalized radial perturbation $C_{(i)}/R$ as a function of $C_{(1,2,3)}$ and the reference radius R. The lines represent surfaces of constant $C_{(i)}/R$. We use a mean Earth radius $R_\oplus = 6.37 \cdot 10^3$ km

This angle may correspond, in principle, to an observable quantity and the relation can be solved for m explicitly,

$$m = \frac{R(3C_{(1)}\pi + \Delta\alpha R)}{3(2C_{(1)}\pi + \Delta\alpha R)}, \tag{33}$$

to obtain an estimate for the relativistic mass monopole of the central object. Thus, the mass can be obtained from the measurement of the deficit angle, assuming the situation can be prepared with initially known reference radius R and radial distance $C_{(1)}$ between both orbits. Also the deviation of a test object from the center of mass within a hollow satellite might be used for such a measurement.

4.1.2 Elliptical Perturbation I

The two parameters $C_{(2,3)}$ cause elliptical perturbations in the (reference orbital plane) if we neglect the influence of all other parameters, i.e. if $C_{(1,4,5,6)} = 0$. If we choose to have $C_{(2)} = 0$, the perturbed orbit is described by

$$r(s) = R + \eta^r(s) = R + C_{(3)} \cos ks = R + C_{(3)} + \mathcal{O}\left(s^2\right), \tag{34a}$$

$$\begin{aligned} t(s) &= \dot{T}s + \eta^t(s) \\ &= \dot{T}s - \frac{2R}{R-2m}\sqrt{\frac{m}{R-6m}}C_{(3)} \sin ks \end{aligned}$$

$$=: \dot{T}s + \delta t_{(3)} \sin ks = (\dot{T} + \delta t_{(3)} k)\, s + \mathcal{O}\left(s^2\right) \tag{34b}$$

$$\varphi(s) = \Omega_\Phi s - 2\sqrt{\frac{R}{R-6m}} \frac{C_{(3)}}{R} \sin ks$$
$$:= \Omega_\Phi s + \delta\omega_{(3)} \sin ks = (\Omega_\Phi + \delta\omega_{(3)} k)\, s + \mathcal{O}\left(s^2\right). \tag{34c}$$

For an elliptically perturbed orbit of this kind, the eccentricity e and the semi major axis a can be linked to the radial perturbation via

$$e = \frac{C_{(3)}}{R}, \quad a = R. \tag{35a}$$

We can as well calculate the distance to the perigee d_p and apogee d_a that are related to the radial perturbation

$$d_p = R - C_{(3)}, \quad d_a = R + C_{(3)}, \tag{35b}$$

and confirm that the semi major axis is half of the sum of the two distances as it should be. Using these relations the spatial shape of the perturbed orbit can be represented in a familiar way as

$$r(s) = a(1 + e\cos ks), \tag{36a}$$

$$\varphi(s) = \Omega_\Phi s - 2\sqrt{\frac{a}{a-6m}}\, e \sin ks. \tag{36b}$$

The eccentricity of the perturbed orbit $e = C_{(3)}/R$ is shown in Fig. 4. If for a specific satellite mission the maximal allowed eccentricity is given, we can read off upper bounds for the parameter $C_{(3)}$ from Fig. 4, or, vice versa: we can model an orbit with a given (small) eccentricity by choosing the necessary value for $C_{(3)}$.

The difference between the effects of the two parameters $C_{(2,3)}$ is just a phase difference, i.e. a spatial rotation of $\pi/2$ of the perturbed orbit within the reference orbital plane. For the case that only $C_{(2)} \neq 0$, the perturbed orbit is described by

$$r(s) = R + \eta^r(s) = R + C_{(2)} \sin ks = R + C_{(2)} ks + \mathcal{O}\left(s^2\right), \tag{37a}$$

$$t(s) = \dot{T}s + \eta^t(s)$$
$$= \dot{T}s + \frac{2R}{R-2m}\sqrt{\frac{m}{R-6m}}\, C_{(2)} \cos ks$$
$$=: \dot{T}s + \delta t_{(2)} \cos ks = \dot{T}s + \delta t_{(2)} + \mathcal{O}\left(s^2\right), \tag{37b}$$

$$\varphi(s) = \Omega_\Phi s + 2\sqrt{\frac{R}{R-6m}} \frac{C_{(2)}}{R} \cos ks$$
$$=: \Omega_\Phi s + \delta\omega_{(2)} \cos ks = \Omega_\Phi s + \delta\omega_{(2)} + \mathcal{O}\left(s^2\right). \tag{37c}$$

Both elliptical orbits are shown in Fig. 3 and the spatial rotation as the difference between the effects of $C_{(2)}$ and $C_{(3)}$ is obvious. These orbits look closed after one reference period, but they are not (recall that the frequencies Ω_Φ and k are just slightly different). After one reference period $s = 2\pi/\Omega_\Phi$ we get

$$r(2\pi/\Omega_\Phi) = R + C_{(3)} \cos\frac{2\pi k}{\Omega_\Phi} \neq r(0)\,, \tag{38a}$$

$$\varphi(2\pi/\Omega_\Phi) = 2\pi + \delta\omega_{(2)} \sin\frac{2\pi k}{\Omega_\Phi} \neq \varphi(0) + 2\pi\,. \tag{38b}$$

However, the difference between the two frequencies is in the range of some 10^{-12} Hz and the periods differ by about 10–25 μs for reference radii in the range from LEO to geostationary orbits. The radial motion has an actual period of $s = 2\pi/k$, i.e. this amount of reference proper time elapses from one perigee to the next. Hence, we get

$$r(2\pi/k) = r(0)\,, \tag{39a}$$

$$\varphi(2\pi/k) = \frac{2\pi\Omega_\Phi}{k} \neq \varphi(0) + 2\pi\,. \tag{39b}$$

Since the increase in the azimuthal angle differs from 2π after one radial period, the perigee of the elliptical orbit will precess. This precession is investigated in the next section in more detail.

4.1.3 Elliptical Perturbation II

Another kind of elliptical orbits can be constructed via the combination with a circular perturbation. For example, we use a combination of both, $C_{(1)}$ and $C_{(3)}$, to arrive at a perturbed orbit that initially fulfills $r(0) = R\,,\ \varphi(0) = \Phi(0) = 0$ and $\dot{r}(0) = 0$. Hence, the perturbed orbit is initially as close as possible to the reference orbit - cf. Fig. 3. To construct it we need to choose $C_{(3)} = -C_{(1)}$. The perturbed orbit is then described by

$$r(s) = R + C_{(1)}(1 - \cos ks)\,, \tag{40a}$$

$$t(s) = (\dot{T} + \delta t_{(1)})\, s + \delta t_{(3)} \sin ks\,, \tag{40b}$$

$$\varphi(s) = (\Omega_\Phi + \delta\omega_{(1)})\, s + \delta\omega_{(3)} \sin ks\,, \tag{40c}$$

$$\vartheta = \Theta \equiv \pi/2\,. \tag{40d}$$

For this orbit type we find the eccentricity and semi major axis to be

$$e = \frac{C_{(1)}}{R + C_{(1)}}\,, \qquad a = R + C_{(1)} \quad \Rightarrow C_{(1)} = ae\,, \tag{41}$$

and this allows to recast the radial motion, again, in the familiar way

$$r(s) = a(1 - e \cos ks) . \tag{42}$$

This radial motion has a period of $2\pi/k$ and we obtain

$$r(2\pi/k) = r(0) , \tag{43a}$$

$$\varphi(2\pi/k) = \frac{2\pi(\Omega_\Phi + \delta\omega_{(1)})}{k} \neq \varphi(0) + 2\pi . \tag{43b}$$

Hence, also this elliptical orbit will precess. The precession is studied in the next section in terms of the perigee advance.

4.1.4 Azimuthal Perturbation

The parameter $C_{(4)}$ causes an offset in the azimuthal motion, i.e. a constant phase difference between the reference and perturbed orbit. When only $C_{(4)} \neq 0$ the perturbed orbit is the same as the reference orbit, but the two test bodies are separated by a constant angle. The perturbed motion is then described by

$$r = R, \tag{44a}$$

$$t(s) = \dot{T}\, s + \eta^t(s) = \dot{T} s + \frac{\sqrt{mR}}{R - 2m} C_{(4)}, \tag{44b}$$

$$\varphi(s) = \Omega_\Phi s + \eta^\varphi(s) = \Omega_\Phi s + \frac{C_{(4)}}{R} , \tag{44c}$$

where the constant azimuthal separation is determined by the value of $C_{(4)}/R$.

4.1.5 Inclined Perturbation

The two parameters $C_{(5)}, C_{(6)}$ incline the orbital plane with respect to the reference plane. If only $C_{(5)} \neq 0$ we obtain a circular orbit with radius $r = R$ and azimuthal motion $\varphi(s) = \Omega_\Phi s$, but with the polar motion given by

$$\vartheta(s) = \frac{\pi}{2} + \frac{C_{(5)}}{R} \cos \Omega_\Phi s . \tag{45}$$

Hence, $C_{(5)}/R$ determines the maximal inclination between the two orbital planes. If only $C_{(6)} \neq 0$ instead, there are just little changes: the $\cos(\Omega_\Phi s)$ becomes $\sin(\Omega_\Phi s)$ and the difference between the effects of these two parameters is simply related to a spatial rotation.

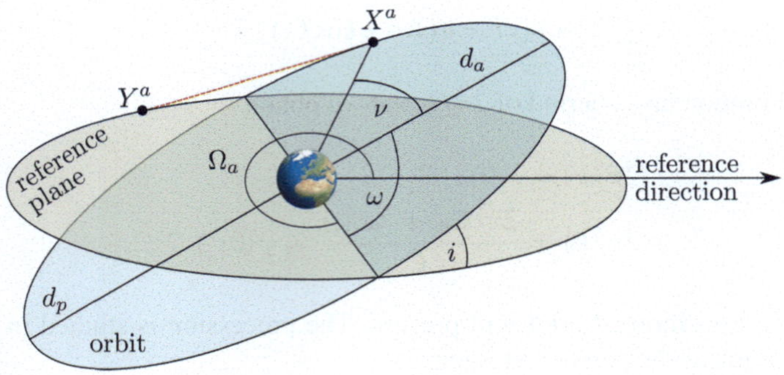

Fig. 5 A sketch of the two orbital planes including the orbital elements of the perturbed orbit X^a

4.2 *The Orbital Elements*

Combining the results of the last section we can link all parameters $C_{(1,\ldots,6)}$ to the initial position and velocity of the test body that follows the perturbed orbit. The initial quantities $r(0)$, $\vartheta(0)$, $\varphi(0)$ and $\dot{r}(0)$, $\dot{\vartheta}(0)$, $\dot{\varphi}(0)$ are summarized in Table 1 together with the resulting orbital elements of the perturbed orbit. The orbital elements are introduced in the sketch shown in Fig. 5.

4.3 *Physical Effects*

As shown before, the parameters $C_{(2,3)}$ lead to an elliptically perturbed orbit if at least one of them does not vanish. For such an orbit the perigee will precess and the orbit is not closed. If either of the parameters $C_{(2,3)} \neq 0$ the radial motion has a period given by $s = 2\pi/k$, but the azimuthal oscillation is advanced already. The difference to a full revolution is then given by

$$\Delta\varphi = \varphi(2\pi/k) - 2\pi = 2\pi\left(\frac{\Omega_\Phi}{k} - 1\right) = 2\pi\left(\sqrt{\frac{a}{a-6m}} - 1\right), \tag{46}$$

where a is the orbit's semi-major axis. Figure 6 shows this precession of the perigee for different reference radii ranging from LEO to geostationary orbits. The result (46) is the same as shown in Ref. [14] and was also derived in [13] as well.[3] Up to linear order in m/a we obtain the well-known result

$$\Delta\varphi = \frac{6\pi m}{a} + \mathcal{O}\left((m/a)^2\right), \tag{47}$$

[3]Note the misprint in Eq. (4.14) in [13], where actually the inverse value of the correct result is shown and we assume this to be simply a typo.

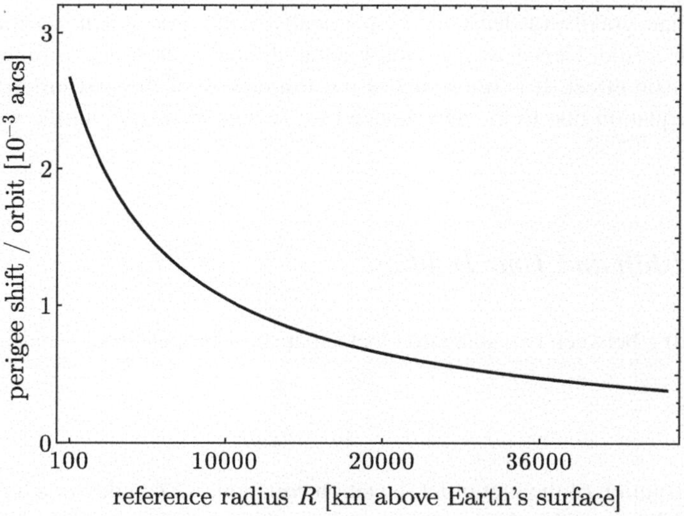

Fig. 6 The perigee shift per orbit in 10^{-3} arcs for the elliptically perturbed orbit with either $C_{(3)} \neq 0$ or $C_{(2)} \neq 0$ (left) This orbit is shown in Fig. 3 in the middle of the upper row

that is the first term in Einstein's result [15]

$$\Delta\varphi = \frac{6\pi G m}{a(1-e^2)} \approx \frac{6\pi m}{a}(1 + e^2 + e^4 + \cdots), \tag{48}$$

for the precession of the perigee in the case of small eccentricities. For the second kind of an elliptically perturbed orbit that is described by Eq. (40) we obtain according to Eq. (43)

$$\begin{aligned} \Delta\varphi &= \frac{2\pi(\Omega_\Phi + \delta\omega_{(1)})}{k} - 2\pi \\ &= 2\pi\left(\sqrt{\frac{R}{R-6m}} - 1\right) - 3\pi\frac{R-2m}{R-3m}\sqrt{\frac{R}{R-6m}}\frac{C_{(1)}}{R}. \end{aligned} \tag{49}$$

The first term resembles the previous result for the perigee precession where the parameter $C_{(1)}$ was set equal to zero and we recover this result in the limit. Up to linear order in e and m/a the result reads

$$\Delta\varphi = \frac{6\pi m}{a}\left(1 - \left(1 + \frac{m}{2a}\right)e\right) + \mathcal{O}\left(e^2, (m/a)^2\right), \tag{50}$$

and depends on the eccentricity, whereas in the first case the result was independent of the perturbation parameters.

It should be mentioned that in the Newtonian solution (12) no perigee precession is present since there is only one frequency, the Keplerian frequency Ω_K, involved.

In Newtonian gravity (at least for a spherically symmetric potential) the Kepler ellipses are closed. Hence, as it is well-known, the precession of an elliptical orbit is a relativistic effect. It is recovered in the framework of the first order geodesic deviation equation due to the appearance of a second frequency in the relativistic solution.

4.4 Redshift and Time Dilation

The redshift z between two standard clocks that show proper times s and $\tilde{s}$ is

$$1 + z = \frac{\nu}{\tilde{\nu}} = \frac{d\tilde{s}}{ds}\,. \tag{51}$$

Using the solution of the first order deviation equation, we can derive a formula for the redshift between the clocks transported along the reference and deviating orbit as follows. Along the orbit $X^a(s)$, the constant of motion E_X related to the energy is

$$E_X = \left(1 - \frac{2m}{r(\tilde{s})}\right)\frac{dt}{d\tilde{s}} = \left(1 - \frac{2m}{r(s)}\right)\frac{dt}{ds}\frac{ds}{d\tilde{s}}\,. \tag{52}$$

Hence, we obtain for the redshift using the solution $t(s)$ and $r(s)$ of the first order deviation equation

$$\begin{aligned} z + 1 = \frac{d\tilde{s}}{ds} &= \left(1 - \frac{2m}{r(s)}\right)\frac{\dot{T} + \dot{\eta}^t(s)}{E_X} \\ &= \left(1 - \frac{2m}{R + g(s)}\right)\frac{\sqrt{\frac{R}{R-3m}} + \frac{\sqrt{m}R}{R-2m}\dot{f}(s)}{E_X}\,, \end{aligned} \tag{53}$$

where the functions $f(s)$ and $g(s)$ are given by Eq. (25) and E_X is fixed by the initial conditions of the deviating orbit, i.e. by the choice of parameters $C_{(i)}$.

$$\begin{aligned} z + 1 = &\frac{1}{E_X}\left(1 - \frac{2m}{R + C_{(1)} + C_{(2)}\sin ks + C_{(3)}\cos ks}\right) \\ &\times\left(\sqrt{\frac{R}{R-3m}} + \lambda_{(1)} + \lambda_{(2)}\sin ks + \lambda_{(3)}\cos ks\right)\,, \end{aligned} \tag{54}$$

where

$$\lambda_{(1)} = -\frac{3}{2}\frac{\sqrt{m}(R-2m)}{R(R-3m)^{3/2}}C_{(1)}\,, \quad \lambda_{(2,3)} = -2\frac{\sqrt{m}R}{(R-2m)\sqrt{R-6m}}C_{(2,3)}\,. \tag{55}$$

This result yields a comparatively simple model for the redshift between the two satellites and is accurate as long as the orbital deviation is small.

For two circular orbits with radii R and $R + C_{(1)}$, we recover the correct result to first order in $C_{(1)}/R$. The redshift becomes

$$z = \frac{C_{(1)}}{R} \frac{3m}{2(R-m)} . \tag{56}$$

Note however, that terms related to Doppler effects are not present here, since we do not consider signals (light rays) send from one orbit to the other. Hence, the formula for the redshift contains only the part related to time dilation effects.

4.5 Accuracy of the First Order Deviation Approach

Exact solutions of the geodesic equation in the Schwarzschild spacetime can be constructed using elliptic functions. To our knowledge, the first work on this is contained in Refs. [16, 17]. The authors used the Jacobi elliptic functions sn, cn, dn to solve the equation of motion. A more recent study of exact orbital solutions in Schwarzschild spacetime (and generalizations for, e.g., Kerr–Newman–deSitter spacetime) can be found in Refs. [18–20], where the Weierstrass elliptic function $\wp$ is used.

Note that especially in the solutions of the geodesic equation that involve the Jacobi elliptic function sn, cn, dn the relation to the solutions (24) of the first order deviation equation is obvious. Take two such exact solutions and construct the deviation between the two orbits as their difference. Then, choosing one of these orbits to be the circular reference geodesic, the linearization of the deviation corresponds to the solution of the first order deviation equation. In Eq. (24) sin and cos terms appear and these are the linearizations of the Jacobi elliptic functions sn and cn.

4.5.1 Circular Orbits

For a circular orbit in the equatorial plane with radius $r = R + C_{(1)}$, the exact azimuthal frequency is given by

$$\omega_\varphi = \sqrt{\frac{m}{(R + C_{(1)})^3}} \sqrt{\frac{R + C_{(1)}}{R + C_{(1)} - 3m}} . \tag{57}$$

Expanding this result w.r.t. the small quantity $C_{(1)}/R \ll 1$ in a Taylor series, we find for the azimuthal frequency

$$\omega_\varphi = \underbrace{\sqrt{\frac{m}{R^3}}\sqrt{\frac{R}{R-3m}}}_{=\Omega_\Phi=:\delta\omega^{(0)}_{(1)}} \underbrace{-\frac{3}{2}\frac{R-2m}{R-3m}\sqrt{\frac{m}{R^3}}\sqrt{\frac{R}{R-3m}}\frac{C_{(1)}}{R}}_{=:\delta\omega^{(1)}_{(1)}}$$
$$\underbrace{+\frac{15}{8}\frac{(R-2m)^2+4/5m^2}{(R-3m)^2}\sqrt{\frac{m}{R^3}}\sqrt{\frac{R}{R-3m}}\left(\frac{C_{(1)}}{R}\right)^2}_{=:\delta\omega^{(2)}_{(1)}}$$
$$+\mathcal{O}\left(\left(C_{(1)}/R\right)^3\right). \tag{58}$$

The quantities $\delta\omega^{(i)}_{(1)}$ are defined as shown above. The superscript denotes the order of expansion and the subscript denotes the connection to the radial perturbation $C_{(1)}$. The 0th order contribution is given by Ω_Φ, the azimuthal frequency for a circular orbit with radius R - cf. Eq. (26). Restricting ourselves to first order contributions we can compare the approximation (58) to the solution of the first order deviation equation (30)

$$\Omega_\Phi + \delta\omega_{(1)} \equiv \Omega_\Phi + \delta\omega^{(1)}_{(1)}. \tag{59}$$

Hence, the approximation up to linear order in $C_{(1)}/R$ is exactly the result that appeared in the solution of the first order deviation equation - cf. Eq. (30). Therefore, the error that we make using this result is dominated by the second order term $\delta\omega^{(2)}_{(1)}$. We show the relative error $\delta\omega^{(2)}_{(1)}/\omega_\varphi$ in Fig. 7 and conclude that even for a radial separation of some 10^2 km between the reference and perturbed orbit this error is less than 0.1%. Using the expansion in Eq. (58), we estimate the error that remains at the next order, i.e. using also second order contributions in the frequency expansion. Then, the error is dominated by the third order term and about two orders of magnitude smaller. One can show that the quantity $\delta\omega^{(2)}_{(1)}/\omega_\varphi$, shown in Fig. 7, is also the dominating error term for the conserved quantity L that corresponds to the angular momentum of the perturbed orbit, since the frequency and angular momentum are related by $L = r^2\omega_\varphi$.

Using the expansion (58), we can write down the solution of the nth order deviation equation for a circular perturbation in the reference orbital plane. The perturbed orbit is then described by

$$r = R + C_{(1)}\,, \tag{60a}$$
$$\varphi = \left(\Omega_\Phi + \delta\omega^{(1)}_{(1)} + \dots + \delta\omega^{(n)}_{(1)}\right) s\,, \tag{60b}$$

where

$$\delta\omega^{(k)}_{(1)} = \frac{1}{k!}\frac{d^k\omega_\varphi}{d(C_{(1)}/R)^k}\left(\frac{C_{(1)}}{R}\right)^k. \tag{60c}$$

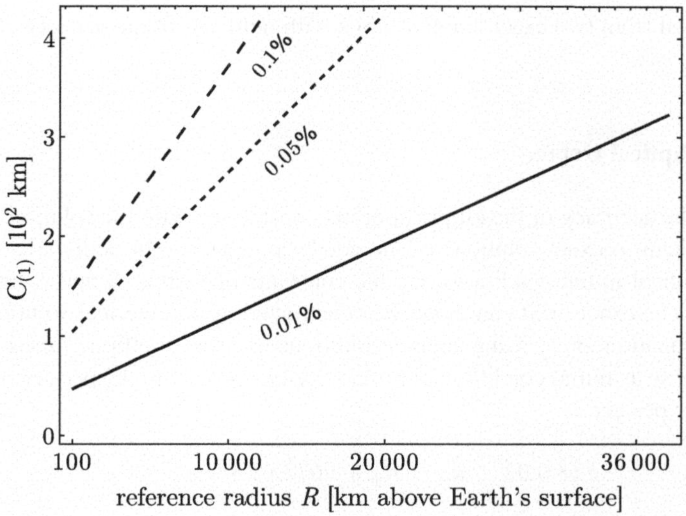

Fig. 7 The dominating term $\delta\omega_{(1)}^{(2)}/\omega_\varphi$ in the frequency error. This relative error is made using the solution of the first order deviation equation to describe a circular perturbed orbit with initial radial distance $C_{(1)}$ to a reference orbit with radius R. Hence, the solution of the first order deviation equation gives the correct result up to a few parts in one hundred

Table 2 Error in the distance between reference and perturbed orbit after one complete reference period. We compare the distance as modeled with Eq. (60), up to 4th order, with the distance constructed from two exact circular orbits with radii and frequencies (R, Ω_Φ) and $(R + C_{(1)}, \omega_\varphi)$. The table shows the values for the two different reference radii, 1000 km and 36000 km above Earth's surface. Bold marked values correspond to cm accuracy level when using the respective approximation order

Radial separation	Error (m)			
$C_{(1)}$	1st	2nd	3rd	4th
	Order			
$R = R_\oplus + 1000$ km				
10 km	159	0.25	**$4 \cdot 10^{-4}$**	**$6 \cdot 10^{-5}$**
50 km	3955	31	0.24	**0.002**
100 km	15700	249	3.8	**0.06**
150 km	35000	535	19	0.43
$R = R_\oplus + 36000$ km				
10 km	27	**0.007**	**10^{-5}**	**10^{-5}**
50 km	690	0.95	**0.001**	**$5 \cdot 10^{-5}$**
100 km	2760	7.6	**0.02**	**$5 \cdot 10^{-5}$**
150 km	6206	26	0.1	**0.0015**

Table 2 shows the error that is made when modeling the distance between the reference orbit and the circular perturbed orbit after one complete reference period. To calculate the error we used the result above and compared it with the distance as calculated from two exact circular orbits with radii and frequencies (R, Ω_Φ) and $(R + C_{(1)}, \omega_\varphi)$.

4.5.2 Elliptical Orbits

To judge the accuracy of the elliptical orbits constructed with the solution (34), we compare them to exact solutions of the geodesic equation. In both cases we have to use identical initial conditions, i.e. the constants of motion E and L have to be the same. The exact orbit can either be constructed as a numerical solution of the geodesic equation, or by using analytic solutions in terms of elliptic functions.

We choose as initial conditions for eccentricity, distance to the perigee and argument of the perigee

$$e = 0.02\,, \quad d_p = R_\oplus + 9672.6\,\text{km}\,, \quad \omega = 0\,. \tag{61}$$

The relative error in the radial and azimuthal motion is shown in Fig. 8 for ten orbital periods. The mean error in the radial deviation is positive, whilst the mean phase error is negative. Lowering the eccentricity by a factor of 10 yields relative errors that decrease two orders of magnitude. The analysis shows that the errors scale roughly as e^2, which is the expected behavior since terms that are quadratic in the eccentricity are neglected in this framework and contribute to second order deviations.

4.5.3 Pendulum Orbits

Pendulum orbit constellations are of special interest for satellite geodesy missions. Hence, we should give an idea of how accurate the modeling of these constellations can be done using the solutions (45) of the first order geodesic deviation equation. Equation (45) describes a circular orbit with unchanged radius that is inclined w.r.t the reference orbit. Thus, all constants of motion are the same but the orbital planes differ. In this solution, the amplitude $C_{(6)}/R$ of the ϑ-oscillation gives the inclination of the perturbed orbit. Since we work with first order perturbations, this amplitude has to be small. For a reference radius of GRACE-type, 500 km above Earth's surface, and an inclination of $1°$ the error is about 100 marcs. For two orbital periods, we show in Fig. 9 the relative error for an even higher inclination of $5°$. As can be seen in the figure, the error in the ϑ-motion is periodic with zero mean value and standard deviation around $0.1\,\%$ for this situation.

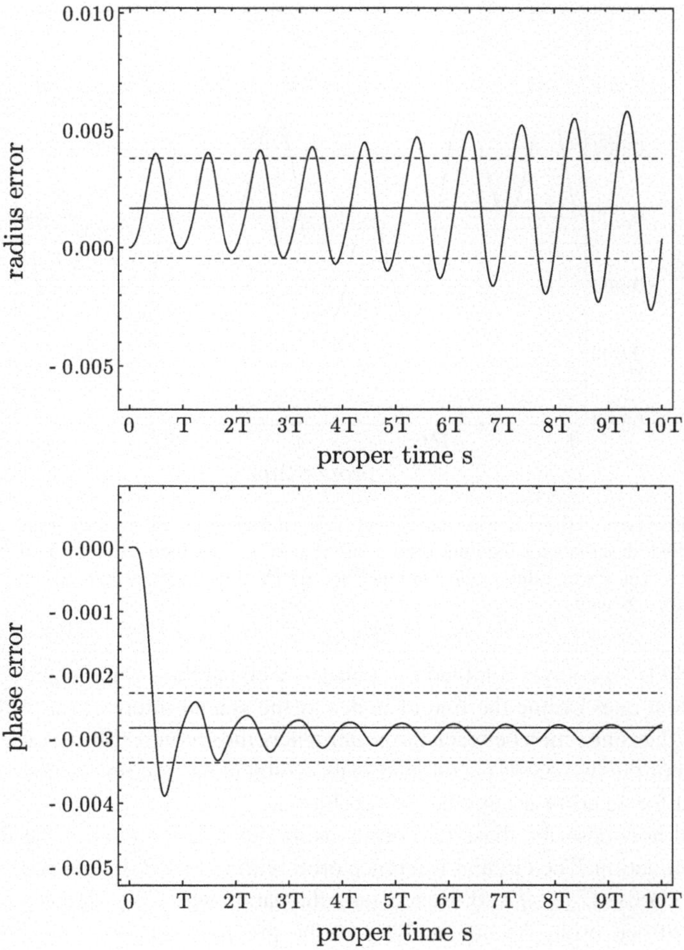

Fig. 8 Relative error in the radial and azimuthal motion compared to the exact solution for 10 periods ($10\,T$). The mean value (solid line) and range of one standard deviation around the mean (dashed lines) are shown

4.6 Other Effects

4.6.1 The Line of Nodes

In Schwarzschild spacetime, due to symmetry, any orbit is confined to one orbital plane that is determined by the initial conditions. Conventionally this plane is then chosen to be the equatorial plane, defined by $\vartheta = \pi/2$, $\dot{\vartheta} = \ddot{\vartheta} = 0$. Let there be two different bound orbits in Schwarzschild spacetime that we describe in the coordinate basis. We define the equatorial plane as the plane of motion for one of them. The

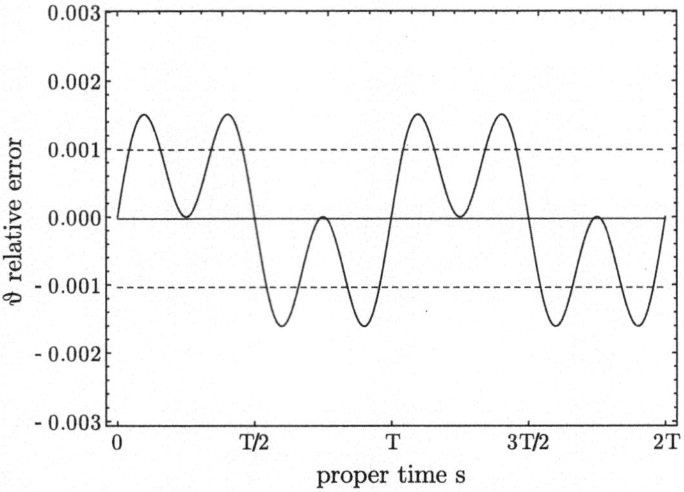

Fig. 9 Relative error in the ϑ-motion that is made using the solution (45) of the first order deviation equation, which describes a pendulum orbit constellation with an inclination of 5° between both orbital planes. The mean value (solid line) and one standard deviation around the mean (dashed lines) are shown as well

second orbit is, of course, confined to a plane as well but this orbital plane is inclined w.r.t. the first one. Define the line of nodes as the spatial intersection of these two planes, i.e. the connection between the points where the second orbit crosses $\theta = \pi/2$. Since each of the two orbits is confined to its orbital plane, the line of nodes remains unchanged for an arbitrary number of revolutions.

We will now describe these two orbits using the solution (24) of the first order deviation equation. The circular reference orbit with radius R defines the equatorial plane. We choose $C_{(2,...,5)} \equiv 0$. Hence, only the parameters $C_{(1,6)}$ describe the shape of the perturbed orbit. As our analysis in the previous sections has shown, $C_{(1)}$ causes a constant radial perturbation and $C_{(6)}$ inclines the perturbed orbital plane. This perturbed orbit is described by

$$r = R + C_{(1)} \,, \tag{62a}$$

$$\varphi(s) = (\Omega_\Phi + \delta\omega_{(1)})\, s \,, \tag{62b}$$

$$\vartheta(s) = \pi/2 + \frac{C_{(6)}}{R} \sin \Omega_\Phi s \,. \tag{62c}$$

The line of nodes is determined by two successive intersections of the ϑ-motion with the equatorial plane. This happens for $s = n\pi/\Omega_\Phi$, $n = 0, 1, 2 \ldots$. For these values we get

$$\vartheta(s = n\pi/\Omega_\Phi) = \pi/2, \tag{63a}$$

$$\varphi(s = n\pi/\Omega_\Phi) = n\pi + \frac{n\pi\delta\omega_{(1)}}{\Omega_\Phi} \,. \tag{63b}$$

Thus, from one orbit to the next the line of nodes shifts by an amount of

$$\Delta n := \frac{2\pi\delta\omega_{(1)}}{\Omega_\Phi} = -3\pi\frac{R-2m}{R-3m}\frac{C_{(1)}}{R}\,. \tag{64}$$

For $C_{(1)}/R = 1\%$ this corresponds to about $-5°$. Since a precession of the line of nodes must not happen in Schwarzschild spacetime, we have to carefully analyze this kind of "effect". As we have shown in the expansion of the exact azimuthal frequency (58) for a circular orbit with radius $R + C_{(1)}$, Ω_Φ is its 0th order and $\delta\omega_{(1)}$ its 1st order approximation. The approximation order in the frequency of the ϑ-motion in (62) is one less than the order in the φ-motion, because the amplitude $C_{(6)}/R$ of the ϑ-motion is already of first order. Hence, when using higher order deviation solutions up to jth order, the orbit will be given by

$$r = R + C_{(1)}\,, \tag{65a}$$

$$\varphi = (\Omega_\Phi + \delta\omega^{(1)}_{(1)} + \cdots + \delta\omega^{(j)}_{(1)})\,s\,, \tag{65b}$$

$$\vartheta = \pi/2 + C_{(6)}\sin(\Omega_\Phi + \delta\omega^{(1)}_{(1)} + \cdots + \delta\omega^{(j-1)}_{(1)})\,s\,, \tag{65c}$$

and subsequently the shift of the line of nodes after one orbit is

$$\Delta n = \frac{2\pi\delta\omega^{(j)}_{(1)}}{\Omega_\Phi + \delta\omega^{(1)}_{(1)} + \cdots + \delta\omega^{(j-1)}_{(1)}}\,. \tag{66}$$

Since $\delta\omega^{(j)}_{(1)}$ is the jth term in the Taylor expansion (58) we notice that $\Delta n \to 0$ for $j \to \infty$. Hence, the shift of the line of nodes vanishes in the limit of infinite accuracy and is simply an artifact of the linearization/approximation (Fig. 10).

4.6.2 Shirokov's Effect Revisited

In [3] a new effect in the context of the standard geodesic deviation equation in Schwarzschild spacetime was reported. This effect was also studied in several follow-up works [21–25], in particular generalizations to other spacetimes than Schwarzschild were given and Shirokov's effect was compared to the influence of the oblateness of the Earth that causes similar perturbations of the reference geodesic.

In the following we critically reassess the derivation, meaning, and physical measurability of Shirokov's effect. Shirokov [3] did only consider periodic solutions of the standard geodesic deviation equation in Schwarzschild spacetime. The solution that is given in the work [3] is obtained in the framework presented here for $C_{(1,3,4,5)} = 0$. Only the parameters $C_{(2,6)}$ are involved. According to the previous analysis of the shape of the perturbed orbit this corresponds to an elliptical orbit in an inclined orbital plane. This orbit is given by, cf. Eq. (21) in [3],

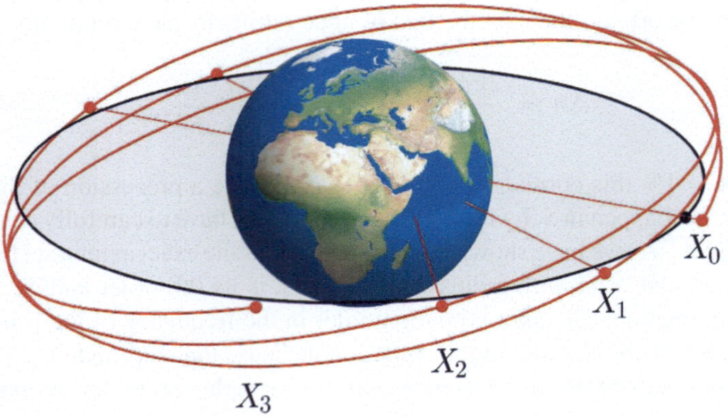

Fig. 10 The reference and perturbed orbit for three reference periods. We have marked one endpoint of the line of nodes X_n after n elapsed reference periods. The precession of the perturbed orbital plane is clearly visible

$$r(s) = R + \eta^r(s) = R + C_{(2)} \sin ks \,, \tag{67a}$$

$$\varphi(s) = \Omega_\Phi s + \eta^\varphi(s) = \Omega_\Phi s + \delta\omega_{(2)} \cos ks \,, \tag{67b}$$

$$\vartheta(s) = \pi/2 + \eta^\vartheta(s) = \pi/2 + \frac{C_{(6)}}{R} \sin \Omega_\Phi s \,. \tag{67c}$$

There are some subtleties that need to be handled with care. First of all, having identified a feature of a physical system within an approximate description, like the first order deviation equation, does not mean that this corresponds to a "real" effect. We have pointed out an example for this with the line of nodes precession that is present in the linearized framework but was shown to decrease with higher order approximations and to vanish in the end. Exactly this artifact of the line of nodes shift is present in Shirokov's solution as well. Since the equatorial plane is intersected by the perturbed orbit for $s = n\pi/\Omega_\Phi$, we get a shift of the line of nodes after one ϑ-period

$$\Delta\varphi = \varphi(s = 2\pi/\Omega_\Phi) - 2\pi = \delta\omega_{(2)} \cos \frac{2\pi k}{\Omega_\Phi} \,. \tag{68}$$

This shift was not mentioned in the work by Shirokov, but is indeed present in the solution. However, Shirokov noticed correctly that in (67) the r, φ and θ oscillations have different frequencies and, thus, different periods

$$T_r = T_\varphi = 2\pi/k \,, \tag{69a}$$

$$T_\vartheta = 2\pi/\Omega_\Phi \,. \tag{69b}$$

Linearized in m/R around the Keplerian value $T_K = 2\pi/\Omega_K = 2\pi\sqrt{R^3/m}$, we obtain

$$T_r = T_\varphi = T_K \left(1 - \frac{3m}{R}\right), \tag{70a}$$

$$T_\vartheta = T_K \left(1 + \frac{3m}{R}\right). \tag{70b}$$

It is by no means obvious why this linearization should be necessary, but it was used in [3]. Of course, for the Earth m/R is a very small quantity when R corresponds to radii above the surface, but the geodesic deviation equation works well even close to black holes, were m/R might be large [12]. However, Shirokov concludes that, due to the different periods of ϑ and r oscillations, the distance $R\,(\vartheta - \pi/2)$ to the equatorial plane, in which the reference orbit lies, is different from 0 after (several) radial oscillations and this is a new effect of GR. Shirokov imagines a satellite that moves on the reference orbit and rotates around the axis perpendicular to the equatorial plane with its orbital period, i.e. $2\pi/\Omega_\Phi$. Placed within this satellite a small test mass shall follow the perturbed orbit and the different frequencies of oscillations in the ϑ and r direction can be observed.

Since each orbit in Schwarzschild spacetime is confined to its orbital plane, the ϑ and φ frequencies have to be equal for exact orbits. Otherwise the line of nodes would shift. The difference in these periods that Shirokov discovered is a result of the approximation. However, the r and φ periods can be different, which leads to the perigee precession for elliptical orbits. Hence, the r and ϑ periods can also be different for elliptical orbits to allow for a perigee precession within an inclined orbital plane. This is exactly what the approximate solution (67a) of the first order deviation equation describes: an elliptical orbit with perigee precession in an inclined orbital plane. The exact solution for this orbit would describe an elliptical orbit in an inclined but fixed orbital plane that shows perigee precession within this orbital plane. The solution (67a) is simply the first order approximation of this situation. Since the radial and the polar period are different, one would observe the object to be above or below the equatorial plane after (several) radial periods. This is nothing but a precessing ellipse in the inclined orbital plane. Hence, Shirokov's "new" effect is not new at all, it is the first order approximation of the well-known perigee precession – discovered by Einstein already in 1916 [15] – in an inclined plane. Furthermore, as we have shown, this is mixed with the artifact of a precessing line of nodes due to the linearized framework.

We conclude that Shirokov's effect is no new effect but the first order description of the perigee shift for an elliptical orbit in an inclined orbital plane.

5 Conclusions

We have shown how to describe orbits using the solution of the first order deviation equation for circular reference geodesics. In particular, we employ the Schwarzschild spacetime as the simplest approximation for the Earth to investigate relativistic satellite orbits and orbit deviations. We describe the shape of all perturbed orbits and connect free parameters in the general solution to the orbital elements of the perturbed orbit. Using this description, one can now apply the solution of the first order deviation equation to model any orbit that is specified in terms of its orbital elements. We have uncovered artificial effects that are due to the linearized framework. For elliptical orbits with small eccentricities the perigee precession was derived as a purely relativistic effect, which is absent in the Newtonian solution of the deviation problem. The solution of the first order deviation problem in Schwarzschild spacetime has shown that such an approximate description must be handled with care. The line of node precession, which is forbidden in Schwarzschild spacetime, will mix in the context of Kerr spacetime with the Lense-Thirring effect that causes a similar behavior.

Reconsidering the so-called Shirokov effect we uncovered its origin. Rather than being a new feature of GR, we identified it as the relict of the approximate description of a well-known perigee precession.

The comparison of perturbed orbits – based on the solution of the geodesic deviation equation – to exact solutions of the underlying geodesic equation has shown, that higher order deviation equations should be used to model modern satellite based geodesy missions.

In (simple) spacetimes, for which analytic solutions of the geodesic equation are available, one can estimate the accuracy of the approximate description to any order. In more realistic spacetimes numerical methods will become necessary.

In conclusion, already at the present level of accuracy applications in geodesy and gravimetry require the use of higher order deviation equations. Such equations will become indispensable for the description of future high precision satellite and ground based measurements.

Acknowledgements The present work was supported by the Deutsche Forschungsgemeinschaft (DFG) through the grant PU 461/1-1 (D.P.), the Sonderforschungsbereich (SFB) 1128 Relativistic Geodesy and Gravimetry with Quantum Sensors (geo-Q), and the Research Training Group 1620 Models of Gravity. We also acknowledge support by the German Space Agency DLR with funds provided by the Federal Ministry of Economics and Technology (BMWi) under grant number DLR 50WM1547. The authors would like to thank V. Perlick, J.W. van Holten, and Y.N. Obukhov for valuable discussions.

A Conventions and Symbols

In the following we summarize our conventions, and collect some frequently used formulas. A directory of symbols used throughout the text can be found in Table 3. The signature of the spacetime metric is assumed to be $(+, -, -, -)$. Latin indices $i, j, k, \dots$ are spacetime indices and take values $0 \dots 3$. For an arbitrary k-tensor $T_{a_1 \dots a_k}$, the symmetrization and antisymmetrization are defined by

Table 3 Directory of symbols

Symbol	Explanation
Geometrical quantities	
g_{ab}	Metric
$A(r)$, $B(r)$	Free metric functions
$\sqrt{-g}$	Determinant of the metric
δ^a_b	Kronecker symbol
x^a, s	Coordinates, proper time
Y^a, X^a	(Reference, perturbed) curve
$\Gamma_{ab}{}^c$	Connection
$R_{abc}{}^d$	Curvature
η^a	Deviation vector
Physical quantities	
G, U	Newtonian gravitational (constant, potential)
R	Reference radius
Ω_K, T_K	Keplerian frequency, period
Ω_Φ, ω_φ	Azimuthal freq. of circ. orbit
k	Second freq. in relativistic solution
T_r, T_φ, T_ϑ	Frequencies of perturbations
$C_{(i)}$	Perturbation parameters, $i = 1 \ldots 6$
$\delta\omega_{(i)}$, $\delta t_{(i)}$	(Azimuthal, temporal) deviations, $i = 1 \ldots 3$
$R_\oplus$	Mean Earth radius $6.37 \cdot 10^3$ km
$\Delta\varphi$, Δn	Angle of (perigee precession, line of nodes shift)
E	Energy
L	Angular momentum
ρ	Matter density
M, m	Mass of the central object (kg), (m)
$f(s)$, $g(s)$, Δ	Abbreviations
Orbital elements	
a	Semi major axis
e	Eccentricity
Ω_a	Ascending node
i	Inclination
d_p, d_a	Distance to (perigee, apogee)
ω	Argument of the perigee
ν	True anomaly
Operators	
∂_i , ∇_i	(Partial, covariant) derivative
$\frac{D}{ds} =$ "·"	Total covariant derivative

$$T_{(a_1...a_k)} := \frac{1}{k!}\sum_{I=1}^{k!} T_{\pi_I\{a_1...a_k\}}, \tag{71}$$

$$T_{[a_1...a_k]} := \frac{1}{k!}\sum_{I=1}^{k!} (-1)^{|\pi_I|} T_{\pi_I\{a_1...a_k\}}, \tag{72}$$

where the sum is taken over all possible permutations (symbolically denoted by $\pi_I\{a_1 \dots a_k\}$) of its k indices.

The covariant derivative defined by the Riemannian connection is conventionally denoted by the nabla or by the semicolon: $\nabla_a =$ "$_{;a}$". Our conventions for the Riemann curvature are as follows:

$$\begin{aligned} 2A^{c_1...c_k}{}_{d_1...d_l;[ba]} &\equiv 2\nabla_{[a}\nabla_{b]}A^{c_1...c_k}{}_{d_1...d_l} \\ &= \sum_{i=1}^{k} R_{abe}{}^{c_i} A^{c_1...e...c_k}{}_{d_1...d_l} - \sum_{j=1}^{l} R_{abd_j}{}^{e} A^{c_1...c_k}{}_{d_1...e...d_l}. \end{aligned} \tag{73}$$

References

1. F. Flechtner, K.-H. Neumayer, C. Dahle, H. Dobslaw, E. Fagiolini, J.-C. Raimondo, A. Guentner, What can be expected from the GRACE-FO laser ranging interferometer for earth science applications? Surv. Geophys. **37**, 453 (2016)
2. B.D. Loomis, R.S. Nerem, S.B. Luthcke, Simulation study of a follow-on gravity mission to GRACE. J. Geod. **86**, 319 (2012)
3. M.F. Shirokov, On one new effect of the Einsteinian theory of gravitation. Gen. Relativ. Gravit. **4**, 131 (1973)
4. D. Philipp, V. Perlick, C. Lämmerzahl, K. Deshpande, On geodesic deviation in Schwarzschild spacetime, in *IEEE Metrology for Aerospace* (2015), p. 198
5. P.J. Greenberg, The equation of geodesic deviation in Newtonian theory and the oblateness of the earth. Nuovo Cimento **24B**, 272 (1974)
6. W.G. Dixon, The new mechanics of Myron Mathisson and its subsequent development, in *Equations of Motion in Relativistic Gravity*, ed. by D. Puetzfeld et al. Fundamental theories of Physics, vol. 179 (Springer, Berlin, 2015), p. 1
7. Y.N. Obukhov, D. Puetzfeld, Multipolar test body equations of motion in generalized gravity theories, in *Equations of Motion in Relativistic Gravity*, ed. by D. Puetzfeld et al. Fundamental theories of Physics, vol. 179 (Springer, Berlin, 2015), p. 67
8. D. Puetzfeld, Y.N. Obukhov, Generalized deviation equation and determination of the curvature in general relativity. Phys. Rev. D **93**, 044073 (2016)
9. H. Fuchs, Conservation laws for test particles with internal structure. Annalen der Physik **34**, 159 (1977)
10. H. Fuchs, Solutions of the equations of geodesic deviation for static spherical symmetric spacetimes. Annalen der Physik **40**, 231 (1983)
11. R. Kerner, J.W. van Holten, R. Colistete Jr., Relativistic epicycles: another approach to geodesic deviations. Class. Quantum Gravity **18**, 4725 (2001)
12. G. Koekoek, J.W. van Holten, Epicycles and Poincaré resonances in general relativity. Phys. Rev. D **83**, 064041 (2011)
13. H. Fuchs, Deviation of circular geodesics in static spherically symmetric space-times. Astron. Nachr. **311**, 271 (1990)

14. C. Misner, K. Thorne, J.A. Wheeler, *Gravitation* (Freeman, San Francisco, 1973)
15. A. Einstein, Die Grundlage der allgemeinen Relativitätstheorie. Annalen der Physik **354**, 769 (1916)
16. A.R. Forsyth, Note on the central differential equation in the relativity theory of gravitation. Proc. R. Soc. Lond. A **97**(682), 145 (1920)
17. W.B. Morton, LXI. The forms of planetary orbits on the theory of relativity. Lond. Edinb. Dublin Philos. Mag. J. Sci. **42**(250), 511 (1921)
18. E. Hackmann, C. Lämmerzahl, Complete analytic solution of the geodesic equation in Schwarzschild (anti-)de Sitter spacetimes. Phys. Rev. Lett. **100**, 171101 (2008)
19. E. Hackmann, C. Lämmerzahl, Geodesic equation in Schwarzschild-(anti-)de Sitter space-times: analytical solutions and applications. Phys. Rev. D **78** (2008)
20. E. Hackmann, V. Kagramanova, J. Kunz, C. Lämmerzahl, Analytic solutions of the geodesic equation in axially symmetric space-times. Europhys. Lett. **88**, 30008 (2009)
21. A. Nduka, On Shirokov's one new effect of the Einsteinian theory of gravitation. Gen. Relativ. Gravit. **8**, 347 (1977)
22. Yu.S. Vladimirov, R.V. Rodichev, Small oscillations of test bodies in circular orbits in the Schwarzschild and Kerr metrics (Shirokov's effect). Sov. Phys. J. **24**, 954 (1981)
23. W. Zimdahl, Shirokov effect and gravitationally induced supercurrents. Exp. Tech. Phys. **33**, 403 (1985)
24. L. Bergamin, P. Delva, A. Hees, Vibrating systems in Schwarzschild spacetime: toward new experiments in gravitation? Class. Quantum Gravity **26**, 18006 (2009)
25. W. Zimdahl, G.M. Gilberto, Temperature oscillations of a gas in circular geodesic motion in the Schwarzschild field. Phys. Rev. D **91**, 024003 (2015)

Measurement of Frame Dragging with Geodetic Satellites Based on Gravity Field Models from CHAMP, GRACE and Beyond

Rolf König and Ignazio Ciufolini

Abstract The experimental measurement of frame-dragging or the Lense-Thirring (LT) effect based on Satellite Laser Ranging (SLR) observations to the LAGEOS satellites was successfully demonstrated with an accuracy of about 10%. Here we look in detail into the effect of the node drift induced by the time variable part of the C(2,0) term of the gravity field model describing the flattening of the Earth. We demonstrate that errors in C(2,0) can effectively be taken care of by analyzing two satellites for the LT measurement. We also adopt some recent gravity field models in order to independently repeat and extend the LT experiments so far. The gravity field models used for this are derived either partly depending on LAGEOS SLR observations or completely independent from LAGEOS, and based on dedicated gravity field satellite missions like CHAMP, GRACE and GOCE. It turns out that from all the gravity field models tested the claimed accuracy of 10% of the LT measurement can be confirmed.

1 Introduction

Frame-dragging, as named by Albert Einstein [1], is an intriguing phenomenon of General Relativity (GR) that predicts that the orbit of particles, the direction determined by gyroscopes, the path followed by photons, and even the time marked by clocks, all are affected by mass-energy currents such as the rotation of a mass. In 1918 Lense and Thirring [2] published the equations for the frame-dragging effect on the orbit of a particle around a rotating central body, e.g. for a satellite orbit-

R. König (✉)
Helmholtz-Zentrum Potsdam Deutsches GeoForschungsZentrum GFZ, c/o DLR Oberpfaffenhofen, Wessling 82234, Germany
e-mail: koenigr@gfz-potsdam.de
URL: http://www.gfz-potsdam.de

I. Ciufolini
Dip. Ingegneria dell'Innovazione, Università del Salento, Lecce, Italy

I. Ciufolini
Centro Fermi, Roma, Italy

D. Puetzfeld and C. Lämmerzahl (eds.), *Relativistic Geodesy*,
Fundamental Theories of Physics 196, https://doi.org/10.1007/978-3-030-11500-5_14

ing the Earth there is a drift of the node and perigee of its orbit due to the rotation of the Earth, this is the so-called Lense-Thirring (LT) effect. Frame-dragging plays a key role in some high-energy astrophysics phenomena, such as the mechanism of emission of jets from active galactic nuclei and quasars [3]. Frame-dragging of the particles of the accretion disk around a central rotating black hole can explain the constant orientation of these jets for periods of time that can reach millions of years. Frame-dragging enters the dynamics of the coalescence of two black holes producing gravitation waves as observed by LIGO in 2015 [4]. The idea to measure frame-dragging by means of two laser-ranged satellites, i.e. tracked by Satellite Laser Ranging (SLR), with supplementary inclinations was first proposed by [5]. The first rough observation [6] was based on the EGM96 gravity field model [7] and SLR tracking to LAGEOS-1 [8] and LAGEOS-2 [9]. Later on [10] presented a breaking result where again SLR data to the satellites LAGEOS-1 and LAGEOS-2 were combined but the analysis was based on the by then newly available gravity field model EIGEN-GRACE02S [11] from the GRACE space mission [12]. Reference [10] estimated an accuracy of the recovery of the node drift due to frame-dragging of about 10%. This accuracy estimate was deduced from a thorough assessment of all possible modelling errors in Precise Orbit Determination (POD).

These results are based on the orbit determination software GEODYN [13] of the National Aeronautics and Space Administration NASA of the United States of America (USA). In order to rule out programming errors, the experiment was repeated by independent orbit determination software packages, one being EPOS-OC [14] by the German Research Centre for Geosciences GFZ, Potsdam, Germany, and one being UTOPIA [15] by the Center for Space Research (CSR) at the University of Texas at Austin, USA. GFZ used some recent GRACE gravity field models to analyze LAGEOS-1 and LAGEOS-2 data over the period 2000–2011 with EPOS-OC [16]. In summary the LT precession is recovered with a mean deviation from the GR prediction of about 8%. CSR tested a multitude of gravity field models using UTOPIA and SLR tracking to LAGEOS-1 and -2 over the years 1992–2006 [17]. They conclude that based on GRACE gravity field models the LT effect can be confirmed at the 8–12% level. So with both software packages the 10% accuracy claim by [10] can be endorsed.

In the following we describe the satellite missions that are involved in this type of analysis, on the one hand the SLR tracked satellites used to measure the LT node drift, on the other hand the satellite missions dedicated to measure the Earth's gravity field which is the pre-requisite for POD of the previous. We then have a glance at the major error source of the LT measurement, the C(2,0) term of the gravity field models. In particular we look into the effect of its variation in time and the impact thereof on the LT measurement by analyzing an extended period of LAGEOS observations with EPOS-OC. Finally we adopt some gravity field models that are generated either with or without LAGEOS data in order to rule out possible influences on the LT measurement by the fact LAGEOS being part of the gravity field model development.

2 Satellite Missions

Geodetic satellites are sphere satellites with low area-to-mass ratio that are tracked by SLR. The most important one being LAGEOS-1 (see Fig. 1), launched in 1976 for geodetic and geophysical applications and his twin, LAGEOS-2, launched in 1992 into a different inclination. In 2012 the LAser RElativity Satellite (LARES) was launched (see Fig. 2). The LARES mission [18] is particularly designed to measure frame dragging with the goal to achieve the 1% accuracy in combination with the LAGEOS satellites. The orbit characteristics of these missions are compiled in Table 1. Table 1 also shows the area-to-mass ratio of the satellites. The smaller this number the smaller the disturbing accelerations by drag of the upper atmosphere, solar radiation pressure and other non-conservative forces. Indeed LARES has the smallest area-to-mass ratio of all artificial satellites ever send into orbit. In the following we are not going to use LARES data as the mission duration is yet too short at the time of conducting this experiment. Later on a very first result with LARES is published by [19] indicating an accuracy of 5%.

Gravity field missions seek for exploring the gravity field of the Earth and its variations in time. The orbit is lower than that of the geodetic satellites, the lower the orbit the more sensitive to the gravity signal. However the lower the orbit, the higher the nuisance perturbations inserted by the non-conservative forces. The first dedicated mission that immediately led to a quantum leap in the accuracy of the derived gravity field models is CHAMP (see Fig. 3) in service from 2000 to 2005. CHAMP was equipped with an on-board two-frequency GPS receiver and a Laser Retro-Reflector (LRR) for POD, a three-axes accelerometer to measure the nuisance

Fig. 1 The LAGEOS satellite (courtesy NASA)

Fig. 2 The LARES satellite (courtesy LARES team)

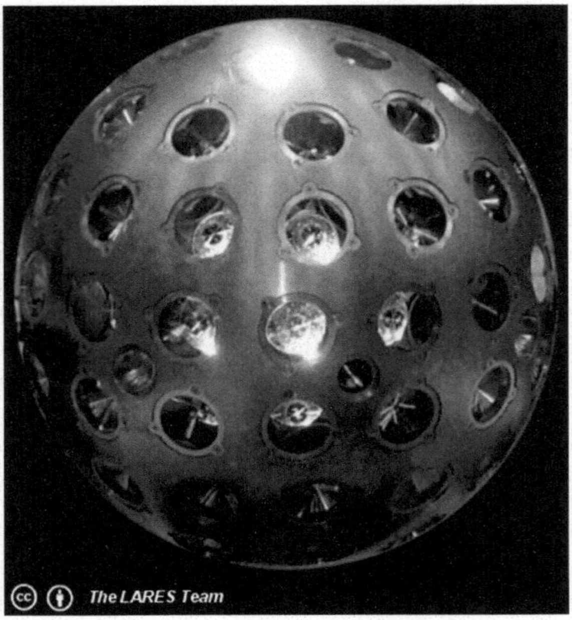

Table 1 Orbit characteristics of geodetic satellites

Satellite	Altitude (km)	Eccentricity	Inclination (deg)	Area-to-Mass Ratio (m^2/kg)
LAGEOS-1	5,900	0.004	109.8	0.000695
LAGEOS-2	5,800	0.014	52.6	0.000697
LARES	1,440	0.001	69.5	0.000269

forces and a star camera to measure the attitude of the satellite for precise reduction of the antenna offsets. Already in 2002 the two GRACE twin satellites (see Fig. 4) were launched. In addition to the equipment as that on CHAMP, the distance between the two satellites was measured with μm precision by an inter-satellite microwave link. This allowed another leap in accuracy, but more important, the recovery of the variations in time of the gravity field which in turn had dramatic impacts on science in the fields of hydrology and cryology. GRACE was designed for a five years lifetime but stayed in duty for impressive 15 years. GOCE (see Fig. 5) was launched in 2010 into the very low altitude of 220 km with the goal of an increased resolution of the static field. The satellite had ion thrusters to keep the spacecraft free of nuisance forces in flight direction. The gravity measurements were done by a three-axes gradiometer, an ensemble of six accelerometers working in differential mode. The gravity campaign took some 6 months only, but it delivered a geoid with 1 cm accuracy at 100 km resolution. The orbit characteristics of these missions are compiled in Table 2.

Fig. 3 CHAMP (courtesy Airbus)

Fig. 4 GRACE (courtesy NASA)

Fig. 5 GOCE (courtesy ESA)

Table 2 Orbit characteristics of gravity field missions

Satellite	Altitude (km)	Eccentricity	Inclination (deg)
CHAMP	470	0.004	87.3
GRACE	490	0.002	89.0
GOCE	290	0.001	96.7

As said, the gravity field missions are necessary to establish the gravity field models that are used in POD of the geodetic satellites for the recovery of the node observations. CHAMP and GRACE are the important missions to establish the long to medium wavelength part or the low to medium degree harmonic coefficients of the gravity field model. GOCE is particularly meant to improve the short wavelength part or the higher degree harmonic coefficients. For POD of the LAGEOS satellites just the coefficients up to degree about 20 play a role. From linear theory just the even degree zonal coefficients influence the node drift, that's why their errors disturb the LT node observation. The by far largest contribution is due to C(2,0), the next one, due to errors in C(4,0), is already smaller by about four orders of magnitude.

3 Measuring Frame Dragging

The idea by [5] to measure the frame dragging effect on the LAGEOS satellites by observing the node drift with the help of SLR tracking is displayed in Fig. 6. It shows the two sources acting on the nodes of two LAGEOS-type satellites with complementary inclinations. The LARES 2 satellite, approved by the Italian Space Agency (ASI), is under construction for a launch in 2019–2020. It will realize the idea of two laser-ranged satellites with supplementary inclinations [20].

The major drift of the node of an orbit is induced by the C(2,0) term of the spherical harmonic expansion of the gravity field. The C(2,0) term is also called J2 term (with $C(2,0) = -J2$) or the quadrupole moment of Earth's inertia. The C(2,0) node drift can be written in first order approximation as:

$$\dot{\Omega}_{C_{2,0}} = \frac{3nC_{2,0}a_e^2}{2a^2(1-e^2)^2}\cos(i) \tag{1}$$

where n is the orbital period, a_e is the Earth's radius, a is the semi-major axis of the orbit, e its eccentricity, and i its inclination. The LT node drift reads:

$$\dot{\Omega}_{LT} = \frac{2J}{a^3(1-e^2)^{3/2}} \tag{2}$$

where J is the angular momentum of the rotating body (the Earth). The sign of the C(2,0) drift depends on the inclination of the orbit. So in Fig. 6 the drifts of the nodes

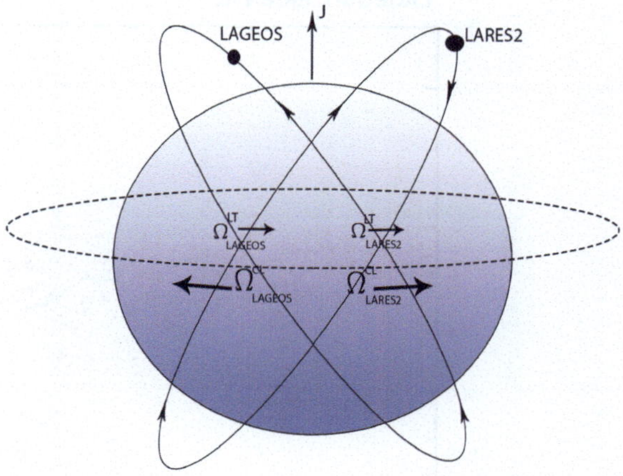

Fig. 6 The nodal drifts due to C(2, 0) and LT for two complementary orbits (adapted from [5])

Table 3 Nodal drifts of the geodetic satellites

Satellite	Lense-Thirring (mas/a)	C(2,0) (mas/a)
LAGEOS-1	30.7	450,000,000
LAGEOS-2	31.5	−830,000,000
LARES	118.4	−2,240,000,000

of the two complementary orbits point in opposite direction. The LT effect however does not depend on inclination, so it points for both orbits in the same direction. Once the node observations of both orbits are combined, the nodal drifts due to C(2,0) will cancel but the drifts due to LT will add. By this not only the comparatively large numbers of the C(2,0) drift vanish but also the errors of the C(2,0) term are ruled out. The drifts induced due to LT and C(2,0) are given for the geodetic satellites in Table 3.

The error of the C(2,0) term of current days gravity field models sizes at about 10^{-8}, so it is subject to erroneous node drifts at the order of the LT effect. Therefore measuring the node drift of one satellite alone can not reveal the LT effect. This is depicted in Figs. 7 and 8 where the node measurements from LAGEOS-1 and -2 are drawn as points. The analysis, done with EPOS-OC, spans the years 2000–2011 and is based on the EIGEN-6C gravity field model [21]. Eventually the node observations are accumulated to yield the node drift drawn as line in each graph. If there was no error in the C(2,0) term, the line should show the LT drift. However as one can see, the drifts behave different for the two satellites and are far from the expected LT signal.

If we now take out the time variable part of C(2,0) in EIGEN-6C, the behaviour of the node changes completely as can be seen in Fig. 9 for LAGEOS-1 and Fig. 10 for LAGEOS-2. The differences of the drifts in Figs. 7 and 9 and those in Figs. 8 and 10 are purely owned to the time variable part of C(2,0) being modeled or not.

The LT drift can finally be observed when the two node observations of the two satellites are combined. Figure 11 displays the combined node observations and the

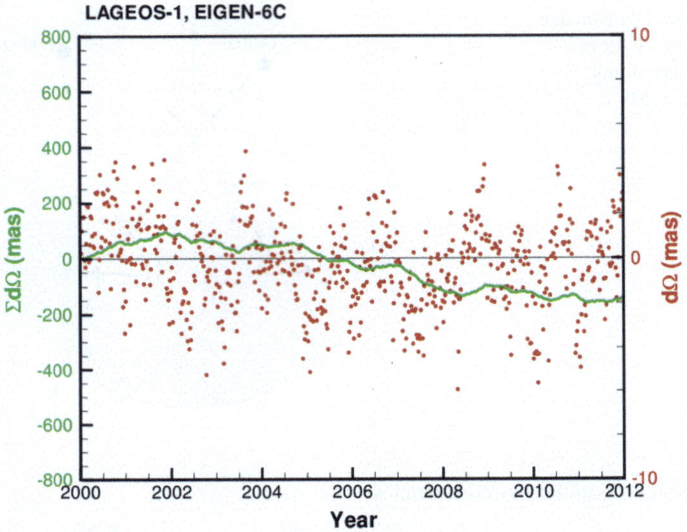

Fig. 7 Node observations and drift from LAGEOS-1, EIGEN-6C, obtained by the use of EPOS-OC

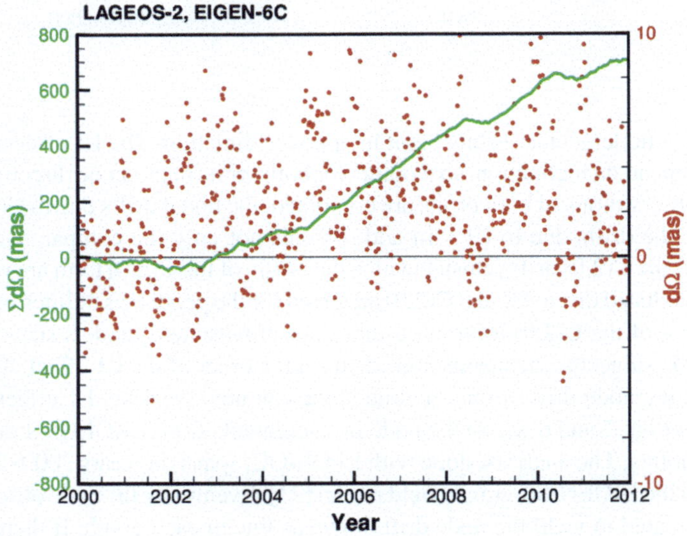

Fig. 8 Node observations and drift from LAGEOS-2, EIGEN-6C, obtained by the use of EPOS-OC

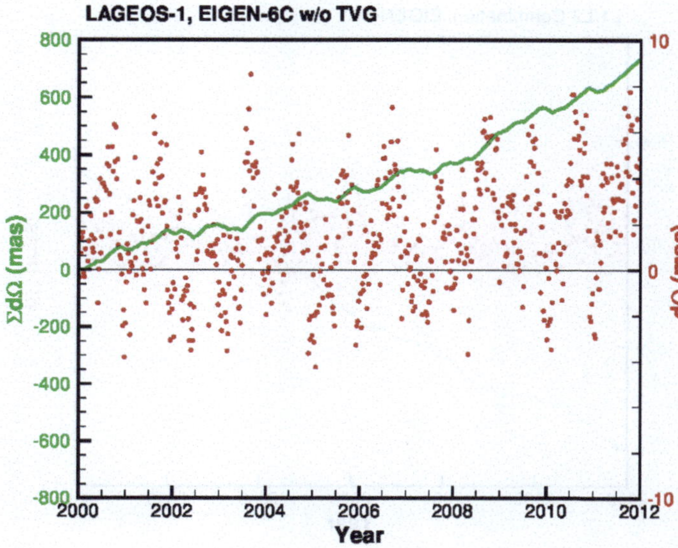

Fig. 9 Node observations and drift from LAGEOS, EIGEN-6C static, obtained by the use of EPOS-OC

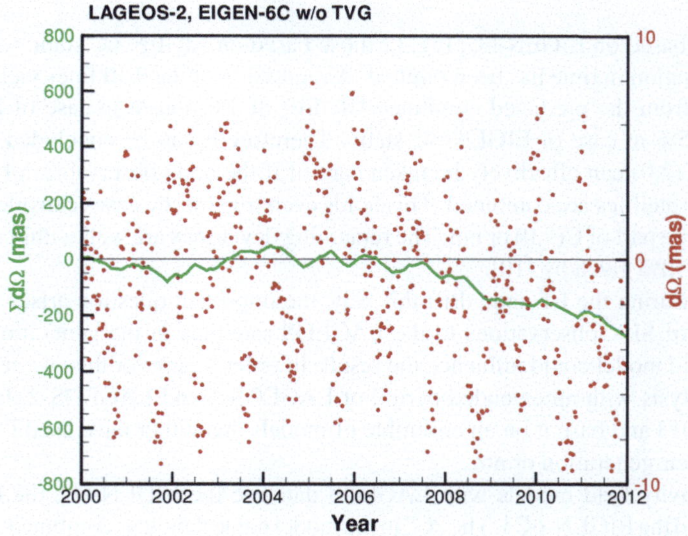

Fig. 10 Node observations and drift from LAGEOS-2, EIGEN-6C static, obtained by the use of EPOS-OC

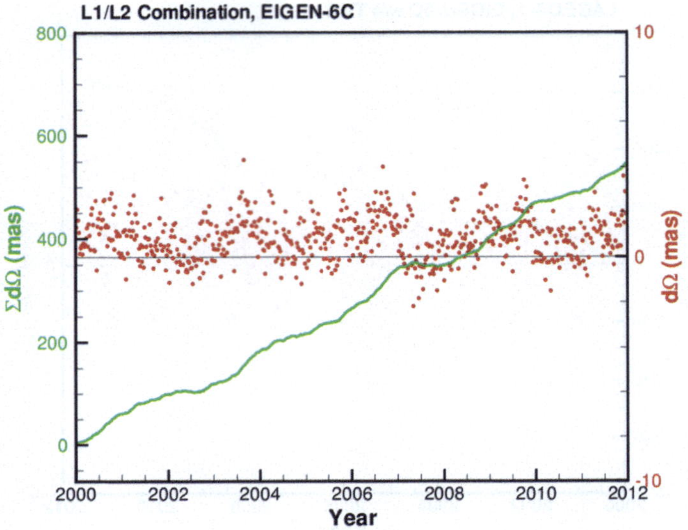

Fig. 11 Combined node observations and drift from LAGEOS-1 and LAGEOS-2, EIGEN-6C, obtained by the use of EPOS-OC

node drift based on EIGEN-6C, Fig. 12 those based on EIGEN-6C static where the C(2,0) variation in time has been omitted. A regression of the drift lines yields 6.9% deviation from the predicted combined LT drift of 48.2 mas/a in case of EIGEN-6C and 3.5% in case of EIGEN-6C static. Therefore it can be concluded that the errors in C(2,0) can effectively be taken care of if the node observables of the two LAGEOS satellites are combined. This holds even true for the cases considering the time variable part of C(2,0) or not. The remaining deviations are well within the error margin of 10% given by [10].

By measuring the LT node drift this way, the important question arises whether the usage of SLR observations to the LAGEOS satellites in the generation of the gravity field model could influence the result. In order to answer this, we repeat the above analysis with an extended period of LAGEOS-1 and LAGEOS-2 data from 2000 to 2013 and base it on an ensemble of models that either have used LAGEOS data for their generation or not.

The gravity field models with LAGEOS data are the EIGEN-6C, the EIGEN-6Sp.34, and the EIGEN-6C3. The "C" in the model name denotes "combined" models meaning satellite data are combined with surface gravity data. So EIGEN-6C is based on LAGEOS data plus GRACE data spanning the period January 2003 to June 2009, and GOCE data from November 2009 to June 2010. Finally the satellite data are combined with DTU2010 global gravity anomalies. The EIGEN-6Sp.34 is the inhouse naming of the official ESA GOCE gravity field model GO_CONS_GCF_2_DIR_R3 and a static, satellite only solution. Besides on LAGEOS it is based on GRACE data from January 2003 to June 2009, and on GOCE data from November 2009 to April 2011. The EIGEN-6C3 relies besides on LAGEOS on GRACE data from February

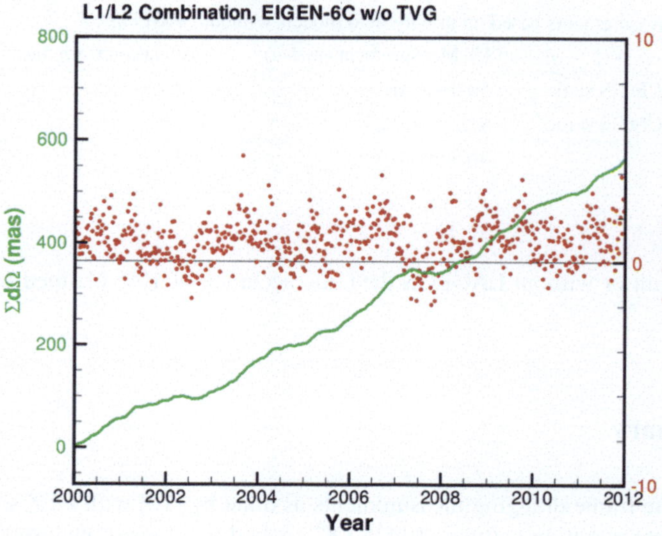

Fig. 12 Combined node observations and drift from LAGEOS-1 and LAGEOS-2, EIGEN-6C static, obtained by the use of EPOS-OC

Table 4 LT measurements based on gravity field models with LAGEOS

Model	LT Measurement (mas/a)	Deviation from theory (%)
EIGEN-6C static	49.2 +− 0.2	2.2
EIGEN-6Sp.34	47.3 +− 0.2	1.8
EIGEN-6C3 static	44.8 +− 0.2	7.3

2003 to September 2012, on GOCE data from November 2009 to May 2013, in combination with DTU2010/12 global gravity anomalies and the DTU12 geoid over the oceans and the EGM2008 geoid over the continents. The LT measurements based on these models are compiled in Table 4. In mean the deviation from the theoretical LT drift is about 4%.

Now we repeat this experiment with gravity field models free of LAGEOS data. These are the EIGEN-GRACE02S, the EIGEN-GRACE03S, and the EIGEN-51C. The EIGEN-GRACE02S is one of the first GRACE-only models with five months of data out of the period August 2002 to August 2003. EIGEN-GRACE03S is also a GRACE-only model with more data out of the time span February 2003 to July 2004. The EIGEN-51C is generated from data of the CHAMP mission and from GRACE data spanning October 2002 to September 2008 combined with DNSC08 global surface gravity anomalies. The LT measurements based on these models are compiled in Table 5. In mean the deviation from the theoretical LT drift is about 10%. In summary, in both trials, measuring the LT effect based on gravity field models

Table 5 LT measurements based on gravity field models without LAGEOS

Model	LT Measurement (mas/a)	Deviation from theory (%)
EIGEN-GRACE02S static	54.0 +− 0.2	12.1
EIGEN-GRACE03S static	41.2 +− 0.2	14.6
EIGEN-51C	46.3 +− 0.3	4.0

generated with or without LAGEOS data, the accuracy of 10% claimed by [10] is confirmed.

4 Summary

We repeat the frame dragging measurements as done by [10] with GFZ's independent orbit determination software EPOS-OC applied to a longer analysis period of LAGEOS-1 and -2 data from 2002 to 2013. Firstly we demonstrate that the LT node drift analyzing just one satellite is hidden behind the drift induced by errors in the C(2,0) term of modern gravity field models. Once the node observations of the two LAGEOS satellites are combined, the LT node drift can clearly be seen. Then we adopt a suite of gravity field models that either had employed LAGEOS data in their generation or not. In summary all results indicate that the LT drift can be recovered with a deviation from the theoretical drift of 10%.

References

1. A. Einstein, Letter to Ernst Mach, 25 June 1913, in *Gravitation* ed. by C. Misner, K.S. Thorne, J.A. Wheeler (Freeman, San Francisco, 1973), p. 544
2. J. Lense, H. Thirring, Über den Einfluss der Eigenrotation der Zentralkörper auf die Bewegung der Planeten und Monde nach der Einsteinschen Gravitationstheorie. Phys. Z. **19**, 156 (1918)
3. K.S. Thorne, R.H. Price, D.A. Macdonald, *The Membrane Paradigm* (Yale University Press, New Haven, 1986)
4. B.P. Abbott et al., Observation of gravitational waves from a binary black hole merger. Phys. Rev. Lett. **116**, 061102 (2016)
5. I. Ciufolini, Measurement of the Lense-Thirring drag on high-altitude, laser-ranged artificial satellites. Phys. Rev. Lett. **56**, 278–281 (1986)
6. I. Ciufolini, E. Pavlis, F. Chieppa, E. Fernandes-Vieira, J. Perez-Mercader, Test of general relativity and measurement of the Lense-Thirring effect with two Earth satellites. Science **279**, 2100–2103 (1998)
7. F.G. Lemoine, S.C. Kenyon, J.K. Factor, R.G. Trimmer, N.K. Pavlis, D.S. Chinn, C.M. Cox, S.M. Klosko, S.B. Luthcke, M.H. Torrence, Y.M. Wang, R.G. Williamson, E.C. Pavlis, R.H. Rapp, and T.R. Olson. The development of the joint NASA GSFC and the National Imagery and Mapping Agency (NIMA) geopotential model EGM96. NASA Technical Paper NASA/TP1998206861, Goddard Space Flight Center, Greenbelt, USA (1998)
8. R.L. Spencer, LAGEOS - a geodynamics tool in the making. J. Geol. Educ. **25**(2), 38–42 (1977)

9. ILRS. Lageos-1/-2, https://ilrs.cddis.eosdis.nasa.gov/missions/satellite_missions/current_missions/lag1_general.html. Accessed 12 March 2018
10. I. Ciufolini, E.C. Pavlis, A confirmation of the general relativistic prediction of the Lense-Thirring effect. Nature **431**, 958–960 (2004)
11. C. Reigber, R. Schmidt, F. Flechtner, R. Koenig, U. Meyer, K.-H. Neumayer, P. Schwintzer, S.Y. Zhu, An Earth gravity field model complete to degree and order 150 from GRACE: EIGEN-GRACE02S. J. Geodyn. **39**(1), 1–10 (2005)
12. B.D. Tapley, S. Bettadpur, M.M. Watkins, Ch. Reigber, The gravity recovery and climate experiment: mission overview and early results. Geophys. Res. Lett. **31**, L09607 (2004)
13. T.V. Martin, W.F. Eddy, D.D. Rowlands, D.E. Pavlis, GEODYN II system description. EG&G Contractor Report, Lanham, MD (1987)
14. S. Zhu, Ch. Reigber, R. Koenig, Integrated adjustment of CHAMP, GRACE, and GPS data. J. Geod. **78**(1–2), 103–108 (2004)
15. B.E. Schutz, B.D. Tapley, UTOPIA: University of Texas Orbit Processor. Inst. for Advanced Study in Orbital Mechanics, University of Texas at Austin, IASOM TR 80-1 (1980)
16. R. Koenig, B. Moreno-Monge, G. Michalak, Some aspects and perspectives of measuring Lense-Thirring with GNSS and geodetic satellites, in *Second International LARES Science Workshop, Accademia dei Lincei, Rome* (2012)
17. J.C. Ries, R.J. Eanes, M.M. Watkins, Confirming the frame-dragging effect with satellite laser ranging, in *Proceedings 16th international workshop on laser ranging*, http://cddis.gsfc.nasa.gov/lw16/docs/presentations/sci_3_Ries.pdf. Accessed 30 April 2018
18. I. Ciufolini, A. Paolozzi, E.C. Pavlis, J. Ries, R. Koenig, R. Matzner, G. Sindoni, The LARES space experiment: LARES orbit, error analysis and satellite structure, in *John Archibald Wheeler and General Relativity*, ed. by I. Ciufolini, R. Matzner (Springer, Berlin, 2010), pp. 371–434
19. I. Ciufolini, A. Paolozzi, E.C. Pavlis, R. Koenig, J. Ries, V. Gurzadyan, R. Matzner, R. Penrose, G. Sindoni, C. Paris, H. Khachatryan, S. Mirzoyan, A test of general relativity using the LARES and LAGEOS satellites and a GRACE Earth gravity model. Eur. Phys. J. C **76**, 120 (2016)
20. I. Ciufolini, A. Paolozzi, E.C. Pavlis, G. Sindoni, R. Koenig, J.C. Ries, R. Matzner, V. Gurzadyan, R. Penrose, D. Rubincam, C. Paris, A new laser-ranged satellite for General Relativity and space geodesy: I. An introduction to the LARES2 space experiment. Eur. Phys. J. Plus **132**, 336 (2017)
21. C. Foerste, S.L. Bruinsma, R. Shako, J.C. Marty, F. Flechtner, O. Abrikosov, C. Dahle, J.M. Lemoine, K.H. Neumayer, R. Biancale, F. Barthelmes, R. Koenig, G. Balmino, EIGEN-6C - A new combined global gravity field model including GOCE data from the collaboration of GFZ-Potsdam and GRGS-Toulouse. Geophysical Research Abstracts, vol. 13, EGU2011-3242-2, EGU General Assembly (2011)

Tests of General Relativity with the LARES Satellites

Ignazio Ciufolini , Antonio Paolozzi, Erricos C. Pavlis, Richard Matzner, Rolf König, John Ries, Giampiero Sindoni, Claudio Paris and Vahe Gurzadyan

Abstract LARES (LAser RElativity Satellite) developed by the Italian Space Agency (ASI) is a laser-ranged satellite successfully launched in February 2012 by ESA (European Space Agency). A second ASI laser-ranged satellite, LARES 2, is scheduled for launch by ESA at the end of 2019. Here we describe the main scientific objectives achieved and achievable by LARES and LARES 2, both in General Relativity and in space geodesy and geodynamics. Among the main tests achieved by LARES is a 5% test of frame-dragging, a fundamental and intriguing prediction of General Relativity. The LARES 2 satellite together with the laser-ranged satellite LAGEOS of NASA, is aimed to provide a 0.2% test of frame-dragging together with other relevant tests and determinations in fundamental physics, space geodesy and geodynamics.

I. Ciufolini (✉)
Dip. Ingegneria dell'Innovazione, Università del Salento, Lecce, Italy
e-mail: ignazio.ciufolini@gmail.com

I. Ciufolini · C. Paris
Centro Fermi - Museo Storico della Fisica e Centro Studi e Ricerche "Enrico Fermi", Rome, Italy

A. Paolozzi · G. Sindoni
Scuola di Ingegneria Aerospaziale, Sapienza Università di Roma, Rome, Italy

E. C. Pavlis
Joint Center for Earth Systems Technology (JCET), University of Maryland, Baltimore County, MD, USA

R. Matzner
Theory Group, University of Texas at Austin, Austin, TX, USA

R. König
Helmholtz-Zentrum Potsdam Deutsches GeoForschungsZentrum GFZ, c/o DLR Oberpfaffenhofen, Wessling 82234, Germany

J. Ries
Center for Space Research, University of Texas at Austin, Austin, TX, USA

V. Gurzadyan
Center for Cosmology and Astrophysics, Alikhanian National Laboratory, Yerevan, Armenia

D. Puetzfeld and C. Lämmerzahl (eds.), *Relativistic Geodesy*,
Fundamental Theories of Physics 196, https://doi.org/10.1007/978-3-030-11500-5_15

1 Introduction: The LARES and LARES 2 Space Missions

LARES (LAser RElativity Satellite) and LARES 2 are two space missions of the Italian Space Agency (ASI) aimed at accurate tests of General Relativity (GR) and fundamental physics and other measurements of space geodesy and geodynamics.

The LARES satellite was launched on 13 February 2012 by the VEGA launcher of the European Space Agency (ESA) [1, 2]. LARES is tracked by the laser ranging stations of the International Laser Ranging Service (ILRS) [3]. It is a spherical satellite, covered with 92 retro-reflectors. The LARES semimajor axis is $\cong$ 7820 km, its orbital eccentricity $\cong$ 0.0008, and its orbital inclination $\cong$ 69.5°. LARES, with LAGEOS (LAser GEOdynamics Satellite) [4], a NASA satellite launched in 1976, and LAGEOS 2, of ASI and NASA, launched in 1992, and with the Earth's gravitational field models by the GRACE mission (2002), of NASA and DLR (German Aerospace Center) [5], has allowed a test of an intriguing prediction of General Relativity, known as frame-dragging (see Sect. 2), with accuracy of about 5%. Such accuracy can be increased with longer periods of Satellite Laser Ranging (SLR) observation of LARES, LAGEOS and LAGEOS 2 to eventually reach an accuracy of approximately 2% in testing frame-dragging. The LARES 2 satellite of ASI, to be launched at the end of 2019-beginning of 2020, is aimed to provide a test of frame-dragging with about 0.2% accuracy.

2 Frame-Dragging

One of the intriguing predictions of General Relativity is the "dragging of inertial frames", or "frame-dragging" [6, 7]. (This is the name that Einstein gave to this GR phenomenon in a letter to Ernst Mach in 1913 [8].) Frame-dragging has a number of fundamental applications in high-energy astrophysics, from the dynamics of coalescing black holes with emission of gravitational waves as observed by LIGO [9] since 2015, to the dynamics of matter falling into spinning black holes with the ejection of huge jets in active galactic nuclei and quasars characterized by a constant direction for emission time scales which may reach millions of years (Bardeen-Petterson effect) [6, 10, 11].

In GR test-gyroscopes determine the axes of the local non-rotating inertial frames where the equivalence principle holds, so that the gravitational field is locally unobservable and all the laws of physics are the laws of special relativity. A test-gyroscope may be realized by a small current of mass in a loop, i.e. using a sufficiently small spinning top. However, GR predicts that a gyroscope has a behavior different from that of classical Galilei–Newton mechanics. In classical mechanics, a torque-free gyroscope always points toward the same distant "fixed" star. On the other hand, in GR a gyroscope is dragged by mass-energy currents, such as a rotating mass, and therefore its orientation will change with respect to a distant "fixed" star. If we rotate with respect to any such test-gyroscope, we would feel centrifugal forces, even though we may not rotate at all with respect to an asymptotic inertial frame, i.e. with respect to the distant "fixed" stars.

In GR, a current of mass-energy generates an additional contribution to the gravitational field, which in weak-field and slow-motion is called gravitomagnetic field because of its formal analogy with electrodynamics. The gravitomagnetic field exerts a torque on a gyroscope in a way formally similar to a magnetic field exerting a torque on a magnetic needle in electrodynamics. Indeed, frame-dragging of a gyroscope has a *formal* analogy with the Larmor effect, i.e. the change of orientation of a magnetic dipole by a magnetic field generated in electrodynamics by an electric current [6, 7, 12].

The Lense-Thirring effect is the frame-dragging perturbation of the orbital elements of a satellite in the weak gravitational field of a slowly rotating body, published in 1918 by Lense and Thirring [13, 14]. The rate of change of the nodal longitude of the satellite is given by $\dot{\Omega} = \frac{2J}{a^3(1-e^2)^{3/2}}$, where J is the angular momentum of the rotating body, Ω is the nodal longitude of the satellite, a its semimajor axis and e its orbital eccentricity.

2.1 Tests of Frame-Dragging

Frame-dragging was observed in 1997–1998 by using the LAGEOS and LAGEOS 2 laser-ranged satellites [15]. It was then measured with accuracy of approximately 10% [16–22] in 2004–2010, using LAGEOS, LAGEOS 2 and the Earth's gravity models obtained by the space geodesy mission GRACE [5]. It was measured with about 5% accuracy in 2016 using LARES and the two LAGEOS satellites (see below) [1]. In 2011 the NASA space mission Gravity Probe B, launched in 2004, also reported a test of frame-dragging with accuracy of about 19%. LARES, using longer periods of SLR observations, should reach a test of frame-dragging with approximately 2% accuracy. LARES 2 of ASI, to be launched at the end of 2019-beginning of 2020, together with LAGEOS, is aimed to provide in a few years a test of frame-dragging with accuracy of a few percent.

3 The LARES Satellite

LARES is a passive, spherical laser-ranged satellite (see Fig. 1). It was designed to approach as closely as possible an ideal test particle [2] and such objective was achieved by minimizing its surface-to-mass ratio; reducing the number of its parts; avoiding any protruding component; using a non-magnetic material; and avoiding the painting of the satellite surface. LARES is made of a non-magnetic tungsten alloy [2], it has a total mass of 386.8 kg and a diameter of 36.4 cm. Its final mean density is 15317 kg/m^3 which makes LARES the single known orbiting object in the solar system with the highest mean density and the artificial satellite with the lowest

Fig. 1 The LARES satellite before launch

surface-to-mass ratio. The satellite was built using one single piece of tungsten alloy, thus reducing thermal contact impedences and the onset of thermal gradients. Indeed, temperature gradients on the LAGEOS satellites produce thermal thrust and a tiny but not negligible orbital perturbation.

4 LARES Results

The measurement of frame-dragging with the two LAGEOS satellites was obtained in 2004–2010 using the two observables provided by the two rates of the nodal longitude of LAGEOS and LAGEOS 2 to determine two unknowns (see: method in [23, 24]): the frame-dragging effect and the uncertainty in the Earth's quadrupole moment, J_2 (the even zonal harmonic of degree two) [25]. The even zonals are the spherical harmonics of the Earth's gravitational potential expansion of even degree and order zero. The largest secular drifts of the nodal longitude of an artificial satellite are due to the Earth's even zonal harmonics and by far the largest node shift is due to the Earth's quadrupole moment, J_2 [25]. To test frame-dragging we either need to perfectly determine the Earth's even zonal harmonics or devise a method to eliminate the propagation of their uncertainties into the satellites' nodal rates, which led to the use of the two LAGEOS satellites to eliminate the uncertainties produced by J_2.

The idea of the LARES space experiment is to use its orbital data together with those of the two LAGEOS satellites to get three observable quantities, i.e. the nodal rates of the three satellites [23, 26]. Then these three observables can be used to solve for the three unknowns: frame-dragging and the two uncertainties in the two even zonal harmonics of lowest degree, i.e. J_2, and J_4 (the spherical harmonics of degree 2 and 4 and order 0). Using this method the two largest sources of uncertainty in the nodal drifts of the satellites are eliminated, thus providing an accurate test of frame-dragging within an uncertainty of a few percent [1].

In [1] we reported our orbital analysis of the laser-ranging data of the LARES, LAGEOS, and LAGEOS 2 satellites from 26 February 2012 to 6 September 2015. These laser-ranging data were collected from more than 30 ILRS stations all over the world. We processed approximately one million normal points of LARES, LAGEOS, and LAGEOS 2. The laser-ranging normal points were processed using NASA's orbital analysis and data reduction software GEODYN II, using the 2016 state-of-the-art Earth gravity field model, GGM05S [27]. GGM05S is an Earth gravity model based on approximately 10 years of GRACE data. It describes the Earth's spherical harmonics up to degree 180. The models in GEODYN II include Earth tides, solar radiation pressure, Earth albedo, thermal thrust, lunar, solar and planetary perturbations, and Earth rotation.

The orbital residuals of a satellite provide a measurement of the orbital perturbations which are not included (unmodeled) or are modeled with some errors (mis-modeled) in the data reduction. They represent the *observed* orbital elements of the satellite minus the *computed* ones. In our analysis, by far the largest part of the residuals of the nodal rates of LARES, LAGEOS and LAGEOS 2 are due to the errors in the Earth's even zonal harmonics and to the Lense-Thirring effect which, in our analysis, was not included in the models of GEODYN II. The theoretical value of the Lense-Thirring effect predicted by General Relativity is about 118.4 milliarcsec/year on the LARES node, and about 30.7 milliarcsec/year and about 31.5 milliarcsec/year, respectively on the LAGEOS and LAGEOS 2 nodes. In our analysis we used three observables, i.e. the three nodal rates of LAGEOS, LAGEOS 2 and LARES, and we were then able to eliminate the uncertainties in their nodal rates due to the errors in the values of even zonal harmonics J_2 and J_4 of the GGM05S model. Our method of analysis was able to eliminate the uncertainties in the nodal rates of the three satellites due to the contribution of the long and medium period tides to the harmonics J_2 and J_4.

We finally fitted our residuals with a secular trend and with the six largest tidal signals of LAGEOS, LAGEOS 2, and LARES and found the result: $\mu = (0.994 \pm 0.002) \pm 0.05$, where μ is our experimental value of frame-dragging normalized to its GR value (i.e. the theoretical prediction of GR is here 1), 0.002 is the formal 1-sigma error and 0.05 is our conservative estimate of the systematic errors due to the uncertainties in the Earth gravity field model GGM05S and to the other error sources.

In Fig. 2 we show our experimental value of frame-dragging obtained by fitting a secular trend together with the six known periodical terms, corresponding to the largest tidal signals observed on the satellite's nodes. In the figure the six fitted peri-

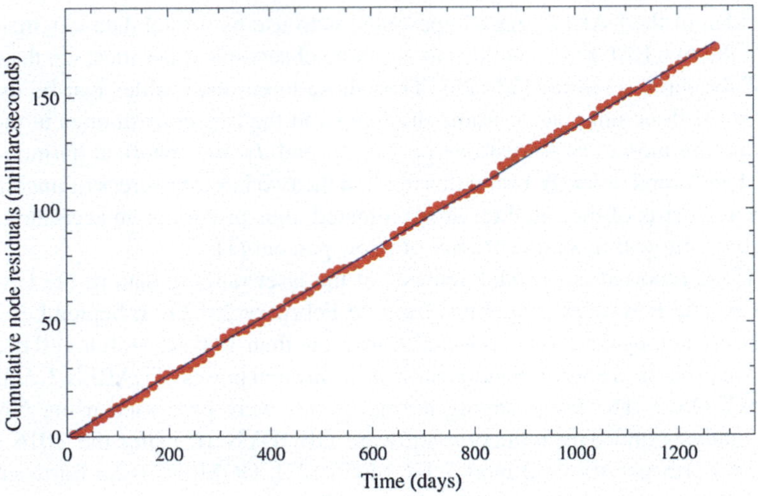

Fig. 2 The solid black line represents the experimental value of frame-dragging obtained by fitting the cumulative combined nodal residuals (the red dots) of LARES, LAGEOS, and LAGEOS 2 with a linear trend and six periodical terms. These six fitted periodical terms, corresponding to the six main tidal perturbations observed in the orbital residuals, are removed in the figure both from the residuals and from the fitting curve where only the secular trend is shown [1]

odical terms are removed both from the residuals and from the fitting curve that shows only the secular trend. The six tidal signals were identified both by analytical calculations [1] of the main tidal perturbations of the nodes of the three satellites and by a Fourier analysis of the observed residuals. Some periodical signals obtained with the Fourier analysis of the residuals of the nodes correspond to the main nongravitational perturbations.

The systematic errors affecting our measurement of frame-dragging with LARES, LAGEOS, and LAGEOS2 are mainly due to the errors in the even zonal harmonics of GGM05S with degree strictly larger than four. To evaluate this source of systematic errors, we tripled the published calibrated errors of each even zonal coefficient of GGM05S, where a calibrated error includes both the statistical and the estimated systematic errors. This technique to multiply by a factor two or three the calibrated errors is a standard method in space geodesy to place an upper bound to the real errors in the Earth's spherical harmonics. We then propagated these tripled errors into the nodal rates of LARES, LAGEOS, and LAGEOS 2 and finally found a systematic error of about 4% in our test of frame-dragging due to the Earth's even zonal harmonics [1, 28, 29].

Other smaller sources of systematic errors are due to long and medium period tidal perturbations and to nongravitational effects either un-modeled or mis-modeled. Nevertheless, we included in our analysis the main tidal and nongravitational perturbations, such as the direct radiation pressure from the Sun and the Earth radiation, i.e. the albedo. However, the systematic errors due to the tidal and nongravitational perturbations, either mis-modeled or un-modeled, are periodical and their effect on the combined nodal residuals is quite small as shown by the Fourier analysis of the post-fit orbital residuals shown in Fig. 2. Furthermore, a number of Monte Carlo simulations and error analyses (see [28, 30] and references in [1]) have shown that the systematic error in the LARES test of frame-dragging due to tides, nongravitational effects and other error sources is at the level of about 3%. In conclusion the total root sum squared (RSS) error in our 2016 LARES test of frame-dragging using the Earth gravity model GGM05S, by including the systematic errors due to the Earth's even zonals, tides and nongravitational perturbations, is approximately 5% of frame-dragging.

However, such 5% test of frame-dragging can be improved using the laser-ranging observations of LARES, LAGEOS, and LAGEOS 2 to decrease the error due the periodical un-modeled or mis-modeled perturbations. Indeed, the extension of the observational time of LARES, LAGEOS, and LAGEOS 2 will improve the modeling of the tidal and nongravitational perturbations and will reduce the total systematic error. Since different Earth gravity models lead to slightly different results, generating a set of solutions with different Earth gravity models and different orbital estimators, such as EPOSOC and UTOPIA, will provide other robust estimates of the systematic errors.

5 LARES 2

5.1 LARES 2 Objectives

The LARES 2 space experiment is aimed at a measurement of frame-dragging with an accuracy of a few parts in a thousand. Indeed, using LARES 2 and LAGEOS, an accuracy of about 0.2% was estimated with detailed error analyses, Monte Carlo simulations and covariance analyses [31–34]. Therefore LARES 2 will improve by about an order of magnitude the best measurements of frame-dragging that could be achieved by the LARES space mission. Furthermore, LARES 2 will provide important contributions in Space Geodesy and Geodynamics.

5.2 Estimated Error Budget in the LARES 2 Experiment

The main sources of systematic errors in the LARES 2 experiment are: orbital injection errors coupled with the uncertainties in the even zonal harmonics; non-zonal Earth harmonics and tides; solar and Earth radiation pressure (albedo), satellites' eclipses, thermal drag and errors in the determination of the satellites' orbital elements [31–34]. Each of these error sources was estimated to produce an error of approximately 0.1%, therefore the overall RSS error was estimated to be at the level

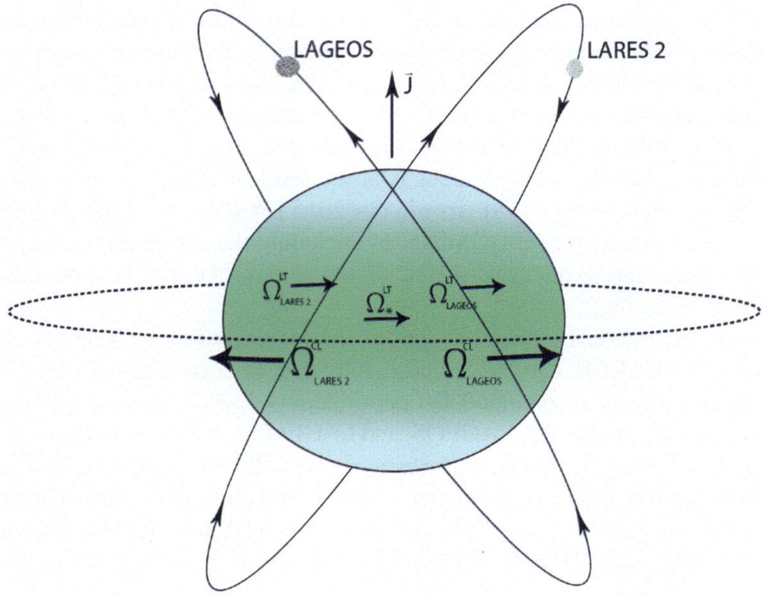

Fig. 3 The concept of the LARES 2 space experiment to test frame-dragging. Two laser-ranged satellites, LARES 2 and LAGEOS, with supplementary inclinations [23, 35–39] (see also: [40]) and the same semimajor axes, will provide the most accurate test of frame-dragging. Indeed, the classical precessions of the nodes of LARES 2 and LAGEOS will be equal and opposite (and thus their modeling uncertainty can be eliminated in the data analysis) whereas frame-dragging of their nodes is equal for the two satellites, both in size and sign

of 0.2%. Other smaller perturbing effects will give a negligible contribution to the RSS error. Such error budget of the LARES 2 experiment was also supported by a number of Monte Carlo simulations and covariance analyses [32]. Here we briefly describe the main error sources, for detailed error analyses we refer to [31–34].

• **Orbital injection errors** The idea of the LARES 2 experiment is to have two laser-ranged satellites with supplementary inclinations to eliminate the error due to the uncertainties in the Earth's even zonal harmonics and thus to accurately test frame-dragging of their nodes (see Fig. 3) [35, 36]. LARES 2 will have an inclination supplementary to that of LAGEOS. However, no launch vehicle can achieve an orbit for LARES 2 with an inclination exactly supplementary to that of LAGEOS and with exactly the same semimajor axis. Thus, since the cancellation of even zonal uncertainties will not be perfect, there will be a systematic error in the measurement of frame-dragging proportional to the uncertainty in the quadrupole moment times the deviations of inclination and semimajor axis from the LARES 2 ideal orbital elements. However, thanks to the space missions GRACE and to the recent GRACE Follow-On, the knowledge of the Earth gravity field has dramatically improved, so a 3-sigma orbital inclination injection error of the new launch vehicle VEGA C will induce a fractional error of only about 10^{-3} in the test of frame-dragging due to the non-perfect supplementary inclination of LARES 2 with respect to LAGEOS. A smaller uncertainty will be induced by the injection error in the semimajor axis of LARES 2. Therefore, the final orbital injection error will be at the level of about $\cong 0.1\%$.

In the following we simply list the other main error sources in the LARES 2 space experiment to test frame-dragging. Detailed analyses of these error sources can be found in [31–34].

• **Non-zonal Earth's harmonics and tides** The non-zonal Earth spherical harmonics do not produce secular nodal rates. They do produce though periodical nodal shifts that may induce a bias in the measurement of frame-dragging [25]. However, the lowest degree spherical harmonics, the ones relevant to the frame-dragging measurement, are very well known thanks to GRACE and GRACE Follow-On. The Earth tidal models are also well known and they will further improve in the near future. Furthermore, in [1], we used a new method that allows us to dramatically reduce the error due to periodic effects with known periods on the orbits of LAGEOS and LAGEOS 2 (tides, non-even zonal harmonics and some nongravitational perturbations.) In conclusion, the bias in the test of frame-dragging with LARES 2 and LAGEOS due to non-even zonals and tides will be at a level of about 0.1%.

• **Nongravitational perturbations** The nongravitational perturbations that will affect the accuracy of the LARES 2 measurement of frame-dragging are direct solar radiation pressure, albedo, satellite eclipses, thermal drag and particle drag [23]. Detailed error analyses are presented in [31, 32, 34], here we just report our estimated systematic error due to each one of these sources: the error due to the modeling of solar radiation pressure will be less than 0.1%; the error due to the modeling of albedo radiation pressure will be about 0.1%; the error due to thermal drag (thrust due to the anisotropic temperature distribution on the satellite) will be about 0.1%; the modeling error of satellite eclipses by the Earth will be about 0.1%; the error due to the nodal shift of LAGEOS and LARES 2 by particle drag will be negligible. In addition we have considered the measurement errors of the orbital parameters. However, Satellite Laser Ranging provides the position of the LAGEOS satellites with a precision of the normal points of less than a millimeter. Such a precision is enough to accurately measure the shift of the node of LAGEOS due to frame-dragging (almost 2 meters per year.) Nevertheless, the other orbital elements of LAGEOS and LARES 2 must also be measured with high accuracy. The semimajor axis and the eccentricity of the LAGEOS satellites are measured with enough accuracy. However the measurement error in the inclination of LAGEOS and LARES 2 may introduce an error in the measurement of frame-dragging (this error in the test of the Lense-Thirring effect with LAGEOS and LARES 2 will be due to the uncertainty in the measurement of the inclination and not to the uncertainty in the prediction of its behavior.) The uncertainty in the determination of the inclination of LAGEOS is mainly due to errors in correcting for atmospheric refraction at low elevations. However, the effect of the errors due to refraction corrections in the measurement of the inclination drops significantly at elevations above 20° and most of the laser-ranging observations are at elevations above 20°. Thus, by keeping in our data analysis only those observations corresponding to elevations above 20°, we will keep the error in the measurement of the inclination to a very small level that, when propagated into the nodes of LARES 2 and LAGEOS, will introduce an error of no more than 0.1% in our test of frame-dragging.

5.3 Error Budget of the LARES 2 Experiment

In conclusion the error sources in the LARES 2 experiment to test frame-dragging are: injection error and even zonal harmonics ≅0.1%, non-zonal harmonics and tides ≅0.1%, albedo ≅0.1%, satellite eclipses ≅0.1%, thermal drag and satellites eclipses ≅0.1%, measurement error of the LARES 2 and LAGEOS orbital parameters ≅0.1% of frame-dragging. Thus, by not including here the errors, listed above, smaller than 0.1% (since they would only affect the next decimal digit of the total RSS error) we estimated the total RSS error in the LARES 2 test of frame-dragging to be at the level of ≅ 0.2% of frame-dragging. This error budget of the LARES 2 space experiment was confirmed by a number of Monte Carlo simulations and covariance analyses [32].

6 LARES 2 Orbital Parameters and Satellite Characteristics

The mass and radius of LARES 2 are chosen to minimize its cross-sectional-to-mass ratio and thus to minimize its nongravitational perturbations. Its mass will be about 300 kg and its radius will be about 21.2 cm. LARES 2 must have the same semimajor axis of LAGEOS but its orbital inclination must be supplementary to that of LAGEOS. The eccentricity of LARES 2 must be zero in order to minimize the nongravitational perturbations of its node. The maximum deviations of the orbital parameters of LARES 2 from these optimal values are determined by the requirement of a 0.1% error (due to the non-perfect elimination of the even zonal harmonics uncertainties.) In summary the LARES 2 orbital parameters must be:

- Semimajor axis 12270 km ± 20 km.
- Inclination 70.16° ± 0.15° (supplementary to that of LAGEOS).
- Orbital eccentricity: between 0 and 0.0025.

7 LARES 2 and Space Geodesy

Satellite Laser Ranging (SLR) provides the most accurate definition of the origin of the International Terrestrial Reference Frame (ITRF) and is comparable with VLBI in the definition of its scale. The orbits of the Global Navigation Satellite System

(GNSS) satellites use reference frames based on the ITRF frame and the ITRF will realize the United Nations (UN), "Global Geodetic Reference Frame (GGRF)", the standard for all applications. This is accomplished with the LAGEOS and LAGEOS 2 SLR satellites, designed for cm level accuracy. (Nevertheless today objectives are 1 mm accuracy and 0.1 mm/yr stability.) LARES has provided the third SLR target, with improved specifications, and has enhanced the previous status. Nevertheless, even with LARES we do not yet have the ideal "SLR Constellation", in comparison with the GNSS one, with a number of LARES-type spacecraft. Indeed, a number of LARES-like satellites, together with LAGEOS, LAGEOS 2 and LARES, will provide a large number of SLR observations with a number of SLR targets above the horizon of each ground station. The forthcoming launch of LARES 2 will give the fourth satellite of the SLR Constellation. LARES 2 was designed to reach submillimeter ranging accuracy. With LARES and LARES 2 added to the SLR constellation we will for instance improve the mean site position and the Earth Orientation Parameters (Polar Motion and Length-of-Day). Finally, in addition to the improved determination of the ITRF, and thus indirectly of the GNSS accuracy, the enhancement of the SLR Constellation with LARES and LARES 2 will improve the estimation of terrestrial geophysical parameters, such as the secular evolution of low degree zonal harmonics, the terrestrial tides and the elastic properties of the Earth.

8 Conclusions

ASI's LARES satellite, launched in February 2012, has already provided tests of General Relativity and fundamental physics, including a 5% test of frame-dragging, a fundamental and intriguing prediction of General Relativity. Using a longer period of SLR observation of LARES, LAGEOS and LAGEOS 2, it will be possible to get a test of frame-dragging with approximately 2% accuracy with relevant implications for fundamental physics too. ASI's LARES 2 satellite is scheduled for launch at the end of 2019-beginning of 2020. Using LARES 2 and LAGEOS, it will be possible to reach a 0.2% test of frame-dragging together with other relevant tests and determinations in fundamental physics, space geodesy and geodynamics.

References

1. I. Ciufolini, A. Paolozzi, E. C. Pavlis, R. Koenig, J. Ries, V. Gurzadyan, R. Matzner, R. Penrose, G. Sindoni, C. Paris, H. Khachatryan, S. Mirzoyan, A test of general relativity using the LARES and LAGEOS satellites and a GRACE Earth gravity model. Measurement of Earth's dragging of inertial frames. Eur. Phys. J. **76**, 120 (2016)
2. A. Paolozzi, I. Ciufolini, LARES successfully launched in orbit: satellite and mission description. Acta Astronautica **91**, 313–321 (2013)
3. M.R. Pearlman, J.J. Degnan, J.M. Bosworth, The international laser ranging service. Adv. Space Res. **30**, 135143 (2002)

4. S.C. Cohen, R.W. King, R. Kolenkiewicz, R.D. Rosen, B.E. Schutz (Eds.), LAGEOS scientific results. J. Geophys. Res. **90**(B11), 9215–9438 (1985)
5. B.D. Tapley, S. Bettadpur, M. Watkins, C. Reigber, The gravity recovery and climate experiment: mission overview and early results. Geophys. Res. Lett. **31**(9) (2004)
6. K.S. Thorne, R.H. Price, D.A. Macdonald, *The Membrane Paradigm* (Yale University Press, NewHaven, 1986)
7. I. Ciufolini, Dragging of inertial frames. Nature **449**, 41 (2007)
8. A. Einstein, Gravitation, in *Letter to Ernst Mach, 25 June 1913*, ed. by C. Misner, K.S. Thorne, J.A. Wheeler (Freeman, San Francisco, 1973), p. 544
9. B.P. Abbott et al. Binary Black Hole Mergers in the first Advanced LIGO Observing Run. Phys. Rev. X **6**, 041015 (2016)
10. J.M. Bardeen, J.A. Petterson, The Lense-Thirring effect and accretion disks around Kerr black holes. ApJ **195**, L65 (1975)
11. C. Nixon, A. King, Do jets precess... or even move at all? ApL **765**, L7 (2013)
12. I. Ciufolini, J.A. Wheeler, *Gravitation and Inertia* (Princeton University Press, Princeton, 1995)
13. J. Lense, H. Thirring, Über den Einfluss der Eigenrotation der Zentralkörper auf die Bewegung der Planeten und Monde nach der Einsteinschen Gravitationstheorie. Phys. Z. **19**, 156–163 (1918)
14. J. Lense, H. Thirring, English translation of [13] by B. Mashhoon, F.W. Hehl, D.S. Theiss. Gen. Relativ. Gravit. **16**, 711 (1984)
15. I. Ciufolini et al., Test of general relativity and measurement of the Lense-Thirring effect with two Earth satellites. Science **279**(5359), 2100–2103 (1998)
16. I. Ciufolini, E.C. Pavlis, A confirmation of the general relativistic prediction of the Lense-Thirring effect. Nature **431**, 958–960 (2004)
17. I. Ciufolini, E.C. Pavlis, R. Peron, Determination of frame-dragging using Earth gravity models from CHAMP and GRACE. New Astron. **11**, 527–550 (2006)
18. I. Ciufolini, E.C. Pavlis, J. Ries, R. Koenig, G. Sindoni, A. Paolozzi, H. Neumayer, Gravitomagnetism and its measurement with laser ranging to the LAGEOS satellites and GRACE Earth gravity models, in *General Relativity and John Archibald Wheeler*, vol. 367 (Springer GmbH, Berlino DEU, 2010), pp. 371434
19. J.C. Ries, R.J. Eanes, M.M. Watkins, Confirming the framedragging effect with satellite laser ranging, in *16th international workshop on laser ranging, Poznan, Poland, 1317 October 2008* (2008)
20. J. Ries, Relativity in satellite laser ranging, in *IAU Symposium 261. Relativity in fundamental astronomy: dynamics, reference frames, and data analysis* (Virginia Beach, VA, USA, 27 April–1 May 2009)
21. R. König, B. Moreno-Monge, G. Michalak, Some aspects and perspectives of measuring Lense-Thirring with GNSS and geodetic satellites, in *Second International LARES Science Workshop, Accademia dei Lincei, Rome* (2012)
22. R. König, I. Ciufolini, Measurement of frame-dragging with geodetic satellites based on gravity field models from CHAMP, GRACE and beyond, in *Current Volume*
23. I. Ciufolini, A comprehensive introduction to the LAGEOS gravimetric experiment. Int. J. Mod. Phys. A **4**, 3083–3145 (1989)
24. I. Ciufolini, On a new method to measure the gravitomagnetic field using two orbiting satellites. Nuovo Cimento A **109**, 1709–1720 (1996)
25. W.M. Kaula, *Theory of Satellite Geodesy* (Blaisdell, Waltham, 1966)
26. I. Ciufolini, On the orbit of the LARES satellite (2006). arXiv:gr-qc/0609081v1
27. J. Ries, S. Bettadpur, R. Eanes, Z. Kang, U. Ko, C. McCullough, P. Nagel, N. Pie, S. Poole, T. Richter, H. Save, B. Tapley, Development and evaluation of the global gravity model GGM05. *CSR-16-02, Center for Space Research*. (The University of Texas at Austin, 2016)
28. I. Ciufolini, A. Paolozzi, E. Pavlis, J. Ries, R. Koenig, R. Matzner, G. Sindoni, The LARES space experiment: LARES orbit, error analysis and satellite structure, in *General Relativity and John Archibald Wheeler*, vol. 367 (Springer-Verlag GmbH, Berlino - DEU, 2010), pp. 467–492

29. I. Ciufolini et al., Reply to "A comment on "*A test of general relativity using the LARES and LAGEOS satellites and a GRACE Earth gravity model*, ed. by I. Ciufolini et al.," Iorio". EPJ C **78**, 880 (2018)
30. I. Ciufolini, B. Moreno Monge, A. Paolozzi, R. König, G. Sindoni, G. Michalak, Monte Carlo Simulations of the LARES space experiment to test General Relativity and fundamental physics. Class. Quantum Gravity **30**, 235009 (2013)
31. I. Ciufolini, A. Paolozzi, E.C. Pavlis, G. Sindoni, R. Koenig, J.C. Ries, R. Matzner, V. Gurzadyan, R. Penrose, D. Rubincam, C. Paris, A new laser-ranged satellite for General Relativity and space geodesy: I. An introduction to the LARES 2 space experiment. Eur. Phys. J. Plus **132**(8), 336 (2017)
32. I. Ciufolini, E.C. Pavlis, G. Sindoni, J.C. Ries, A. Paolozzi, R. Matzner, R. Koenig, C. Paris, A new laser-ranged satellite for General Relativity and space geodesy: II. Monte Carlo simulations and covariance analyses of the LARES 2 experiment. Eur. Phys. J. Plus **132**(8), 337 (2017)
33. I. Ciufolini, R. Matzner, V.G. Gurzadyan, R. Penrose, A new laser-ranged satellite for General Relativity and space geodesy: III. De Sitter effect and the LARES 2 space experiment. Eur. Phys. J. C **77**, 819 (2017)
34. I. Ciufolini, R.A. Matzner, J. Feng, D.P. Rubincam, E.C. Pavlis, J.C. Ries, G. Sindoni, A. Paolozzi, C. Paris, A new laser-ranged satellite for General Relativity and Space Geodesy IV. Thermal drag and the LARES 2 space experiment. Eur. Phys. J. Plus **133**, 333 (2018)
35. I. Ciufolini, Measurement of the Lense-Thirring drag on high-altitude laser-ranged artificial satellites. Phys. Rev. Lett. **56**, 278–281 (1986)
36. I. Ciufolini, *Theory and Experiments in General Relativity and other Metric Theories*. Ph.D. Dissertation, Univ. of Texas, Austin, Pub. Ann Arbor, Michigan, 1984
37. B. Tapley, J.C. Ries, R.J. Eanes, M.M. Watkins, NASA-ASI Study on LAGEOS III. *Center for Space Research Report CSR-89-03, The University of Texas at Austin* (1989)
38. I. Ciufolini et al., *ASI-NASA Study on LAGEOS III* (CNR, Rome, Italy, 1989)
39. J.C. Ries. *Simulation of an experiment to measure the Lense-Thirring precession using a second LAGEOS satellite*. Ph.D. Dissertation, Center for Space Research Report CSR-89-05, The University of Texas at Austin (1989)
40. G.E. Peterson. *Estimation of the Lense-Thirring precession using laser-ranged satellites*. Ph.D. Dissertation, The University of Texas at Austin (1997)

The manufacturer's authorised representative in the EU is Springer Nature Customer Service Centre GmbH, Europaplatz 3, 69115 Heidelberg, Germany. If you have any concerns regarding our products, please contact ProductSafety@springernature.com

Printed and bound by CPI Group (UK) Ltd, Croydon, CR0 4YY
30/09/2026
02225443-0018